SPECIFIC HEAT OF SOLIDS

CINDAS DATA SERIES ON MATERIAL PROPERTIES

Y. S. Touloukian* and C. Y. Ho, Series Editors

Group I   **Theory, Estimation, and Measurement of Properties**
Volume I-1   Transport Properties of Fluids: Thermal Conductivity, Viscosity, and Diffusion Coefficient (1988)
Volume I-2   Specific Heat of Solids (1988)

Group II   **Properties of Special Materials**
Volume II-1   Thermal Accommodation and Adsorption Coefficients of Gases (1981)
Volume II-2   Physical Properties of Rocks and Minerals (1981)

Group III   **Properties of the Elements**
Volume III-1   Properties of Selected Ferrous Alloying Elements (1981)
Volume III-2   Properties of Nonmetallic Fluid Elements (1981)

Group V   **Properties of Fluids and Fluid Mixtures**
Volume V-1   Properties of Inorganic and Organic Fluids (1988)

VOLUMES IN PREPARATION

Group I   **Theory, Estimation, and Measurement of Properties**
Volume I-3   Transport Properties of Solids: Thermal Conductivity, Electrical Resistivity, and Thermoelectric Properties
Volume I-4   Thermal Expansion of Solids
Volume I-5   Thermal Radiative Properties of Solids

Group II   **Properties of Special Materials**
Volume II-3   Thermoelectric Power of Selected Metals and Binary Alloy Systems
Volume II-4   Optical Properties of Optical Materials

Group IV   **Properties of Alloys**
Volume IV-1   Properties of Stainless Steels

Group V   **Properties of Fluids and Fluid Mixtures**
Volume V-2   Properties of Commercial Refrigerants and Fluid Mixtures

Group VI   **Properties of Oxides and Oxide Mixtures**
Volume VI-1   Properties of Rare-Earth Oxides and Actinide Oxides

Additional titles will be listed and announced as the preparation of other volumes is well in progress.

*Deceased in 1981.

CONTRIBUTING AUTHORS

A. C. Anderson
Chapter 2
Department of Physics
University of Illinois
Urbana, Illinois

D. W. Bonnell
Chapter 7
Inorganic Materials Division
National Bureau of Standards
Gaithersburg, Maryland

C. R. Brooks
Chapter 5 and Appendix A
Chemical, Metallurgical and
*   Polymer Engineering Department*
University of Tennessee
Knoxville, Tennessee

Ared Cezairliyan
Chapter 9
Thermophysics Division
National Bureau of Standards
Gaithersburg, Maryland

M. G. Chasanov
Chapter 12
Chemical Engineering Division
Argonne National Laboratory
Argonne, Illinois

D. A. Ditmars
Chapter 6
Chemical Thermodynamics Division
National Bureau of Standards
Gaithersburg, Maryland

D. F. Fischer
Chapter 12
Chemical Engineering Division
Argonne National Laboratory
Argonne, Illinois

R. K. Kirby
Appendix B
Office of Standard Reference
*   Materials*
National Bureau of Standards
Gaithersburg, Maryland

Ya. A. Kraftmakher
Chapter 8
Institute of Inorganic Chemistry
USSR Academy of Sciences
Novosibirsk, USSR

L. Leibowitz
Chapter 12
Chemical Engineering Division
Argonne National Laboratory
Argonne, Illinois

C. Loriers-Susse
Chapter 10
CNRS-Laboratoire de Bellevue
Bellevue, France

J. L. Margrave
Chapter 7
Department of Chemistry
Rice University
Houston, Texas

Douglas L. Martin
Chapter 3
Division of Physics
National Research Council of
*   Canada*
Ottawa, Canada

A. P. Miiller
Chapter 1
Thermophysics Division
National Bureau of Standards
Gaithersburg, Maryland

R. L. Montgomery
Chapter 7
Department of Chemistry
Rice University
Houston, Texas

S. C. Mraw
Chapter 11
Exxon Research and Engineering
*   Co.*
Annandale, New Jersey

E. E. Stansbury
Chapter 5 and Appendix A
Chemical, Metallurgical and
*   Polymer Engineering Department*
University of Tennessee
Knoxville, Tennessee

B. Stephenson
Chapter 7
Department of Chemistry
Rice University
Houston, Texas

P. C. Sundareswaran
Chapter 7
Department of Chemistry
Rice University
Houston, Texas

E. F. Westrum, Jr.
Chapter 4 and Appendix C
Department of Chemistry
University of Michigan
Ann Arbor, Michigan

CINDAS Data Series on Material Properties
Volume I-2

SPECIFIC HEAT OF SOLIDS

Edited by

C. Y. Ho

Director, Center for Information and Numerical Data Analysis and Synthesis
Purdue University

Authored by

Ared Cezairliyan

Senior Author and Volume Coordinator

and

A. C. Anderson, D. W. Bonnell, C. R. Brooks, M. G. Chasanov, D. A. Ditmars,
D. F. Fischer, R. K. Kirby, Ya. A. Kraftmakher, L. Leibowitz, C. Loriers-Susse,
J. L. Margrave, D. L. Martin, A. P. Miiller, R. L. Montgomery, S. C. Mraw,
E. E. Stansbury, B. Stephenson, P. C. Sundareswaran, and E. F. Westrum, Jr.

HEMISPHERE PUBLISHING CORPORATION
A member of the Taylor & Francis Group

New York   Washington   Philadelphia   London

SPECIFIC HEAT OF SOLIDS

1 2 3 4 5 6 7 8 9 0   B C B C   8 9 8

Library of Congress Cataloging-in-Publication Data

Cezairliyan, A.
    Specific heat of solids / edited by C. Y. Ho ; authored by
Ared Cezairliyan and A. C. Anderson . . . [et al.].
        p. cm.—(CINDAS data series on material properties. Group
I, Theory, estimation, and measurement of properties ; v. I-2)
    Includes bibliographies and index.
    1. Solids—Thermal properties—Measurement—Handbooks, manuals,
etc. 2. Specific heat—Measurement—Handbooks, manuals, etc.
3. Calorimeters and calorimetry—Handbooks,manuals, etc. I. Ho,
C. Y. (Cho Yen), date. II. Anderson, A. C. (Ansel Cochran),
date. III. Title. IV. Series: CINDAS data series on material
properties. Group I, Theory, estimation, and measurement of
properties ; vol. I-2.
QC176.8.T4C49 1988
536′.63—dc19   88-604
                                                            CIP

ISBN 0-89116-834-6

Contents

Foreword to the Series

The years since 1955 have seen an explosive growth in scientific research throughout the world. While this growth has been modulated at times by economic problems, the trend is still upward. The increased number of people involved in research and development, combined with the power of modern instrumentation, has led to this striking expansion of scientific knowledge. At the same time, public concern over government expenditures for research has led to the demand that this knowledge produce more tangible public benefits.

The familiar "information explosion" has inspired much rhetoric and a certain amount of action. Considerable effort has been put into utilizing modern computer technology to establish better monitoring of the scientific and technical literature. The operations of the abstracting and indexing services have been revolutionized by the computer, leading to more efficient methods of searching the literature and retrieving documents of potential interest. The computer has made it possible for these services to keep up with the exponential growth in journal articles, reports, and conference proceedings that describe the fruits of research and development projects. On-line searching of bibliographic files is growing rapidly, and retrieval techniques are becoming more sophisticated.

There is increasing recognition, however, that the retrieval of pertinent documents is only the first step in achieving optimum utilization of the results of scientific research. The hard information content, exemplified most directly by the numerical data contained in a scientific report, is the commodity that gives any document its real value. Experience has shown that data reported in the research literature are often contradictory, incomplete, and poorly documented. The refereeing process tends to focus on the scientific insights and conclusions presented by the author, rather than on the supporting data. Better quality control of these data is essential to realizing the full potential value of the research.

The Center for Information and Numerical Data Analysis and Synthesis (CINDAS), which is an outgrowth of the Thermophysical Properties ResearchCenter (TPRC) founded by Professor Y. S. Touloukian in 1957, provides one response to the problem of how to utilize published scientific information most effectively. CINDAS has attempted to capture the world literature pertinent to its defined scope, to organize and index this literature, to extract the data contained therein, and to subject the compiled data to critical evaluation and careful analysis. Its ultimate objective is to present the most reliable values possible for the physical properties in question and to use these data as a basis for extrapolation and prediction of properties in regimes where direct measurements have not been reported. In this way CINDAS serves to compact a vast amount of scientific information into a convenient and reliable form.

CINDAS is one component of the National Standard Reference Data System (NSRDS), which is coordinated by the Office of Standard Reference Data of the National Bureau of Standards as a nationwide effort to provide reliable chemical and physical reference data. The NSRDS program seeks to enlist the help of the entire scientific community in carrying out critical evaluation of data in all branches of the physical sciences. We are proud of the long association of CINDAS with the NSRDS and recognize its many contributions to science and technology in the United States.

The *CINDAS Data Series on Material Properties* is a major undertaking to present the fruits of CINDAS's efforts to the technical public in a convenient format. This series represents the essential quantitative results from tens of thousands of research projects throughout the world. By screening and organizing the raw data, carrying out critical evaluation where possible, and extrapolating from the resulting data base, CINDAS performs an invaluable service to the scientific and engineering community. In so doing it contributes in an important way to the utilization of science and technology for the public benefit.

David R. Lide, Jr.
Director
Standard Reference Data
National Bureau of Standards

Preface to the Series

The *CINDAS Data Series on Material Properties* is an updated, expanded, and completely reorganized successor of the earlier *TPRC Data Series* entitled *Thermophysical Properties of Matter.* The reorganization is intended to bring into each volume of the series a complete summary of the properties of a group of related materials, within the scope of thermophysical and electronic properties assigned to CINDAS as a National Data Center.

Unlike the practice of most data handbooks, this series presents critically evaluated data and incorporates a brief discussion of the state of knowledge in a given case. Whenever possible, an indication is given as to how these values would change under certain experimental conditions. In those instances where no data evaluation was possible, a table of selected representative data from the literature, with specimen specifications and remarks, is included together with a figure presenting the data in order to assist the user in exercising some judgment in data selection. In addition, a number of single-valued physical constants and, whenever relevant, room-temperature values of a number of mechanical properties are also reported to enhance the value of the volume to the user interested in material selection and material characterization.

The mission of CINDAS is to search for, evaluate, analyze, reconcile, correlate, and synthesize all available data for a large range of thermophysical and electronic properties of materials in order to obtain the best possible consistent sets of property values. In such work, gaps in the data often occur for ranges of temperature, composition, etc. Whenever feasible, such gaps are filled by methods ranging from empirical procedures to detailed theoretical calculations. Depending on the information available for a particular property and material, the end product may vary from a simple listing of isolated values to detailed tabulations with generating functions and accompanying plots showing the degree of concordance of the different values and, in some cases, values in a range of parameters presently unexplored in the laboratory.

The volumes in the *CINDAS Data Series* are organized into a number of groups. Those in Group I deal with the theory, measurement, and estimation of the major properties covered by the other groups. Written by invited experts in the field, these volumes constitute major works in their own right. The volumes in Group II differ from those in both Group I and all other groups in that they relate to special materials and both deal with the theories of properties and provide comprehensive compendia and evaluation of the available data in given fields. More typical of the *Data Series* are the volumes beginning with Group III. In each of these volumes, Chapter 1 presents pertinent information on the presentation of data, methods used in data evaluation, unit conversion tables, etc. Each of the succeeding chapters deals with a particular material, beginning with a description of that material and a table of selected physical constants and room-temperature values of a number of selected mechanical properties. This is followed by sections presenting data on a given property of the material. Each section includes a discussion of the available data and information on the property, a discussion of the considerations involved in arriving at the final assessment and recommendation, a discussion of any anomalous behavior of the property, a comparison of the recommended values with the experimental data, and a statement of the estimated uncertainty in the recommended values. This is accompanied by a figure showing the recommended values along with some selected experimental data, a table of recommended values, and a list of references. Each volume concludes with a detailed index to properties and materials.

The volumes in the *CINDAS Data Series* constitute an invaluable source of data for engineers and scientists; here for the first time they have access to a wealth of information heretofore not known or readily available. It should be emphasized that new data are being compiled and evaluated at CINDAS on a continuing basis. The user is encouraged to contact CINDAS for the latest information that may be available in any given area covered by this and other volumes.

It is indeed a pleasure to acknowledge with gratitude the multisource financial assistance received from over fifty past and present sponsors of CINDAS, which has made the continued generation of these reference works possible. In particular, I wish to single out the sustained major support being received from the Department of Defense in the operation of its *Thermophysical and Electronic Properties Information Analysis Center* (TEPIAC), and from the Office of Standard

Reference Data, National Bureau of Standards, for continuing support in data evaluation and critical analysis. The outputs of these two major programs have made possible the preparation of the majority of the volumes in this Data Series.

While the preparation and continued maintenance of this work is the responsibility of CINDAS's Reference Data Division, it would not have been possible without the direct input of CINDAS's Scientific Documentation Division. The contributing authors of the various volumes are primarily the senior staff members at CINDAS in responsible charge of the work. It should be clearly understood, however, that many others have contributed over the years, and I wish to take this opportunity to personally thank these members of the CINDAS staff, past and present: research assistants and graduate research assistants, as well as computer, graphics, and technical typing support personnel, without whose diligent and painstaking efforts this work could not have materialized.

Lastly, it is inherent to the character of such a work that we have drawn heavily upon the scientific and technical literature and want to express our debt of gratitude to the authors of the referenced articles as well as the many not cited herein. While their often discordant results have given rise to difficulties in reconciling their findings, we consider this to be our challenge and contribution.

Y. S. Touloukian*
Founding Director
Center for Information
  and Numerical Data Analysis and Synthesis
Purdue University

*Deceased in 1981.

Introduction

In 1980, Professor Y. S. Touloukian had discussed the desirability of the preparation of a comprehensive volume on the specific heat of solids to include theory and measurement techniques for the temperature range from below 1 K to several thousands of degrees as a part of the CINDAS Data Series on Material Properties. The objective was to have an authoritative book to serve both the researchers and the students in the field of calorimetry. It was also the intent to provide guidelines in assessing the quality of data reported in the literature and in estimating specific heat of substances in the absence of reliable experimental data.

Although some aspects of both the theory and measurement of specific heat date back over 100 years, they have been continuously improved and extended over the variables temperature, pressure, etc. A major surge in calorimetry came about in the late 1950's and extended over approximately two decades. During this period, several conventional calorimetry techniques, such as adiabatic and drop, were perfected to their practical limits. General interest in cryogenics stimulated development of accurate specific heat measurement techniques and generation of a large amount of data at low temperatures. Also during this period, several specialized calorimetry techniques, such as levitation, modulation, and pulse, were developed for the measurements of specific heat at high temperatures and under other extreme conditions (high pressures, etc.). Almost all the calorimetry techniques have benefited immensely from the advances during the last three decades in electronics and computer technologies. As a result, improvements in the temperature measurement and power generation and control as well as improvements in the measuring instruments and data reduction techniques contributed to the increased accuracy and range in specific heat determinations.

It may be said that our understanding of the specific heat of solids and the techniques for its measurement have reached a state of maturity and are ready to be applied successfully to emerging new challenges. As examples, it can be pointed out that the sharp increase in specific heat at high temperatures is not satisfactorily explained yet, and that the accurate measurement of the specific heat under combined high-temperature and high-pressure conditions remains to be conquered.

This volume is comprised of 12 chapters and three appendices; its contents may be grouped into the following three categories: (1) theory, (2) measurement techniques, and (3) general interest items. The theory of the specific heat of solids is discussed in detail in Chapter 1. Also, as a part of this chapter, methods for the estimation of specific heat are presented. The remaining chapters are devoted to measurement techniques from cryogenic to very high temperatures. The calorimetry techniques in the range from below 1 K to 350 K are covered in Chapters 2–4. Chapters 5 and 6 present adiabatic and drop calorimetry above 300 K and up to moderately high temperatures. Levitation, modulation, and pulse calorimetry for measurements at very high temperatures (up to several thousands of degrees) are discussed in Chapters 7–9. Calorimetry techniques for measurements under very high pressures are described in Chapter 10. Immergence of differential scanning calorimetry as a means for reliable specific heat measurements is discussed in Chapter 11. Specialized calorimetry techniques applicable to measurements on radioactive substances are presented in Chapter 12. Finally, several items of general interest in calorimetry are given as appendices; they include: materials of construction in high temperature calorimetry, reference materials for calorimetry, and presentation of thermophysical data in the scientific and technological literature. The emphasis in this volume has been placed on discussing the well-established techniques for the measurement of the specific heat of solids. Techniques in which specific heat is obtained only as a by-product of measurements of primarily other properties are not included.

On this occasion, it is a great pleasure to acknowledge the contributions of the following individuals: the authors of the chapters for their enthusiastic cooperation and patience throughout the preparation of this volume; the reviewers: A. C. Anderson, G. Bäckström, C. R. Brooks, K. Churney, J. H. Colwell, D. A. Ditmars, D. L. Martin, R. A. MacDonald, A. P. Miiller, and W. W. Rodenburg for their constructive comments; A. P. Miiller for editorial assistance during the early stage of the preparation of the volume; C. Y. Ho and the members of CINDAS at Purdue University for their support and assistance throughout the preparation of the volume; specifically, D. M. Lenartz for her effective coordination of the efforts and for her competent preparation of the camera-ready copy of the entire volume.

This volume would not have existed if it were not for the vision and planning of the late Professor Y. S. Touloukian. It is with great respect and gratitude that we wish to dedicate this volume to his memory.

A. Cezairliyan

CHAPTER 1

Theory of Specific Heat of Solids

A. P. MIILLER

1.1. INTRODUCTION

The interaction between theory and experiment in the study of specific heat has contributed significantly to our knowledge of the properties of substances in their solid state. Any theory describing an atomic or molecular phenomenon in a solid generally leads to a set of energy levels which the particles of the system can occupy. The levels may be associated with a particular mode of energy such as the vibrational, rotational, electronic, or magnetic energy of the constituent particles. By suitable statistical methods, one can evaluate the average energy of the system and from it, the contribution of the particular mode to the specific heat. A comparison of the calculated values with those observed experimentally provides a direct test of the validity of the theoretical model employed. The observed specific heat, however, often contains contributions from more than one mode of energy. In such instances, theory can provide a means of identifying the contributing modes and a method of estimating and, hence, separating their specific heats.

In principle, any temperature-dependent phenomenon can contribute to the specific heat of a solid, and as a result the range of topics related in some way to the study of specific heat is indeed broad. Clearly, it is not possible to discuss all topics thoroughly within the confines of a single chapter and, therefore, our description of theory must, of necessity, be selective. In this chapter, the discussion encompasses three major areas of study: the specific heat in solids arising from the lattice vibrations, the conduction electrons, and from magnetic excitations. The aim of the present review is twofold: (1) to introduce students of the field to basic principles underlying the theory of specific heat, and (2) to provide the potential thermodynamic data user with a reasonably comprehensive coverage of important theoretical models in the above-mentioned areas of study.

There are a number of reviews of the theory of specific heat given in the literature. The earlier studies [1-3] deal primarily with the role played by harmonic lattice vibrations in determining the observed specific heat. Since then, the development of inelastic neutron scattering methods has enabled a detailed comparison between experiment and harmonic lattice theory, and has stimulated the study of anharmonic behavior in solids. These topics are included in our discussion of lattice specific heat given in Section 1.2. In more recent surveys of theory [4-6], considerable emphasis is placed on the effects of various phenomena on the specific heat at low temperatures. Information given in these reviews, suitably updated by more recent works, forms the

1

basis of our discussions on electronic and magnetic specific heats in Sections 1.3 and 1.4, respectively. A brief summary of the theory described in the present chapter is presented in Section 1.5.

Some discussions of estimation and correlation of specific heat contributions are given throughout the main text of this chapter in connection with comparisons between theory and experiment. These discussions, suitably amplified, are presented again in the Appendix (Section 1.6) for the reader who is not concerned with details of the various theoretical models, but is interested in the application of theoretical results to the analysis of specific heat data.

1.1.1. Definition of Specific Heat

The heat capacity of a system of arbitrary mass may be defined in terms of the following limit

$$C \equiv \lim_{\Delta T \to 0} \left(\frac{\Delta Q}{\Delta T} \right) \tag{1}$$

where ΔQ is the quantity of heat that must be added to the system to raise its temperature by an amount ΔT. In order to obtain a quantity that is independent of mass, Eq. (1) is divided by the system mass m to yield the specific heat capacity or, more simply, the specific heat

$$c \equiv \frac{C}{m} = \frac{đq}{dT} \tag{2}$$

where $đq$ is the quantity of heat required to raise the temperature of a unit mass of the system by an amount dT; the symbol '$đ$' refers to an infinitesimal amount rather than an exact differential change as denoted by the symbol 'd'. In general, the required quantity of heat will depend upon the temperature of the system as well as changes that may occur in other physical properties of the system during the temperature rise. However, the physical properties (or state variables) which define the equilibrium state of a thermodynamic system are not totally independent but rather are related by an equation of state. For an ideal fluid or a solid subject to a uniform hydrostatic pressure, the equation of state has the functional form $f(P,V,T) = 0$ and involves only three variables: the volume V and temperature T of the substance and the hydrostatic pressure P. Since one variable can always be expressed in terms of the other two, only two variables can have arbitrary values at the same time. Therefore, as T is increased, only one of the two remaining variables, P or V, can be kept constant. Consequently, there are two principal specific heats, one defined at constant pressure, the other at constant volume:

$$c_p \equiv \left(\frac{đq}{dT} \right)_p \qquad c_v \equiv \left(\frac{đq}{dT} \right)_v . \tag{3}$$

In most theoretical calculations, the natural quantity to calculate is the 'heat capacity per mole' since this refers to a fixed number of particles. This quantity is also a 'specific' heat capacity and, consequently, is also referred to in the literature as the specific heat. (The term 'heat capacity' is also used by some authors when referring to this quantity.) By convention, the principal (molar) specific heats are denoted by upper case symbols and are defined as in Eq. (3) by

$$C_p \equiv \left(\frac{đQ}{dT}\right)_p \qquad C_v \equiv \left(\frac{đQ}{dT}\right)_v \tag{4}$$

where $đQ$ is the quantity of heat required to raise the temperature of one mole of the substance by an amount dT under conditions of constant pressure or constant volume. In the present chapter, the term 'heat capacity' will be used when referring to an arbitrary amount of a substance, whereas the term 'specific heat' will refer to a unit amount of the substance; the precise meaning of 'specific heat' will be clear from the symbols or context used.

In recent years, the SI system of units [7,8] has become more widely used in the scientific literature for expressing values of specific heat. The SI units for specific heat are $J \cdot kg^{-1} \cdot K^{-1}$ for c_p or c_v and $J \cdot mol^{-1} \cdot K^{-1}$ for C_p or C_v. In SI, the mole (mol) is defined as 'the amount of substance of a system which contains as many elementary entities (atoms, molecules, etc.) as there are atoms in 0.012 kg of carbon-12.' The number of elementary entities is known as Avogadro's constant and is equal to 6.022045×10^{23} mol^{-1}. In the earlier literature, specific heats were usually expressed in other units such as $cal \cdot g^{-1} \cdot K^{-1}$, $cal \cdot (g\text{-}atom)^{-1} \cdot K^{-1}$ or $cal \cdot (g\text{-}molecule)^{-1} \cdot K^{-1}$ where the terms 'gram-atom' (gram atomic weight) and 'gram-molecule' (gram molecular weight) were used as synonyms for 'mole'; the present conversion factor of 1 calorie = 4.1840 J was not always used in determining specific heat values from the measurements. The units of $BTU \cdot lb^{-1} \cdot F^{-1}$ are still commonly used to express specific heats in the engineering literature.

All numerical values of and relationships among the physical quantities in this chapter, except where explicitly noted otherwise, are based on the SI system of units.

1.1.2. Relation of Specific Heat to Other Thermodynamic Quantities

According to the first law of thermodynamics [9-12], when a quantity of heat $đQ$ is added to a substance, part of the heat energy is converted into an increase in the internal energy U of the system and part is utilized in performing external work $đW$, that is,

$$đQ = dU + đW. \tag{5}$$

If the external work consists only of work done by the system in increasing its volume against an external pressure, then

$$đW = PdV. \tag{6}$$

It follows from Eqs. (4-6) that the specific heat at constant volume may be expressed as

$$C_v = \left(\frac{dU}{dT}\right)_v. \tag{7}$$

The quantity C_v is of particular theoretical interest since a change in the internal energy of a substance with temperature can, in principle, be related by statistical methods to changes in the translational, vibrational, rotational, electronic, and/or magnetic energy of its atoms.

Other properties of specific heat follow from the second law of thermodynamics which, for a reversible heat exchange, enables the calculation of the change in entropy from the relation

$$đQ = TdS. \tag{8}$$

The condition of reversibility is equivalent to saying that the system remains in equilibrium with its surroundings throughout the change of state. Equations (4) and (8) allow the principal specific heats to be expressed as

$$C_p = T\left(\frac{\partial S}{\partial T}\right)_p \qquad C_v = T\left(\frac{\partial S}{\partial T}\right)_v. \tag{9}$$

The change in the internal energy of the system may also be written as

$$dU = TdS - PdV \tag{10}$$

by combining Eqs. (5), (6), and (8). In certain cases, it is convenient to introduce other principal thermodynamic functions; namely, the Helmholtz free energy F, the enthalpy H, and the Gibbs free energy G which, in differential form, are given by

$$dF = d(U-TS) = -SdT - PdV \tag{11}$$

$$dH = d(U+PV) = TdS + VdP \tag{12}$$

$$dG = d(U-TS+PV) = SdT + VdP. \tag{13}$$

The four functions are essentially measures of the energy content of the system under various conditions and, in reversible thermodynamic processes, their changes depend only on the initial and final states of the system.

The Helmholtz free energy is particularly useful in discussing the statistical thermodynamic behavior of systems of particles [12,13] since it can be obtained directly from the partition function Z by the relation

$$F = -k_B T \ln Z \tag{14}$$

where k_B is Boltzmann's constant ($1.381 \times 10^{-23} J \cdot K^{-1}$). The partition function may be evaluated for a given theoretical model of the system from

$$Z = \sum_i \exp(-E_i/k_B T) \tag{15}$$

where the sum extends over all possible states of the system and E_i is the system energy corresponding to a given state. The system entropy can be readily obtained from Eq. (11) as $S = -(\partial F/\partial T)_v$ and from it, the specific heat at constant volume

$$C_v = T\left(\frac{\partial S}{\partial T}\right)_v = -T\left(\frac{\partial^2 F}{\partial T^2}\right)_v. \tag{16}$$

Thus, the specific heat at constant volume can be evaluated from the temperature dependence of either U or F; the choice will depend upon the nature of the given theoretical model, as seen in Sections 1.2-1.4.

As mentioned earlier, a comparison of the calculated values of specific heat with those determined experimentally provides a quantitative test of the validity of a given theory. In the study of solids, however, it is very difficult to measure C_v directly and, thus, C_p is the quantity that is normally determined by experiment.

Under isobaric conditions, the change in enthalpy is given by $dH = TdS = đQ$ and, therefore, it follows from the definition of specific heat that

$$C_p \equiv \left(\frac{đQ}{dT}\right)_p = \left(\frac{\partial H}{\partial T}\right)_p. \tag{17}$$

Experimentally, C_p is often derived from measurements of the rates of heat input to and temperature rise of the specimen, that is, $đQ/dT = (đQ/dt)(dT/dt)^{-1}$, or from measurements of the change in enthalpy (or heat content) of the specimen with changing temperature. The different measurement methods are discussed in detail in the succeeding chapters of this volume.

The third law of thermodynamics [9] requires that, as the temperature of a substance approaches absolute zero, its specific heats (C_p and C_v) must tend to zero at least as rapidly as the first power of T. At finite temperatures, however, C_p is always greater than C_v. The reason for this is easy to see. Heating a substance at constant pressure through a temperature rise dT requires sufficient heat not only to increase its internal energy but also to enable the substance to do work in expanding against an external pressure. Under the restraint of constant volume, no external work is done by the substance and less heat is required to obtain the same temperature rise. For this reason, the difference C_p-C_v is often referred to as the dilation contribution to the observed specific heat.

The general relation between C_p and C_v is derived in most texts on thermodynamics [9–12] and is given by

$$C_p - C_v = \frac{V_m \beta^2 T}{K_T} \tag{18}$$

where V_m is the molar volume, $\beta \equiv V^{-1}(\partial V/\partial T)_p$ is the coefficient of volumetric expansion, and $K_T \equiv -V^{-1}(\partial V/\partial P)_T$ is the isothermal compressibility. For isotropic solids, β is related to the coefficient of linear thermal expansion α by

$$\beta = \frac{1}{\ell^3}\left(\frac{\partial \ell^3}{\partial T}\right)_p = \frac{3}{\ell}\left(\frac{\partial \ell}{\partial T}\right)_p = 3\,\alpha \tag{19}$$

where ℓ is the length of a side of a representative cube of the solid.

Thus, measured values of C_p can be converted to C_v if values of β (or α) and K_T are known at the desired temperatures. Unfortunately, for many solids values of K_T are usually available in the literature only at near-room temperatures and, consequently, some method of estimating C_p-C_v must often be employed. Such methods are described in the Appendix (Section 1.6) of this chapter.

In many solids, the difference C_p-C_v is only about 1% of C_p at $\theta_D/3$ and decreases to about 0.1% of C_p at $\theta_D/6$, where θ_D is the Debye characteristic temperature (Section 1.2.2). Clearly, values of C_v derived from the observed C_p rapidly become insensitive to the uncertainty in estimating the dilation contribution as temperature is decreased. The largest uncertainty in C_v may be expected at the melting point, where C_p-C_v is usually a maximum and is of the order of 10% of C_p; in such cases an error of 10%, for example, in estimating the dilation term would contribute an uncertainty of the order of 1% in the derived value of C_v.

1.1.3. Historical Background

The subject of specific heat studies is much older than is commonly realized, dating back to the work by Dulong and Petit [14] in 1819. Their measurements of specific heat of 13 solid elements near room temperature revealed that the product of the specific heat per unit mass at constant pressure and the atomic weight is approximately a constant, about 6 cal·(g-atom)$^{-1}$·deg^{-1}. The implication of this observation, known as the 'Dulong-Petit law,' is that the heat capacity per atom is about the same for the different elements. For this reason, the heat capacity per gram atomic weight of an element was sometimes referred to as the 'atomic heat.' An important extension of the Dulong-Petit rule became evident during the period 1840 to 1860 as a result of experiments by several investigators on chemical compounds. Their observations led to a generalization known as the 'Kopp-Neumann law' which states that the 'molecular heat' (heat capacity per gram molecular weight) of a compound is equal to the sum of 'atomic heats' of the constituent elements. In other words, diatomic and triatomic solids were expected to have specific heats of approximately 12 and 18 cal·(g-molecule)$^{-1}$·deg^{-1}, respectively.

A theoretical explanation of Dulong and Petit's law was first given in 1871 by Boltzmann [15] on the basis of his theorem on the equipartition of energy. A detailed development of the theorem and its application to a classical solid may be found in texts on statistical mechanics [13]. In brief, the theorem states that, for a system in thermal equilibrium, each degree of freedom contributes $1/2 \, k_B T$ to the average energy of the system. For example, a linear harmonic oscillator of mass m and (angular) frequency ω has a total energy given by the sum $p^2/(2m) + m\omega^2 x^2/2$ where the momentum p and displacement x (from equilibrium) are the two independent variables needed to describe the motion of the system. Therefore, the linear oscillator is said to have two degrees of freedom and, indeed, the theorem of equipartition of energy yields a value $k_B T$ for the average energy of the oscillator. A three-dimensional oscillator requires six independent variables (p_x, p_y, p_z, x, y, z) to describe its motion; the six degrees of freedom in this case yield an average energy of $3k_B T$. A simple model of a solid is to view the atoms oscillating about their lattice sites as a system of (3-dimensional) oscillators. Thus, in one gram atomic weight (one mole) of a monatomic solid the total (average) energy of the N_A atoms is given by

$$U = 3N_A k_B T = 3RT \tag{20}$$

where N_A is Avogadro's constant and $R = N_A k_B$ is the universal gas constant. It follows from Eq. (7) that the specific heat of the solid at constant volume is given by

$$C_v = 3R. \tag{21}$$

The quantity 3R = 5.9616 cal·mol^{-1}·K^{-1} = 24.9432 J·mol^{-1}·K^{-1} is often referred to as the 'Dulong-Petit' value of specific heat. For a compound containing n atoms per molecule, the above argument is readily extended to yield $C_v = 3nR$ as the specific heat. Clearly, the empirical generalization of Dulong and Petit and others appeared to be justified on the basis of classical statistical mechanics. Nevertheless, exceptions to the Dulong-Petit rule were known to exist, notably silicon, boron, and diamond which had specific heats of about 20, 11, and 8 J·mol^{-1}·K^{-1}, respectively, near 300 K. In 1875, measurements by Weber [16] on diamond over a wide temperature range (200 to

1300 K) suggested that, at sufficiently high temperatures, the specific heats of all substances approach the Dulong-Petit value.

During the next few decades, experimental studies at low temperatures showed that, in general, the specific heats of solids decrease rapidly with decreasing temperature. This led Nernst [17] and others to suggest that the specific heat of solids must tend towards zero as the absolute zero of temperature is approached.

The failure of classical theory to predict the behavior of the specific heat of solids at low temperatures initiated considerable theoretical and experimental effort into studies on the dynamical behavior of atoms in a solid. It was known (from mineralogical studies, later confirmed by X-ray diffraction measurements) that the atoms in a crystalline solid are arranged in a regular geometrical array or lattice. Hence, this field of study which began in the early 1900's became known as 'lattice dynamics.' The different lattice dynamical models which were subsequently proposed to explain the contribution of the lattice vibrations to the specific heat are discussed in Section 1.2.

Around the turn of the century it also became evident that the high electrical and thermal conductivities which characterized metals were related to the movement of electrons through the solid. The first successful theoretical description of metallic conductivity was given by Drude [18] in 1900. By treating the kinetic behavior of the electrons as that of a classical ideal gas, he was able to derive the Wiedemann-Franz relation between electrical and thermal conductivity which had been discovered empirically [19] nearly 50 years earlier. However, the classical electron gas model also predicted an excessively large contribution from the conduction electrons to the specific heat: on the basis of the equipartition theorem, each (free) electron contributes $3(1/2\ k_BT)$ to the internal energy and, hence, for a monovalent metal, the electrons should contribute $(3/2)R$ to the specific heat in addition to $3R$ from the lattice. This result was in complete disagreement with the available calorimetric data (near room temperature) which indicated that the observed specific heat was entirely accounted for by the atomic vibrations.

With the development of quantum mechanics in the 1920's, Sommerfeld [20,21] was able to explain the unexpectedly small magnitude of the electronic contribution by applying quantum statistics to the problem. The next decade brought about the rapid development of a reasonably complete description of metallic properties in terms of nearly-free electron behavior [22]. By the 1960's it became clear that a nearly-free electron picture, which had seemed rather naive, gave a relatively accurate description of the electronic structure of simple metals [23]. A description of the electronic specific heat in terms of free electron behavior and deviations therefrom is discussed in Section 1.3.

1.2. LATTICE SPECIFIC HEAT

The basis of all lattice dynamic theories is the harmonic approximation in which thermal vibrations of the atoms about their equilibrium positions are treated as a superposition of motions of harmonic oscillators, according to the laws of quantum statistics. This approximation is mathematically equivalent to expanding the interatomic potential as a Taylor series in atomic displacements and discarding all terms higher than quadratic. High-order terms may then be treated as perturbations to describe deviations from harmonic behavior. As will become clear in the following discussion of lattice models, this approach has been remarkably successful in describing the lattice specific heat of solids.

1.2.1. The Einstein Model

In 1907, Einstein [24] formulated the first successful dynamical model of a solid on the basis of Planck's quantum hypothesis. The solid is treated as a collection of uncoupled linear harmonic oscillators each vibrating with the same frequency ν_E. The number of oscillators is chosen to equal the number of degrees of freedom in the problem. With these few simple assumptions, Einstein demonstrated that the rapid reduction of specific heat with decreasing temperature was indeed a manifestation of quantization of the mechanical oscillators; that is, a quantization of the atomic vibrations.

According to quantum theory, the energy of a linear harmonic oscillator of frequency ν is quantized in units of $h\nu$ according to

$$\varepsilon_n = (n + 1/2)\ h\nu \tag{22}$$

where $n = 0, 1, 2, 3, \ldots$ are the quantum numbers corresponding to energy states ε_n, h is Planck's constant $(6.626 \times 10^{-34}\text{J}\cdot\text{s})$ and $1/2\ h\nu$ is the zero point energy. The collective properties of an ensemble of such oscillators can be determined by recourse to statistical methods [12,24-26]. For a given oscillator in thermal equilibrium at temperature T, the probability that it has energy ε_n is proportional to the Boltzmann factor $\exp(-\varepsilon_n/k_B T)$, and so one may express the average energy as

$$\langle \varepsilon \rangle = \frac{\sum_n \varepsilon_n \exp(-\varepsilon_n/k_B T)}{\sum_n \exp(-\varepsilon_n/k_B T)} = (\langle n \rangle + 1/2)\ h\nu \tag{23}$$

where

$$\langle n \rangle = \frac{\sum_n n \exp(-\varepsilon_n/k_B T)}{\sum_n \exp(-\varepsilon_n/k_B T)} \tag{24}$$

is the average quantum number of the oscillator. The identity $y \equiv \exp(-h\nu/k_B T)$ and the property

$$\sum_n y^n = (1-y)^{-1}$$

enable one to write

$$\langle n \rangle = \frac{\sum_n n\, y^n}{\sum_n y^n} = y\,\frac{d}{dy}\left(\ell n \sum_n y^n\right) = -y\,\frac{d}{dy}\,\ell n(1-y) = \frac{y}{1-y}\ . \tag{25}$$

Therefore,

$$\langle n \rangle = \frac{1}{\exp(h\nu/k_B T) - 1}\ . \tag{26}$$

Consideration of Eqs. (23) and (26) shows that, at high temperatures ($T \gg h\nu/k_B$), the average $\langle n \rangle$ varies as $k_B T/h\nu$ and, hence, the average oscillator energy approaches the classical value of $k_B T$. In the limit as temperature approaches absolute zero, $\langle n \rangle$ approaches zero exponentially and, therefore, $\langle \varepsilon \rangle$ approaches $1/2 \, h\nu$, the energy held by each oscillator at zero kelvin.

In the Einstein model, the total vibrational energy of N_A atoms is equal to the sum of the average energies of $3N_A$ oscillators vibrating with a common frequency ν_E, that is,

$$U = 3N_A \langle \varepsilon \rangle = 3N_A (\langle n \rangle + 1/2) \, h\nu_E. \tag{27}$$

Therefore, the contribution of the atomic vibrations to the specific heat of the solid is

$$C_v = \left(\frac{\partial U}{\partial T}\right)_v = \frac{3 \, N_A k_B (h\nu_E/k_B T)^2 \, \exp(h\nu_E/k_B T)}{[\exp(h\nu_E/k_B T) - 1]^2}. \tag{28}$$

The ratio $h\nu_E/k_B$ acts as a scaling factor for the temperature and, thus, is defined as the characteristic temperature θ_E, known as the 'Einstein temperature.' This enables the specific heat to be conveniently expressed as

$$C_v = \frac{3R(\theta_E/T)^2 \, \exp(\theta_E/T)}{[\exp(\theta_E/T) - 1]^2}. \tag{29}$$

Tabulations of the Einstein specific heat function $C_v(\theta_E/T)$ may be obtained from a number of sources [5,27,28], including the Appendix (Table A2) of the present chapter. At high temperatures ($T \gg \theta_E$) the exponential terms may be replaced by their series expansions so that Eq. (29) becomes

$$C_v = 3R \left[1 - \frac{1}{12} \left(\frac{\theta_E}{T}\right)^2 + \ldots\right]. \tag{30}$$

At sufficiently low temperatures ($T \ll \theta_E$), the specific heat is approximated by

$$C_v \simeq 3R \, (\theta_E/T)^2 \, \exp(-\theta_E/T). \tag{31}$$

Therefore, Einstein's theory of specific heat yields the classical Dulong-Petit value 3R in the high temperature limit and also predicts that the specific heat of a solid should decrease to zero (exponentially) as the temperature approaches zero kelvin.

The publication of Einstein's theory in 1907 provided considerable stimulus to the development of low temperature calorimetry. Within a few years Nernst and co-workers [29] had measured the specific heats of a number of substances at temperatures as low as ~20 K. Their results showed that while the low-temperature behavior predicted by the Einstein model was qualitatively correct, the specific heat of real solids did not decrease as rapidly with decreasing temperature. It was recognized by Einstein [30] and others that in a more realistic model of a solid the atoms would be highly coupled and would vibrate collectively at many different frequencies. This work inspired investigations of more sophisticated models in which the single Einstein frequency was replaced by a frequency spectrum. Models of this type were proposed simultaneously by Debye [31] and by Born and von Kármán [32] in 1912, using rather different approaches. The much simpler Debye model is considered first in the next section.

In spite of its limitations, Einstein's theory clearly demonstrated the need to take quantum-mechanical considerations into account in any dynamical model of a solid and, thus, provided a basis for the more sophisticated lattice models which followed. The Einstein model is particularly useful for determining the vibrational specific heat in systems characterized by one or more discrete frequencies [33] and is sometimes used to represent the lattice contribution in the analysis of specific heat in complex solids [34].

1.2.2. The Debye Model

On the basis of quantum statistics, the internal energy of a solid at low temperatures is determined primarily by the low frequency modes of vibration. These 'acoustic' oscillations have wavelengths (λ) considerably greater than the interatomic distances and, therefore, propagate through the solid as if it were an elastic continuum. The fundamental assumption in the Debye model is that the solid can be treated as an isotropic elastic continuum for all possible vibrational modes. This simplification eliminates the need for making detailed assumptions about the coupling of the atomic motions. The discreteness of the solid enters the theory only in normalizing the total number of vibrational modes to $3N_A$, thus limiting the range of frequencies to some maximum or cut-off frequency ν_D. Since $3N_A$ is very large ($\sim 10^{24}$), it is convenient to treat the vibrational levels as continuous and, therefore, to define the number of modes in frequency range ν to $\nu + d\nu$ as $g(\nu) \, d\nu$; the normalization condition then becomes

$$\int_0^{\nu_D} g(\nu) \, d\nu = 3N_A. \tag{32}$$

The derivation of the frequency distribution function $g(\nu)$ is based on the assumption that the relation between frequency ν and wave vector $q = 2\pi/\lambda$ is given by the well-known linear relation for ordinary sound waves; that is,

$$\nu = c \, q/2\pi = c/\lambda \tag{33}$$

where c, the velocity of sound, is a constant independent of the frequency. A convenient device for enumerating the vibrational modes is the method of periodic boundary conditions [26] which assumes the solid to be infinite in extent but requires that the elastic (traveling) waves have solutions that are periodic over some finite large distance L. In this way, consideration of the surface of the solid is avoided without changing the essential physics of the problem, at least for large systems. If L is taken as the dimension of a representative cube of the continuum, then one may express the boundary condition as

$$\exp[i\{xq_x + yq_y + zq_z\}] = \exp[i\{(x + L)q_x + (y + L)q_y + (z + L)q_z\}]. \tag{34}$$

Therefore, the allowed values for the components of $\vec{q}$ are given by

$$q_x, \; q_y, \; q_z = 0; \; \pm \frac{2\pi}{L}; \; \pm \frac{4\pi}{L}; \; \cdots \tag{35}$$

Since the volume in $\vec{q}$-space associated with each allowed $\vec{q}$ is $(2\pi/L)^3$, it follows that number of allowed values of $\vec{q}$ per unit volume of $\vec{q}$-space is $(L/2\pi)^3 = V/8\pi^3$ where $V = L^3$ is a representative volume of the solid. In practical cases, the number of allowed values of $\vec{q}$ (that is, allowed modes of vibration) is so large that $q = |\vec{q}| = [q_x^2 + q_y^2 + q_z^2]^{1/2}$ may be taken as a continuous variable. Therefore, the number of modes, N, with a wavevector equal to q or less is simply given by $V/8\pi^3$ times the volume of a sphere of radius q; that is,

$$N = \frac{V}{8\pi^3} \cdot \frac{4}{3} \pi q^3 = \frac{V}{6\pi^2} q^3. \tag{36}$$

With the aid of Eq. (33), one may express the number of modes per unit frequency interval as

$$g(\nu) = \frac{dN}{d\nu} = \frac{4\pi V}{c^3} \nu^2. \tag{37}$$

In a real solid, the elastic waves associated with a given wavevector q are of three types: a longitudinal compression wave and two transverse shear waves, each with its own velocity of sound. Therefore, the total number of vibrational modes per unit frequency range is actually equal to the sum of the contributions from each type of wave; that is,

$$g(\nu) = \left(\frac{12\pi V}{c_m^3}\right) \nu^2 \tag{38}$$

where c_m is a 'mean' velocity of sound obtained by averaging the reciprocals of the sound velocities for the three wave types over the crystallographic directions (see Eq. (47) below); in an isotropic solid the mean sound velocity is determined by the relation $3 c_m^{-3} = c_L^{-3} + 2 c_T^{-3}$, where c_L and c_T are the respective sound velocities of the longitudinal and transverse waves. Although Eq. (38) was derived for a representative cube of the continuum, this result is not significantly changed by consideration of large systems of other shapes [35,36].

The cut-off frequency for the vibrational spectrum in Eq. (38) is determined from the normalization condition of Eq. (32) as

$$\nu_D = (3N_A/4\pi V)^{1/3} c_m. \tag{39}$$

The quantity $(3N_A/4\pi V)^{-1/3}$ has the dimensions of length and, thus, may be regarded as the effective (minimum) wavelength corresponding to the maximum frequency ν_D. In a typical solid this quantity is ~4 $\overset{\circ}{A}$, which is of the same order as the interatomic spacings.

The total vibrational energy of the solid can now be obtained by summing the average energy of each vibrational mode over the frequency spectrum:

$$U = \int_0^{\nu_D} g(\nu) \; (\langle n \rangle + 1/2) \; h\nu \; d\nu. \tag{40}$$

The contribution of the lattice vibrations to the specific heat of the solid is then determined from Eqs. (26), (38), and (40) as

$$C_v = \left(\frac{\partial U}{\partial T}\right)_v = 9R \left(\frac{T}{\theta_D}\right)^3 \int_0^{\theta_D/T} \frac{x^4 e^x dx}{(e^x - 1)^2} \tag{41}$$

where $x = h\nu/k_B T$ and the Debye characteristic temperature θ_D is defined in terms of the cut-off frequency by

$$\theta_D = h\nu_D/k_B. \tag{42}$$

At temperatures $T \gg \theta_D$, the dimensionless quantity $x \ll 1$ so that e^x may be replaced in Eq. (41) by its series expansion, yielding

$$C_v = 3R \left[1 - \frac{1}{20} \left(\frac{\theta_D}{T} \right)^2 + \ldots \right], \tag{43}$$

in agreement with classical theory in the high temperature limit. At very low temperatures ($T \ll \theta_D$), the upper limit of integration in Eq. (41) may be replaced by infinity. The value of the resulting integral is found in standard tables as $12\pi^4/45$ and accordingly one obtains

$$C_v = \frac{12}{5} \pi^4 R \left(\frac{T}{\theta_D} \right)^3, \tag{44}$$

which is a good mathematical approximation of C_v for temperatures below $\theta_D/10$. Figure 1.1 clearly illustrates the T^3-variation in the specific heat of NaF [37] at low temperatures.

Specific heat measurements have shown that the range of strict validity of the T^3-law is often limited to temperatures below about $\theta_D/50$. The departure from a T^3-dependence for $T > \theta_D/50$ is the result of deviations of $g(\nu)$ of the real solid from the assumed ν^2-dependence. At these temperatures a more accurate representation of the spectrum is given by its low-frequency expansion [38]

$$g(\nu) = \alpha_1 \nu^2 + \alpha_2 \nu^4 + \alpha_3 \nu^6 + \ldots . \tag{45}$$

This leads to a corresponding series for the specific heat

$$C_v = \beta_1 T^3 + \beta_2 T^5 + \beta_3 T^7 + \ldots \tag{46}$$

which provides a convenient means of representing the lattice contribution to the specific heat at low temperatures.

At moderate temperatures, the integral in Eq. (41) must be evaluated numerically in order

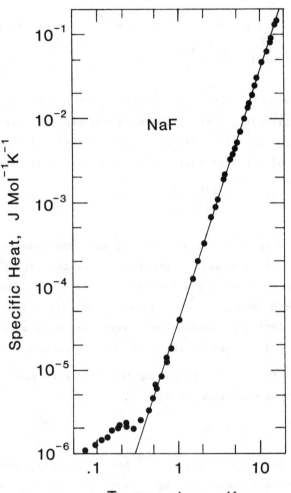

Figure 1.1. Specific heat of NaF plotted as a function of temperature on a log-log scale. The straight line represents the equation $C_p = 3.84 \times 10^{-5} T^3$, which corresponds to $\theta_D = 466$ K. The deviation from a straight line at the lowest temperatures is attributed to a Schottky anomaly arising from impurities in the specimen (after Harrison et al. [37]).

to obtain C_v. Values of the Debye specific heat function $C_v(\theta_D/T)$ have been tabulated in the literature [5,39,40] for a wide range of θ_D/T and are also given in the Appendix (Table A3) of the present chapter. The variation of $C_v(\theta_D/T)$ with changing temperature is illustrated in dimensionless form in Fig. 1.2 as a plot of $C_v/3R$ versus T/θ_D; a similar plot for the Einstein model is given for comparison. The physical significance of the Debye characteristic temperature becomes apparent: the Debye θ is the temperature which approximately marks the beginning of the rapid reduction in C_v with decreasing temperature; that is, the approximate temperature below which the high-frequency vibrations begin to 'freeze out' and no longer contribute to the internal energy nor the specific heat.

Simple Debye theory suggests that by suitable selection of a single parameter θ_D the specific heat of any solid should fall on a 'universal' curve as given in Fig. 1.2. However, it is well known that in order to accurately fit the Debye model to the observed C_p data (suitably corrected for C_p-C_v), the required value of θ_D must be permitted to vary with temperature. In spite of this shortcoming, the Debye theory is still useful for describing the specific heat of a solid, usually in the form of a θ_D versus T plot. In this way, one can clearly elucidate small differences in the temperature variation of specific heat for different solids. In general, the $\theta_D(T)$ functions for different solids exhibit a number of similar features (see Figs. 1.3 and 1.7). At temperatures below $\theta_D/50$, the Debye characteristic temperature becomes essentially constant and is approximately equal to its limiting value θ_o^c at absolute zero. At very high temperatures, all vibrational modes are excited and, thus, on the basis of harmonic theory, θ_D is expected to be constant. In many solids, this is approximately true at temperatures above about $\theta_D/2$. At intermediate temperatures, however, θ_D varies somewhat, often passing through a minimum. The variation of θ_D from its mean value is usually less than 10%, although variations of 20% or more have been observed in some cases [41].

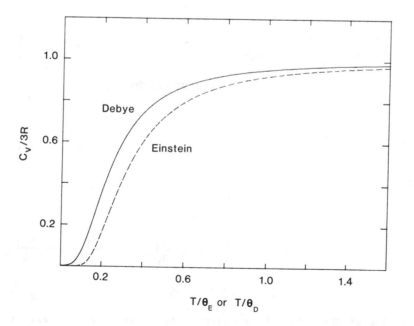

Figure 1.2. Temperature variation of the specific heat in the Einstein and Debye models.

The connection between the thermal and elastic properties of a solid is inherent in Debye's continuum model, at least at very low temperatures. Equations (39) and (42) show that a characteristic temperature θ can also be determined from the velocities of the longitudinal and transverse sound waves. In general, these velocities depend upon the direction of propagation through the crystalline lattice. Therefore, it is convenient to define a mean velocity c_m such that

$$3 \; c_m^{-3} = \int \sum_{j=1}^{3} c_j^{-3} \; d\Omega/4\pi \qquad (47)$$

where index j refers to the vibration polarization (longitudinal or transverse) and c_j are values of the principal sound velocities in the element of solid angle $d\Omega$. The methods for determining average velocities from elastic constant data have been reviewed in the literature [2,42]. It is expected that, in the low-temperature limit, the characteristic temperature θ_o^{el} determined from elastic constant values should be the same as θ_o^c determined from calorimetric measurements. Indeed, this equality has been shown formally to hold even if the lattice vibrations are anharmonic [43,44]. A comparison of θ_o^{el} and θ_o^c is presented in Table 1.1 for selected substances. The agreement for most substances is found to lie within the experimental uncertainties arising from measurements of the specific heats and the elastic constants. At higher temperatures, θ values obtained from elastic constant data generally do not agree with those obtained from calorimetric

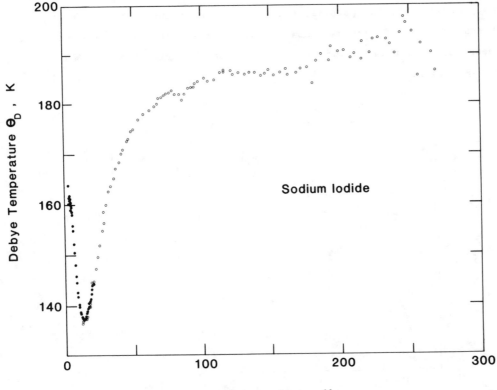

Figure 1.3. Variation of the Debye characteristic temperature of NaI with temperature (after Berg and Morrison [41]).

Table 1.1. A Comparison of θ_o^c and θ_o^{e1} for Selected Substances [42]

Substance	θ_o^c (in K)	θ_o^{e1} (in K)
Ag	226.6[a]	226.4
Al	427.7	430.6
Au	161.8[a]	161.6
Cd	219.5	213.4
Cu	344.3[a]	344.4
Fe	445	478.1
Ge	371	374.3
In	109	111.3
Mo	470	474.0
Ni	441	476.0
Nb	238	276.2
Pb	106.7	105.3
Pd	275	275.5
Sn	200	201.4
Ta	240	263.8
W	378	384.3
CaF_2	508	514.1
InSb	200	205.8
KCl	235.1	236.2
LiF	736	734.3
NaI	163.2	167.5

[a]Values taken from more recent measurements [45].

data. This is not unexpected since the elastic constants are determined only at very low frequencies where $g(\nu) \propto \nu^2$ whereas the specific heat values involve contributions from higher frequencies in the vibrational spectrum where $g(\nu)$ departs significantly from a ν^2-behavior (Section 1.2.5).

Characteristic temperatures can also be determined from other properties of a solid such as compressibility and melting temperature, thermal expansion, infrared absorption, etc. Details regarding the determination of these θ-values have been reviewed by a number of authors [2,5,46]. The important point to note here is that there is no common theoretical basis for the exact equality of these θ-values with either θ_o^c or θ_o^{e1}. Their usefulness is limited only to obtaining rough estimates of the lattice specific heat in the absence of calorimetric or elastic constant data.

The inability of Debye's continuum model to accurately predict the specific heat over the entire temperature range can be attributed to the fact that real solids are composed of discrete atoms. However, when explicit consideration is given to the interatomic forces, the enumeration of the normal-mode frequencies becomes a difficult task, as will become clear in the following sections (1.2.3-1.2.5). Because of this, the Debye theory has been and continues to be used extensively in the description of the thermal behavior of solids, particularly at low temperatures.

1.2.3.   Theory of Harmonic Lattice Dynamics

The basis of the modern theory of lattice dynamics is the three-dimensional model of a crystalline solid proposed by Born and von Kármán [32] in 1912. Later, extensions of this model to a general theory of lattice vibrations were developed by Born and his collaborators, resulting in what is now known as the Born-von Kármán theory. The theory has been discussed in the literature by a number of authors [2,47-49].

Unfortunately, because of the general nature of the Born-von Kármán model, the effect of interatomic coupling on the vibrational modes of a crystal is not readily apparent from the formalism. Therefore, it is instructive to defer the discussion of the Born-von Kármán theory (Section 1.2.3.2) until we have considered a simple model for which there is an explicit solution. One such model often discussed in introductory accounts of lattice dynamics [3,5,25,26] is the linear diatomic chain of atoms.

1.2.3.1.   Dynamics of a One-Dimensional Diatomic Lattice

For simplicity, assume that the two kinds of atoms are spaced a distance d apart. Atoms of mass m are located at the even-numbered lattice sites 2n, 2n+2, ..., whereas atoms of mass M are located at the odd sites 2n-1, 2n+1, ... Furthermore, assume that each atom interacts only with its two nearest neighbors so that a relative displacement $u_{2n+1} - u_{2n}$ between atoms at lattice sites 2n+1 and 2n gives rise to a force $\beta (u_{2n+1} - u_{2n})$ on atom 2n; here u_{2n} denotes the (small) displacement of atom 2n from its equilibrium position. It follows that the classical equations of motion for the atoms at these two sites may be written as

$$m \, \ddot{u}_{2n} = \beta \, (u_{2n+1} - u_{2n} + u_{2n-1} - u_{2n})$$
$$M \, \ddot{u}_{2n+1} = \beta \, (u_{2n+2} - u_{2n+1} + u_{2n} - u_{2n+1})$$

(48)

where the notation $\ddot{u}_{2n}$ denotes the second time derivative of u_{2n}. The solutions may be taken in the form for traveling waves; that is,

$$u_{2n} = \xi_m \exp\{i[\omega t + 2nqd]\}$$
$$u_{2n+1} = \xi_M \exp\{i[\omega t + (2n+1)qd]\}$$

(49)

which, when substituted into Eq. (48), lead to

$$-\omega^2 m \, \xi_m = \beta \, \xi_M \, (e^{iqd} + e^{-iqd}) - 2\beta\xi_m$$
$$-\omega^2 M \, \xi_M = \beta \, \xi_m \, (e^{iqd} + e^{-iqd}) - 2\beta\xi_M.$$

(50)

The condition for a nontrivial solution to this set of homogeneous equations is that the determinant of the coefficients of ξ_m, ξ_M must vanish:

$$\begin{vmatrix} 2\beta - m\omega^2 & -2\beta \cos qd \\ -2\beta \cos qd & -2\beta - M\omega^2 \end{vmatrix} = 0.$$

(51)

On expanding the determinant and solving in terms of the lattice constant a = 2d, one obtains

$$\omega^2 = \beta \left(\frac{m+M}{mM}\right) \pm \beta \left[\left(\frac{m+M}{mM}\right)^2 - \frac{4}{mM} \sin^2 \frac{qa}{2}\right]^{1/2}.$$

(52)

The variation of ω with q in Eq. (52) is shown in Fig. 1.4(a) for M > m. It can be seen that the two roots correspond to two branches of the frequency-wave vector relation $\omega(q)$. The nature of the two branches can be understood by considering the motion of the atoms for waves of small q (long wavelength); in this case the roots are

(i) $\omega^2 = \frac{\beta a^2}{2(m+M)} q^2$ and   (ii) $\omega^2 = 2\beta \left(\frac{m+M}{mM}\right).$

If root (i) is used in Eq. (50), one obtains $\xi_m \simeq \xi_M$; that is, the atoms (and their center of mass) move together as in acoustical vibrations. Thus, this branch is called the 'acoustical branch.' If root (ii) is used in Eq. (50), one obtains $\xi_m \simeq -(M/m)\xi_M$; that is, the atoms vibrate against each other with their center of mass fixed. If the two types of atoms are oppositely charged ions, this motion can be excited by electric fields as, for example, light waves. Therefore, this branch is called the 'optical branch.'

In the case of a linear monatomic chain (m=M), the pairs of relations in Eqs. (48-50) become degenerate and yield only the acoustic branch as a solution:

$$\omega^2 = \frac{4\beta}{M} \sin^2 \frac{qa}{2}.$$

(53)

The dependence of ω on q is shown in Fig. 1.4(b) for the positive branch of $\omega(q)$.

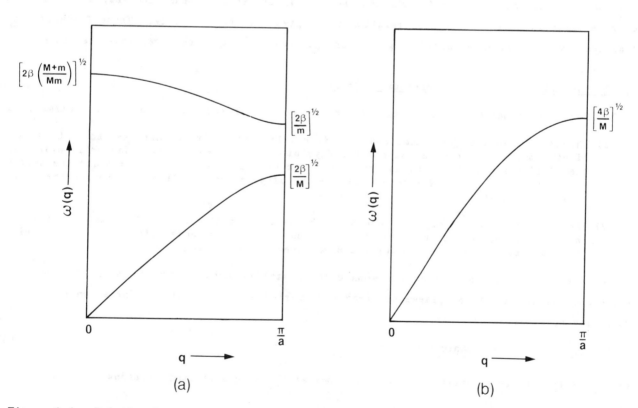

(a)   (b)

Figure 1.4. Relation between frequency $\omega(q)$ and wave vector q for vibrations in (a) a linear diatomic lattice and (b) a linear monatomic lattice.

The values of ω given by Eqs. (52) and (53) are clearly periodic in q. However, it can be shown [50] that the atomic displacements ... u_{2n-1}, u_{2n}, u_{2n+1}, ... associated with $|q| > \pi/a$ are indistinguishable from those corresponding to smaller values of q and that all physical vibrational modes can be accounted for by the sole consideration of wave vectors in the range $|q| \leq \pi/a$. Within the fundamental interval, called the first Brillouin zone [50], the allowed values of q are determined by the boundary conditions of the problem. By an argument similar to that used for the elastic continuum, one can apply periodic boundary conditions to a representative length L >> a of the linear crystal containing N atoms. The allowed values of q are $\pm 2\pi L$, $\pm 4\pi/L$, ..., $\pm(N/2)(2\pi/L)$, ... for a total of N values which, in turn, yield a frequency distribution function expressed as

$$g(\omega) = (L/2\pi)/(d\omega/dq). \tag{54}$$

The quantity dω/dq, which is the slope of the ω(q) relation, is the group velocity [25,26] of the lattice waves. As can be seen in Fig. 1.4, the group velocity approaches zero in the limiting cases of $|q| \to 0$ (optical branch only) and π/a, thereby giving rise to singularities in the frequency spectrum g(ω).

The one-dimensional lattice models considered here exhibit many of the general features of vibrations observed in real crystals. These include (a) the existence of dispersion in the ω(q) or ν(q) relation, (b) the presence of acoustical modes of vibration in all crystals and additional types of vibration called optical modes in diatomic (and polyatomic) crystals, (c) the existence of a fundamental volume in q-space containing the wave vectors associated with all possible vibrational modes of the crystal, and (d) the presence of singularities in the frequency spectrum which are characteristic of the dispersion relation for a given crystal geometry. These features are not as readily apparent when considering the more general case of three-dimensional lattices.

1.2.3.2. Dynamics of a Three-Dimensional Lattice

The Born-von Kármán theory of lattice dynamics is based on two assumptions or approximations:

(1) The oscillations of the atoms about their equilibrium positions are assumed to be sufficiently small that one can expand the crystal potential energy as a Taylor's series in the displacements of the atoms from their equilibrium positions and truncate the series expansion after terms of second order in the displacements. This is known as the 'harmonic approximation.'

(2) The electrons in the crystal move so much faster than the nuclei (because of the mass ratio) that they are assumed to follow the nuclear motions adiabatically. This is called the 'adiabatic approximation' or the Born-Oppenheimer theorem.

For simplicity we consider only a monatomic crystal of cubic symmetry in the following discussion. If the atomic displacements from equilibrium are small, the potential energy of the crystal may be expanded as

$$\Phi = \Phi_0 + \Phi_1 + \Phi_2 + \text{(anharmonic terms)} \tag{55}$$

where Φ_0 is the static potential energy (all atoms at their equilibrium positions),

$$\Phi_1 = \sum_{\ell,\alpha} \frac{\partial \Phi}{\partial u_\alpha^\ell} u_\alpha^\ell \tag{56}$$

and

$$\Phi_2 = \frac{1}{2} \sum_{\ell,\alpha} \sum_{\ell',\beta} \frac{\partial^2 \Phi}{\partial u_\alpha^\ell \partial u_\beta^{\ell'}} u_\alpha^\ell u_\beta^{\ell'}. \tag{57}$$

The Greek subscripts represent the cartesian components (x,y,z) and $\vec{u}^\ell$ is the (small) displacement from equilibrium of atom ℓ. The derivatives are evaluated at the equilibrium positions of the atoms. The cubic (Φ_3), quartic (Φ_4), and higher-order terms in the expansion are neglected in the harmonic approximation. The term Φ_1 is equal to zero since the sum in Eq. (56) involves the quantities $\partial\Phi/\partial u_\alpha^\ell$; that is, the forces acting on the atoms at their equilibrium positions, which must be zero. For convenience, the quantities involving the second derivative of the potential energy in Eq. (57) are relabeled as

$$\phi_{\alpha\beta}^{\ell\ell'} = \frac{\partial^2 \Phi}{\partial u_\alpha^\ell \partial u_\beta^{\ell'}}. \tag{58}$$

On the basis of the approximations above, the Hamiltonian of the crystal can now be written as

$$H = \sum_{\ell,\alpha} \frac{(p_\alpha^\ell)^2}{2M} + \Phi_0 + \frac{1}{2} \sum_{\ell,\alpha} \sum_{\ell',\beta} \phi_{\alpha\beta}^{\ell\ell'} u_\alpha^\ell u_\beta^{\ell'} \tag{59}$$

where $\vec{p}^\ell$ denotes the momentum of atom ℓ and M is the atomic mass. To obtain the equation of motion for atom ℓ, one can proceed formally with Hamilton's equations [51], $\dot{u}_\alpha = \partial H/\partial p_\alpha$ and $\dot{p}_\alpha = -\partial H/\partial u_\alpha$, which yield

$$M\ddot{u}_\alpha^\ell = - \sum_{\ell',\beta} \phi_{\alpha\beta}^{\ell\ell'} u_\beta^{\ell'}, \quad \text{(all } \alpha, \ell \text{)}. \tag{60}$$

Physically, $-\phi_{\alpha\beta}^{\ell\ell'} u_\beta^{\ell'}$ is interpreted as the force on atom ℓ in the α-direction when atom ℓ' is displaced a distance $u_\beta^{\ell'}$ in the β-direction. Thus, each coefficient $\phi_{\alpha\beta}^{\ell\ell'}$ plays the role of an 'interatomic force constant.'

The simplest traveling wave solution of Eq. (60) is given by

$$u_\alpha^\ell = \xi_\alpha \exp[i(\vec{q}\cdot\vec{R}_\ell - \omega t)] \tag{61}$$

where $\vec{\xi}$ is a polarization vector describing the direction and amplitude of the lattice vibration and $\vec{R}_\ell$ is the position vector of atom ℓ. Upon substitution, one obtains

$$M\omega^2 \xi_\alpha = \sum_\beta D_{\alpha\beta}(\vec{q}) \xi_\beta \tag{62}$$

where

$$D_{\alpha\beta}(\vec{q}) = \sum_{\ell'} \phi_{\alpha\beta}^{\ell\ell'} \exp(-i\vec{q}\cdot\vec{R}_{\ell\ell'}) \text{ and } \vec{R}_{\ell\ell'} = \vec{R}_\ell - \vec{R}_{\ell'}. \tag{63}$$

The general equation of motion can be written more concisely in matrix notation as

$$M \omega^2 \tilde{\xi} = \tilde{D} \tilde{\xi} \tag{64}$$

where $\tilde{\xi}$ is a column matrix and $\tilde{D}$ is a 3 x 3 square matrix. The condition for a nontrivial solution to this set of equations then becomes

$$\det |\tilde{D} - M\omega^2| = 0. \tag{65}$$

For each value of $\vec{q}$, the determinant yields three distinct frequencies which one can formally identify by setting $\omega = \omega_j(\vec{q})$, where $j = 1,2,3$ is the branch index.

For a polyatomic crystal containing n different types of atoms, the analysis is similar to that given above except that the quantity M is replaced by a diagonal matrix $\tilde{M}$ containing the different masses [48,49]. The dynamical matrix $\tilde{D}$ is now of the order 3n x 3n, thereby yielding a total of 3n distinct frequencies for each $\vec{q}$. It can be shown [47] by the 'method of long waves' that three of the $\omega_j(\vec{q})$ always tend to zero in the limit $\vec{q} \to 0$ and, thus, are associated with the acoustical vibrations of the lattice.

As in the case of the one-dimensional lattice considered earlier, all distinct values of frequency can be accounted for by consideration of a finite range of wave vectors which, in three dimensions, lie in a volume in $\vec{q}$-space known as the 'first Brillouin zone' [50]. The geometry of the Brillouin zone is determined by the crystal structure under consideration.

The vibrational frequencies of the lattice are further specified by the 'allowed' values of $\vec{q}$ which, in turn, are fixed by the boundary conditions. Again, periodic boundary conditions are applied to a representative volume (containing N atoms) to avoid explicit consideration of the atoms at the crystal surface. As a result, for large N, the distinct values of $\vec{q}$ may be regarded as continuously and uniformly distributed throughout the first Brillouin zone so that the total number of normal-mode frequencies is equal to 3N.

By a suitable transformation of coordinates, it can be shown formally [47] that the equations of motion for a lattice of N atoms become equivalent to those for a system of 3N independent harmonic oscillators with frequencies given by the eigenvalues of Eq. (64). The formalism and results are discussed in a number of texts [52,53]. For our purposes we need only note that the oscillator energies are quantized in units of $h\nu$ known as 'phonons' and that the thermodynamic properties of the crystal can be derived by well-known quantum statistical methods (Section 1.2.6).

Phonon propagation along directions of high symmetry in the crystal (such as the [100], [110], and [111] directions) is of particular interest since the polarization $\vec{\xi}$ of the lattice vibration is fixed by the symmetry of the crystal and is either purely longitudinal or transverse. For a particular branch in a particular symmetry direction, one may express Eq. (62) as

$$M\omega^2 = \sum_n \Phi_n (1 - \cos n\pi\zeta) \tag{66}$$

where the Φ_n are linear combinations of the $\phi_{\alpha\beta}^{\ell\ell'}$ for which $\vec{q} \cdot \vec{R}_{\ell\ell'}$ is a constant. The quantity ζ

is the 'reduced wave vector' and is defined as the ratio q/q_m, where q_m is the (maximum) q at the Brillouin zone boundary. Since $\vec{q} \cdot \vec{R}_{\ell \ell'} =$ constant defines a plane of atoms in the crystal, Φ_n may be interpreted as the force between an atom and the two planes of atoms perpendicular to $\vec{q}$ and n planes away. For this reason the Φ_n are generally called interplanar force constants [54]. If the dispersion relation is measured along the major symmetry directions, one can calculate the interplanar force constants and, hence, the interatomic force constants by making a linear least-squares fit to the observed frequencies. The values of $\phi_{\alpha\beta}^{\ell \ell'}$ constitute a 'force constant model' of the crystal and become the basis for computing the phonon spectrum (Section 1.2.5).

It is clear from the above discussion that, in its most general form, the Born-von Kármán theory is highly phenomenological in that no assumptions are made about the nature of the inter-atomic forces. During the last few decades, considerable effort has been devoted to semiphenomen-ological approaches in describing the dynamics of certain classes of solids, such as metallic, rare gas, and ionic crystals.

In metals, the conduction electrons are not confined to the vicinity of the ion cores and, thus, are not expected to follow the ionic motions adiabatically, as is assumed by the Born-von Kármán model. However, it has been shown [55,56] that, in a metal, the ion plus its screening charge may be treated as a single entity called a 'neutral pseudo-atom.' The interactions between pseudo-atoms are described by an 'effective' crystal potential consisting of a direct bare ion Coulomb term, a term involving the indirect interaction through the electrons and a repulsion or 'overlap' term. The potential may then be used within the framework of the Born-von Kármán theory.

In rare gas solids, the interatomic force system in its simplest form is described by two-body pair potentials of the Lennard-Jones type. The crystal potential energy Φ is obtained by summing the pair potentials over the crystal and the interatomic force constants $\Phi_{\alpha\beta}$ are then evaluated from the appropriate derivatives. A detailed discussion of the lattice dynamics of rare gas solids has been given in a recent book [57].

The simplest model of an ionic solid is the 'rigid ion' model [47-49] in which the crystal potential energy is derived from long-range Coulomb interactions between ions and the short-range overlap forces between nearest-neighbors; the ions are treated as point charges, which neglects the ionic polarizability (hence, the label 'rigid ion'). This model yields a general equation of motion similar to Eq. (64) except that the dynamical matrix $\tilde{D}$ consists of a Coulomb and a repul-sive part. A more realistic description of an ionic solid is provided by the 'shell' model [58-60] and its related formulations [61-63]. In addition to Coulomb interactions and short-range overlap forces, these formulations include the polarizability of each ion by a model in which a rigid shell of electrons (taken to have zero mass) is allowed to move with respect to the massive ionic core. Differences between the various formulations (manifested by differences among terms in $\tilde{D}$) have been reviewed in the literature [49,60].

For the purpose of studying specific heat the importance of force-constant models lies in their ability to parameterize the measured dispersion relations so that accurate frequency spectra may be computed.

1.2.4. Determination of Phonon Dispersion Relations

The earliest measurements of phonon dispersion in single crystals were performed by the method of thermal diffuse scattering of X-rays [64-66]. However, the method was rapidly superseded by the development of a significantly more accurate technique, the method of inelastic scattering of 'slow' neutrons.

Both X-rays and slow neutrons have wavelengths of the same order as the interatomic spacings (≈ 1 Å), making them invaluable for diffraction studies of the crystalline structure of solids. In the study of lattice vibrations, however, slow neutrons are vastly superior since their energies (~0.025 eV) are comparable to the vibrational energies of the lattice, in contrast to X-rays which have much larger energies (~5000 eV). Thus, measurements of energy transfer to a crystal become feasible in neutron scattering experiments. Indeed, the interpretation of neutron scattering results is straightforward and reliable since one can determine unambiguously both the frequency and the wave-vector of a phonon using only the laws of conservation of energy and momentum, respectively [67,68].

The most significant advance in neutron spectroscopy was the development of the triple-axis crystal spectrometer [67,68] with its special modes of operation such as 'constant-momentum transfer' and 'constant-energy transfer.' These techniques allowed a preselection of each phonon to be studied, thereby enabling the experimenter to map out any desired part of $\nu_j(\vec{q})$ with a high degree of accuracy (now typically 1% or better). As a result of these developments, the dispersion relations for a great many crystalline solids have now been studied in detail and, in some cases, at more than one temperature; extensive lists may be found in the literature [67-70].

The dispersion curves for copper at 296 K [71], shown in Fig. 1.5(a), are a good example of results that are obtained for lattice vibrations in a monatomic crystal. The general appearance of dispersion curves is, to a considerable extent, determined by crystal symmetry and so the phonon dispersion curves for substances of a given crystal structure are found to be rather similar. Beyond consideration of a given crystal structure, further resemblances are found among substances that are chemically alike, such as elements in the same column of the periodic table. For example, nickel [72] and palladium [73] have dispersion curves which differ by little more than a frequency scale factor given approximately by the inverse square root of their mass ratio. This striking similarity is not unexpected since the lattice dynamics of metals are dominated by large, almost-central, first-neighbor forces; we have seen earlier that the acoustical frequencies of a linear lattice (with first-neighbor forces only) vary as the inverse square root of the atomic mass (see Fig. 1.4).

The results for metals have been analyzed, for the most part, in terms of Born-von Kármán models. Elastic constant data, when available, are often included in the least-squares fit to the observed frequencies. In all cases, systems of long range forces are found, extending to at least several neighbors and sometimes much further. Details of the weak longer range forces vary from one metal to another, presumably the result of differences in electronic structure.

The dispersion relation for KCl at 115 K [74] is shown in Fig. 1.5(b) as an illustrative example of results obtained for diatomic crystals. The comments above regarding the influence of

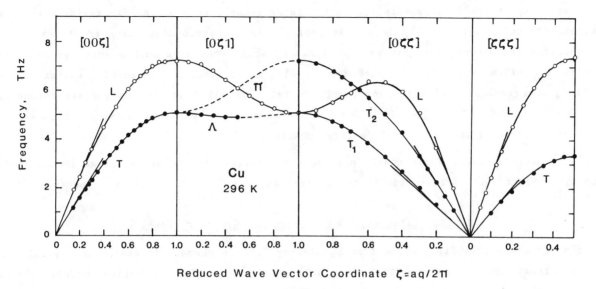

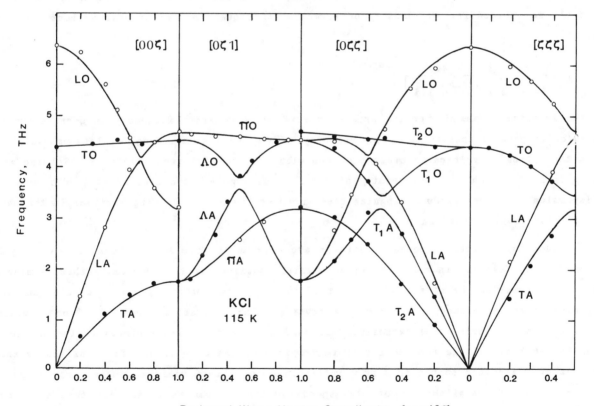

Figure 1.5.   Frequency/wave vector dispersion relations along the major symmetry directions in (a) copper at 296 K (after Svensson et al. [71]) and (b) potassium chloride at 115 K (after Copley et al. [74]).   The solid curves represent force-constant model fits to the observed frequencies of longitudinal (open symbols) and transverse (solid symbols) modes.

crystal symmetry and chemical similarity on the general appearance of the dispersion curves among polyatomic crystals apply equally here. In general, the measured dispersion relations for the alkali halides are described rather well in terms of shell models which, in turn, yield elastic constants and optical properties (high-frequency dielectric constant and small-$\vec{q}$ optical frequencies) in good agreement [74] with experimental data. For other insulators and semiconductors (e.g., UO_2 [75], GaAs [76]), good fits can also be obtained by the shell model, but many of the parameters no longer have their usual physical meaning.

As mentioned earlier, force models provide a convenient interpolation scheme for determining the possible vibrational frequencies of the crystal and, in turn, the frequency spectrum.

1.2.5. Calculation of Vibrational Frequency Spectra

Prior to the availability of neutron scattering data on phonon dispersion in crystals, considerable effort had been devoted to finding analytic forms of $g(\nu)$ for various models, the Debye model being the most successful example. It was recognized that in a real crystal the dispersion relation would be highly anisotropic with respect to the direction of $\vec{q}$ and, thus, the surfaces of constant frequency would no longer be spheres in $\vec{q}$-space as in the case of the Debye isotropic continuum but, rather, their shape would depend on the direction of $\vec{q}$. Consequently, a more general result was needed for the density of normal-mode frequencies, given here without proof as [e.g., 26]

$$g(\nu) = \frac{V}{(2\pi)^3} \sum_j \int \frac{dS}{|\mathrm{grad}_{\vec{q}}\nu_j(\vec{q})|} \tag{67}$$

where the integral is taken over the area of a surface of constant frequency in $\vec{q}$-space and the denominator is proportional to the group velocity. This expression, though of less practical value, has theoretical significance because of contributions to $g(\nu)$ from points at which the group velocity is zero. Such critical points (at the Brillouin zone boundaries or at saddle points in the dispersion relation) produce singularities known as Van Hove [77] singularities in the phonon frequency-distribution function.

In principle, constant-frequency surfaces are obtainable from Eq. (65), given an analytic form for $\nu_j(\vec{q})$. However, in practice, $g(\nu)$ is rarely obtained in this manner; rather, semianalytic or wholly numerical methods are usually used in conjunction with a given force-constant model to approximate the frequency spectrum of the crystal. One approach has been the 'root-sampling' method in which the force model parameters are used to generate a large number of frequencies by solving Eq. (65) at a large number of uniformly-spaced values of $\vec{q}$ in the first Brillouin zone. The values of ν are then sorted in narrow intervals $d\nu$ or 'bins' yielding a histogram representation of $g(\nu)$. The availability of high-speed electronic computers has made this approach particularly attractive since clearly the accuracy of $g(\nu)$ improves with the number of frequencies computed. The most accurate of the computational schemes is the semianalytic method developed by Gilat and Raubenheimer originally for cubic crystals [78], later extended to hexagonal [79] and tetragonal [80] crystals. A detailed discussion of the 'root-sampling' and other approximate techniques is given in Ref. [49].

Typical frequency distributions are shown in Fig. 1.6 for (a) copper at 296 K [71] and for (b) potassium chloride at 115 K [74]; the smooth curves are drawn through the histograms calculated with 'bin-widths' of 0.01 THz and 0.05 THz, respectively. Many of the critical points in the distributions can be correlated with the stationary values and cross-over points of the dispersion relations illustrated in Fig. 1.5.

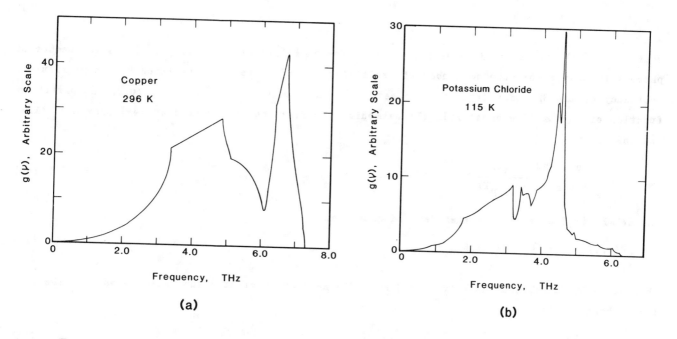

(a)   (b)

Figure 1.6.   Frequency distribution for the lattice vibrations in (a) copper at 296 K (after Svensson et al. [71]) and (b) potassium chloride at 115 K (after Copley et al. [74]).

It should be noted that frequency distributions are usually calculated from force models based solely on measurements of $\nu_j(\vec{q})$ along directions of high symmetry. This implicity assumes that the models are 'perfect' in that they correctly predict the phonon frequencies for the 'off-symmetry' values of $\vec{q}$. Of course, the validity of such force models can be checked by measurements at a sufficient number of wave vectors along off-symmetry directions. Examples of 'off-symmetry' measurements have been reported in the literature [e.g., 73,81] in cases where accuracy of $g(\nu)$ was of particular concern. The requirements for validity of a given force model are not as stringent as it might first appear since thermodynamic properties arising from lattice vibrations are determined by averages over the frequency spectrum and, therefore, it is probably sufficient that force models predict the correct frequencies 'on average.'

1.2.6.   Thermodynamic Properties in the Harmonic Approximation

From a thermodynamic point of view, a crystal can be described by its volume V and temperature T; the natural thermodynamic function corresponding to these parameters is the Helmholtz free energy F(V,T), which has a simple and direct relation to the partition function (Eq. 14). The partition function, in turn, is related to a sum over all appropriate energy levels (vibrations) in the crystal (Eq. 15). For these reasons, $F(V,T)$ is often the starting point in many calculations in statistical thermodynamics.

As mentioned earlier in Section 1.2.3.2, the motions of N atoms in the crystal are equivalent to the vibrations of a system of 3N oscillators. The total vibrational energy of the crystal in state i is, therefore, given by a sum over all oscillator energies:

$$E_i = \sum_s (n_s + 1/2) \, h\nu_s \tag{68}$$

where $s \equiv (\vec{q}, j)$ labels the 3N vibrational modes (or oscillators), $n_s = 0, 1, 2, \ldots$ is the number of phonons in mode s (or quantum number of oscillator s) and state i is specified by the particular occupancy of the 3N modes (that is, the set $\{n_s\} = n_{s_1}, n_{s_2}, n_{s_3}, \ldots$). Therefore, the partition function of a vibrating crystal in the harmonic approximation can be derived [49] from Eqs. (15) and (68) as

$$Z = \prod_s \frac{\exp(-h\nu_s/2k_BT)}{1 - \exp(-h\nu_s/k_BT)} . \tag{69}$$

The Helmholtz free energy is then derived from Z as

$$F(V,T) = k_BT \sum_s \ln(2 \sinh x_s) \tag{70}$$

where $x_s = h\nu_s/2k_BT$, and the entropy and specific heat at constant volume associated with the lattice vibrations follow:

$$S = -\left(\frac{\partial F}{\partial T}\right)_V = k_B \sum_s [x_s \coth x_s - \ln(2 \sinh x_s)] \tag{71}$$

$$C_v = T\left(\frac{\partial S}{\partial T}\right)_V = -T\left(\frac{\partial^2 F}{\partial T^2}\right)_V = k_B \sum_s x_s^2 \operatorname{cosech}^2 x_s. \tag{72}$$

Thus, in the harmonic approximation, the thermodynamic functions are additive functions of the normal mode frequencies and, consequently, can be expressed as averages over the frequency spectrum. Of particular interest to us is the contribution of the lattice vibrations to the specific heat of the crystal, which may be obtained from Eq. (72) as

$$C_v = k_B \int_0^{\nu_m} \frac{(h\nu/k_BT)^2 \, \exp(h\nu/k_BT)}{[\exp(h\nu/k_BT) - 1]^2} \, g(\nu) \, d\nu \tag{73}$$

where the frequency distribution is normalized according to Eq. (32) with ν_m the maximum frequency.

It is rather straightforward to calculate C_v once $g(\nu)$ is known. The values of C_v are usually expressed in terms of a Debye equivalent temperature by comparison with Eq. (41) and the results given as a plot of Debye θ versus temperature in order to maximize the sensitivity to details of $g(\nu)$. As an example, the results for copper calculated [71] from $g(\nu)$ at room temperature [Fig. 6(a)] are presented as a dashed curve in Fig. 1.7; the solid and open circles represent values of the Debye temperature [45,82] determined from calorimetric measurements of C_p after suitable corrections for C_p–C_v and the electronic specific heat. The agreement is reasonable considering that a room temperature $g(\nu)$ was used to calculate the characteristic temperature at all T without corrections for anharmonic effects such as the volume-dependence of the frequencies. The most significant feature of the dashed curve is that it has essentially the correct shape, approaching a constant value at higher temperatures as expected from harmonic theory.

The description of the lattice specific heat on the basis of the harmonic approximation is expected to become more accurate with decreasing temperature, simply on the grounds that vibrational amplitudes will become smaller and that the volume coefficient of expansion approaches zero in the limit T → 0. For example, the frequency spectrum of copper at absolute zero was determined by 'rescaling' the room temperature $g(\nu)$ on the basis of measurements of average shifts in the frequencies with changing temperature and the 'harmonic' lattice specific heat was then calculated [73]. The corresponding values of the characteristic temperature are presented as a solid curve in Fig. 1.7. As can be seen, the agreement between these values and those derived from calorimetric data is rather good, except at very low temperatures where differences may be attributed to inaccuracies in the lowest-frequency part of the $g(\nu)$ histogram. It should be noted that harmonic theory is not strictly valid even at absolute zero because of zero-point vibrations [83,84], though anharmonicity here is usually assumed to be negligibly small.

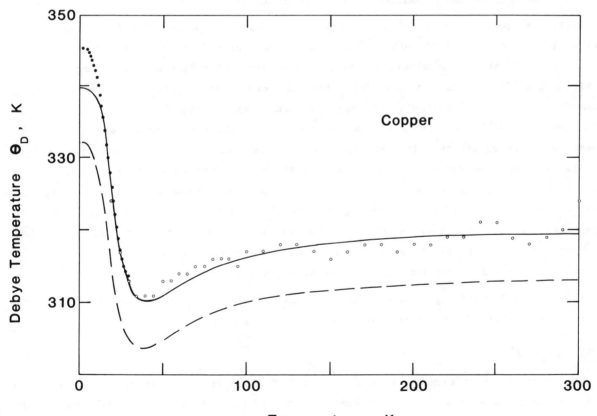

Figure 1.7. Temperature variation of the Debye characteristic temperature for copper as derived from (a) calorimetric data [45,82]: solid and open symbols, (b) a frequency distribution at 296 K [71]: dashed curve, and (c) frequency distribution extrapolated to zero kelvin [73]: solid curve.

During the past two decades, the dispersion relations for many crystalline substances have been studied in sufficient detail to yield accurate frequency spectra [70]. In most instances, the dispersion relations are not strongly temperature-dependent, particularly at low temperatures. Experience has shown that as long as the temperature is only a small fraction (~10%) of the melting point, the results are essentially equivalent to those at absolute zero in the sense that the corrections necessitated by the finite measurement temperature lie within the measurement errors; in such cases, a reasonable approximation of the harmonic lattice specific heat may be obtained. In addition, a number of crystals have now been studied at more than one temperature and corrections to the results for the effects of anharmonicity have been estimated.

1.2.7. Thermodynamic Properties of Lattice Vibrations in Anharmonic Crystals

An approach most often used in the study of anharmonic effects in crystals is to separate the thermodynamic properties into quasiharmonic and explicitly anharmonic contributions. Another approach involves the total (harmonic plus anharmonic) lattice entropy or lattice specific heat, which can be derived from neutron spectroscopic data if the temperature dependence of the frequencies is known. These methods of analysis are discussed briefly below.

1.2.7.1. Quasiharmonic and Explicitly Anharmonic Contributions

The vibrational properties of solids are often analyzed in terms of the 'quasiharmonic' approximation [83] in which the effects of the higher-order terms ($\Phi_3+\Phi_4+...$) neglected by harmonic theory are, to a first approximation, taken into account indirectly by evaluating the second derivatives $\Phi_{\alpha\beta}$ (Eq. 58) at the mean positions which the atoms actually occupy at temperature T; these volume-dependent force constants are then used within the framework of harmonic theory to obtain the 'quasiharmonic' frequency spectrum and then the quasiharmonic values of F, S, and C_v via the usual harmonic equations. In this approximation the vibrational frequencies of the crystal are still assumed to be harmonic though they are allowed to depend on volume. Since volume-dependence of the frequencies is itself an anharmonic effect, this approach does not correspond to any self-consistent model. However, it does give a valid first approximation for the effect of crystal expansion on the vibrational spectra and enables the deviations from quasiharmonic behavior (i.e., the explicit temperature dependence of the frequencies) observed in crystals at higher temperatures to be treated as perturbations to the basic harmonic problem.

In cases where anharmonicity is small enough to be treated as a perturbation (probably true for most simple crystals), the Helmholtz free energy may be written as

$$F(V,T) = F^{qh} + F^{a} \tag{74}$$

where F^{qh} represents the quasiharmonic contribution and F^{a} the explicitly anharmonic contribution. The entropy and specific heat at constant volume are related to derivatives of F(V,T) and, therefore, may be expressed in similar form:

$$S(V,T) = S^{qh} + S^{a} \tag{75}$$

$$C_v(V,T) = C_v^{qh} + C_v^{a} . \tag{76}$$

The explicitly anharmonic contributions, F^a, S^a, and C_v^a are directly related to the high-order terms $\Phi_3 + \Phi_4 + \cdots$ in Eq. (55) and are derivable, in principle, if the interatomic potential is known [83].

Formal expressions for the two lowest-order terms of the Helmholtz energy, commonly known as the cubic (F_3) and quartic (F_4) terms (arising from Φ_3 and Φ_4, respectively), have been derived by Leibfried and Ludwig [83], Maradudin et al. [85], and others [86,87] using the low-order perturbation theory. The expressions, which involve multiple summations of Fourier transforms of the third- and fourth-rank tensor atomic force constants ($\Phi_{\alpha\beta\gamma}$ and $\Phi_{\alpha\beta\gamma\delta}$) over the phonon wave vectors of the Brillouin zone, are exceedingly complex and require extensive machine calculations for their evaluation, even for the simplest of force models. As a consequence, many of the calculations of F_3 and F_4 during the past two decades have dealt with rare gas crystals in which force systems may be adequately approximated by nearest-neighbor central force models; for an extensive review of the calculations for rare gas solids, see Ref. [57].

In the high-temperature limit ($T > \theta_D$), the explicit anharmonic contributions to $C_v(V,T)$ arising from F_3 and F_4 are linear in T:

$$C_v^a = -T \frac{\partial^2}{\partial T^2} (F_3 + F_4) = 3RAT \tag{77}$$

where A is a volume-dependent coefficient whose value is determined by the particular force model selected for the anharmonic calculations. The anharmonic correction C_v^a is somewhat sensitive to the form of the interatomic potential because of the large degree of cancellation that often occurs between F_3 and F_4 (which have opposite signs).

In recent years, a number of calculations of F_3 and F_4, hence C_v^a, have been reported for a number of metallic crystals principally by Shukla, MacDonald, and collaborators: for example, copper [88-90], sodium and potassium [91], rubidium [92,93], and lead, silver, nickel, aluminum, calcium, and strontium [88,90]. A comparison of values of the anharmonic coefficient A obtained from theory with those determined from experiment (see discussion below) is given in Table 1.2 for selected metals. The theoretical and experimental values of A are often of the same sign and, in some cases, quantitative agreement is observed. The agreement is reasonably good considering that C_v^a (hence A) is rather sensitive to the approximations involved in the theoretical models and to uncertainties in the experimental data.

As the melting temperature (T_m) is approached, low-order perturbation theory (LOPT) may become a poor approximation because of possible contributions to C_v^a from higher-order terms in the expansion of the crystal potential energy. The general treatment of crystal anharmonicity [83] shows that at $T \gg \theta_D$ C_v^a is of the form $3R[AT + BT^2 + \cdots]$. The range of validity of LOPT depends upon the degree of anharmonicity exhibited by the crystal. For example, in a weakly anharmonic crystal such as copper, good results [89,90] are obtained at temperatures up to at least 0.5 T_m; in sodium chloride, the range of validity appears to extend to 0.7 T_m [94]. However, in rare gas solids, the second-order perturbation corrections become significant at temperatures as low as 0.3 T_m [95].

Table 1.2.   Theoretical and Experimental Values for the Anharmonic Coefficient A in the Explicitly Anharmonic Contribution $C_v^a = 3RAT$ for Selected Metals

Metal	Theory		Experiment	
	A $(10^{-4} \cdot K^{-1})$	Ref.	A $(10^{-4} \cdot K^{-1})$	Ref.
Al	0.09	[88]	0.2^a	[99]
	-0.39^b	[90]	-0.19, -0.35	[97]
Cu	0.06	[88]	0.1^c	[98]
	-0.17^b	[90]		
Pb	-0.02	[88]	-1.0^d, -1.6^d	[97]
	0.06^b	[90]		
K	0.9	[91]	1.2^e	[100]
	0.30^f			
Na	0.4	[91]	1.7^e	[100]
	-0.19^f			
Rb	1.1	[93]	1.3^e	[100]
	0.36^f			

[a] Value of $C_v^a/3RT$ determined at $T \simeq 0.7\, T_m$ where T_m is the melting temperature.

[b] Values correspond to theoretical results described in Ref. [90] and were kindly provided by one of the authors (R.A. MacDonald).

[c] Value of $C_v^a/3RT$ determined at $T \simeq 0.5\, T_m$ where T_m is the melting temperature.

[d] Reported values of A as in $C_v^a/3R = AT + 2BT^2$; the corresponding values of B are ~6 x 10^{-8} and ~8 x $10^{-8} K^{-2}$, respectively.

[e] Reported values of A derived from $C_v - C_v^h = 3RAT$ where C_v^h is the harmonic lattice contribution.

[f] Values were kindly provided by R.A. MacDonald from theoretical results to be published.

Alternative formulations have been used for the calculation of thermodynamic properties of anharmonic crystals, among these the self-consistent phonon theory (SC) [95] and the Monte Carlo method [96]. A comparison of results obtained by LOPT with those obtained by SC and by Monte Carlo methods are given in Refs. [89] and [93], respectively.

The determination of C_v^a from experiment requires an accurate evaluation of the various contributions to the total specific heat of the solid. The procedure involves converting the experimental C_p to C_v by means of data on volume expansivity and isothermal compressibility (Eqs. 18, 19, and in the Appendix, Eqs. A2-A5). For metals, corrections to C_v must also be made for the electronic specific heat, which is usually assumed to contribute linearly with temperature (Section 1.3). By subtracting the quasiharmonic contribution C_v^{qh} [97] or the harmonic lattice contribution C_v^h [98,99] from the resulting values of C_v one obtains the so-called 'excess' specific heat. In the high-temperature limit $(T > \theta_D)$ both C_v^{qh} and C_v^h rapidly approach 3R; thus, the excess specific heats as determined by $C_v - C_v^{qh}$ and $C_v - C_v^h$ become equal. As the melting temperature T_m is approached, consideration must also be given to the contribution from the formation of equilibrium lattice defects, primarily monovacancies (Section 1.2.8). The resulting values of excess specific heat are then equated to C_v^a. At intermediate temperatures ($T \sim 0.5\ T_m$), the experimental values of C_v^a are approximately linear in temperature and, thus, can be described by the anharmonic coefficient $A = C_v^a/3RT$; experimental values of A are given in Table 1.2 for selected metals. At higher temperatures, empirical values of C_v^a often become nonlinear in temperature, an effect which is usually attributed to higher-order anharmonic effects and/or uncertainties in evaluating lattice vacancy contributions.

1.2.7.2. Total (Harmonic Plus Anharmonic) Contribution

Barron [84] has shown that within LOPT the entropy of the lattice vibrations in an anharmonic crystal is correctly determined by replacing the harmonic frequencies in the usual harmonic formula (Eq. 71) with frequencies determined by inelastic neutron scattering experiments. This kind of replacement is not valid for other thermodynamic functions derived from harmonic theory such as free energy and internal energy. The identical result was subsequently derived by Werthamer [101] and by Hui and Allen [102] on the basis of self-consistent phonon theory.

Since the vibrational frequencies are generally measured at constant pressure (essentially zero pressure), the total (harmonic plus anharmonic) lattice specific heat of the real crystal at constant pressure is then derived from Eq. (71) as follows:

$$C_p^\ell = T\left(\frac{\partial S}{\partial T}\right)_p = k_B \sum_s x_s^2 \operatorname{cosech}^2 x_s \left(1 - \frac{T}{\nu_s}\frac{d\nu_s}{dT}\right) \tag{78}$$

where $x_s = h\nu_s/2k_BT$. Since it is impractical to measure the temperature-dependence of all $3N$ normal modes, some interpolation scheme is required.

One approach is to assume that the fractional change in frequency of a mode with temperature is independent of the mode [73,92,102]. Then C_p^ℓ becomes an additive function of the normal-mode frequencies and can be expressed as an average over the temperature-dependent frequency spectrum:

$$C_p^\ell = k_B \int_0^{\nu_m} \frac{(h\nu/k_BT)^2 \exp(h\nu/k_BT)}{[\exp(h\nu/k_BT) - 1]^2} \left(1 - \frac{T}{\nu}\frac{d\nu}{dT}\right) g(\nu)\ d\nu. \tag{79}$$

By measuring the (average) change of the frequencies with temperature, one can then determine the temperature dependence of $g(\nu)$, hence of C_p^ℓ. Figure 1.8 illustrates the results obtained for copper [73]. The dashed line gives the harmonic lattice specific heat (obtained by extrapolating the frequency shifts to absolute zero) whereas the solid line gives the total lattice specific heat of the anharmonic crystal; that is, C_p^ℓ. The solid symbols, which represent the calorimetric measurements of C_p [82,103], lie above the solid line by an amount which agrees well with the expected electronic contribution.

For weakly anharmonic crystals, such as copper, the above approach may be expected to give good results for C_p^ℓ at temperatures as high as perhaps 70% of the melting point. However, in more strongly anharmonic crystals, the approximation that all modes have the same fractional change in frequency with temperature is not as well justified. For example, in rubidium the fractional frequency shifts for the transverse modes are measurably larger than those for the longitudinal modes and, consequently, the above approximation yields less satisfactory results for C_p^ℓ [92].

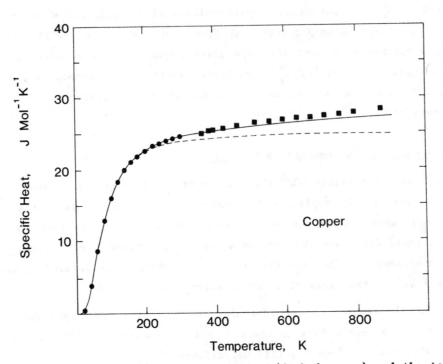

Figure 1.8. Comparison of the harmonic contribution (dashed curve) and the total harmonic plus
anharmonic contribution (solid curve) of the lattice vibrations to the specific heat
of copper as determined from neutron scattering measurements with calorimetric measurements of C_p (solid circles [82], solid squares [103]). The difference between the
total lattice contribution and the calorimetric data is well accounted for by the
contribution from the conduction electrons (after Miiller and Brockhouse [73]).

1.2.8. Contribution of Lattice Vacancies

An anomalous rapid rise in specific heat prior to melting has been observed in numerous
solids [97-99, 104-107] and is particularly large in the refractory metals (see Fig. 1.9). This
effect is attributed to high-order anharmonic contributions and/or to contributions from the lattice vacancies which are expected to be present in solids near their melting point.

It can be shown [108] from elementary statistical mechanics that the fractional concentration
of monovacancies in a solid increases rapidly with temperature according to

$$n_v = A \exp[-E_f/k_B T] \tag{80}$$

where E_f is the vacancy formation energy; the pre-exponential term A is given by $\exp[S_f/k_B]$ where
S_f is the entropy associated with vacancy formation. The increase in internal energy arising from
the formation of vacancies is $n_v N_A E_f$ and their contribution to the specific heat is given by

$$C_{vac} = \frac{d}{dT}(n_v N_A E_f) = R(E_f/k_B T)^2 A \exp[-E_f/k_B T] = R(E_f/k_B T)^2 n_v. \tag{81}$$

In principle, once C_{vac} is obtained from experiment, E_f may be determined from the slope of
$\ln(C_{vac} T^2)$ versus T^{-1}, which should be a straight line; with the value of E_f known, n_v may also
be determined. In practice, however, the separation of C_{vac} from other contributions to the observed values of C_p, in particular the anharmonic contributions, is an exceedingly difficult
problem.

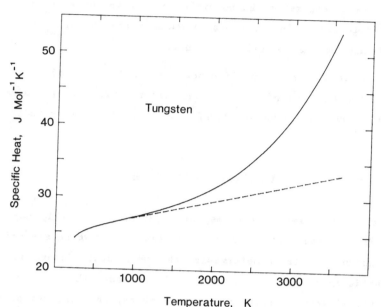

Figure 1.9. Specific heat of tungsten in the range 300 to 3600 K (solid curve) as given by $C_p =$ $3R - 7.72 \times 10^4 T^{-2} + 2.33 \times 10^{-3}T + 1.18 \times 10^{-13}T^4$. The dashed curve represents C_p when only the first three terms are included (after Cezairliyan and McClure [104]).

One approach [106,107] to the problem involves representing the dilation, electronic, and lowest-order anharmonic contributions to C_p by a term linear in temperature; this term is fitted to measurements at intermediate temperatures ($T \sim 0.3\ T_m$) and extrapolated to higher temperatures. The additional or 'excess' specific heat associated with the anomalous rise in C_p is then attributed entirely to lattice vacancies, enabling values to be obtained for E_f and n_v as described above. The reliability of the results, of course, depends on the validity of the linear approximation for anharmonic effects. This method of analysis yields values for n_v that are typically one or two orders of magnitude larger than those obtained by other independent measurements (see below).

A second method [97-99] of analyzing the high-temperature C_p data has been mentioned earlier in Section 1.2.7.1: the dilation, electronic, and harmonic (or quasiharmonic) lattice contributions to C_p are evaluated and the remaining or 'excess' specific heat is analyzed in terms anharmonic and vacancy contributions. Values of E_f and n_v, obtained from independent measurements (see below), are used to calculate C_{vac}, thus yielding the anharmonic contribution. Clearly the accuracy of the analysis depends upon, among other things, the reliability of E_f and n_v.

Consistent and reliable values of E_f can be determined by a variety of experimental techniques; for example, simultaneous measurement of lattice parameter and expansion [109], electrical resistivity of rapidly quenched metals [110], tracer diffusion [111], etc. However, estimates of vacancy concentration appear less certain. The most direct method of determining n_v is by the simultaneous measurement of lattice parameter and expansion [109]. These values, however, are too low by at least an order of magnitude to account for the anomalous rise in the high-temperature specific heat. Estimates of n_v obtained from rapid quenching experiments [110] are also low;

these results are less certain because of the possible loss of vacancies during the rapid quench. Values of n_v can also be determined, in principle, from positron annihilation studies in metals [112] but, as yet, this technique has not been fully developed.

The problem of the anomalous rise in specific heat near melting is of current theoretical and experimental interest. The resolution of this problem may require a better understanding of the full anharmonic contribution before the magnitude of vacancy contributions to specific heat can be reliably determined.

1.3. ELECTRONIC SPECIFIC HEAT

The conduction electrons in normal metals may be regarded as a highly degenerate Fermi gas and, as such, they obey Fermi–Dirac statistics. These statistics, which are briefly described in Section 1.3.1, play a fundamental role in determining the temperature dependence of the electronic contribution to the specific heat. In Section 1.3.2, it is shown that the electronic specific heat is proportional to the electronic density of states which, in turn, may be determined from a given model of electronic structure. Finally, a discussion of the specific heat of superconductors is given in Section 1.3.3.

1.3.1. Statistics of an Electron Fermi Gas

As shown in several texts [12,25,26], the quantum statistical behavior of an electron gas is a direct result of the Pauli exclusion principle, which allows only one electron in any given quantum state. The appropriate statistics are described by the Fermi–Dirac (F–D) distribution function

$$f(\varepsilon) = \frac{1}{\exp[(\varepsilon-\varepsilon_F)/k_B T] + 1} \tag{82}$$

where $f(\varepsilon)$ gives the probability that an electronic state with energy ε is occupied and ε_F, the so-called 'Fermi energy,' is selected so that the total number of electrons is accounted for correctly. The shape of the F–D function is shown in Fig. 1.10 at various temperatures defined in terms of the Fermi temperature, $T_F = \varepsilon_F/k_B$. At absolute zero, all energy levels with $\varepsilon < \varepsilon_F$ are fully occupied whereas all those with $\varepsilon > \varepsilon_F$ are vacant. At a finite temperature T, the shape of $f(\varepsilon)$ reflects the fact that only those electrons with energies within approximately $k_B T$ of ε_F have sufficient thermal energy to become excited to vacant higher energy levels.

The variation of $f(\varepsilon)$ with temperature qualitatively explains the relatively small contribution of the conduction electron gas to the specific heat of metals at ordinary temperatures ($T \ll T_F$). Since the fraction of electrons which gain energy $\sim k_B T$ is of the order of T/T_F, the increase in internal energy per mole is approximately $N_A(T/T_F) k_B T$ and, hence, the electronic specific heat $C_e \simeq 2RT/T_F$. In metals, T_F is typically in the range 10^4–10^5K [26] which means that, at room temperature, C_e is about $10^{-2}R$ or approximately 1% of the lattice contribution to the specific heat. A more detailed analysis of the problem, given in Section 1.3.2, shows that the variation of C_e with T is indeed linear.

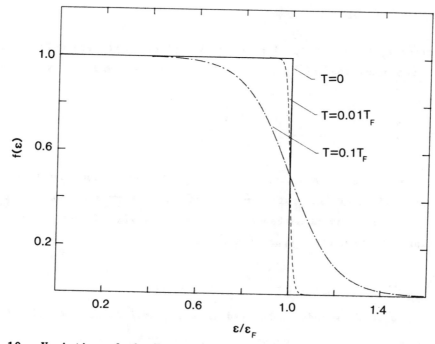

Figure 1.10. Variation of the Fermi-Dirac function (Eq. 82) as a function of energy for various temperatures.

1.3.2. Specific Heat of Electrons in Normal Metals

For a quantitative evaluation of C_e, consideration must be given to the number of energy levels near ε_F. Since the number of conduction electrons (hence, number of quantum states) is typically of the order of N_A, it is convenient to define the total density of states (for both directions of spin) $n(\varepsilon)$ such that $n(\varepsilon)d\varepsilon$ is the number of energy levels in the interval ε and ε + $d\varepsilon$. Thus, the internal energy of a system of N electrons may be expressed as

$$U = \int_0^\infty \varepsilon\ f(\varepsilon)n(\varepsilon)d\varepsilon \qquad (83)$$

with

$$N = \int_0^\infty f(\varepsilon)n(\varepsilon)d\varepsilon. \qquad (84)$$

The specific heat of the electrons is then obtained from

$$C_e = \frac{\partial U}{\partial T} = \int_0^\infty \varepsilon\ \frac{\partial f}{\partial T}\ n(\varepsilon)d\varepsilon. \qquad (85)$$

At low temperatures $(T \ll T_F)$, C_e may be evaluated [26] by differentiating Eq. (84) with respect to T, and multiplying the result by ε_F,

$$0 = \varepsilon_F\ \frac{\partial N}{\partial T} = \int_0^\infty \varepsilon_F\ \frac{\partial f}{\partial T}\ n(\varepsilon)d\varepsilon, \qquad (86)$$

and then subtracting Eq. (86) from (85) to get

$$C_e = \int_0^\infty (\varepsilon - \varepsilon_F) \frac{\partial f}{\partial T} n(\varepsilon) d\varepsilon. \tag{87}$$

From Fig. 1.10 it is clear that, for $T \ll T_F$, the derivative $\partial f/\partial T$ is nonzero only in the vicinity of ε_F. Thus, to a good approximation, $n(\varepsilon)$ may be evaluated at the Fermi energy and taken outside the integral so that

$$C_e \simeq n(\varepsilon_F) \int_0^\infty (\varepsilon - \varepsilon_F) \frac{\partial f}{\partial T} d\varepsilon = n(\varepsilon_F) k_B^2 T \int_{x_0}^\infty \frac{x^2 e^x dx}{(e^x + 1)^2} \tag{88}$$

where $x = (\varepsilon - \varepsilon_F)/k_B T$ and $x_0 = -T_F/T$. Since the integrand in Eq. (88) becomes negligibly small as x approaches x_0, the lower limit of integration may be replaced by $-\infty$ with negligible error; the resulting integral is given in standard mathematical tables as $\pi^2/3$. Therefore, the electronic specific heat may be expressed to a first approximation as

$$C_e = \frac{\pi^2}{3} n(\varepsilon_F) k_B^2 T \equiv \gamma T. \tag{89}$$

A linear variation of C_e with T is indeed observed in all normal metals at sufficiently low temperatures.

For a metal in which $n(\varepsilon)$ is a rapidly varying function near the Fermi level, the replacement of $n(\varepsilon)$ by $n(\varepsilon_F)$ in Eq. (87) becomes a poor approximation as temperature is raised. Stoner [113] has shown that, to the next order of approximation, the electronic specific heat depends also upon the variation of $n(\varepsilon)$ at the Fermi level according to

$$C_e = \gamma T \left\{ 1 - \frac{\pi^2 k_B^2 T^2}{2} \left[\left(\frac{n'(\varepsilon_F)}{n(\varepsilon_F)} \right)^2 - \frac{7 n''(\varepsilon_F)}{5 n(\varepsilon_F)} \right] \right\} \tag{90}$$

where the primes indicate derivatives with respect to energy ε.

The exact form of C_e as a function of T depends upon the functional dependence of $n(\varepsilon)$ on ε over the entire energy range, and the calculations have to be carried out numerically by means of Eq. (85). However, since the melting points of metals are considerably less than the respective values of T_F, the electronic specific heat of most metallic solids can be adequately represented by either Eq. (89) or (90).

From the above discussion, it becomes evident that the problem of evaluating C_e is one of determining the density of states in the vicinity of the Fermi energy. Except for the simplest of models, however, the calculation of $n(\varepsilon)$ generally involves rather formidable computations.

1.3.2.1. The Free Electron Model

The earliest calculation [20,21] of density of states, hence specific heat, of the conduction electrons was carried out in terms of the free electron model. This calculation, which is described in numerous texts [12,25,26], provides a good insight into the specific heat of electrons and is outlined briefly here.

In this model, the effective potential seen by an electron is assumed to be zero throughout the volume of the metal; that is, the electrons move as free particles. As a result, the wave functions which satisfy the free-particle Schrodinger equation are of the form of traveling plane waves, yielding the following relation for the electron energy:

$$\varepsilon = \frac{\hbar^2}{2m} k^2 = \frac{\hbar^2}{2m} [k_x^2 + k_y^2 + k_z^2] \tag{91}$$

where $\hbar = 1.054589 \times 10^{-34}$ J·s is Planck's constant divided by 2π, and m and $\vec{k} \equiv [k_x, k_y, k_z]$ are the mass and wave vector of the free electron, respectively. The values of the wave vector are restricted, however, by the boundary conditions of the problem. As described in Section 1.2.2, traveling waves readily satisfy periodic boundary conditions over a representative cube of the system (with side L) provided that each component of the wave vector is restricted to values given by $2n\pi/L$, where n = 0, ±1, ±2, ... The values of n associated with k_x, k_y, and k_z, i.e., n_x, n_y, and n_z, are quantum numbers which, along with quantum number m_s for spin direction, define the possible states of the system.

The allowed wave vectors form a closely-spaced grid in $\vec{k}$-space with one allowed wave vector for each cell of size $(2\pi/L)^3 = 8\pi^3/V$; thus, the number of allowed wave vectors per unit volume of $\vec{k}$-space is $V/8\pi^3$. Since the energy of each state is given by $\hbar^2 k^2/2m$, the lowest energy state of the system is obtained by placing pairs of electrons in the states of smallest $\vec{k}$, thus occupying all states within a sphere of radius k_F large enough to accommodate all electrons. For a system of N electrons, radius k_F is

$$\left(\frac{4}{3} \pi k_F^3\right) \cdot \frac{V}{8\pi^3} \cdot 2 = N \tag{92}$$

where the factor 2 allows for both directions of spin. Solving for k_F, one obtains

$$k_F = \left(\frac{3\pi^2 N}{V}\right)^{1/3}. \tag{93}$$

The surface of this sphere, known as the Fermi surface, separates the occupied and unoccupied states in wave vector space. The energy at the surface, the Fermi energy, is then given by

$$\varepsilon_F = \frac{\hbar^2}{2m} \left(\frac{3\pi^2 N}{V}\right)^{2/3}. \tag{94}$$

From Eqs. (93) and (94) it may be seen that the Fermi sphere radius and Fermi energy do not depend upon the size of the system, only upon the electron density N/V which, for a metal, is of the order of 10^{23} cm^{-3}. Correspondingly, $k_F \approx 1.4$ Å^{-1} and $\varepsilon_F \approx 7.9$ eV (1.3×10^{-18} J). In addition, the Fermi temperature may also be estimated: $T_F = \varepsilon_F/k_B \approx 90,000$ K. A list of calculated free-electron Fermi surface parameters is given in Ref. [26] for a number of the simpler metals.

Since Eq. (94) relates ε_F to the number of states N with energies $\leq \varepsilon_F$, it follows that the number of states N with energies $\leq \varepsilon$ may be expressed as

$$N = \frac{V}{3\pi^2} \left(\frac{2m\varepsilon}{\hbar^2}\right)^{3/2}. \tag{95}$$

Therefore, the density of states for the free electron gas may be expressed as

$$n(\varepsilon) = \frac{dN}{d\varepsilon} = \frac{V}{2\pi^2} \left(\frac{2m}{\hbar^2}\right)^{3/2} \varepsilon^{1/2},$$

(96)

which is a simple parabolic function of electron energy.

Of particular interest is the density of states at ε_F which, from Eqs. (94) and (96), may be expressed in the form

$$n(\varepsilon_F) = \frac{m}{\pi\hbar^2} \left(\frac{3zN_A V_m^2}{\pi}\right)^{1/3}$$

(97)

where $z = N/N_A$ is the conduction electron/atom ratio and V_m is the molar volume. Therefore, the electronic specific heat in the free electron approximation may be obtained from Eq. (90) as

$$C_e = \gamma_o T \left[1 - \frac{3\pi^2}{10} \left(\frac{T}{T_F}\right)^2 \right]$$

(98)

where

$$\gamma_o = \frac{m}{\hbar^2} (z\,N_A)^{1/3} \left(\frac{\pi V_m}{3}\right)^{2/3} k_B^2.$$

(99)

Since $T_F \sim 10^5 K$, the values of C_e given by free-electron theory are essentially linear in T over the entire temperature range associated with metallic solids.

For a 'typical' metal, say copper, $z = 1$ and $V_m \simeq 7\ cm^3 \cdot mol^{-1}$, which corresponds to $\gamma_o = 0.50$ $mJ \cdot mol^{-1} \cdot K^{-2}$. This is in approximate agreement with the empirical value $\gamma = 0.695\ mJ \cdot mol^{-1} \cdot K^{-2}$ [26]. In general, the free electron model gives values of γ of the right order of magnitude, as seen in the next section.

1.3.2.2. Experimental Specific Heat at Low Temperatures

For normal metals at temperatures in the liquid helium range (~2–4 K), the specific heat at constant volume may be expressed in the form

$$C_v = \gamma T + \beta T^3$$

(100)

where the coefficients γ and β are defined by Eqs. (89) and (44), respectively. At these temperatures there is no significant difference between C_v and C_p, the directly observable quantity. Thus, experimental results are usually analyzed by fitting the values of $C_p\, T^{-1}$ versus T^2 to a straight line by the method of least-squares, yielding values for slope β (hence, θ_o) and intercept γ (see Fig. 1.11). The values of θ_o for a number of metals were given earlier in Table 1.1. Typical values of γ for selected metals are presented in Table 1.3. An extensive review of the observed values of θ_o and γ reported in the literature is given in Ref. [6].

For the simple metals, such as those exhibiting nearly-free electron behavior (Section 1.3.2.3), the difference between γ and the corresponding free electron value γ_o is often described in terms of an 'effective' electron mass m* defined by

$$\frac{m^*}{m} = \frac{\gamma}{\gamma_0} \, .$$

(101)

This definition follows from the proportional relationships $\gamma \propto n(\varepsilon_F)$ and $n(\varepsilon_F) \propto m$ expressed by Eqs. (89) and (97), respectively. Values for m^*/m are given in Table 1.3 for nontransition metals. In the case of transition metals, the details of electronic structure are quite complicated and, thus, a comparison with the free electron model has little physical significance.

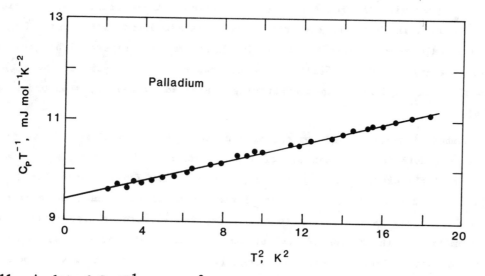

Figure 1.11. A plot of $C_p T^{-1}$ versus T^2 below 4.2 K for palladium (after Veal and Rayne [114]).

Table 1.3. Values of the Low-Temperature Specific Heat Coefficient γ for Selected Metals. Values of m^*/m are also Given for Nontransition Metals [26].

Metal	z (electrons/atom)	γ (mJ·mol^{-1}·K^{-2})	$\frac{m^*}{m}$	Metal	z (electrons/atom)	γ (mJ·mol^{-1}·K^{-2})	$\frac{m^*}{m}$
Na	1	1.38	1.26	Fe	8	4.98	
K	1	2.08	1.25	Ru	8	3.3	
Rb	1	2.41	1.26	Os	8	2.4	
Be	2	0.17	0.34	Co	9	4.73	
Mg	2	1.3	1.3	Rh	9	4.9	
Ca	2	2.9	1.9	Ir	9	3.1	
Sc	3	10.7		Ni	10	7.02	
Y	3	10.2		Pd	10	9.42	
Ti	4	3.35		Pt	10	6.8	
Zr	4	2.80		Cu	1	0.695	1.38
Hf	4	2.16		Ag	1	0.646	1.00
V	5	9.26		Au	1	0.729	1.14
Nb	5	7.79		Zn	2	0.64	0.85
Ta	5	5.9		Cd	2	0.688	0.73
Cr	6	1.40		Al	3	1.35	1.48
Mo	6	2.0		In	3	1.69	1.37
W	6	1.3		Tl	3	1.47	1.14
Re	7	2.3		Pb	4	2.98	1.97

The departure of m*/m from unity arises from three separate effects: the interaction of the conduction electrons (1) with the ions at their lattice sites, (2) with the phonons, and (3) with each other. As seen in the next section, calculations which include these effects can give a reasonable account of the experimental values of m*/m.

1.3.2.3. The Effect of Electron Band Structure and Many-Body Enhancement

In solids, the valence (or conduction) electrons are subject to the influence of the periodic potential field associated with the ion cores of the lattice. As a result, each atomic valence level broadens into a band in which the electron energy ε is a continuous function of wave vector; however, at the Brillouin zone boundaries the bands themselves are separated by gaps in which there are no allowed energy levels [115,116]. The existence of energy gaps in the $\varepsilon(\vec{k})$ relation readily explains the wide difference in conductivity exhibited by metals, semiconductors, and insulators [25,26].

There are a number of methods used for calculating the electron band structure; that is, the $\varepsilon(\vec{k})$ relation. In all methods, a common set of approximations reduces the complete many-body problem to that of a single electron in a periodic potential. The first of these is the adiabatic approximation (Section 1.2.3.2), in which motion of the electrons is assumed to be independent of motions of the nuclei. This is equivalent to neglecting the electron-phonon interaction. The second approximation is that the motion of each electron in the system is considered independently of the motions of the other electrons and that each electron is treated as moving in the average periodic field of all other electrons and the nuclei. This is equivalent to neglecting direct electron-electron interactions. The problem then becomes one of solving the single-particle Schrodinger equation for a given periodic potential to obtain $\varepsilon(\vec{k})$. The various methods for calculating $\varepsilon(\vec{k})$ involve different assumptions about the form of the periodic potential and the nature of the electron wave functions. The most powerful of the methods is one based on fitting empirical pseudopotentials to experimental Fermi surface dimensions. The details concerning the development of these methods and their application to different solids are described in a vast literature on the subject and clearly must be left to special reviews [117-119]. However, to illustrate the influence of band structure on electronic specific heat, two groups of metals, namely the nearly-free electron (NFE) metals and the transition metals, are considered here briefly.

For the NFE metals, the $\varepsilon(\vec{k})$ relation is well approximated by the free electron model except at the Brillouin zone boundaries, where the major aspects of the energy bands are determined almost entirely by crystal structure. The simplest examples of this group are the alkali metals (Li, Na, K, Rb, and Cs) which contribute one electron per atom to their respective conduction bands. In each case, the available states in the conduction (or s-) band are only half-filled and the Fermi surface is a sphere, contained entirely within the first Brillouin zone. Thus, the variation of the resultant density of states $n(\varepsilon)$ with ε is expected to be essentially parabolic (see Eq. 96) and, consequently, the variation of C_e with T should be very nearly linear over the entire temperature range of the solid phase. However, even in such simple cases the effective mass of the electrons must be substantially increased from the 'free-particle' value in order to account for the magnitude of the observed specific heat (see Table 1.3). A number of polyvalent metals also exhibit nearly-free electron behavior; these include the divalent metals Be, Mg, Zn, Cd, and Hg, the trivalent metals Al, Ga, In, and Tl, and also Pb, which contributes four electrons per atom to its conduction band. The Fermi surfaces of many of the polyvalent metals can be represented reasonably accurately by geometrical constructions, depending only on the crystal structure and the valence of the material in question [116,120]. As the valence (i.e., electron density) is increased, ε_F increases and the Fermi sphere expands. In part, this is accomplished by an increase in the Fermi radius, but for certain energies the gaps at the Brillouin zone boundaries cause the disappearance and reappearance at higher energies of parts of the Fermi surface. The latter changes cause a distortion of the Fermi surface and give rise to discontinuous changes in the slope of the electron density of states $n(\varepsilon)$. Since the behavior of the electronic specific heat of individual polyvalent NFE metals will depend sensitively on the details of $n(\varepsilon)$ in the vicinity of the Fermi level, no generalizations about the temperature dependence of C_e can be made for these metals.

The transition metals are characterized by an electron band structure in which there is considerable overlap in energy of the d-band and s-band electron states [121]. Since the d-electrons are more highly localized about their ion cores than are the s-electrons, there is less overlap among the d-state wave functions than among the s-state wave functions. As a result, the spread in energy of the d-band, in which there are 10 electron states per atom, is much less than the broad range of energies associated with the s-band, which contains only two electron states per atom. Consequently, the density of d-states will be large, resulting in a large peak in the total density of states function $n(\varepsilon)$. Therefore, one may expect the relative position of the Fermi level ε_F and the large peak in $n(\varepsilon)$ to strongly influence the temperature dependence of C_e in transition metals. If, for example, ε_F coincides with the maximum of the large peak, $n(\varepsilon_F)$ and, hence, γ will be relatively large (see Eq. 89). Consideration of Eq. (90) shows that the slope of $C_e(T)$ will decrease rapidly from its initial value of γ as temperature is raised. A particularly striking example of this behavior is provided by palladium: experimental results yield an unusually large value of γ (see Table 1.3) as well as a 'saturation' of C_e at temperatures above about 200 K [73,122–124]. Recent band structure calculations [125] have indeed confirmed the existence of a large peak in the density of states near ε_F (see Fig. 1.12).

The density of states determined from $\varepsilon(\vec{k})$ is called the 'band structure' density of states $n_b(\varepsilon)$, and may be expressed as

$$n_b(\varepsilon) = \frac{2V}{(2\pi)^3} \int \frac{dS}{|\text{grad}_{\vec{k}}\, \varepsilon(\vec{k})|} \tag{102}$$

where the integral is taken over the area of a constant energy surface in $\vec{k}$-space. These surfaces often have complicated shapes and, therefore, in general, Eq. (102) is not solved analytically but by various semianalytic methods. The approach now most commonly used for calculating $n_b(\varepsilon)$ is one based on the Gilat-Raubenheimer scheme [78] developed earlier for phonon spectra. Modifications [126,127] of this scheme in recent years have considerably increased the accuracy of histogram representations of $n_b(\varepsilon)$.

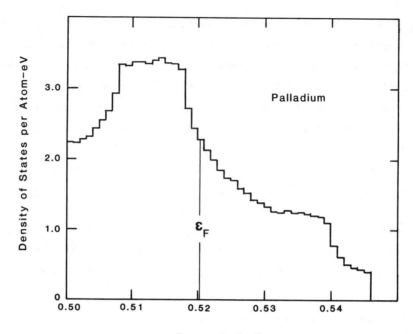

Figure 1.12. Histogram representation of the electronic density of states for palladium in the
immediate vicinity of the Fermi energy (after Mueller et al. [125]).

Although careful energy band calculations have been made for a number of metals [119], many
of these studies, unfortunately, do not include a calculation of the density of states. In cases
where accurate theoretical values of $n_b(\varepsilon_F)$ are available, it is generally found that these values
are smaller than the density of states $n(\varepsilon_F)$ required to account for the experimentally observed γ
[125,128,129]. This difference is customarily attributed to the enhancement effects of electron-
phonon and direct coulomb electron-electron interactions. The enhancement effects are usually
assumed to be additive [6], such that

$$n(\varepsilon_F) = (1 + \lambda + \lambda_c)\, n_b(\varepsilon_F) \tag{103}$$

where λ and λ_c are the phonon and coulomb enhancement factors, respectively. Details concerning
the evaluation of λ and λ_c are quite involved and, therefore, must be left to discussions given in
the literature [130-134].

In order to facilitate a comparison between theory and experiment, it is convenient to introduce the 'band' effective mass m_b, defined so that m_b/m is equal to the ratio of $n_b(\varepsilon_F)$ to the
corresponding density of states for the free electron gas [131,132]. Thus, the theoretical effective mass is obtained from Eq. (103) as

$$\frac{m^*}{m} = (1 + \lambda + \lambda_c)\, \frac{m_b}{m} . \tag{104}$$

A comparison of theoretical and experimental values of m^*/m, as defined by Eqs. (104) and (101),
respectively, is presented for several metals in Table 1.4. It can be seen that, even for the
simple case of the alkali metals, the band structure alone cannot account for the observed

electronic specific heat. However, when many-body enhancement effects are taken into account, the examples given as well as other available data in the literature [132] suggest that the agreement between theory and experiment is within the uncertainties in the calculated values for λ, λ_c, and m_b/m. Except in the case of alkali metals, λ_c is of the order of or smaller than the uncertainties in λ and is, therefore, often neglected in comparisons between theory and experiment.

Table 1.4.   Comparison of Theoretical Values of m_b/m and m^*/m for Selected Metals with Observed Values of m^*/m Obtained from Specific Heat Measurements

Metal	Theory				Experiment
	m_b/m	λ	λ_c	m^*/m	m^*/m
Na	1.00[a]	0.18	0.06	1.24[c]	1.26[b]
		0.13	0.06	1.20[d]	
K	0.99[a]	0.15	0.11	1.25[e]	1.25[b]
		0.12	0.1	1.22[d]	
Mg	1.01[a]	0.39	0.01	1.41[d]	1.3[b]
Al	1.04[a]	0.49	−0.01	1.54[c]	1.48[b]
		0.53	−0.01	1.58[d]	
Pb	0.86[a]	1.05	0.0	1.76[c]	1.97[b]
		1.26	0.0	1.94[d]	

[a] Values of band efective mass from Ref. [129].

[b] Values of observed effective mass from Ref. [26].

[c] Values of $m^*/m = (1+\lambda+\lambda_c)\, m_b/m$ from Ref. [131].

[d] Values of $m^*/m = (1+\lambda)(1+\lambda_c)\, m_b/m$ from Ref. [133].

[e] Values of $m^*/m = (1+\lambda+\lambda_c)\, m_b/m$ from Ref. [134].

1.3.2.4.   The Rigid-Band Model

The rigid-band model was originally applied to calorimetric studies of alloys, particularly of transition metals, in order to obtain the variation of $n(\varepsilon_F)$ with z (valence electron/atom ratio) and, thus, the $n(\varepsilon)$ function for the pure host metal. In this model, it is assumed that in the process of alloying the band structure remains 'rigid' and only z is altered. This assumption is expected to be reasonably good for very dilute alloys, particularly of elements of near atomic number which readily go into solid solution without a change in crystal structure.

Consider, for example, palladium and its neighboring elements rhodium and silver. According to the rigid-band model, the addition of silver to palladium yields one more electron for each silver atom and, consequently, the conduction band is filled to a higher energy level. Conversely, a small amount of rhodium added to palladium lowers the average z and hence the Fermi level. A measurement of the low-temperature specific heat coefficient γ for each alloy determines $n(\varepsilon_F)$ at a different Fermi energy (Eq. 89). In this way, the variation of $n(\varepsilon)$ with ε in the vicinity of ε_F for palladium can be determined from measurements on a suitable range of alloys. Such measurements [122–124], conducted nearly two decades ago, provided the earliest evidence of the large peak near ε_F in the electronic density of states for palladium. This observation was subsequently confirmed when accurate band structure calculations became available [125].

In recent years, many results on the energy bands of transition metals have become available. The calculations show a great deal of similarity between the band structures of different elements with the same crystal structure [116,119]. In fact, metals with the same z tend to have the same crystal structure, so that a strong correlation of $n(\varepsilon_F)$ or γ with z can be expected. Such a correlation is clearly evident among the empirical values of γ given in Table 1.3. These properties provide some justification of the rigid-band model for determining the relation between γ and z in transition metals and their alloys.

Attempts to interpret the γ values for alloys of nontransition metals in terms of the rigid-band model have been less successful. In part, this is due to the fact that changes in γ with z are much smaller (see Table 1.3) than for transition metals and, hence, are more difficult to measure accurately. For example, most measurements on the α-phase alloys of noble metals seem to suggest that γ increases with z, in contradiction to the behavior expected from a simple consideration of changes in the Fermi surface area with z [6]. Several explanations have been offered for the observed deviation from rigid-band behavior and these have been discussed in detail in Ref. [6].

In the case of alloys that undergo superconducting transitions, the variation of the electron-phonon enhancement λ with z can often be determined from electron tunneling experiments. In such cases, a direct comparison can be made between the z dependence of $n(\varepsilon_F)$ determined from alloying data and that of $n_b(\varepsilon_F)$ derived from a rigid-band model based on the band structure for the pure host metal. Extensive investigations of the PbTl and PbBi alloy systems indicate that theory and experiment are in good agreement [6].

1.3.3. Specific Heat of Superconductors

Much of the theoretical understanding of thermal properties of (type I) superconductors has been derived from classical thermodynamics and, more recently, from the Bardeen-Cooper-Schrieffer (BCS) theory; the corresponding theoretical results are discussed in Sections 1.3.3.1 and 1.3.3.2, respectively. A brief description of the behavior of specific heat in type II superconductors is given in Section 1.3.3.3. For detailed expositions of the theory and other topics related to superconductivity, the reader is referred to several recent books [135-137] and the references cited therein.

1.3.3.1. Specific Heat According to Classical Thermodynamics

For a decade or so after the discovery of superconductivity in 1911 by Kamerlingh Onnes [138], measurements showed no striking difference between the specific heats in the normal and superconducting phases. With improvements in thermometry, Keesom and Andrews [139] were able to measure the sudden increase in specific heat of a superconductor as it cooled through the transition temperature T_c. Although thermodynamic arguments were successfully applied to the superconducting transition, it was not clear that the use of thermodynamics was fully justified until the reversible magnetization measurements of Meissner and Ochsenfeld [140] in 1933. Their measurements showed that when placed in a small magnetic field the superconductor is perfectly diamagnetic; that is, it completely expels the magnetic flux from its interior. Furthermore, it was found that the final state of the superconductor was independent of whether it was cooled through T_c and then placed in a field or vice versa, since the flux was expelled from the super-conductor in either case. This phenomenon is now known as the Meissner effect.

For the purpose of the present simple discussion of specific heats, the demagnetization effects dependent on specimen shape [135-137] are assumed to be negligible; this condition may be realized for specimens in the form of a thin rod parallel to the applied field, H. Furthermore, it is assumed that the Gibbs free energy of the normal phase, G_n, is independent of the applied field; that is, the specimen in its normal state is virtually nonmagnetic. In the superconducting state, however, the specimen experiences an induced magnetization (magnetic moment per unit volume) of magnitude M = -H because of the perfect diamagnetism. In this case, the Gibbs free energy per mole of specimen may be expressed in SI units as [137]

$$G_s(H) = G_s - \mu_o V_m \int_0^H M\, dH = G_s + 1/2\, \mu_o\, H^2\, V_m \tag{105}$$

where G_s is the Gibbs free energy in the superconducting state at zero field, μ_o is the magnetic constant ($4\pi \times 10^{-7}$ H·m^{-1}) and V_m is the molar volume. If one defines the 'thermodynamic' critical field H_c as the value of applied field for which $G_s(H_c) = G_n$, then the difference in free energy between the normal and superconducting states in an applied magnetic field is given by

$$G_n - G_s(H) = 1/2\, \mu_o\, (H_c^2 - H^2)\, V_m. \tag{106}$$

For superconductors which exhibit a complete Meissner effect (i.e., type I superconductors), H_c has the following physical significance: for $H > H_c$ the perfect diamagnetism and perfect conductivity are destroyed and the normal state of the metal is restored, whereas for $H < H_c$ the properties of the superconductive state are retained. For many common (type I) superconductors such as cadmium, lead, tantalum, tin, and vanadium the critical field has a temperature dependence which can be approximated by an empirical parabolic law,

$$H_c = H_o \left[1 - \left(\frac{T}{T_c}\right)^2\right] \tag{107}$$

where H_o is the value of critical field at zero kelvin.

Since $S = -(\partial G/\partial T)_{H,P}$ the entropy difference, at constant pressure and field, between the normal and superconducting states may be expressed as

$$S_n - S_s = -\mu_o\, H_c \left(\frac{\partial H_c}{\partial T}\right) V_m. \tag{108}$$

From Eq. (107), it may be seen that $\partial H_c/\partial T$ is always negative and, therefore, $S_s \leq S_n$; that is, the superconducting state is more highly ordered than the normal state. Since $C_p = T(\partial S/\partial T)_p$, the difference in specific heats of the two states may be expressed as

$$C_n - C_s = -\mu_o T \left[H_c \frac{\partial^2 H_c}{\partial T^2} + \left(\frac{\partial H_c}{\partial T}\right)^2\right] V_m. \tag{109}$$

If the specimen is heated in the absence of a magnetic field, the transition from superconducting to normal state takes place at T_c with no change in entropy (since $H_c = 0$) but with a discontinuity in specific heat given by

$$(C_n - C_s)_{T_c} = -\mu_o T_c \left(\frac{\partial H_c}{\partial T}\right)^2 V_m, \tag{110}$$

a formula often called 'Rutger's relation.' If the heating is performed in a constant applied field of magnitude H, then the transition from the superconducting state takes place at a temperature $T_1 < T_c$ such that $H_c(T_1) = H$; this results in a sudden increase in entropy (Eq. 108) and the absorption of a latent heat given by

$$L = -\mu_o T H_c \frac{\partial H_c}{\partial T} V_m. \tag{111}$$

Thus, the transition implied by these relations is first order in a magnetic field and second order in the absence of a magnetic field [9].

In comparing the predictions of theory with specific heat measurements, the usual method of evaluating the data is to separate the observed specific heat into electronic and lattice contributions:

$$C_n = \gamma T + C_{\ell n} \tag{112}$$

$$C_s = C_{es} + C_{\ell s} \tag{113}$$

where C_{es} is the electronic specific heat in the superconducting state, and $C_{\ell n}$ and $C_{\ell s}$ are the lattice specific heats in the normal and superconducting states, respectively. The observed specific heat is generally obtained below T_c for both the superconducting state and the normal state, which can be restored by an applied field greater than H_c (see Fig. 1.13). The analysis is simplified by the assumption that $C_{\ell n} = C_{\ell s}$, which is reasonable since studies by X-ray crystallography reveal no change in crystal structure or lattice spacing at the transition. This leads to the relations

$$C_n - C_s = \gamma T - C_{es} \tag{114}$$

$$C_{es} = C_s - C_{\ell n}. \tag{115}$$

At sufficiently low temperatures, C_{es} may be readily obtained as a function of temperature from the observed values of C_s, since $C_{\ell n} = \beta T^3$, where $\beta = 12\pi^4 R/5\theta_o^3$.

Prior to about 1950, the values of C_{es} obtained by experiment appeared to be of the form aT^3. This is consistent with a parabolic variation for $H_c(T)$, as may be seen from Eqs. (107) and (109), which then yield

$$C_n - C_s = 2\mu_o \left(\frac{H_o}{T_c}\right)^2 V_m T \left[1 - 3\left(\frac{T}{T_c}\right)^2\right]. \tag{116}$$

A comparison with Eq. (114) yields

$$\gamma = 2\mu_o \left(\frac{H_o}{T_c}\right)^2 V_m \quad \text{or} \quad \frac{\gamma T_c^2}{\mu_o H_o^2 V_m} = 2, \tag{117}$$

and

$$C_{es} = \left(\frac{6\,\mu_o H_o^2 V_m}{T_c^4}\right) T^3. \tag{118}$$

The relation given in Eq. (117) suggests an alternate method of determining γ from measurements of H_o and T_c. Recent studies [6], however, show that γ values obtained from this relation may be in error by as much as 6% because of the nonparabolic behavior of $H_c(T)$ in real superconductors.

Although deviations of C_{es} from a T^3-law had been observed earlier, it was not until 1954 that improvements in low-temperature measurement techniques were sufficient to reveal an exponential temperature dependence of C_{es} at the very lowest temperatures [142]. An exponential variation suggested the existence of an energy gap in the electron excitation spectrum, since the number of electrons excited across such a gap would vary exponentially with temperature.

Another experimental result which was important in the development of the theory of superconductivity was the discovery of the isotope effect in 1950. The transition temperature T_c was

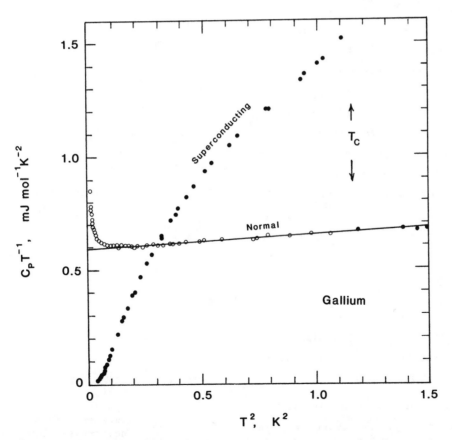

Figure 1.13. Specific heat of gallium in its normal and superconducting states. At temperatures below T_c, the normal state is restored by an applied magnetic field (after Phillips [141]).

found to vary with the isotopic mass M approximately as $T_c \sim M^{-1/2}$. This suggested a fundamental connection between the electrons in the superconducting state and the phonons of the lattice, since phonon frequency is also approximately proportional to $M^{-1/2}$ (see Sections 1.2.3 and 1.2.4).

These empirical observations provided important insights into the microscopic mechanism of superconductivity and ultimately led to formulation of a quantum theory of superconductivity by Bardeen et al. [143] in 1957.

1.3.3.2. Thermodynamic Properties According to BCS Theory

The fundamental basis of superconductivity, according to the BCS theory [143–145], is the indirect attraction between pairs of conduction electrons caused by virtual phonon exchange. The electron–phonon coupling is assumed to be independent of electron energy and crystal direction, approximations which are valid in the weak coupling limit ($\lambda \ll 1$). The attraction between electron pairs, known as 'Cooper' pairs, leads to a temperature-dependent energy gap $2\Delta(T)$ in the electron energy spectrum at the Fermi level. The energy gap is a maximum at absolute zero, where it is simply related to the critical temperature by

$$2\Delta(0) = 3.52\ k_B T_c. \tag{119}$$

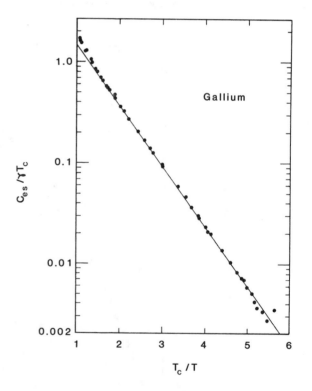

As temperature is increased, $\Delta(T)$ remains nearly constant up to about $0.5\ T_c$ and then it begins to decrease rapidly, becoming zero at T_c. At temperatures below about $0.5\ T_c$ the energy required to break up a Cooper pair is approximately $2\Delta(0)$ and, therefore, the number of pairs broken up is proportional to $\exp[-2\Delta(0)/k_B T]$. This leads to an exponential temperature variation for the electronic specific heat, a behavior which is observed in experiments at sufficiently low temperatures (see Fig. 1.14). The result for C_{es} from BCS theory may be approximated in limited temperature ranges by

$$\frac{C_{es}}{\gamma T_c} = a\ \exp\left(-b\ \frac{T_c}{T}\right) \tag{120}$$

where $a = 8.5$, $b = 1.44$ for $2.5 < T_c/T < 6$, and $a = 26$, $b = 1.62$ for $7 < T_c/T < 12$. Values of $C_{es}/\gamma T_c$ have been tabulated in the literature [146] for the range $1 < T_c/T < 7$.

The thermodynamics of the transition is determined by the behavior of the energy gap in the vicinity of T_c. As temperature is increased through T_c there is an abrupt fall in specific

Figure 1.14. A semi-log plot of the electronic specific heat of gallium in its superconducting state as a function of T_c/T. The straight line represents the equation $C_{es}/\gamma T_c = 7.46$ $\exp(-1.39\ T_c/T)$ (after Phillips [141]).

heat because the contribution due to the splitting up of electron pairs vanishes at T_c. In addition, the transition involves no latent heat, since the total energy of the electrons as T approaches T_c from below approaches the same value as when T approaches T_c from above. The combination of a discontinuity in the specific heat and the absence of latent heat is characteristic of a second-order phase transition [9]. The BCS result for the discontinuity in the electronic specific heat is given by

$$\frac{C_{es}(T_c) - \gamma T_c}{\gamma T_c} = 1.43.$$

(121)

Another simple result predicted by BCS theory is a relation involving the critical field at absolute zero (H_o), the critical temperature (T_c), and the coefficient of electronic specific heat in the normal state (γ):

$$\frac{\gamma T_c^2}{\mu_o H_o^2 V_m} = 2.14.$$

(122)

This relation is equivalent to that in Eq. (117), which was derived on the basis of a parabolic variation for $H_c(T)$. The BCS values for $H_c(T)$ exhibit a negative deviation from a reference parabola with a maximum deviation of almost -4% near $(T/T_c)^2 = 0.5$; the values may be approximated at low temperatures ($T \ll T_c$) by

$$H_c \simeq H_o \left[1 - 1.07 \left(\frac{T}{T_c} \right)^2 \right].$$

(123)

The properties of the superconducting state as expressed in Eqs. (119) to (123) form a 'law of corresponding states' in the sense that the properties, when expressed in terms of certain reduced variables, are the same function of the variables for all superconductors. It may be seen in Table 1.5 that many widely different superconductors do obey quite closely the BCS model for corresponding states.

The law of corresponding states breaks down for strong-coupling superconductors such as mercury or lead (see Table 1.5). Strong-coupling superconductors are characterized by relatively large ratios of energy gap to T_c, low values for θ_o, and positive deviations of $H_c(T)$ from a parabolic temperature dependence. The effects of strong coupling on various thermodynamic properties such as the quantities $2\Delta(0)/k_B T_c$, $H_c(T)$, and T_c have been studied theoretically [148-150].

Deviations from simple BCS theory may also arise if the energy gap is anisotropic over the Fermi surface. For example, measurements on lead [151] have shown two rather distinct slopes on a semilog plot (as in Fig. 1.14) of C_{es} versus T_c/T, suggesting the presence of two distinct energy gaps. Specific heat measurements performed on high-purity niobium, tantalum, and vanadium [152] also seem to provide evidence for the existence of two gaps. The origins of (and even the validity of the evidence for) energy gap anisotropy has been the subject of recent theoretical studies [153].

Table 1.5.   A Comparison of Empirically-Derived Values of the Properties $2\Delta(0)/k_BT_c$, $(C_{es}(T_c) - \gamma T_c)/\gamma T_c$, and $\gamma T^2_c/\mu_o H^2_o\ V_m$ for Different Superconductors [147] with the Values Predicted by BCS Theory

Superconductor	$\dfrac{2\Delta(0)}{k_BT_c}$	$\dfrac{C_{es}(T_c)-\gamma T_c}{\gamma T_c}$	$\dfrac{\gamma T^2_c}{\mu_o H^2_o}$
Al	3.37	1.29–1.59	2.15
Cd	3.2	1.40	2.22
Ga	3.50	1.44	2.12
Hg	4.6	2.37	1.68
In	3.45	1.73	1.97
Pb	4.38	2.71	1.68
Sn	3.46	1.60	2.02
Ta	3.60	1.59	2.02
Tl	3.57	1.50	2.02
V	3.4	1.49	2.14
Zn	3.2	1.30	2.22
BCS	3.52	1.43	2.14

1.3.3.3. Specific Heat of Type II Superconductors

Superconductors may be broadly classified as type I or type II according to their behavior in a magnetic field, as illustrated in Fig. 1.15. Type I superconductors exhibit a complete Meissner effect and, apart from deviations noted in the previous section, are reasonably well described by BCS theory. In type II superconductors, however, the Meissner effect is incomplete. The magnetic flux begins to penetrate the specimen at a lower critical field $H_{c1} < H_c$ causing the magnetization to decrease rapidly in magnitude until, at an upper critical field $H_{c2} > H_c$, flux penetration is complete and the specimen is returned to its normal state. At fields less than H_{c1}, the properties of type II superconductors are the same as those of type I in the superconducting state. In the region between H_{c1} and H_{c2} the superconductor is said to be in a 'mixed' state. According to Abrikosov [154,155], the mixed state consists of a regular two-dimensional array of normal filaments embedded in a superconductive medium. These filaments, or fluxoids, are quantized and increase in number with magnetic field until at H_{c2} they fill the entire volume of the specimen.

The basic theory of type II superconductivity is that developed by Ginsburg and Landau [156], Abrikosov [154,155], and Gor'kov [157,158]. Various aspects of the GLAG theory are described in several chapters of Ref. [135]. In the present work, we consider only the implications of the theory on the behavior of specific heat in type II superconductors.

According to the GLAG theory, there is no entropy change during either transition -- that is, between the superconducting and mixed states or between the mixed and normal states. Since no latent heat is involved both transitions are of the second order. It may be shown [135] that such transitions are accompanied by discontinuities in the specific heat given by:

$$C_s - C_m = T_1 \left(\frac{dH_{c1}}{dT}\right)^2 \left[\left(\frac{\partial M_s}{\partial H}\right)_T - \left(\frac{\partial M_m}{\partial H}\right)_T\right] V_m \qquad (124)$$

$$C_m - C_n = T_2 \left(\frac{dH_{c2}}{dT} \right)^2 \left[\left(\frac{\partial M_m}{\partial H} \right)_T - \left(\frac{\partial M_n}{\partial H} \right)_T \right] V_m \qquad (125)$$

where the subscripts s, m, and n refer to the superconducting, mixed, and normal states, respectively, and T_1 and T_2 are the temperatures at which the superconducting-to-mixed and the mixed-to-normal transitions occur. The nature of the specific heat discontinuities may be determined by considering a 'typical' measurement.

Specific heat measurements are usually performed in a constant applied field H as a function of increasing temperature. This corresponds to traversing a horizontal line in the H–T plane as shown in Fig. 1.16(a). At temperature T_1, $H = H_{c1}(T_1)$ and the specimen passes from the superconducting to the mixed state. In the Abrikosov model, the slope of the magnetization in the mixed state $(\partial M_m/\partial H)$ becomes infinite at H_{c1} [see Fig. 1.15(b)]. Consideration of Eq. (124) suggests that a singularity in specific heat will appear at T_1, yielding a 'λ-type' specific heat anomaly. The sharp peak observed at T_1 in Fig. 1.16(b) is consistent with such an anomaly. Unusual behavior may also be expected at T_2 where $H = H_{c2}(T_2)$. From Fig. 1.15(b), it can be seen that as H_{c2} is approached from below and from above, $\partial M_m/\partial H > 0$ and $\partial M_n/\partial H = 0$, respectively. Thus, Eq. (125) indicates that there will be sudden drop in specific heat at T_2 as the specimen undergoes a transition from the mixed to the normal state. Such a drop in specific heat is indeed observed at T_2 in Fig. 1.16(b).

The appearance of the specific heat versus temperature curve for a type II superconductor is, of course, dependent on the strength of the applied field relative to $H_{c1}(0)$ and $H_{c2}(0)$. A good example of the effect of magnetic field strength on specific heat data is provided by the detailed measurements [160] on niobium. An extensive review of these and other measurements on the specific heat of superconductors of both types is given in Ref. [6].

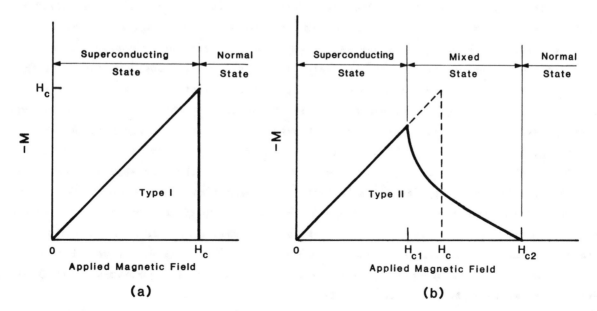

Figure 1.15. Magnetization versus applied magnetic field for (a) a type I superconductor and (b) a type II superconductor.

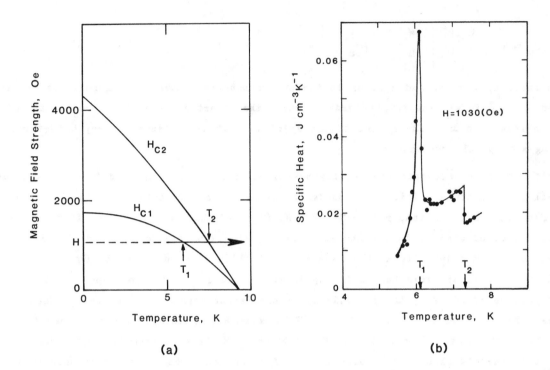

Figure 1.16.   (a) Heating a type II superconductor in a constant magnetic field corresponds to traversing a horizontal line in the H–T plane.   (b) The observed specific heat of niobium exhibits anomalies at T_1 and T_2 which correspond to second–order transitions from the superconducting to the mixed state and from the mixed to the normal state, respectively (after McConville and Serin [159]).

1.4.   MAGNETIC SPECIFIC HEAT

The various forms of magnetism [161] are the result of different electronic configurations within solids.   Solids in which the electron shells of the constituent atoms are filled have no permanent magnetic moments because of zero electron spin and orbital angular momenta.   Even in these cases, however, a magnetic moment can be induced by an externally applied magnetic field. The interaction of the electrons with the field gives rise to a (Larmor) precession of their orbital motions and, hence, to a magnetic moment opposite in direction to that of the field.   This phenomenon, known as diamagnetism, occurs to some extent in all substances.   All other types of magnetic behavior are associated with permanent magnetic moments which result primarily from the intrinsic spin of the electrons.   Substances in which the spin moments of the constituent atoms are randomly oriented are said to be paramagnetic, in the sense that the spin moments can be preferentially oriented by an applied field to yield a net magnetization parallel to the field direction.   In some substances the electron spins are strongly coupled, making an ordered arrangement of the elementary magnetic moments possible.   Magnetic ordering was originally attributed to phenomenological internal magnetic fields by Weiss [162] in 1907.   In 1928, however, Heisenberg [163] was able to show that the ordered magnetic state was the result of a quantum–mechanical exchange interaction between neighboring electron spins.

The ordered arrangements of the electron spins in simple ferromagnetic, antiferromagnetic, and ferrimagnetic substances are illustrated schematically in Fig. 1.17. Other more complicated types of ordering in which the spins are canted or arranged in a helical pattern are also possible. Except for simple antiferromagnets, all other spin arrangements have a spontaneous magnetization even under conditions of zero applied field. The ordered arrangement among spins begins to break down as temperature is raised because of thermal agitation, until at some critical temperature T_c the spins are completely disordered; T_c is known as the Curie temperature in ferro- and ferrimagnets and as the Néel temperature in antiferromagnets. Thus, the thermal energy required to raise the temperature of these substances will include a temperature-dependent contribution associated with spin randomization, and an observable effect may be expected in the specific heat.

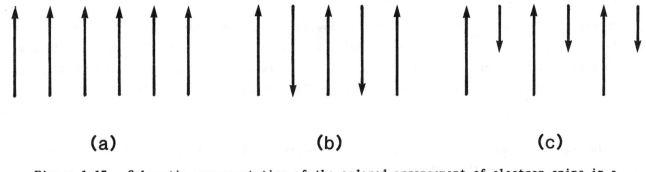

(a)   (b)   (c)

Figure 1.17. Schematic representation of the ordered arrangement of electron spins in a simple (a) ferromagnet, (b) antiferromagnet, and (c) ferrimagnet.

1.4.1. Spin-Wave Specific Heat

Complete order among atomic spins exists only at zero kelvin, as required by the third law of thermodynamics. As the temperature is raised slightly, sufficient thermal energy becomes available to disturb a spin from perfect alignment with its neighbors; the disturbance then propagates through the lattice because of the exchange coupling between neighboring spins. At low but finite temperatures, the resulting oscillations in the relative orientations of spins may be regarded as a superposition of spin waves.

The basis of spin-wave theory, originally developed by Bloch [164] in 1932, is the Heisenberg exchange Hamiltonian

$$H = -2J \sum_{i,j} \vec{S}_i \cdot \vec{S}_j \qquad (126)$$

where J is the isotropic exchange integral and is related to the overlap of the charge distributions of atoms i and j, bearing spins $\vec{S}_i$ and $\vec{S}_j$, respectively. The summation extends over all atoms i in the crystal and all j of their nearest neighbors. The Heisenberg model treats magnetism as coming entirely from unpaired electron spins localized at the atomic lattice sites and, therefore, is expected to be valid for magnetic insulators. These include all ferrimagnets and most antiferromagnetic materials. Only a few nonmetallic ferromagnets (e.g., $CrBr_3$) are known to exist. The extent to which the model can be generalized to include itinerant effects of magnetic electrons in metals is of considerable interest and has been discussed in the literature [165, 166]. For the purpose of our simple discussion of spin waves, it is sufficient to note that neutron spectroscopic data on magnetic materials (see below) seem to support the general conclusions of the Heisenberg model, even for metallic ferromagnets.

The usual approach in spin-wave theory is to approximate the magnetic spin system by a system of linear harmonic oscillators. As a result, the spin waves are quantized in energy units of $\hbar\omega$, known as magnons, and are characterized by a frequency/wave vector dispersion relation $\omega(\vec{q})$. Various thermal properties, such as the exchange specific heat at low temperature, can then be calculated on the basis of the same quantum statistics (Bose-Einstein) used for phonons. A detailed description of spin-wave theory lies outside the scope of the present chapter and is left to specialized reviews of the subject [165,167].

An insight into the nature of spin waves, however, can be obtained by considering the spin oscillations along a linear ferromagnet [5,26]. In its ground state, the ferromagnet will have all spins aligned parallel, say, along the z-axis. At finite temperatures the excitation energy associated with the spin at lattice site ℓ is given by (see Eq. 126)

$$-2\,J\,\vec{S}_\ell \cdot (\vec{S}_{\ell-1} + \vec{S}_{\ell+1}). \tag{127}$$

Each spin has an associated magnetic moment given by $\vec{\mu}_\ell = \gamma\hbar\,\vec{S}_\ell$, where γ is the gyromagnetic ratio. Therefore, Eq. (127) may be expressed as

$$-(2J/\gamma\hbar)\,\vec{\mu}_\ell \cdot (\vec{S}_{\ell-1} + \vec{S}_{\ell+1}). \tag{128}$$

Since the energy of a magnetic moment $\vec{\mu}_\ell$ in a magnetic field $\vec{B}$ is of the form* $-\vec{\mu}_\ell\cdot\vec{B}$, Eq. (128) suggests that the exchange coupling of $\vec{S}_\ell$ with its neighbors can be described in terms of the interaction of $\vec{\mu}_\ell$ with an effective magnetic field or exchange field given by

$$\vec{B}_e = (2J/\gamma\hbar)\,(\vec{S}_{\ell-1} + \vec{S}_{\ell+1}). \tag{129}$$

In the exchange field, spin $\vec{S}_\ell$ will experience a torque $\vec{\mu}_\ell \times \vec{B}_e = \gamma\hbar\,\vec{S}_\ell \times \vec{B}_e$ which, according to classical mechanics, is equal to the time rate of change of the angular momentum:

$$\frac{d(\hbar\vec{S}_\ell)}{dt} = \gamma\hbar\,\vec{S}_\ell \times \vec{B}_e \qquad \text{or} \qquad \frac{d\vec{S}_\ell}{dt} = \frac{2J}{\hbar}\,\vec{S}_\ell \times (\vec{S}_{\ell-1} + \vec{S}_{\ell+1}). \tag{130}$$

The differential equations represented by Eq. (130) contain products of the spin components (S_ℓ^x, S_ℓ^y, S_ℓ^z) and, therefore, are nonlinear. The equations become approximately linear, however, for small spin deviations such that $S_\ell^z \simeq S$ and S_ℓ^x, $S_\ell^y \ll S$; in this approximation, Eq. (130) may be expressed as

$$\frac{dS_\ell^x}{dt} = (2JS/\hbar)\,(2S_\ell^y - S_{\ell-1}^y - S_{\ell+1}^y) \tag{131a}$$

---

*The magnetic field may be described either by its magnetic field strength $\vec{H}$ (in $A\cdot m^{-1}$) or by an equivalent magnetic flux density $\vec{B}$ (in T) where, in SI units, $\vec{B} = \mu_0\vec{H}$. The use of $\vec{B}$ here avoids the appearance of the magnetic constant μ_0 in the equations.

$$\frac{dS_\ell^y}{dt} = -(2JS/\hbar) \ (2S_\ell^x - S_{\ell-1}^x - S_{\ell+1}^x) \qquad (131b)$$

$$\frac{dS_\ell^z}{dt} = 0. \qquad (131c)$$

The analogy between spin oscillations and lattice vibrations (Section 1.2.3) suggests that Eq. (131) will have traveling wave solutions of the form

$$S_\ell^x = \xi_x \ \exp[i(\omega t + \ell qa)]$$
$$S_\ell^y = \xi_y \ \exp[i(\omega t + \ell qa)] \qquad (132)$$

where a is the lattice constant. Upon substitution into Eq. (131) one obtains

$$-i\omega \ \xi_x + (4JS/\hbar) \ (1-\cos qa) \ \xi_y = 0$$
$$(4JS/\hbar) \ (1-\cos qa) \ \xi_x + i\omega \ \xi_y = 0. \qquad (133)$$

Nontrivial solutions for ξ_x and ξ_y exist only if the determinant of their coefficients is zero, a condition which leads to the spin–wave dispersion relation:

$$\omega = (4JS/\hbar) \ (1-\cos qa). \qquad (134)$$

The above derivation of $\omega(q)$ yields the following classical picture of a spin wave in a linear ferromagnet: (a) Components S_ℓ^x and S_ℓ^y of a given spin are out of phase by 90°, as can be seen from Eqs. (133) and (134), which yield $\xi_x = i \ \xi_y$. This corresponds to a precession of each spin about the z-axis. (b) Equation (132) shows that the precession phase of each successive spin is advanced by a constant amount qa, giving the coupled spin oscillations the appearance of a traveling wave as shown in Fig. 1.18.

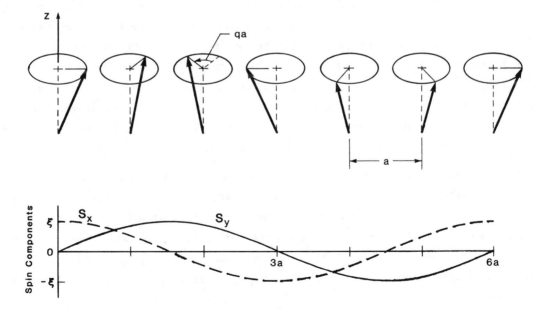

Figure 1.18. Schematic representation of a spin wave in a linear ferromagnet. The phase of each successive spin is advanced by amount qa.

The result in Eq. (134) was derived in the limit of small spin deviations. This is expected to be a reasonable approximation at the lowest temperatures, where the spin system energy is dominated by the low-frequency or long-wavelength spin waves. In the long-wavelength limit $qa \ll 1$, the quantity $(1-\cos qa) \simeq 1/2\ qa^2$ and the magnon dispersion relation in Eq. (134) simplifies to

$$\omega = (2JSa^2/\hbar)\ q^2. \tag{135}$$

The theory of spin waves in a three-dimensional ferromagnetic crystal [165] yields essentially the same dispersion law in the long-wavelength (small q) limit:

$$\omega_q = \alpha_f\ (2JSa^2/\hbar)\ q^2. \tag{136}$$

The quantity α_f is a constant which depends upon details of the crystal structure. The same result is obtained for ferrimagnetic crystals except that S must be replaced by $2S_1S_2/(S_1-S_2)$, where S_1 and S_2 are the magnitudes of the antiparallel spins. In the case of antiferromagnets, special considerations are required to derive the dispersion law which, in the absence of anisotropy effects, has the long-wavelength limiting form [165]

$$\omega_q = \alpha_a\ (2J'Sa^2/\hbar)\ q \tag{137}$$

where J' is the magnitude of the exchange constant. In this case the dispersion relation is doubly-degenerate; that is, there are two spin-wave modes with frequency ω for each q.

Magnon dispersion relations have now been studied in a large number of magnetic systems by means of inelastic neutron scattering (see Section 1.2.4). An extensive bibliography of relevant work may be found in Ref. [70]. Materials that have been studied include the ferromagnets Co, Fe, and Ni [168] and $CrBr_3$ [169], the ferrimagnet Fe_3O_4 [170], and the antiferromagnets MnO [171] and $MnCO_3$ [172]. The results for the ferro- and ferrimagnetic insulators are reasonably well described by the isotropic Heisenberg model for exchange coupling. Although the dispersion relations in the metallic ferromagnets Co, Fe, and Ni exhibit some departure from Heisenberg theory as q is increased, they appear to be well approximated by Eq. (136) for small q. The magnetic behavior of the antiferromagnets, however, is complicated by anisotropy effects, and the Heisenberg model must be extended to include additional exchange parameters in order to obtain a good fit to the data [171,172]. As a result, the double-degeneracy in the dispersion relation predicted by the isotropic model (Eq. 137) is removed and only one of the branches of ω_q approaches zero in the limit $q \rightarrow 0$. This branch, which corresponds to the lowest energy magnons, appears to follow a dispersion law of the form $\omega \propto q$ in the long-wavelength limit.

As mentioned earlier, the spin-wave model of a magnetic system treats the elementary excitations as a superposition of independent harmonic oscillators. Therefore, the application of quantum statistics to the problem of calculating low-temperature thermal properties of the system follows in a relatively straightforward manner. It was seen in Section 1.2.1 that, apart from the zero-point energy, the average energy of an oscillator with frequency ω_q is $\langle n \rangle \hbar \omega_q$ where $\langle n \rangle$ is the Bose-Einstein population factor given by Eq. (26). It follows that the total energy of the spin system is given by

$$U = \sum_q \frac{\hbar\omega_q}{\exp(\hbar\omega_q/k_BT) - 1} \tag{138}$$

where the sum extends over the allowed modes q (the first Brillouin zone). The number of modes with wave vectors between q and q + dq is readily determined from Eq. (36) as

$$n(q) \; dq = \left(\frac{dN}{dq}\right) dq = \frac{V}{2\pi^2} q^2 dq. \tag{139}$$

The sum in Eq. (138) can then be replaced by an integral:

$$U = \frac{V}{2\pi^2} \int_0 \frac{\hbar\omega_q q^2 dq}{\exp(\hbar\omega_q/k_BT) - 1} \; . \tag{140}$$

The upper limit of integration over q cannot be specified without detailed analysis. However, at low temperatures only low frequency (small q) modes contribute significantly to U, and the upper limit may be taken as infinity without serious error. Given the dispersion relation ω_q, one can then evaluate U and from it the low-temperature contribution of the spin waves to the specific heat.

1.4.1.1. Ferromagnetic and Ferrimagnetic Systems

The energy of the spin system in a ferromagnet may be derived from Eqs. (136) and (140) and expressed in the form

$$U = \frac{V}{2\pi^2} b^{-3/2} (k_BT)^{5/2} \int_0^\infty \frac{x^4 dx}{e^{x^2} - 1} \tag{141}$$

where $x^2 = q^2b/k_BT$ and $b = 2\alpha_f JSa^2$. The integral may then be evaluated from standard tables and the magnetic specific heat expressed as [165]

$$C_M = \left(\frac{dU}{dT}\right)_V = c_f N_A k_B \left(\frac{k_BT}{2JS}\right)^{3/2} \tag{142}$$

where c_f is a geometrical factor which depends on the type of lattice. For example, $c_f \simeq 0.113$ for a simple cubic spin structure. The same $T^{3/2}$ law also describes the low-temperature specific heat of ferrimagnets, except that S is replaced by $2S_1S_2/(S_1-S_2)$.

In ferrimagnets and nonmetallic ferromagnets, the observed specific heat consists of contributions only from the lattice vibrations and the spin waves. At liquid helium temperatures, $C_p=C_v$ and, thus, the observed specific heat will vary as

$$C_p = \beta T^3 + \delta T^{3/2}. \tag{143}$$

Therefore, a plot of $C_p T^{-3/2}$ versus $T^{3/2}$ should be a straight line with slope β and intercept δ. The results for magnetite (Fe_3O_4) [173], presented in Fig. 1.19, clearly show the presence of a magnetic $T^{3/2}$ term.

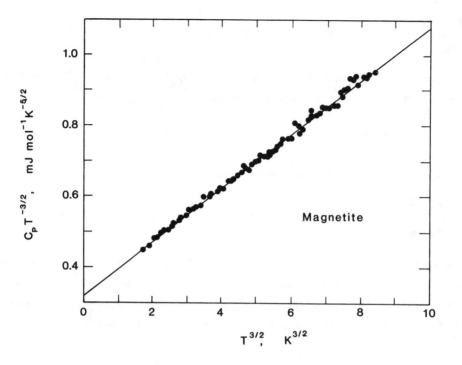

Figure 1.19. Temperature variation of the specific heat of magnetite (Fe_3O_4) observed at low temperatures (after Dixon et al. [173]).

The situation is less favorable in metallic ferromagnets because of the additional contribution of the conduction electrons to the specific heat. In this case,

$$C_p = \gamma T + \beta T^3 + \delta T^{3/2}. \tag{144}$$

The term with the lowest power of T, the electronic term, dominates the temperature variation of C_p and makes it extremely difficult to resolve the spin-wave term. In certain instances, for example the Ni-Fe alloy system [174], the spin-wave contribution is sufficiently large (~5% of C_p) that it can be determined by fitting Eq. (144) to the observed data.

1.4.1.2. Antiferromagnetic Systems

In the absence of anisotropy effects, the spin system energy in an antiferromagnet may be determined from Eqs. (137) and (140) as

$$U = 2 \cdot \frac{V}{2\pi^2} b^{-3} (k_B T)^4 \int_0^\infty \frac{x^3 dx}{e^x - 1} \tag{145}$$

where the initial factor of 2 allows for the double degeneracy of the spin-wave modes, $x = qb/k_B T$, and $b = 2a_a J'Sa^2$. A suitable evaluation of the integral enables the magnetic specific heat to be expressed in the form

$$C_M = c_a N_A k_B \left(\frac{k_B T}{2J'S}\right)^3. \tag{146}$$

The geometrical constant c_a has been evaluated for several crystal structures [165].

In real antiferromagnets, the effect of exchange-coupling anisotropy may spoil the rigor of the T^3-law for C_M. If the degeneracy of the spin-wave modes is lifted, the observed specific heat at the lowest temperature will contain a T^3-contribution only from the lowest-frequency branch of ω_q. As temperature is raised, the modes belonging to the second branch will begin to contribute to C_p as well. Some anisotropic exchange models [171] also predict a small energy gap at q = 0 which, if present, would result in an initial exponential variation of specific heat with temperature near zero kelvin.

The T^3-law for spin-wave specific heat in antiferromagnets is of the same form as the T^3-law for lattice specific heat at low temperatures. This makes it virtually impossible to resolve the magnetic contribution from the total specific heat observed in metallic antiferromagnets. However, in antiferromagnet insulators the separation of the spin-wave and lattice contributions may become possible, particularly if the Néel temperature is much less than the Debye temperature. This condition is satisfied by, for example, the carbonates of Mn, Co, and Fe. Extensive calorimetric measurements on these materials [175] have shown that the T^3 spin-wave term is approximately an order of magnitude larger than the T^3 lattice term. The magnetic specific heat was resolved from the measurements in each case by equating the lattice contribution to the specific heat observed in nonmagnetic $CaCO_3$. The results, presented in Fig. 1.20 as a plot of $C_M T^{-1}$ versus T^2, clearly demonstrate the T^3-variation of C_M by the linear behavior of the curves. The changes in slope observed at higher temperatures in the cases of $MnCO_3$ and $CoCO_3$ are attributed to the excitation of spin-wave modes belonging to the higher-frequency branch of ω_q. This is consistent with neutron scattering measurements on $MnCO_3$ [172], which clearly show that the degeneracy of the branches of ω_q has been removed.

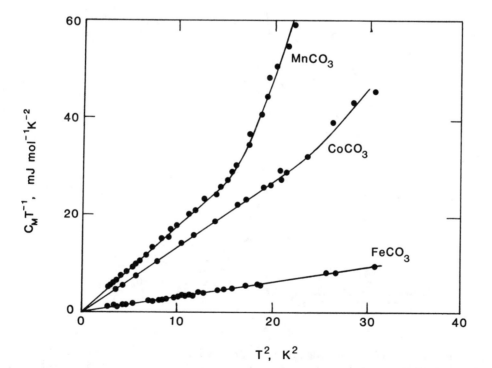

Figure 1.20. Magnetic specific heat of the antiferromagnets $MnCO_3$, $CoCO_3$, and $FeCO_3$ (after Kalinka [175]).

1.4.2. Specific Heat in the Critical Temperature Region

At sufficiently high temperatures, the ordered ferromagnetic, ferrimagnetic, and antiferromagnetic states are transformed into the disordered paramagnetic state by the randomizing effects
of thermal agitation on spin alignment.   In the vicinity of the critical temperature T_c, the
order-to-disorder transition is accompanied by rapid changes in a number of thermophysical properties.   For example, the spontaneous magnetization of ferromagnetic materials decreases rapidly as
T_c is approached from below, and is zero for temperatures $T \geq T_c$. The specific heat, on the other
hand, increases rapidly and approaches infinity as T_c is approached from below and from above (see
Fig. 1.21).   The λ-type singularity in the specific heat is a characteristic of second-order phase
transitions and is observed in a large variety of cooperative transitions, generally known as
critical phenomena.   Typical of these are the order-disorder transitions in binary alloys and in
ferroelectrics, as well as the magnetic transitions already mentioned.

The subject of critical phenomena is a vast one and clearly cannot be discussed in depth in
the present chapter. A reader who is interested in the details of the theory is referred to a
number of reviews [177,178] and books [179-182]. In the following sections, we highlight some of
the important results for the form of the specific heat in the critical region, using the ferromagnetic-paramagnetic transition as the illustrative example.

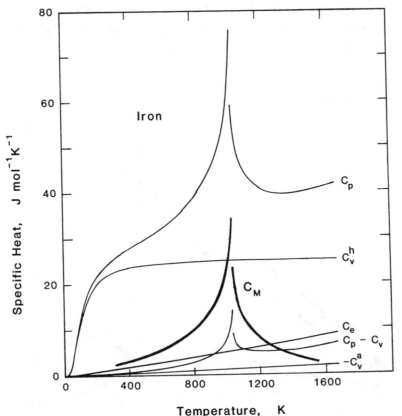

Figure 1.21.   The magnetic specific heat C_M of iron derived from measurements of C_p and estimates
of the harmonic and anharmonic lattice terms, C^h_v and C^a_v, and of the electronic C_e
and dilation (C_p-C_v) contributions (after Kollie [176]).

1.4.2.1. Approximate Methods: The Weiss Molecular Field Model

A number of approximate methods have been developed for the study of magnetic transitions and other order-disorder problems. Examples of these are the first treatment of ferromagnetic-paramagnetic transitions given by Weiss [162] and a similar treatment used by Bragg and Williams [183] for order-disorder transitions in alloys. A brief description of the Weiss molecular field model is given below; discussions of the Bragg-Williams model may be found in Refs. [13,179].

In the Weiss model for the ferromagnetic state, each electron spin is assumed to experience a local magnetic field B consisting of a phenomenological molecular field B_m due to the neighboring spins, plus the externally applied field B_a (if any):

$$B = B_m + B_a. \tag{147}$$

The molecular field is assumed to be a function of the average magnetic moment of all spins, i.e., the magnetization M, and is expressed as

$$B_m = \lambda M \tag{148}$$

where λ is a constant, independent of temperature. The magnitude of B_m may be as large as 10^7 gauss (10^3 tesla) which is considerably larger than magnetic fields normally produced in the laboratory.

At temperatures above T_c, the spontaneous magnetization vanishes and the system can be magnetized only by an external field. From the theory of paramagnetism [26,161], it is known that the net magnetization of a system of particles, each with spin 1/2 and magnetic moment μ, may be written as

$$M = N\mu \tanh \frac{\mu B}{k_B T} \tag{149}$$

where N is the number of particles per unit volume. By combining Eqs. (147) to (149), one obtains the familiar Curie-Weiss law for the paramagnetic susceptibility:

$$\chi = \mu_o \left(\frac{M}{B_a} \right) \frac{C}{T - T_{cw}} \tag{150}$$

where the Curie-Weiss value for the critical temperature is related to the Curie constant C by $T_{cw} = \lambda C = \lambda N\mu^2/k_B$. At temperatures below T_c, the spontaneous magnetization (in the absence of an external field) is implicitly given by Eqs. (147) to (149) as

$$M_s = N\mu \tanh \frac{M_s T_{cw}}{N\mu T}. \tag{151}$$

Except for temperatures very near T_c, Eqs. (150) and (151) give a reasonable description of the variation of χ and M_s with T for many ferromagnetic materials (e.g., iron, nickel, cobalt) if μ is set equal to the Bohr magneton. This suggests that ferromagnetism in these materials arises primarily from the electron spin rather than the electron orbital angular momentum.

The spontaneous magnetization provides an additional contribution to the internal energy which, in the absence of an external field, may be expressed as [6,161]

$$U_m = -V_m \int_0^{M_s} B_m dM = -1/2 \ \lambda V_m M_s^2 \tag{152}$$

where V_m is the molar volume. The additional contribution to the specific heat then follows as

$$C_M = -1/2 \ \lambda V_m \frac{dM_s^2}{dT} . \tag{153}$$

The magnetic specific heat predicted by the Weiss molecular field model is illustrated in Fig. 1.22. A comparison with experimental values of C_M, as seen in Fig. 1.21, shows that the theoretical and experimental curves are somewhat similar. However, the theoretical curve exhibits a maximum of 3/2 R at $T = T_{cw}$ instead of diverging to infinity and drops discontinuously to zero at $T >$ T_{cw} in contrast with the characteristic 'high-temperature tail' seen experimentally. The latter is due to the presence of short-range order among the electron spins at temperatures above T_{cw} which is not taken into account by the Weiss theory.

More sophisticated phenomenological theories (e.g., the Bethe–Peierls model of order–disorder transitions [13,179]) give a small 'tail' to the specific heat curve, but the agreement with experiment in the critical region is not much improved. The value of phenomenological theories lies in their simplicity and their ability to give a reasonable description of the overall behavior. However, if interested in details, particularly of the critical behavior near T_c, one must take into account the statistical nature of the problem.

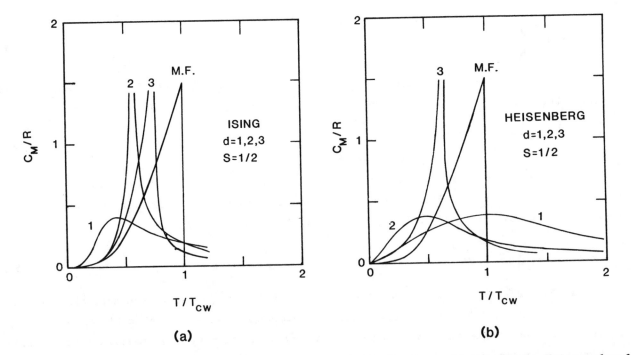

Figure 1.22.   Comparison of theoretical magnetic specific heats predicted by the Weiss molecular field model (MF) and by (a) the Ising model and (b) the Heisenberg model for 1, 2, and 3-d lattices with S = 1/2. The quantity T_{cw} is the Curie-Weiss transition temperature predicted by MF theory (after DeJongh and Miedema [184]).

1.4.2.2. Statistical Methods: The Heisenberg and Ising Models

The statistical approach to the study of critical behavior involves the evaluation of the partition function Z of the system (Eq. 15). This requires a knowledge of the total energy (E_i) or Hamiltonian which, in the case of magnetic transitions, consists only of the potential energies of interaction. Once Z has been calculated, the free energy and other thermodynamic quantities such as the specific heat can be obtained from standard relations (Eqs. 14 and 16).

Although straightforward in principle, such calculations become mathematically feasible only for models in which the magnetic exchange interaction is more or less simplified. Consider, for example, an interaction Hamiltonian of the form

$$H = -2J \sum_{i,j} [a \, S_i^z S_j^z + b \, (S_i^x S_j^x + S_i^y S_j^y)]$$ (154)

where the sum extends over nearest neighbor spins and J is the exchange constant. If a = b = 1, the Hamiltonian corresponds to the isotropic Heisenberg model mentioned earlier (Section 1.4.1). With a = 0 and b = 1, we have the planar Heisenberg model (or XY model) if one requires that the spins are constrained to lie in the xy plane. The third case of a = 1 and b = 0 (spins aligned only along the z-axis) is the Ising model.

Of the model systems that have been studied, the Ising model is the simplest one which appears to reproduce many of the features observed experimentally. Exact solutions of this model have been obtained for one-dimensional [185] and two-dimensional [186] spin lattices, but as yet a rigorous solution of the three-dimensional problem has not been achieved. However, a number of powerful approximate procedures have been developed [177] and successfully applied to both Ising and Heisenberg model systems of various dimensionality [184]. Examples of results obtained for the specific heat are shown in Fig. 1.22. As may be seen, only the two- and three-dimensional Ising models and the three-dimensional Heisenberg model exhibit singularities in their specific heat curves.

Much of the recent interest in model systems has been directed towards the study of critical exponents [177-182] which describe the behavior of various thermodynamic quantities near the critical point. An important quantity in this regard is the so-called order parameter, which is a measure of the degree of ordering in the system below T_c. In magnetic systems, the spontaneous magnetization is taken as the order parameter and has the form

$$M_s \sim (T_c - T)^\beta$$ (155)

where β is called a critical exponent. In the case of the magnetic specific heat, the variation near T_c exhibits an inverse power-law singularity of the form*

$$C_M \sim A \, (T-T_c)^{-\alpha} \text{ for } T > T_c$$

and   (156)

$$C_M \sim A' \, (T_c-T)^{-\alpha'} \text{ for } T < T_c$$

---

*This is the general form for the specific heat near second-order phase transitions as predicted by the modern theory of critical phenomena [179-182].

where α and α' are the critical exponents. With the development of the scaling hypothesis and the use of renormalization-group techniques [177-182,187,188], a number of relationships (scaling laws) among the various critical exponents have become well established. This has resulted in the reduction of the number of independent exponents to only two fundamental unknowns. In particular, the scaling laws predict that the specific heat exponents α and α' are equal. Another consequence of the scaling hypothesis is the concept of universality which states that transitions may be classified simply according to the dimensionality d of the lattice and the number of degrees of freedom n of the order parameter. Thus, the critical behavior of systems belonging to the same class (same d and n) is expected to be similar; that is, described by the same set of critical exponents.

Experimental verification of the scaling result $\alpha = \alpha'$ has been recently reported for a number of Heisenberg magnetic systems (ferromagnets Fe, Ni, and EuO and antiferromagnet $RbMnF_3$) [189]. The results show that the ferromagnets all belong to the same universality class with a mean value $\alpha = \alpha' = 0.10 \pm 0.01$, which is in good agreement with the estimate of $\alpha = \alpha' = 0.10$ (n=d=3) based on scaling and renormalization-group theory [187-189]. However, the antiferromagnet $RbMnF_3$ failed to conform to the universal behavior of the ferromagnets, yielding the result $\alpha = \alpha' = 0.14 \pm 0.01$.

1.4.3. The Schottky Anomaly

In the study of systems with discrete energy levels, the specific heat can provide useful information about the levels, particularly when their number is small. Consider, for example, a simple system in which the particles can exist only in two states, with energies $\varepsilon_0 = 0$ and $\varepsilon_1 = \Delta$, and with corresponding degeneracies g_0 and g_1. Since the probability of a particle occupying a given level is proportional to $g_i \exp(-\varepsilon_i/k_BT)$, the average energy of N_A particles in the two-level system may be expressed as

$$U = \frac{N_A g_1 \, \Delta \, \exp(-\Delta/k_BT)}{g_0 + g_1 \, \exp(-\Delta/k_BT)} \, . \tag{157}$$

The specific heat then follows directly from dU/dT:

$$C_s = R \, \frac{g_0}{g_1} \left(\frac{T_\Delta}{T}\right)^2 \frac{\exp(T_\Delta/T)}{[1 + (g_0/g_1) \, \exp(T_\Delta/T)]^2} \tag{158}$$

where $T_\Delta = \Delta/k_B$ is the level separation measured in degrees kelvin. As can be seen in Fig. 1.23, the specific heat $C_s(T)$ exhibits a peak at a temperature approximately equal to $0.5 \, T_\Delta$ and the magnitude of the peak depends upon the ratio of degeneracies for the levels. Peaks of this type are known as Schottky anomalies [190]. Information about the separation and degeneracies of the levels can also be determined from $C_s(T)$ in the limit of either low or high temperature. From Eq. (158) it is easy to show that

$$C_s = R \, \frac{g_1}{g_0} \left(\frac{T_\Delta}{T}\right)^2 \exp\left(\frac{-T_\Delta}{T}\right) \quad \text{for } T \ll T_\Delta \tag{159}$$

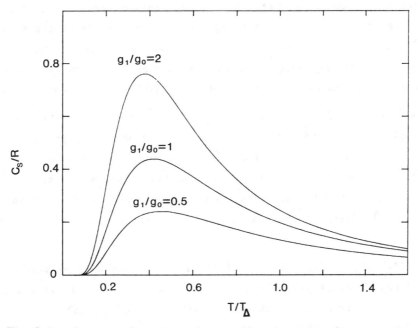

Figure 1.23. The Schottky specific heat of a two-level system for several values of g_1/g_0.

and

$$C_s = R \frac{g_0 g_1}{(g_0 + g_1)^2} \left(\frac{T_\Delta}{T}\right)^2 \quad \text{for } T \gg T_\Delta .$$ (160)

As the number of levels capable of thermal excitation is increased, the analysis of the problem rapidly becomes more complex. The specific heat must be derived from a more general expression for U or, alternatively, from the free energy associated with the discrete levels, namely, $F_s = -RT\ln z$, where z is the single-particle partition function defined by

$$z = \sum_i g_i \exp(-\varepsilon_i/k_B T).$$ (161)

The Schottky specific heat then follows from the standard relations in Eqs. (14) and (16) as

$$C_s = \frac{R}{(k_B T)^2} (\langle \varepsilon^2 \rangle - \langle \varepsilon \rangle^2)$$ (162)

where the angular brackets denote thermal averages of the form

$$\langle x \rangle = \frac{1}{z} \sum_i g_i x_i \exp(-\varepsilon_i/k_B T).$$ (163)

Implicit in this derivation [and that of Eq. (158)] is the assumption that the particles are statistically independent. As in the case of the simple two-level system, $C_s(T)$ increases with increasing temperature as $\exp(-\text{constant}/T)$ at temperatures small compared with the separation of levels and decreases as T^{-2} at temperatures large compared with the separation. There is no simple result, in the multi-level case, for the variation of $C_s(T)$ in the vicinity of the characteristic peak.

In general, a Schottky anomaly, if present at liquid–helium temperatures (~2-4 K), will dominate the variation of the specific heat of a solid since the magnitude of the peak is of the order of R (see Fig. 1.23) whereas the lattice and electronic contributions are typically $10^{-2}R$. Thus, at these temperatures, the Schottky specific heat can be accurately resolved even when there are relatively large uncertainties in evaluating the lattice (βT^3) and electronic (γT) terms.

The analysis is simplified further when only the high–temperature 'tail' of C_s is observed. The total specific heat of a nonconductor at or below liquid–helium temperatures will be of the form $C_p = \beta T^3 + DT^{-2}$ and a plot of $C_p T^2$ versus T^5 should be a straight line with slope β and intercept D. At sub–kelvin temperatures, the total specific heat of a metal will be determined primarily by the electronic and Schottky terms; that is, $C_p = \gamma T + DT^{-2}$. The terms can easily be separated by a plot of $C_p T^2$ versus T^3, which should be linear if other contributions are negligible (see Fig. 1.24).

The problem of resolving C_s from other contributions rapidly becomes more difficult as temperature is increased above the liquid–helium range because of the rapid increase in the lattice term. The usual approach is to assume that the contributions other than C_s are equal to the specific heat observed for another solid of similar composition and same crystal structure but which does not exhibit the anomaly.

In many solids, Schottky contributions to the specific heat arise from the splitting of atomic and nuclear energy levels by internal and/or externally applied fields; examples of systems which exhibit such anomalies are given below. Other Schottky systems, not described herein, include certain molecular solids (e.g., CH_4) in which discrete energy levels arise from quantum–mechanical tunneling between orientational energy states.

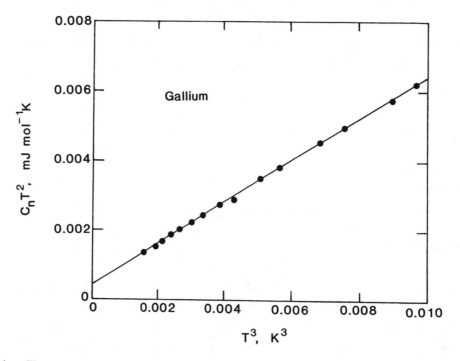

Figure 1.24.   The normal state specific heat of gallium at T < 0.21 K as given by $C_p = 4.3$ x $10^{-4}T^{-2} + 0.596$ T. At these temperatures, the specific heat is dominated by the nuclear ($\propto T^{-2}$) and electronic ($\propto T$) terms (after Phillips [141]).

1.4.3.1. Schottky Specific Heat in Paramagnetic Systems

It is well known that the ground state of a free atom with spin quantum number J is (2J+1)-fold degenerate because of the possible orientations of the spin; in a magnetic field, the degeneracy is lifted and the atomic level is split into 2J+1 equally-spaced levels. In any solid, the degeneracies of some of the atomic levels are removed, even in the absence of external magnetic fields, by internal crystalline electric fields through the familiar Stark effect; however, no simple results can be given since the crystal field can split the 2J+1 levels of the ground state in many different ways [191,192]. Other effects such as magnetic dipole and exchange interactions between neighboring ions may also contribute to the splitting of the levels.

Specific heat, being an integrated quantity, cannot provide any direct information about the separation and degeneracies of the energy levels. This must come from other theoretical and experimental studies such as those on magnetic susceptibility, magnetic resonance, and optical spectroscopy [191,192]. However, empirical values of C_s can provide a valuable means of checking the validity of a proposed energy-level scheme.

A necessary requirement for applying Schottky theory is that the ions should be independent of one another. Since the strength of magnetic interactions decreases slowly with increasing ion separation, the condition of statistical independence can normally be approximated only in paramagnetic systems that are sufficiently 'dilute.' For this reason, the anomalous contributions of magnetic impurities to the specific heat are often of the Schottky-type. For example, in a recent study of 0.1 and 1.0% Fe substitutional impurities in a crystal of ZnS [193], the impurity (Schottky) contributions to the specific heat were determined as a function of temperature by comparison with similar measurements on pure ZnS. The contributions were compared with those calculated for the crystal-field splitting of the Fe^{2+} ground state (based on spectroscopic data) and quantitative agreement was obtained for the 0.1% Fe specimen; the lack of agreement for the 1% Fe specimen was attributed to possible direct interactions between the impurity ions.

The required 'dilution' of magnetic interactions can be achieved in some paramagnetic salts by the presence of water molecules around each ion, as in the case of $\alpha NiSO_4 \cdot 6H_2O$. A calorimetric study of this salt [194] has shown that a peak (~0.8 R and at 2.6 K) in the specific heat curve can be well fitted by a Schottky term, yielding the separation in energy between the three lowest spin states of the Ni^{2+} ion. On the basis of the level separations, other properties such as susceptibility and magnetization were calculated and found in reasonable agreement with observed values.

In certain paramagnetic salts, for example, $PrCl_3$ and $NdCl_3$, the Schottky (noncooperative) contribution to the specific heat can be separated from the (cooperative) contribution associated with direct ion-ion interactions [195]. The resolution of the magnetic contribution into cooperative and noncooperative parts is made possible by the fact that cooperative magnetic ordering in these salts takes place at temperatures below 5 K whereas the Schottky peaks are observed at temperatures above 50 K. The Schottky contributions, obtained by subtracting the specific heat of diamagnetic $LaCl_3$ from those of the paramagnetic salts, were found in good agreement with specific heats derived from energy levels based on spectroscopic data.

In a number of nearly-ferromagnetic alloys, there is a tendency for the magnetic atoms to form large superparamagnetic clusters of various sizes and magnetic moments; the spin moments can be preferentially oriented by an applied magnetic field to yield a magnetization parallel to the field direction (hence, the term 'paramagnetic'). The magnetic levels of a cluster with spin S can also be split by internal crystalline fields or by externally applied fields. Hence, anomalies in the specific heat can be expected. According to the theory of specific heat of superparamagnetic clusters [196-198] the limit $S \to \infty$ (an infinitely large cluster) yields an anomaly that approaches a constant value (k_B or $1/2\ k_B$ per cluster depending upon assumptions) at the lowest temperatures and that falls off as T^{-2} at high temperatures. In the more realistic case of finite S (hence, a finite number of levels), the anomaly is of the Schottky-type considered earlier. In recent studies of several alloys [199,200], anomalies in the specific heats were successfully interpreted in terms of multi-level Schottky functions from which the number of levels and cluster concentrations were determined.

1.4.3.2. Nuclear Specific Heat

A Schottky contribution to the specific heat also arises when an interaction removes the degeneracies of the nuclear levels. The energy-level splitting in this case may be produced by externally applied magnetic fields, by effective (hyperfine) magnetic fields arising from the orbital and conduction electrons and, in noncubic crystals, by electric field gradients.

As a result of the interaction of a nuclear dipole moment with a magnetic field, B, a nucleus with spin quantum number I will have $2I+1$ equally-spaced levels given by

$$\varepsilon_m = -\mu_I\ Bm_I/I \tag{164}$$

where μ_I is the maximum observable component of the nuclear magnetic moment and m_I is the azimuthal quantum number with allowed values I, I-1, ..., -I. The quantity μ_I is often written as $\mu_I \equiv g_N I \mu_N$ where g_N is the 'nuclear g factor' and μ_N is the nuclear magneton. Nuclear 'magnetic moments' are usually tabulated in units of μ_N as the dimensionless numbers $g_N I$ (see Ref. [26]).

For a nucleus with an electric quadrupole moment, the $2I+1$ levels arising from an electric field gradient may be expressed in the form [5,6]

$$\varepsilon_m = K\ [3m^2 - I\ (I + 1)] \tag{165}$$

where K is a coupling constant related to the quadrupole moment and field gradient. Nuclear quadrupole contributions to the specific heat only occur for $I \geq 1$, since there is no quadrupole moment associated with $I = 1/2$.

Measurements of nuclear specific heats have been used to determine the hyperfine magnetic fields and the quadrupole coupling constants in a number of metals and alloys. Extensive reviews are given in the literature [6,201]. Owing to the smallness of the nuclear moments, nuclear Schottky peaks usually occur at temperatures of the order of 10^{-2} K. Consequently, in the temperature range of most measurements the nuclear specific heat can be represented by $C_N = DT^{-2}$ (see Fig. 1.24). In certain cases, the data are analyzed in terms of the high-temperature expansion of Eq. (162):

$$C_N = D_2 T^{-2} + D_3 T^{-3} + D_4 T^{-4} + \dots \; . \tag{166}$$

Expressions for the coefficients in terms of I are given in the literature [202].

1.4.4. Other Magnetic Contributions to Specific Heat

Other specific heat contributions of magnetic origin have been studied in recent years. A brief discussion of selected topics is given below.

1.4.4.1. Spin Fluctuations

In certain nearly-ferromagnetic systems (notably Pd) the exchange interaction, although not strong enough to produce an ordered ferromagnetic state, strongly enhances the paramagnetic susceptibility. This exchange enhancement is characterized by persistent spin fluctuations or critically-damped spin waves. The effect of persistent spin fluctuations on the specific heat has been considered by a number of authors.

Doniach and Engelsberg [203] and Berk and Schrieffer [204] have shown that the absorption and re-emission of spin fluctuations lead to an enhancement of the effective mass of the conduction electrons. This mass enhancement is analogous to that arising from the electron-phonon interactions (Section 1.3.2.3). The effect of the spin fluctuations also manifests itself in the temperature dependence of the electronic specific heat which, at low temperatures ($T \ll T_s$), is approximately given by [203]

$$C_e = \gamma T + \alpha \; (T/T_s)^3 \; \ln \; (T/T_s). \tag{167}$$

The coefficient γ includes the many-body enhancements of both phonons and spin fluctuations, T_s is a temperature characteristic of the spin-fluctuation energy, and α is a constant. A similar expression for C_e has been derived by Bennemann [205] for alloys with compositions on both sides of the ferromagnetic-paramagnetic phase boundary. Theoretical calculations by Brinkman and Engelsberg [206] and by Béal-Monod et al. [207] show that the spin-fluctuation contribution to specific heat is nearly insensitive to the presence of external magnetic fields.

Quantitative comparisons of theory with experiment have been largely unsuccessful for the exchange-enhanced pure metals, primarily because of their high spin-fluctuation temperatures ($T_s \sim$ 250 K for Pd). In such cases, the $T^3 \ln (T/T_s)$ term is spread out over a large temperature region and cannot be empirically separated from the much larger lattice contribution [208].

A spin fluctuation contribution was recently observed in the specific heat of the paramagnetic alloy UAl_2 [208]. The measurements revealed a nearly field-independent upturn in $C_p T^{-1}$ as temperature was decreased below 10 K. The anomalous behavior was well described by a $T^3 \ln (T/T_s)$ term with $T_s = 12$ K. Similar anomalies have also been observed in the specific heat of a number of 'nearly-ferromagnetic' alloys. In most cases, the anomalies have been attributed to the Schottky effect arising from superparamagnetic clustering (Section 1.4.3.1). In the NiRh system the observed upturns in $C_p T^{-1}$ were fitted by $T^3 \ln (T/T_s)$ terms and attributed to spin fluctuations [209]. In subsequent measurements [210], however, the anomalous specific heat in the NiRh alloys was shown to be strongly dependent on magnetic field, a behavior consistent with magnetic clustering but not with spin-fluctuation effects.

1.4.4.2. The Kondo or Single-Impurity Effect

In dilute solid solutions of nonmagnetic metals containing magnetic impurities, the exchange coupling between the magnetic ions and the conduction electrons has important consequences [6,211, 212]. In the 'single-impurity' limit, the impurity ion magnetizes the surrounding conduction electron gas so that the effective impurity spin becomes zero. At a finite concentration of impurities, the induced magnetization causes an indirect coupling between pairs of impurity spins known as the Rudermann-Kittel-Kasuya-Yosida (RKKY) interaction. The RKKY mechanism plays a fundamental role in the so-called 'spin-glass' effect (see next section).

The effect of the magnetic impurity/conduction electron coupling on physical properties has been studied theoretically for the idealized 'single-impurity' case. Kondo [212] has shown that, at low temperatures, the exchange coupling can lead to a rapid increase in the impurity scattering contribution to resistivity as temperature is decreased. Since phonon resistivity increases as T^5 at low temperature, there will be a minimum in the resistivity-temperature curve, a behavior often referred to as the Kondo effect. From Kondo's theory it can be shown [6] that, at high temperatures ($T \gg T_K$), each magnetic impurity contributes a term to the specific heat which varies with temperature as $[A + B \ln (CT/T_K)]^{-4}$ where T_K is the Kondo characteristic temperature. Bloomfield and Hamann [213] have shown that the single-impurity contribution to the specific heat exhibits a peak of the order of k_B with a maximum at $T \simeq T_K/3$. The theoretical situation is less clear at low temperatures ($T \ll T_K$), where different models yield different results for the temperature dependence of the specific heat. A detailed review of the different theoretical models is given in Ref. [6].

Single impurity effects have been identified in a number of dilute magnetic alloys, a criterion being the linear dependence of the measured properties on impurity concentration [211]. For example, measurements performed on dilute CuCr alloys [214] reveal a peak in ΔC_p, the specific heat in excess of that for pure Cu. The values of ΔC_p are proportional to the impurity concentration (0.002 to 0.005 at.%) and are in good agreement with the Bloomfield-Hamann theory.

1.4.4.3. The Spin-Glass Effect

In dilute magnetic systems, the indirect interaction between impurity spins via the RKKY mechanism may be interpreted as the effect of internal fields which tend to align the associated magnetic moments [6,215]. Because of the random distribution of impurities in real dilute alloys, some of the moments are located in positions of low fields and, below some 'critical' temperature, they are frozen into thermal equilibrium orientations with no long-range order. The thermal excitation of these moments gives rise to the so-called 'spin-glass' contribution to the specific heat.

A quantitative theoretical treatment of spin-glass systems is made difficult by the random nature of the impurity distributions and the long range of the RKKY interactions. In early theoretical models, as given by Marshall [216] and by Klein and Brout [217], the effect of the distribution of impurity configurations was approximated by a distribution (in magnitude) of effective internal fields. These models were able to explain the earliest experimental results in CuMn alloys [218,219], which had suggested that the spin-glass contribution to the specific heat at low temperatures is approximately linear in temperature and independent of impurity concentration. In subsequent calculations, Sherrington and Kirkpatrick [220] considered an Ising model in which the spins are coupled by infinite-ranged random interactions following a Gaussian distribution law. Their model yielded a 'spin-glass' phase in which the leading term in the specific heat at low temperatures is also linear in temperature.

Recent computer simulation studies by Walker and Walstedt [221], using a RKKY coupling scheme with spins randomly distributed on an fcc lattice, suggest that the form of the spin-glass specific heat is more complicated at the lowest temperatures, becoming nonlinear near T = 0. Other theoretical calculations [222] based on the Sherrington-Kirkpatrick model yield a spin-glass contribution of the same general form as that predicted by the numerical simulations. These theoretical developments are supported by recent experimental work on spin-glass alloys of Cu containing 0.08 to 0.9 at.% Mn [223] and 1.0 at.% Fe [224]. The data show that the spin-glass contributions have an initial positive curvature leading to a region linearly dependent on temperature which extrapolates back to zero at a positive temperature.

1.5. SUMMARY

In the preceding sections, the theory describing the contributions of the lattice vibrations, conduction electrons, and various magnetic excitations to the specific heat of solids was discussed.

The lattice contribution which is present in all solids can be determined from a number of models based on the harmonic approximation, the simplest being the Einstein and Debye models. Although both models give the correct high-temperature limiting value (3R for a monatomic solid), only the latter model correctly predicts the behavior of the specific heat at low temperatures ($C_v \propto T^3$). The major shortcoming of the Debye model is reflected in the fact that θ_D is not constant over the entire temperature range. The variation of θ_D is understood in terms of the Born-von Kármán model, which can be used, at the expense of considerable computational effort, to calculate a more realistic vibrational spectrum and, hence, specific heat from neutron scattering data. The anharmonicity of the lattice vibrations in real solids becomes appreciable as temperature is raised and, to first order, contributes a term to the specific heat which varies linearly with temperature at $T > \theta_D$. High-order anharmonic terms may become important as the melting point is approached but are exceedingly difficult to calculate for even the simplest of lattice models. The form of the specific heat is further complicated at these temperatures by the contribution arising from the formation of lattice vacancies.

In electrical conductors, the conduction electrons in their normal state contribute to the specific heat at low temperatures a term linear in temperature, $C_e = \gamma T$. Although the free electron model gives values of γ of about the right order of magnitude, reasonable agreement with experiment can be achieved only if the effects of electron band structure and of electron-phonon and electron-electron coupling on the electron density of states are taken into account. As temperature is increased, C_e becomes sensitive to any rapid variation of the electron density of states which may occur in the vicinity of the Fermi level; in such cases (e.g., Pd, Cr), C_e will deviate significantly from the initial linear trend. The thermodynamic properties of the conduction electrons in their superconducting state are reasonably well described by the Bardeen-Cooper-Schrieffer theory, which predicts an exponential variation of the electronic specific heat at the lowest temperatures.

Additional specific heat contributions also arise from various magnetic excitations in solids. At low temperatures, the thermal excitation of spin waves yields a specific heat which, according to the isotropic Heisenberg model, varies as $T^{3/2}$ in ferromagnets (and ferrimagnets) and as T^3 in antiferromagnets. At a higher temperature, these magnetic systems undergo an order-disorder transition which is accompanied by a λ-type anomaly in the specific heat at the critical temperature. The form of the anomaly is characteristic of a second-order phase transition and, in the critical region, can be described by an inverse-power law singularity. Other anomalies, known as Schottky anomalies, appear in the specific heat of certain paramagnetic systems, arising from the splitting of atomic levels by internal and/or externally applied fields. These anomalies manifest themselves as peaks in the specific heat curve with a maximum often of the order of R and, when observed at low temperatures, they can be readily separated from the much smaller lattice and electronic contributions ($\sim 10^{-2}R$). Similar anomalies also arise from the splitting of levels associated with the magnetic dipole and electric quadrupole moments of the nucleus. However, since nuclear Schottky peaks usually occur at $T \lesssim 10^{-2}$ K, only the high-temperature 'tail' is observed in the temperature range of most measurements and the nuclear contribution to the specific heat can often be represented by the leading term; that is, $C_N = DT^{-2}$.

Other specific heat contributions of magnetic origin may become important at low temperatures. These include contributions arising from persistent spin fluctuations in certain 'nearly-ferromagnetic' systems and, in dilute magnetic alloys, from the so-called 'single-impurity' and 'spin-glass' effects.

1.6. APPENDIX: ESTIMATION AND CORRELATION OF SPECIFIC HEAT

One of the earliest methods used to estimate specific heat, in particular that of compounds, was based on an empirical generalization known as the 'Kopp-Neumann law' (Section 1.1.3). This empirical rule is still sometimes used to predict the specific heat of alloys (e.g., [225]) in which case it may be expressed as

$$c_p = \sum_i f_i c_{pi} \tag{A1}$$

where f_i and c_{pi} are the mass fraction and specific heat*, respectively, of the i-th constituent element.

---

*In this chapter, lower-case symbols (c_p, c_v) refer to the specific heat per unit mass whereas the molar specific heat is denoted by upper-case symbols (C_p, C_v).

The validity of the Kopp–Neumann rule has been investigated for several alloy systems. For example, Jaeger and co-workers have reported that, on the basis of their measurements on crystals of Ag and Au [226] and the alloy Ag$_3$Au [227], the values of specific heat calculated by this additivity rule are in good agreement with those observed for the alloy at temperatures between 0 and 400°C but, at higher temperatures (up to 800°C), the calculated values for the alloy are lower than the observed values by as much as 2%. Recent measurements by Cezairliyan and co-workers on the refractory metals Nb [228], Ta [229], and W [104], and the alloys Ta–(10 mass %)W [230] and Nb–(10 mass %)Ta–(10 mass%)W [231], in the temperature range 1500 to about 3000 K, show that the alloy specific heats calculated by Eq. (A1) are about 2–3% lower than those observed experimentally. In the above investigations, the differences between the calculated and measured specific heats of a given alloy were always less than the combined experimental uncertainties in the data for the same alloy and its constituents, though the differences tended to be of the same sign. This suggests that the Kopp–Neumann rule may underestimate the specific heats of these alloys by an amount of the order of the experimental errors.

As mentioned earlier, theoretical models for the different modes of energy (vibrational, electronic, magnetic, etc.) in a solid provide a means of estimating and, hence, separating their contributions to the observed specific heat. Theoretical values for specific heat, however, generally refer to conditions of constant volume whereas, for practical reasons, the specific heat of a solid is ordinarily measured at constant pressure. The difference, $(C_p - C_v)$, is associated with the thermal expansion of the solid and, therefore, is often referred to as the dilation contribution to the measured specific heat. From classical thermodynamics, it can be shown [9–12] that the dilation contribution is given by

$$C_p - C_v = \frac{V_m \beta^2}{K_T} T \tag{A2}$$

where V_m is the molar volume, T is the absolute temperature, and $\beta \equiv V^{-1}(\partial V/\partial T)_p$ and $K_T \equiv -V^{-1}(\partial V/\partial P)_T$ are the coefficient of volumetric thermal expansion and the isothermal compressibility, respectively. For isotropic solids, the coefficient β is simply related to the coefficient of linear thermal expansion, $\alpha \equiv \ell^{-1}(\partial \ell/\partial T)_p$, by $\beta = 3\alpha$ [see Eq. (19)].

Although data on linear thermal expansion are available in the literature for many solids over a broad range of temperature, experimental values of compressibility, if available, are usually known only near room temperature. Consequently, one of two methods is usually employed to estimate the dilation contribution at other temperatures.

The first of these is based on Gruneisen's theory [232], which relates the thermal expansion of solid to its specific heat at constant volume according to

$$\beta = G \frac{K_T C_v}{V_m} \tag{A3}$$

where the Gruneisen parameter G is assumed to be independent of temperature. This enables Eq. (A2) to be written as

$$C_p - C_v = G \, C_v \beta T. \tag{A4}$$

Detailed theoretical analyses [233–236] have shown that significant deviations from a constant G can be expected at temperatures below about one-third the Debye characteristic temperature θ_D. These conclusions are supported by considerations of empirical data on thermal expansion, compressibility, and specific heat of several metals in the temperature range 300 K and below [237–239]. However, since the dilation contribution is typically less than about 1% of C_p at temperatures below $\theta_D/3$, the uncertainty in (C_p-C_v) arising from the nonconstancy of G should not introduce serious error in the comparison between theory and experiment.

A more frequently used method to estimate (C_p-C_v) involves the semiempirical Nernst–Lindemann relation [240]

$$C_p - C_v = A C_p^2 T \tag{A5}$$

where the parameter $A(=V_m \, \beta^2 \, K_T^{-1} \, C_p^{-2})$ is taken as a temperature-independent constant; A is calculated from values of V_m, β, K_T, and C_p at one temperature (often room temperature) and is then used to calculate (C_p-C_v) over the desired temperature range. It is known experimentally that parameter A is nearly constant over a wide range of temperatures in many solids. For example, in copper, ultrasonic measurements have been used to determine values for the compressibility at a number of temperatures which, when combined with data on thermal expansion and specific heat, yield the values $A = 4.0 \times 10^{-6}$ mol·J^{-1} at 100 K and 3.9×10^{-6} mol·J^{-1} at 800 K [241]. A similar study conducted on tungsten [242] shows that the values of A vary between 3.5×10^{-6} and 3.8×10^{-6} mol·J^{-1} for temperatures in the range 300–1600 K.

The general expression for the specific heat of a solid at constant pressure may be written as a sum of several contributions:

$$C_p = (C_p-C_v) + C_\ell + C_e + C_M + C' \tag{A6}$$

where (C_p-C_v) is the dilation contribution and C_ℓ, C_e, and C_M are the specific heats at constant volume arising from the lattice vibrations, the conduction electrons, and the magnetic excitations, respectively. The quantity C' refers to the remaining contributions not included by the other terms (e.g., lattice vacancies, order–disorder transformations, etc.). The form of the $C_p(T)$ function is, of course, determined by those terms which contribute significantly to the total specific heat and this, in turn, depends upon the nature of the solid (i.e., whether it is electrically conducting or nonconducting, magnetic or nonmagnetic, etc.) and its temperature. Theoretical expressions for the different specific heat terms are summarized in Table A1 according to three temperature ranges: low, $T \lesssim \theta_D/50$; moderate, $\theta_D/50 \lesssim T \lesssim \theta_D$; and high, $T \gtrsim \theta_D$, where θ_D is the Debye temperature. It should be emphasized that these expressions for specific heat are only approximations based on an assumed model for a given contribution and are subject to a number of qualifications which are described in detail in the appropriate sections of this chapter.

The correlation of specific heat contributions and methods of determining (or estimating) their magnitudes in the above-mentioned temperature ranges are briefly described below. Some discussion of these topics has already been presented earlier in this chapter in connection with

Table A1. A Summary of Theoretical Results for the Functional Form of Various Contributions to the Specific Heat of a Solid

Source of Specific Heat	Sect.[a]	Functional Form of the Specific Heat[b]		
		Low Temperatures[b]	Moderate Temperatures[b]	High Temperatures[b]
Dilation Contribution	1.1.2		$C_p - C_v = \dfrac{V_m \beta^2}{K_T} T$	
Lattice Vibrations: (1) Harmonic Contribution[c]	1.2.2	$C_v = \beta T^3$	$C_v(\theta_D/T) = 9R\left(\dfrac{T}{\theta_D}\right)^3 \displaystyle\int_0^{\theta_D/T} \dfrac{x^4 e^x\,dx}{(e^x-1)^2}$	$C_v^h = 3R\left[1 - \dfrac{1}{20}\left(\dfrac{\theta_D}{T}\right)^2 + \cdots\right]$
(2) Anharmonic Contribution	1.2.7.1			$C_v^a = 3RAT$
Lattice Monovacancies	1.2.8			$C_{vac} = R(E_f/k_B T)^2\, A\,\exp[-E_f/k_B T]$
Conduction Electrons: (1) Normal State[d]	1.3.2		$C_e = \gamma T$	$C_e = \gamma T\left[1 - \dfrac{3\pi^2}{10}\left(\dfrac{T}{T_F}\right)^2\right]$
(2) Superconducting State	1.3.3.2	$C_{es} = a\gamma T_c\,\exp[-bT_c/T]$ for $T < T_c$		
Spin Waves: (1) Ferromagnets or Ferrimagnets	1.4.1.1	$C_M = \delta T^{3/2}$		
(2) Antiferromagnets	1.4.1.2	$C_M = \delta' T^3$		
Second-Order Phase Transitions: Critical Region	1.4.2.2		$C \sim A'\,(T_c - T)^{-\alpha'}$ for $T < T_c$	$C \sim A\,(T-T_c)^{-\alpha}$ for $T > T_c$
Nuclear Schottky Effect	1.4.3.2	$C_N = DT^{-2}$		

[a] Refers to the section of this chapter in which a discussion of the corresponding theoretical model is given.

[b] Temperature ranges may be expressed in terms of the Debye characteristic temperature approximately as: low, $T \lesssim \theta_D/50$; moderate, $\theta_D/50 \lesssim T \lesssim \theta_D$; high, $T \gtrsim \theta_D$.

[c] Based on the Debye model.

[d] The form at high temperatures is based on the free electron model.

comparisons between theory and experiment. However, for the sake of completeness, much of this information is included here again but in summarized form.

1.6.1. Specific Heat at Low Temperatures

In this temperature range (typically $T \lesssim 4$ K), there is no significant difference between the specific heats at constant pressure and constant volume; that is, $(C_p - C_v) \simeq 0$. The lattice contribution, which is present in the specific heat of all solids, follows a Debye T^3-law. Thus, in the absence of magnetic effects, the specific heat of an electrical insulator is simply given by

$$C_p = \beta T^3 \qquad \qquad (A7)$$

where β is related to θ_O, the low-temperature limit of the Debye temperature (Section 1.2.2). A log-log plot of C_p versus T is sometimes used to present data taken over a large temperature span in order to accommodate the several orders of magnitude change in C_p (see Fig. 1.1)*.

In a normal metal, the conduction electrons contribute an additional term, linear in T, to the total specific heat and, therefore,

$$C_p = \gamma T + \beta T^3 . \qquad \qquad (A8)$$

The separation of the electronic and lattice contributions from the measured values of C_p is easily achieved by plotting $C_p T^{-1}$ versus T^2, which should yield a straight line with slope β and intercept γ if other contributions are indeed negligible (see Fig. 1.11).

In a superconducting metal, the conduction electrons undergo a transition from their normal state to a superconducting state as the metal is cooled below a certain critical temperature T_c. As a result, the electronic contribution C_{es} departs significantly from γT, varying with temperature approximately as $a\gamma T_c \exp[-bT_c/T]$. However, the normal state can be restored at temperatures below T_c by applying an external magnetic field of sufficient strength. Therefore, by measuring the specific heat of the metal below T_c in both superconducting and normal states (see Fig. 1.13) and by assuming the lattice contribution to be the same in both states, one can determine the values of C_{es} as a function of temperature (see Eqs. 112–115). Since $\ln(C_{es}/\gamma T_c) = -b(T_c/T) + \ln a$, a semilog plot of $C_{es}/\gamma T_c$ versus T_c/T should yield a straight line with slope $-b$ and intercept $\ln a$ if the values of C_{es} do, in fact, vary exponentially with temperature (see Fig. 1.14).

In magnetic materials, an additional contribution to the specific heat arises at low temperatures from the thermal excitation of spin waves. The spin-wave contribution varies as $T^{3/2}$ in ferromagnets and ferrimagnets, but in antiferromagnets the temperature dependence follows a T^3-law. Most ferromagnets are good electrical conductors (e.g., Ni, Fe, Co), in which case the total specific heat may be written as the sum of electronic, lattice, and spin-wave terms:

$$C_p = \gamma T + \beta T^3 + \delta T^{3/2} . \qquad \qquad (A9)$$

---

*All figures mentioned in the Appendix are presented in the main text of this chapter.

Since the variation of C_p with temperature is dominated by the term with the lowest power of T, namely the electronic term, it is usually very difficult to resolve the spin-wave term in these materials. Ferrimagnets, on the other hand, are electrical insulators and, therefore, their total specific heat consists only of lattice and spin-wave terms:

$$C_p = \beta T^3 + \delta T^{3/2}. \tag{A10}$$

The lattice and spin-wave contributions can be resolved by a plot of $C_p T^{-3/2}$ versus $T^{3/2}$, which should give a straight line with slope β and intercept δ (see Fig. 1.19). In antiferromagnetic materials the temperature dependence of the spin-wave term is of the same form as the Debye T^3-law for the lattice contribution, making it virtually impossible to resolve the magnetic term; a notable exception is provided by the carbonates of Mn, Co, and Fe (see Fig. 1.20).

The splitting of atomic energy levels in certain paramagnetic salts [194,195] and in super-paramagnetic alloys [199,200] by internally or externally applied fields gives rise to a magnetic contribution to C_p known as a Schottky anomaly (Section 1.4.3.1). These anomalies manifest themselves as peaks in the specific heat curve with a maximum often of the order of R; when observed at low temperatures such peaks can be easily separated from the electronic and lattice contributions, which are of the order of 10^{-2}R.

Similar splitting of the nuclear levels (Section 1.4.3.2) also gives rise to Schottky anomalies though, in the temperature range of most measurements, only the high-temperature 'tail' of the peak is seen in the C_p data. This nuclear contribution is usually represented by a T^{-2}-term in the expression for the total specific heat which, in the case of an electrical insulator, may be written as

$$C_p = \beta T^3 + DT^{-2}. \tag{A11}$$

The lattice and nuclear terms are easily separated by a plot of $C_p T^2$ versus T^5, which should yield a straight line with slope β and intercept D. In metals the nuclear specific heat can be easily resolved at sub-kelvin temperatures, where the total specific heat is determined primarily by the electronic and nuclear terms; that is,

$$C_p = \gamma T + DT^{-2}. \tag{A12}$$

A straight line with slope γ and intercept D is obtained by plotting $C_p T^2$ versus T^3 (see Fig. 1.24).

Other specific heat contributions of magnetic origin, which may become important in certain materials at low temperatures, are described in Section 1.4.4.

1.6.2. Specific Heat at Moderate Temperatures

The dominant contribution to the specific heat in this temperature range is the lattice term, which increases with temperature by more than three orders of magnitude. The dilation contribution increases rapidly as well, and becomes a non-negligible fraction of C_p. In copper, for example, (C_p-C_v) is only 0.1% of C_p at 50 K ($\sim\theta_D/6$) but increases rapidly to about 3% of C_p at 300 K ($\sim\theta_D$).

The harmonic lattice contribution can be determined from a number of models, the simplest being the Einstein and Debye models. The Einstein model (Section 1.2.1) gives the correct high-temperature limit, that is, $C_v = 3R$ (24.9432 $J \cdot mol^{-1} \cdot K^{-1}$)*, but underestimates the specific heat values at low temperatures (see Fig. 1.2). Even so, because of its simplicity, this model is still used in certain studies (e.g., [34]) to approximate the lattice contribution in terms of a single adjustable parameter, namely, the Einstein temperature θ_E. The Debye model (Section 1.2.2), on the other hand, correctly predicts the behavior of the lattice contribution at low temperatures ($C_v \propto T^3$) as well as in the high-temperature limit. The shortcoming of this model is that θ_D for a real solid is not constant, as is implied by the model, but must be allowed to vary in order to accurately describe the lattice specific heat over a wide temperature range (see Figs. 1.3 and 1.7). The variation of θ_D is understood in terms of the Born-von Kármán model, which can be used to calculate a more realistic vibrational spectrum and, hence, the lattice specific heat from neutron scattering data (Sections 1.2.3-1.2.6). The major drawback of this approach lies in the considerable computations that are required in any Born-von Kármán analysis. Consequently, the Debye model has been and continues to be extensively used in studies concerned with the specific heat of solids.

The specific heats at constant volume as given by the Einstein and Debye models are tabulated in dimensionless form in Tables A2 and A3, in terms of the Einstein specific heat function

$$\frac{C_v(\theta_E/T)}{3R} = \frac{(\theta_E/T)^2 \exp(\theta_E/T)}{[\exp(\theta_E/T) - 1]^2} \tag{A13}$$

and the Debye specific heat function

$$\frac{C_v(\theta_D/T)}{3R} = 3 \left(\frac{T}{\theta_D}\right)^3 \int_0^{\theta_D/T} \frac{x^4 e^x dx}{(e^x-1)^2} , \tag{A14}$$

respectively. The functions are tabulated at intervals of 0.1 over the range $\theta_E/T \leq 20$ and $\theta_D/T \leq 20$. A linear interpolation gives an inaccuracy of 1 to 2 units in the fourth significant figure over most of this range. It should be noted that values of C_v derived from these tables become increasingly sensitive to the value used for θ_E or θ_D as T is decreased. In the Debye model, for example, a 2% uncertainty in θ_D at $T = \theta_D/2$ corresponds to an uncertainty in C_v of only 0.7%, but at $T = \theta_D/20$ the corresponding uncertainty in C_v is about 6%.

At $T \leq \theta_D/50$, the values of θ_D are essentially equal to their low-temperature limiting value θ_o. For many solids, θ_D also becomes approximately constant at temperatures above about $\theta_D/2$. However, at intermediate temperatures θ_D varies by as much as 20% in some substances, often passing through a minimum (see Figs. 1.3 and 1.7). Values of θ_o, which may be determined from either calorimetric or elastic constant measurements, are presented in Table 1.1 (Section 1.2.2) for selected substances; more extensive lists are given in the literature [6,26,42]. Characteristic temperatures have been derived from other properties of a solid [2,5,46] such as

---

*For polyatomic solids, the high-temperature limit is 3nR where n is the number of different atomic species.

Table A2. Tabulation of the Einstein Specific Heat Function $C_v(\theta_E/T)/3R$ Versus θ_E/T

θ_E/T	$10^n \cdot C_v(\theta_E/T)/3R$										
	0	1	2	3	4	5	6	7	8	9	
0.0	1.0000	0.9992	0.9967	0.9925	0.9868	0.9794	0.9705	0.9601	0.9483	0.9351	
1.0	0.9207	0.9050	0.8882	0.8703	0.8515	0.8318	0.8114	0.7903	0.7687	0.7466	
2.0	0.7241	0.7013	0.6783	0.6552	0.6320	0.6089	0.5859	0.5631	0.5405	0.5182	$n=0$
3.0	0.4963	0.4747	0.4536	0.4330	0.4129	0.3933	0.3743	0.3558	0.3380	0.3207	
4.0	0.3041	0.2881	0.2726	0.2578	0.2436	0.2300	0.2170	0.2046	0.1928	0.1815	
5.0	0.1707	0.1605	0.1508	0.1416	0.1329	0.1246	0.1168	0.1094	0.1025	0.9588	
6.0	0.8968	0.8383	0.7833	0.7315	0.6828	0.6371	0.5942	0.5539	0.5162	0.4808	$n=1$
7.0	0.4476	0.4166	0.3876	0.3605	0.3351	0.3115	0.2894	0.2687	0.2495	0.2316	
8.0	0.2148	0.1993	0.1848	0.1713	0.1587	0.1471	0.1362	0.1261	0.1168	0.1081	
9.0	0.9999	0.9249	0.8554	0.7909	0.7311	0.6756	0.6243	0.5767	0.5326	0.4918	$n=2$
10.0	0.4540	0.4191	0.3867	0.3568	0.3292	0.3036	0.2800	0.2581	0.2380	0.2193	
11.0	0.2021	0.1862	0.1715	0.1580	0.1455	0.1340	0.1233	0.1135	0.1045	0.9616	
12.0	0.8848	0.8140	0.7487	0.6886	0.6333	0.5823	0.5353	0.4921	0.4523	0.4157	$n=3$
13.0	0.3820	0.3510	0.3225	0.2962	0.2721	0.2499	0.2294	0.2107	0.1934	0.1776	
14.0	0.1630	0.1496	0.1373	0.1260	0.1156	0.1060	0.9728	0.8923	0.8184	0.7506	
15.0	0.6883	0.6311	0.5786	0.5305	0.4863	0.4458	0.4086	0.3744	0.3431	0.3144	$n=4$
16.0	0.2881	0.2639	0.2418	0.2215	0.2029	0.1858	0.1702	0.1559	0.1427	0.1307	
17.0	0.1196	0.1095	0.1003	0.9179	0.8402	0.7690	0.7038	0.6441	0.5894	0.5393	
18.0	0.4935	0.4515	0.4130	0.3778	0.3456	0.3162	0.2892	0.2645	0.2419	0.2212	$n=5$
19.0	0.2023	0.1849	0.1691	0.1546	0.1413	0.1292	0.1181	0.1080	0.9870	0.9021	
20.0	0.8245										$n=6$

At lower temperatures, that is, $\theta_E/T > 20$, $C_v(\theta_E/T)/3R = (\theta_E/T)^2 \exp(-\theta_E/T)$.

Table A3. Tabulation of the Debye Specific Heat Function $C_v(\theta_D/T)/3R$ Versus θ_D/T

θ_D/T	$10^n \cdot C_v(\theta_D/T)/3R$										
	0	1	2	3	4	5	6	7	8	9	
0.0	1.0000	0.9995	0.9980	0.9955	0.9921	0.9876	0.9822	0.9759	0.9687	0.9607	
1.0	0.9517	0.9420	0.9316	0.9204	0.9085	0.8960	0.8829	0.8692	0.8550	0.8404	
2.0	0.8254	0.8100	0.7944	0.7784	0.7622	0.7459	0.7294	0.7128	0.6961	0.6794	
3.0	0.6628	0.6461	0.6296	0.6132	0.5969	0.5807	0.5647	0.5490	0.5334	0.5181	$n=0$
4.0	0.5031	0.4883	0.4738	0.4595	0.4456	0.4320	0.4187	0.4057	0.3930	0.3807	
5.0	0.3686	0.3569	0.3455	0.3345	0.3237	0.3133	0.3031	0.2933	0.2838	0.2745	
6.0	0.2656	0.2569	0.2486	0.2405	0.2326	0.2251	0.2177	0.2107	0.2038	0.1972	
7.0	0.1909	0.1847	0.1788	0.1730	0.1675	0.1622	0.1570	0.1521	0.1473	0.1426	
8.0	0.1382	0.1339	0.1297	0.1257	0.1219	0.1182	0.1146	0.1111	0.1078	0.1046	
9.0	0.1015	0.9847	0.9558	0.9280	0.9011	0.8751	0.8500	0.8259	0.8025	0.7800	
10.0	0.7582	0.7372	0.7169	0.6973	0.6784	0.6601	0.6424	0.6253	0.6087	0.5928	
11.0	0.5773	0.5624	0.5479	0.5339	0.5204	0.5073	0.4946	0.4824	0.4705	0.4590	
12.0	0.4478	0.4370	0.4265	0.4164	0.4066	0.3970	0.3878	0.3788	0.3701	0.3617	
13.0	0.3535	0.3455	0.3378	0.3303	0.3230	0.3160	0.3091	0.3024	0.2960	0.2896	
14.0	0.2835	0.2776	0.2718	0.2661	0.2607	0.2553	0.2501	0.2451	0.2402	0.2354	$n=1$
15.0	0.2307	0.2262	0.2218	0.2174	0.2132	0.2092	0.2052	0.2013	0.1975	0.1938	
16.0	0.1902	0.1867	0.1832	0.1799	0.1766	0.1734	0.1703	0.1673	0.1643	0.1614	
17.0	0.1586	0.1558	0.1531	0.1505	0.1479	0.1454	0.1429	0.1405	0.1382	0.1359	
18.0	0.1336	0.1314	0.1293	0.1272	0.1251	0.1231	0.1211	0.1192	0.1173	0.1154	
19.0	0.1136	0.1118	0.1101	0.1084	0.1067	0.1051	0.1035	0.1019	0.1004	0.9888	
20.0	0.9741										$n=2$

At lower temperatures, that is, $\theta_D/T > 20$, $C_v(\theta_D/T)/3R = 4\pi^4 (T/\theta_D)^3/5$.

compressibility and melting temperature, thermal expansion, infrared absorption, etc. However, there is no common theoretical basis for an exact equality between these θ-values and θ_D, and thus, their usefulness is limited to obtaining rough estimates of the lattice specific heat in the absence of calorimetric (or elastic constant) data.

The anharmonic lattice contribution is expected to increase with temperature but in any case is small in the present temperature range. For example, detailed analyses of the C_p data for aluminum and lead [97] have shown that the anharmonic contribution in each case is $\lesssim 0.01(3R)$ for $T \lesssim \theta_D$.

The electronic contribution to the specific heat of a metal is comparable in magnitude to the lattice contribution at temperatures near the low end of the present range but, because of the rapid increase in the lattice term with temperature, γT is only a few percent of C_p at $T \simeq \theta_D$. Values of γ derived from low-temperature specific heat measurements are given in Table 1.3 (Section 1.3.2.2) and elsewhere in the literature [5,6,26].

1.6.3. Specific Heat at High Temperatures

Above the Debye temperature, the specific heat of a solid at constant volume arises primarily from the lattice vibrations. The harmonic contribution may be represented by the high-temperature expansion of the Debye specific heat function; that is,

$$C_v^h = 3R \left[1 - \frac{1}{20} \left(\frac{\theta_D}{T} \right)^2 + \ldots \right] \tag{A15}$$

and to first order in temperature the anharmonic contribution may be expressed as

$$C_v^a = 3RAT \tag{A16}$$

where the anharmonic coefficient A is typically $\sim 10^{-4} K^{-1}$ (see Table 1.2). It may be seen from Eqs. (A4, A15, and A16) that, to a first approximation, the variation of the dilation contribution is also linear in T.

In metals, the electronic contribution may depart from the initial linear trend in T when the Fermi level coincides with a large maximum or minimum in the electron density of states $n(\varepsilon)$ (see Eq. 90); significant departure at temperatures as low as $\sim\theta_D$ have been observed in certain transition metals (e.g., palladium [73,122-124], chromium [243]). However, in many other metals, in particular the alkali and noble metals, the form of the $n(\varepsilon)$ function near the Fermi level may be adequately approximated by the free electron model which, to next order of approximation, yields

$$C_e = \gamma T \left[1 - \frac{3\pi^2}{10} \left(\frac{T}{T_F} \right)^2 \right] \tag{A17}$$

where T_F is the Fermi temperature. Since T_F is typically in the range 10^4-10^5 K [26], the values of C_e for such metals are essentially linear in T over the entire temperature range up to their melting points.

An additional contribution to specific heat arises from the thermal generation of lattice vacancies at temperatures approaching the melting point. The fractional concentration of vacancies increases exponentially with temperature according to

$$n_v = A \exp[-E_f/k_B T] \tag{A18}$$

where E_f is the vacancy formation energy and A is the pre-exponential term. It may be shown (Section 1.2.8) that the vacancy contribution to the specific heat is given by

$$C_{vac} = R(E_f/k_B T)^2 \, A \, \exp[-E_f/k_B T]. \tag{A19}$$

In principle, values of C_{vac} and, hence, E_f and n_v can be determined directly from the measurement of specific heat [106,107]. However, in practice, the resolution of C_{vac} from the C_p data is an extremely difficult problem because of the uncertainties involved in determining the other contributions to C_p near the melting point. Other empirical methods for obtaining values for E_f and/or estimates of n_v include: simultaneous measurement of lattice parameter and expansion [109], rapid quenching experiments [110], tracer diffusion [111], and positron annihilation [112].

The total specific heat measured at temperatures above θ_D may be expressed in the form

$$C_p = a - \frac{b}{T^2} + cT + \Delta C \tag{A20}$$

where the constant term is 3R (24.9432 J·mol^{-1}·K^{-1}) for a monatomic solid, the term in T^{-2} corresponds to the second term in the expansion of the Debye function, the linear term in T represents (to first order) the dilation, anharmonic, and electronic contributions, and ΔC represents the excess specific heat not accounted for by the first three terms.

An anomalous rise in the excess specific heat has been observed in numerous solids prior to melting, an effect which is particularly large in the refractory metals. For example, the quantity ΔC can be represented by a T^4-term in molybdenum [244] and tungsten [104] and by a T^5-term in niobium [228] and tantalum [229]. Figure 1.9 (Section 1.2.8) illustrates the specific heat of tungsten when fitted to the form given by Eq. (A20). Estimates based on literature values of E_f and n_v for these refractory metals indicate that the vacancy contribution to specific heat is, in each case, less than 10% of the quantity ΔC near the melting point. This suggests that higher-order terms from the lattice anharmonicity and the electronic specific heat may contribute significantly at high temperatures. Unfortunately, theoretical evaluations of the higher-order terms are exceedingly difficult and as yet the problem of the anomalous rise in specific heat prior to melting is unresolved.

At a sufficiently high temperature, the ordered state of various magnetic materials is transformed into the disordered paramagnetic state. The transformation is accompanied by a λ-type singularity in the specific heat curve at some critical temperature T_c, as shown in Fig. 1.21 for iron [176]. This form of singularity is a characteristic of second-order phase transitions, which include other examples such as order-disorder transitions in binary alloys and ferroelectrics. The specific heat in the critical region can be described by an inverse-power law singularity (see Table A1) and may be expressed in the form

$$C_p = A'(T_c-T)^{-\alpha'} + B' \quad \text{for } T < T_c$$
$$\quad = A(T-T_c)^{-\alpha} + B \quad\quad \text{for } T > T_c \tag{A21}$$

where α' and α are the critical exponents describing the specific heat anomaly. The quantities A', A, B', B, and the critical exponents are treated as parameters in fitting the C_p data to Eq. (A21). Certain situations may require that B and B' be functions of T in order to account for the variation of the other 'background' contributions to C_p with temperature [189].

1.7.   REFERENCES

1. Blackman, M., in Reports on Progress in Physics (Mann, W.B., Editor), Vol. VIII, The Physical Society, London, England, 11-30, 1942.

2. Blackman, M., in Handbuch der Physik (Flugge, S., Editor), Vol. VII, Springer-Verlag, Berlin, 325-82, 1955.

3. DeLaunay, J., in Solid State Physics (Seitz, F. and Turnbull, D., Editors), Vol. 2, Academic Press, New York, NY, 219-303, 1956.

4. Parkinson, D.H., in Reports on Progress in Physics (Strickland, A.C., Editor), Vol. XXI, The Physical Society, London, England, 226-70, 1958.

5. Gopal, E.S.R., Specific Heats at Low Temperatures, Plenum Press, New York, NY, 240 pp., 1966.

6. Phillips, N.E., in Critical Reviews in Solid State Sciences (Schuele, D.E. and Hoffman, R.W., Editors), Vol. 2, Chemical Rubber Co., Cleveland, OH, 467-553, 1972.

7. National Bureau of Standards, The International System of Units (SI), NBS Special Publ. 330, 41 pp., 1977.

8. National Physical Laboratory, SI Units in Electricity and Magnetism, Her Majesty's Stationary Office, London, England, 23 pp., 1970.

9. Pippard, A.B., The Elements of Classical Thermodynamics, University Press, Cambridge, England, 165 pp., 1957.

10. Zemansky, M.W., Heat and Thermodynamics, McGraw-Hill, New York, NY, 484 pp., 1957.

11. Swalin, R.A., Thermodynamics of Solids, Wiley, New York, NY, 343 pp., 1962.

12. Kittel, C., Thermal Physics, Wiley, New York, NY, 418 pp., 1969.

13. Huang, K., Statistical Mechanics, Wiley, New York, NY, 470 pp., 1963.

14. Dulong, P.L. and Petit, A.T., Ann. Chim. Phys., 10, 395-413, 1819.

15. Boltzmann, L., Adad. Wissen. Wien., Sitzungsber., 63, 679-732, 1871.

16. Weber, H.F., Philos. Mag., 49, 161-83, 1875.

17. Nernst, W., Akad. Wiss. Berlin, Sitzber., 52, 933-40, 1906.

18. Drude, P., Ann. Physik, 1, 566-613, 1900.

19. Wiedemann, G. and Franz, R., Ann. Physik, 89(2), 497-531, 1853.

20. Sommerfeld, A., Z. Phys., 47, 1-32, 1928.

21. Sommerfeld, A. and Bethe, H., in Handbuch der Physik (Geiger, H. and Scheel, K., Editors), Vol. XXIV/2, Springer-Verlag, Berlin, 333-622, 1933.

22. Mott, N.F. and Jones, H., Theory of the Properties of Metals and Alloys, Clarendon Press, Oxford, 326 pp., 1936.

23. Harrison, W.A. and Webb, M.B. (Editors), The Fermi Surface, Wiley, New York, NY, 356 pp., 1960.

24. Einstein, A., Ann. Physik, 22, 180-90, 1907.

25. Kittel, C., Introduction to Solid State Physics, 2nd Ed., Wiley, New York, NY, 617 pp., 1956.

26. Kittel, C., Introduction to Solid State Physics, 5th Ed., Wiley, New York, NY, 599 pp., 1976.

27. Sherman, J. and Ewell, R.B., J. Phys. Chem., 46, 641-62, 1942.

28. Hilsenrath, J. and Ziegler, G.G., Tables of Einstein Functions, NBS Monograph 49, 258 pp., 1962.

29. Nernst, W. and Lindemann, F.A., Z. Elektrochem., 17, 817-27, 1911.

30. Einstein, A., Ann. Physik, 35, 679-94, 1911.

31. Debye, P., Ann. Physik, 39, 789-839, 1912.

32. Born, M. and von Kármán, T., Phys. Z., 13, 297-309, 1912.

33. Schröder, K., J. Appl. Phys., 32, 880-2, 1961; see also Ref. [196].

34. Fink, J.K., Int. J. Thermophys., 3, 165-200, 1982.

35. Ledermann, W., Proc. R. Soc. London, A182, 362-77, 1944.

36. Peierls, R.E., Proc. Natl. Inst. Sci. India, 20A, 121-6, 1954.

37. Harrison, J.P., Lombardo, G., and Peressini, P.P., J. Phys. Chem. Solids, 29, 557-9, 1968.

38. Barron, T.H.K. and Morrison, J.A., Can. J. Phys., 35, 799-810, 1957.

39. Beattie, J.A., J. Math. and Phys., 6, 1-32, 1926.

40. Landolt-Börstein Physikalich Chemische Tabellen, Vol. 2, Part 4, Springer-Verlag, Berlin, 742-9, 1961.

41. Berg, W.T. and Morrison, J.A., Proc. R. Soc. London, A242, 467-77, 1957.

42. Alers, G.A., in Physical Acoustics (Mason, W.P., Editor), Vol. IIIB, Academic Press, New York, NY, 1-42, 1965.

43. Barron, T.H.K. and Klein, M.L., Phys. Rev., 127, 1997-8, 1962.

44. Feldman, J.L., Proc. Phys. Soc., 84, 361-9, 1964.

45. Martin, D.L., Phys. Rev., B8, 5337-60, 1973.

46. Herbstein, F.H., Advan. Phys., 10, 313-55, 1961.

47. Born, M. and Huang, K., Dynamical Theory of Crystal Lattices, Clarendon Press, Oxford, 420 pp., 1954.

48. Cochran, W., in Reports on Progress in Physics (Strickland, A.C., Editor), Vol. XXVI, The Physical Society, London, England, 1-45, 1963.

49. Maradudin, A.A., Montroll, E.W., Weiss, G.H., and Ipatova, I.P., Solid State Physics (Ehrenreich, H., Seitz, F., and Turnbull, D., Editors), Suppl. 3, 2nd Ed., Academic Press, New York, NY, 708 pp., 1971.

50. Brillouin, L., Wave Propagation in Periodic Structures, 2nd Ed., Dover, New York, NY, 255 pp., 1953.

51. Goldstein, H., Classical Mechanics, 2nd Ed., Addison-Wesley, Reading, MA, 672 pp., 1980.

52. Kittel, C., Quantum Theory of Solids, Wiley, New York, NY, 435 pp., 1963.

53. Taylor, P.L., A Quantum Approach to the Solid State, Prentice-Hall, Englewood Cliffs, NJ, 322 pp., 1970.

54. Foreman, A.J.E. and Lomer, W.M., Proc. Phys. Soc. London, B70, 1143-50, 1957.

55. Ziman, J.M., Advan. Phys. (Phil. Mag. Suppl.), 13, 89-138, 1964.

56. Cochran, W., in Inelastic Scattering of Neutrons, Vol. I, IAEA, Vienna, 3-23, 1965.

57. Klein, M.L. and Venables, J.A. (Editors), Rare Gas Solids, Vol. I, Academic Press, New York, NY, 607 pp., 1976.

58. Dick, B.J. and Overhauser, A.W., Phys. Rev., 112, 90-103, 1958.

59. Woods, A.D.B., Cochran, W., and Brockhouse, B.N., Phys. Rev., 119, 980-99, 1960.

60. Cowley, R.A., Cochran, W., Brockhouse, B.N., and Woods, A.D.B., Phys. Rev., 131, 1030-9, 1963.

61. Mashkevich, V.S. and Tolpygo, K.B., Sov. Phys.-JETP, 5, 435-9, 1957.

62. Hardy, J.R., Philos. Mag., 7, 315-36, 1962.

63. Schröder, U., Solid State Commun., 4, 347-8, 1966.

64. Cole, H. and Warren, B.E., J. Appl. Phys., 23, 335-40, 1952.

65. Jacobsen, E.H., Phys. Rev., 97, 654-9, 1955.

66. Walker, C.B., Phys. Rev., 103, 547-57, 1956.

67. Brockhouse, B.N., in Inelastic Scattering of Neutrons, IAEA, Vienna, 113-57, 1961.

68. Brockhouse, B.N., in Phonons in Perfect Lattices and Lattices with Point Imperfections (Stevenson, R.W.H., Editor), Oliver and Boyd, London, England, 110-52, 1966.

69. Dolling, G. and Woods, A.D.B., in Thermal Neutron Scattering (Egelstaff, P.A., Editor), Academic Press, New York, NY, 193-249, 1965.

70. Larose, A. and Vanderwal, J., Solid State Physics Literature Guides (Connolly, T.F., Editor), Vol. 7, IFI/Plenum, New York, NY, 527 pp., 1974.

71. Svensson, E.C., Brockhouse, B.N., and Rowe, J.M., Phys. Rev., 155, 619-32, 1967.

72. Birgeneau, R.J., Cordes, J., Dolling, G., and Woods, A.D.B., Phys. Rev., A136, 1359-65, 1964.

73. Miiller, A.P. and Brockhouse, B.N., Can. J. Phys., 49, 704-23, 1971.

74. Copley, J.R.D., MacPherson, R.W., and Timusk, T., Phys. Rev., 182, 965-72, 1969; see erratum in Phys. Rev., B1, 4193, 1970.

75. Dolling, G., Cowley, R.A., and Woods, A.D.B., Can. J. Phys., 43, 1397-413, 1965.

76. Dolling, G. and Waugh, J.L.T., in Lattice Dynamics (Wallis, R.F., Editor), Pergamon Press, London, England, 19-32, 1965.

77. Van Hove, L., Phys. Rev., 89, 1189-93, 1953.

78. Gilat, G. and Raubenheimer, L.J., Phys. Rev., 144, 390-5, 1966.

79. Raubenheimer, L.J. and Gilat, G., Phys. Rev., 157, 586-99, 1967.

80. Kam, Z. and Gilat, G., Phys. Rev., 175, 1156-63, 1968.

81. Stedman, R. and Nilsson, G., Phys. Rev., 145, 492-500, 1966.

82. Martin, D.L., Can. J. Phys., 38, 17-24, 1960.

83. Leibfried, G. and Ludwig, W., in Solid State Physics (Seitz, F. and Turnbull, D., Editors), Vol. 12, Academic Press, New York, NY, 275-444, 1961.

84. Barron, T.H.K., in Lattice Dynamics (Wallis, R.F., Editor), Pergamon Press, London, England, 247-54, 1965.

85. Maradudin, A.A., Flinn, P.A., and Coldwell-Horsfall, R.A., Ann. Phys., 15, 337-59, 1961.

86. Cowley, R.A., Advan. Phys., 12, 421-80, 1963.

87. Shukla, R.C. and Muller, E.R., Phys. Status Solidi B, 43, 413-22, 1971.

88. Shukla, R.C., Int. J. Thermophys., 1, 73-82, 1980.

89. Cowley, E.R. and Shukla, R.C., Phys. Rev., B9, 1261-7, 1974.

90. MacDonald, R.A. and MacDonald, W.M., Phys. Rev., B24, 1715-24, 1981.

91. Shukla, R.C. and Taylor, R., Phys. Rev., B9, 4116-20, 1974.

92. Copley, J.R.D., Can. J. Phys., 51, 2564-86, 1973.

93. MacDonald, R.A., Mountain, R.D., and Shukla, R.C., Phys. Rev., B20, 4012-6, 1979.

94. Cowley, E.R., J. Phys. C, 4, 988-97, 1971.

95. Werthamer, N.R., in Rare Gas Solids (Klein, M.L. and Venables, J.A., Editors), Vol. 1, Academic Press, New York, NY, 265-300, 1976.

96. Metropolis, N., Rosenbluth, A.W., Rosenbluth, M.N., Teller, A.H., and Teller, E., J. Chem. Phys., 21, 1087-92, 1953.

97. Leadbetter, A.J., J. Phys. C, 1, 1481-504, 1968.

98. Brooks, C.R., J. Phys. Chem. Solids, 29, 1377-85, 1968.

99. Brooks, C.R. and Bingham, R.E., J. Phys. Chem. Solids, 29, 1553-60, 1968.

100. Martin, D.L., Phys. Rev., A139, 150-60, 1965.

101. Werthamer, N.R., Phys. Rev., B1, 572-81, 1970.

102. Hui, J.C.K. and Allen, P.B., J. Phys. C, 8, 2923-35, 1975.

103. Pawel, R.E. and Stansbury, E.E., J. Phys. Chem. Solids, 26, 607-13, 1965.

104. Cezairliyan, A. and McClure, J.L., J. Res. Natl. Bur. Stand., 75A, 283-90, 1971.

105. Cezairliyan, A., High Temp. Sci., 13, 117-33, 1980.

106. Kraftmakher, Ya.A. and Strelkov, P.G., Sov. Phys.-Solid State, 8, 838-41, 1966.

107. Kraftmakher, Ya.A., in Proceedings of the 7th Symposium on Thermophysical Properties (Cezairliyan, A., Editor), ASME, New York, NY, 160-8, 1977.

108. Damask, A.C. and Dienes, G.J., Point Defects in Metals, Gordon and Breach, New York, NY, 314 pp., 1963.

109. Simmons, R.O. and Balluffi, R.W., Phys. Rev., 129, 1533-44, 1963.

110. Berger, A.S., Seidmann, D.N., and Balluffi, R.W., Acta Metall., 21, 123-48, 1973.

111. Batra, A.P. and Slifkin, L.M., J. Phys. Chem. Solids, 38, 687-92, 1977.

112. Fluss, M.J., Smedskjaer, L.C., Chason, M.K., Legnini, D.G., and Siegel, R.W., Phys. Rev., B17, 3444-55, 1978.

113. Stoner, E.C., Proc. R. Soc. London, A154, 656-78, 1936.

114. Veal, B.W. and Rayne, J.A., Phys. Rev., A135, 442-6, 1964.

115. Altmann, S.L., Band Theory of Metals, The Elements, Pergamon Press, Oxford, 250 pp., 1970.

116. Harrison, W.A., Electronic Structure and the Properties of Solids, Freeman and Co., San Francisco, CA, 582 pp., 1980.

117. Ziman, J.M., Advan. Phys., 13, 89-138, 1964.

118. Cohen, M.L. and Heine, V., in Solid State Physics (Ehrenreich, H., Seitz, F., and Turnbull, D., Editors), Vol. 24, Academic Press, New York, NY, 37-248, 1970.

119. Dimmock, J.O., in Solid State Physics (Ehrenreich, H., Seitz, F., and Turnbull, D., Editors), Vol. 26, Academic Press, New York, NY, 103-274, 1971.

120. Harrison, W.A., Phys. Rev., 118, 1190-208, 1960.

121. Mott, N.F., Advan. Phys., 13, 325-422, 1964.

122. Shimizu, M., Takahashi, T., and Katsuki, A., J. Phys. Soc. Jpn., 18, 240-8, 1963.

123. Kimura, H., Katsuki, A., and Shimizu, M., J. Phys. Soc. Jpn., 21, 307-12, 1966.

124. Hindley, N.K. and Rhodes, P., Proc. Phys. Soc., 81, 717-25, 1963.

125. Mueller, F.M., Freeman, A.J., Dimmock, J.O., and Furdyna, A.M., Phys. Rev., B1, 4617-35, 1970.

126. Rath, J. and Freeman, A.J., Phys. Rev., B11, 2109-17, 1975.

127. Gilat, G. and Bharatiya, N.R., Phys. Rev., B12, 3479-81, 1975.

128. Anderson, O.K. and Mackintosh, A.R., Solid State Commun., 6, 285-90, 1968.

129. Weaire, D., Proc. Phys. Soc., 92, 956-61, 1967.

130. Rice, T.M., Ann. Phys., 31, 100-29, 1965.

131. Ashcroft, N.W. and Wilkins, J.W., Phys. Lett., 14, 285-7, 1965.

132. Allen, P.B. and Cohen, M.L., Phys. Rev., 187, 525-38, 1969.

133. Janak, J.F., Phys. Lett. A, 27, 105-6, 1968.

134. Ashcroft, N.W., Phys. Rev., A140, 935-40, 1965.

135. Parks, R.D. (Editor), Superconductivity, Vols. 1 and 2, Marcel Dekker, Inc., New York, NY, 1412 pp., 1969.

136. Grassie, A.D.C., The Superconducting State, Sussex University Press, Sussex, 135 pp., 1975.

137. Rose-Innes, A.C. and Rhoderick, E.H., Introduction to Superconductivity, 2nd Ed., Pergamon Press, New York, NY, 237 pp., 1978.

138. Omnes, H. Kamerlingh, Akad. van Wetenschappen (Amsterdam), 14, 113-5, 1911.

139. Keesom, W.H. and Andrews, D.H., Verslag. Akad. Weten. Amsterdam, 36, 52-61, 1927.

140. Meissner, W. and Ochsenfeld, R., Naturwissenschaften, 21, 787-8, 1933.

141. Phillips, N.E., Phys. Rev., A134, 385-91, 1964.

142. Corak, W.S., Goodman, B.B., Satterthwaite, C.B., and Wexler, A., Phys. Rev., 96, 1442-4, 1954.

143. Bardeen, J., Cooper, L.N., and Schrieffer, J.R., Phys. Rev., 108, 1175-204, 1957.

144. Bardeen, J. and Schrieffer, J.R., Prog. Low Temp. Phys., 3, 170-287, 1961.

145. Schrieffer, J.R., Theory of Superconductivity, Benjamin, New York, NY, 282 pp., 1965.

146. Mühlshlegel, B., Z. Phys., 155, 313-27, 1959.

147. Meservey, R. and Schwartz, B.B., in Superconductivity (Parks, R.D., Editor), Marcel Dekker, Inc., New York, NY, 117-91, 1969.

148. Swihart, J.C., Scalapino, D.J., and Wada, Y., Phys. Rev. Lett., 14, 106-11, 1965.

149. McMillan, W.L., Phys. Rev., 167, 331-44, 1968.

150. Carbotte, J.P. and Dynes, R.C., Phys. Rev., 172, 476-84, 1968.

151. van der Hoeven, B.J.C., Jr. and Keesom, P.H., Phys. Rev., A137, 103-7, 1965.

152. Shen, L.Y.L., Senozan, N.M., and Phillips, N.E., Phys. Rev. Lett., 14, 1025-6, 1965.

153. Weber, H.W. (Editor), Anisotropy Effects in Superconductors, Plenum Press, New York, NY, 316 pp., 1977.

154. Abrikosov, A.A., Sov. Phys.-JETP, 5, 1174-82, 1957.

155. Abrikosov, A.A., J. Phys. Chem. Solids, 2, 199-208, 1957.

156. Ginsburg, V.L. and Landau, L.D., Zh. Eksp. Teor. Fiz., 20, 1064-82, 1950.

157. Gor'kov, L.P., Sov. Phys.-JETP, 10, 593-9, 1960.

158. Gor'kov, L.P., Sov. Phys.-JETP, 10, 998-1004, 1960.

159. McConville, T. and Serin, B., Phys. Rev., A140, 1169-77, 1965.

160. Ferreira da Silva, J., Burgemeister, E.A., and Dokupil, Z., Physica, 41, 409-39, 1969.

161. Morrish, A.H., The Physical Principles of Magnetism, Wiley, New York, NY, 680 pp., 1965.

162. Weiss, P., J. Phys. (Paris), 6, 661-90, 1907.

163. Heisenberg, W., Z. Phys., 49, 619-36, 1928.

164. Bloch, F., Z. Phys., 74, 295-335, 1932.

165. Van Kranendonk, J. and Van Vleck, J.H., Rev. Mod. Phys., 30, 1-23, 1950.

166. Herring, C., Magnetism (Rado, G.T. and Suhl, H., Editors), Vol. IV, Academic Press, New York, NY, 407 pp., 1966.

167. Keffer, F., in Handbuch der Physik (Flügge, S., Editor), Vol. XVIII, Springer-Verlag, Berlin, 1-273, 1966.

168. Shirane, G., Minkiewicz, V.J., and Nathans, R., J. Appl. Phys., 39, 383-90, 1968.

169. Samuelsen, E.J., Silberglitt, R., and Shirane, G., Phys. Rev., B3, 157-66, 1971.

170. Alperin, H.A., Steinsvoll, O., Nathans, R., and Shirane, G., Phys. Rev., 154, 508-14, 1967.

171. Pepy, G., J. Phys. Chem. Solids, 35, 433-44, 1974.

172. Holden, T.M., Svensson, E.C., and Martel, P., Can. J. Phys., 50, 687-91, 1972.

173. Dixon, M., Hoare, F.E., and Holden, T.M., Phys. Lett., 14, 184-5, 1965.

174. Dixon, M., Hoare, F.E., and Holden, T.M., Proc. R. Soc. London, A303, 339-54, 1968.

175. Kalinkina, I.N., Sov. Phys.-JETP, 16, 1432-8, 1963.

176. Kollie, T.G., Univ. Tennessee, Ph.D. Thesis, 254 pp., 1969.

177. Fisher, M.E., in Reports on Progress in Physics (Strickland, A.C., Editor), Vol. XXX, The Physical Society, London, England, 615-730, 1967.

178. Kadanoff, L.P., Götze, W., Hamblen, D., Hecht, R., Lewis, E.A.S., Palciauskas, V.V., Martin, R., and Swift, J., Rev. Mod. Phys., 39, 395-431, 1967.

179. Rao, C.N.R. and Rao, K.J., Phase Transitions in Solids, McGraw-Hill, New York, NY, 330 pp., 1978.

180. Stanley, H.E., Introduction to Phase Transitions and Critical Phenomena, Oxford University Press, Oxford, 308 pp., 1971.

181. Patashinskii, A.Z. and Pokrovskii, V.L., Fluctuation Theory of Phase Transitions, Pergamon Press, New York, NY, 321 pp., 1979.

182. Ma, S.K., Modern Theory of Critical Phenomena, Benjamin, London, England, 561 pp., 1976.

183. Bragg, W.L. and Williams, E.J., Proc. R. Soc. London, A145, 699-730, 1934.

184. DeJongh, L.J. and Miedema, A.R., Advan. Phys., 23, 1-260, 1974.

185. Ising, E., Z. Phys., 31, 253-8, 1925.

186. Onsager, L., Phys. Rev., 65, 117-49, 1944.

187. Wilson, K.G., Phys. Rev. Lett., 28, 548-51, 1972.

188. Wilson, K.G. and Kogut, J., Phys. Repts., 12C, 75-199, 1974.

189. Lederman, F.L., Salamon, M.B., and Shacklette, L.W., Phys. Rev., B9, 2981-8, 1974.

190. Schottky, W., Phys. Z., 23, 448-55, 1922.

191. Rosenberg, H.M., Low Temperature Solid State Physics, Clarendon Press, Oxford, 420 pp., 1963.

192. Griffith, J.S., The Theory of Transition-Metal Ions, University Press, Cambridge, England, 455 pp., 1961.

193. Sheard, F.W., Smith, T.F., White, G.K., and Birch, J.A., J. Phys. C, 10, 645-55, 1977.

194. Stout, J.W. and Hadley, W.B., J. Chem. Phys., 40, 55-63, 1964.

195. Sommers, J.A. and Westrum, E.F., J. Chem. Thermodyn., 8, 1115-36, 1976.

196. Livingston, J.D. and Bean, C.P., J. Appl. Phys., 32, 1964-6, 1961.

197. Schröder, K. and Cheng, C.H., J. Appl. Phys., 31, 2154-5, 1960.

198. Schröder, K., J. Appl. Phys., 32, 880-2, 1961.

199. Falge, R.L., Jr. and Wolcott, N.M., J. Low Temp. Phys., 5, 617-50, 1971.

200. Flotow, H.E., Kuentzler, R., and Osborne, D.W., Phys. Status Solidi A, 34, 291-5, 1976.

201. Lounasmaa, O.V., in Hyperfine Interactions (Freeman, A.J. and Frankel, R.B., Editors), Academic Press, New York, NY, 467-96, 1967.

202. Bleaney, B. and Hill, R.W., Proc. Phys. Soc. London, 78, 313-5, 1961.

203. Doniach, S. and Engelsberg, S., Phys. Rev. Lett., 17, 750-3, 1966.

204. Berk, N.F. and Schrieffer, J.R., Phys. Rev. Lett., 17, 433-5, 1966.

205. Bennemann, K.H., Phys. Rev., 167, 564-72, 1968.

206. Brinkmann, W.F. and Engelsberg, S., Phys. Rev., 169, 417-31, 1968.

207. Béal-Monod, M.T., Ma, S.K., and Fredkin, D.R., Phys. Rev. Lett., 20, 929-32, 1968.

208. Trainor, R.J., Brodsky, M.B., and Culbert, H.V., Phys. Rev. Lett., 34, 1019-22, 1975.

209. Bucher, E., Brinkman, W.F., Maita, J.P., and Williams, H.J., Phys. Rev. Lett., 18, 1125-7, 1967.

210. Triplett, B.B. and Phillips, N.E., Phys. Lett. A, 37, 443-4, 1971.

211. Rizzuto, C., in Reports on Progress in Physics (Ziman, J.M., Editor), Vol. 37, The Institute of Physics, London, England, 147-229, 1974.

212. Kondo, J., in Solid State Physics (Seitz, F., Turnbull, D., and Ehrenreich, H., Editors), Vol. 23, Academic Press, New York, NY, 183-281, 1969.

213. Bloomfield, P.E. and Hamann, D.R., Phys. Rev., 164, 856-65, 1967.

214. Triplett, B.B. and Phillips, N.E., Phys. Rev. Lett., 27, 1001-4, 1971.

215. Blandin, A., J. Phys. (Paris), 39(C6), 1499-516, 1978.

216. Marshall, W., Phys. Rev., 118, 1519-23, 1960.

217. Klein, M.W. and Brout, R., Phys. Rev., 132, 2412-25, 1963.

218. Zimmerman, J.E. and Hoare, F.E., J. Phys. Chem. Solids, 17, 52-6, 1960.

219. Crane, L.T. and Zimmerman, J.E., J. Phys. Chem. Solids, 21, 310-3, 1961.

220. Sherrington, D. and Kirkpatrick, S., Phys. Rev. Lett., 35, 1792-6, 1975.

221. Walker, L.R. and Walstedt, R.E., Phys. Rev. Lett., 38, 514-8, 1977.

222. Thouless, D.J., Anderson, P.W., and Palmer, R.G., Philos. Mag., 35, 593-601, 1977.

223. Martin, D., Phys. Rev., B20, 368-75, 1979.

224. Martin, D., Phys. Rev., B21, 1906-10, 1980.

225. Touloukian, Y.S. (Editor), Thermophysical Properties Research Center, TPRC Rept, 16, 323-46, 1966.

226. Jaeger, F.M., Rosenbohm, E., and Bottema, J.A., Proc. Akad. van Weten. Amsterdam, 35, 763-79, 1932.

227. Bottema, J.A. and Jaeger, F.M., Proc. Akad. van Weten. Amsterdam, 35, 929-31, 1932.

228. Cezairliyan, A., J. Res. Natl. Bur. Stand., 75A, 565-71, 1971.

229. Cezairliyan, A., McClure, J.L., and Beckett, C.W., J. Res. Natl. Bur. Stand., 75A, 1-13, 1971.

230. Cezairliyan, A., High Temp.-High Pressures, 4, 541-50, 1972.

231. Cezairliyan, A., J. Chem. Thermodyn., 6, 735-42, 1974.

232. Gruneisen, E., in Handbuch der Physik (Henning, F., Editor), Vol. X, Springer-Verlag, Berlin, 1-59, 1926.

233. Barron, T.H.K., Philos. Mag., 46, 720-34, 1955.

234. Barron, T.H.K., Ann. Phys. (Paris), 1, 77-90, 1957.

235. Blackman, M., Proc. Phys. Soc. London, B70, 827-32, 1957.

236. Blackman, M., Philos. Mag., 3, 831-8, 1958.

237. Bijl, D. and Pullan, H., Philos. Mag., 45, 290-4, 1954.

238. Bijl, D. and Pullan, H., Physica, 21, 285-98, 1955.

239. Fraser, D.B. and Hallett, A.C.H., in Proceedings of the VIIIth International Conference on Low Temperature Physics (Graham, G.M. and Hallett, A.C.H., Editors), Univ. of Toronto Press, Toronto, 689-92, 1961.

240. Nerst, W. and Lindemann, F.A., Z. Elektrochem., 17, 817-27, 1911.

241. Chang, Y.A. and Hultgren, R., J. Phys. Chem., $\underline{69}$, 4162-5, 1965.

242. Singh, R.P. and Verma, G.S., Solid State Commun., $\underline{6}$, 113-4, 1968.

243. Shimizu, M., Takahashi, T., and Katsuki, A., J. Phys. Soc. Jpn., $\underline{17}$, 1740-6, 1962.

244. Cezairliyan, A., Morse, M.S., Berman, H.A., and Beckett, C.W., J. Res. Natl. Bur. Stand., $\underline{74A}$, 65-92, 1970.

CHAPTER 2

Calorimetry Below 1 K

A. C. ANDERSON

2.1. INTRODUCTION

Calorimetric measurements have been extended into the temperature range below 1 K only in the recent past. A 1961 review [1] of cryogenic calorimetry barely discusses the temperature range below 1 K. Nevertheless, 1958 was a turning point in the history of low-temperature calorimetry. By 1958, pure gaseous ^{3}He had become available as a result of the military nuclear programs of a few nations. The higher vapor pressure of liquid ^{3}He provided temperatures to roughly 0.3 K by evaporative cooling. The calorimetric techniques which had been used in the liquid ^{4}He temperature range above 1 K were rapidly adapted to the range 0.3–1 K, and numerous papers have been published. These techniques are not discussed in the present chapter, but rather in the next chapter of the present book, 'Calorimetry in the Range 0.3–30 K.' The present chapter deals primarily with alternative techniques applicable to temperatures below 0.1 K, but which may be utilized to temperatures as high as ≈1 K. The number of calorimetric measurements which have been made in the temperature range below 0.3 K is still relatively small.

The change in appropriate experimental techniques at a temperature near 0.3 K is not arbitrary, but stems from several constraints. For the purpose of this introduction, three constraints will be mentioned. First, evaporative cooling can be used only to ≈0.3 K. For lower temperatures either adiabatic demagnetization or dilution refrigeration is generally required. Second, an international temperature scale does not extend below ≈0.3 K. Each laboratory has had to establish a scale as close as possible to the thermodynamic scale. Third, thermal isolation of the heat-capacity sample is successfully provided by a mechanical heat switch for temperatures above 0.3 K. At much lower temperatures, alternative approaches to thermal isolation have had to be established.

The problems encountered at temperatures below 0.3 K are discussed in Section 2.2 of this chapter. The discussions are brief, and are intended to provide the reader with the flavor of low-temperature experimentation. Frequent reference will be made to the literature, especially to the review paper [2] entitled 'Instrumentation at Temperatures Below 1 K' and its extensive bibliography. Section 2.3 of this chapter discusses several examples of calorimetric techniques utilized at temperatures below 0.3 K.

2.2. PROBLEMS ENCOUNTERED IN THE TEMPERATURE RANGE 0.01-1 K

2.2.1. Refrigeration

With few exceptions, temperatures below 0.3 K are available only through the use of adiabatic demagnetization of paramagnetic salts or the use of ^{3}He-^{4}He dilution refrigerators [2]. Temperatures to 0.01 K are most easily and inexpensively obtained by adiabatic demagnetization. The paramagnetic crystals are readily available, and a superconducting solenoid to supply the magnetic field may be made in the laboratory or purchased from commercial vendors. Unfortunately, the presence of a magnetic field may be deleterious in some situations, as in calorimetric measurements of superconducting materials.

Most laboratories now use ^{3}He-^{4}He dilution refrigerators, probably because the refrigerator is also used for measurements other than heat capacity. A measurement of thermal conductivity, for example, requires a refrigerator which can accept the applied measuring power for an extended period of time without warming. This is a feature provided only by the dilution refrigerator. Dilution refrigerators may be constructed, or purchased, which provide reliable and useful refrigeration to below 0.01 K. But these are expensive and complex in construction. Dilution refrigerators of more simple design provide temperatures of ≈ 0.04 K.

2.2.2. Temperature Scale and Standards

A refrigerator can operate at less than optimum performance and still permit specific heat measurements to be obtained over a more limited temperature range. Thermometers, on the other hand, must always be accurate and reliable or the data are less than worthless. Thus, thermometry is a serious problem. The actual thermometers are discussed in the next section; here we are concerned with the establishment of a temperature scale.

Prior to 1979, nearly all laboratories based their temperature scale on the vapor pressures of liquid ^{4}He or liquid ^{3}He, since there was international agreement as to a relation between temperature and vapor pressure. Liquid ^{3}He is to be preferred as it has a higher vapor pressure and, therefore, is both easier to measure and useful to lower temperatures. Liquid ^{3}He is also not a superfluid at temperatures above 0.3 K, and, thus, avoids the subtle problems associated with the superfluid-film flow of liquid ^{4}He at temperatures T below 2 K.

The vapor pressure of liquid ^{3}He is useful for thermometry, in a practical sense, only for $T \gtrsim 0.5$ K. Thus, a means is required to extrapolate the vapor pressure scale by a factor of 50 to 0.01 K. This has commonly been done using a material which is accurately paramagnetic (that is, the magnetic susceptibility, χ, varies as $1/T$). The susceptibility may be measured with an a.c. mutual-inductance bridge which provides a reading $M = M_O + GT^{-1}$. The constants M_O and G are established by a calibration against a helium vapor-pressure thermometer. With appropriate precautions [2] in the preparation of the paramagnet and the construction of the measuring apparatus, the extrapolation of the calibration to low temperatures can provide an accuracy of $\approx 1\%$ relative to the vapor-pressure scales. Other methods of establishing a low-temperature scale have been used, such as nuclear magnetic resonance, γ-ray anisotropy, and Johnson noise.

Two important changes have occurred since, roughly, 1979. First, a revised international temperature scale has been proposed which should more closely approximate the absolute thermodynamic scale. The differences between this EPT 76 scale and the helium vapor pressure scales are a fraction of one percent [3]. The second change is the availability of a set of superconducting thermometric fixed points [4,5] for temperatures ranging from 0.015 to 7 K. At temperatures above 0.5 K the fixed points are based on the EPT 76 scale. Since EPT 76 does not extend to lower temperatures, the fixed points for T < 0.5 K have been calibrated against 'absolute' thermometric techniques such as Johnson-noise thermometry. Although the fixed points below 0.5 K have not received international endorsement, they do provide a means for the comparison of working scales between various laboratories, and between those laboratories and the National Bureau of Standards. Even with the use of fixed points, some means of interpolation between the fixed-point temperatures is required. This may be done, for example, with the paramagnetic thermometers discussed above. Proper use of fixed points does avoid the expense, time, and care otherwise devoted to reliable vapor-pressure measurements.

It is also significant that reliable doped-germanium resistance thermometers have become available. With proper precautions, to be discussed in the next section, germanium thermometers have been found to be highly reproducible. Hence, a single, careful calibration can be maintained indefinitely on these thermometers. Nevertheless, it would be prudent to periodically check the calibration of a resistance thermometer using superconducting fixed points.

Calibrated germanium resistance thermometers may be purchased. However, prior to 1979, the author found no calibrated thermometer which was not in gross error at temperatures below ≈ 0.3 K. Although the situation has improved since 1979, it is advisable to check a commercial calibration against a set of superconducting fixed points.

At temperatures above 1 K it has been the practice to use pure copper as a calorimetric standard to ascertain the accuracy of both the calorimeter and the working temperature scale. This approach would also be helpful at temperatures below 1 K, and pure copper has been used for this purpose. However, several measurements on copper at $T \lesssim 0.3$ K have found a specific heat in excess of that generally attributed to conduction electrons and phonons. That is, if the temperature dependence established at $T \gtrsim 1$ K for the specific heat of copper is extrapolated to 0.03 K, the extrapolation is in error by $\approx 100\%$ [6]. The excess specific heat is probably caused by an unidentified impurity in the copper.

2.2.3. Thermometry

In measuring the temperature, T, and the temperature change, ΔT, of a calorimeter, any type of thermometer can be used including the paramagnetic thermometer discussed in Section 2.2.2. However, the resistance thermometer is used most frequently, either a doped-germanium unit or one fashioned from a carbon resistor [2]. The germanium resistor has the advantage of a reproducible temperature calibration, but the disadvantage of a large added heat capacity contributed by its encapsulation. The carbon thermometer is small in heat capacity and very inexpensive, but the reproducibility of its temperature calibration cannot be relied upon.

For either germanium or carbon resistance thermometers, the most serious problem is the dissipation of electrical power within the thermometer. This causes the thermometer to indicate a temperature higher than the sample to which it is attached. Thus, the electronics used to measure resistance must operate with a measuring current such that the power dissipated in the thermometer is very small. The problem becomes more severe with decreasing temperatures, varying as T^{-4} or T^{-5}. At temperatures near 0.02 K, the measuring power must be $\leq 10^{-14}$ W. Other electrical currents must also be kept small. These spurious sources include residual d.c. currents from the electronics, thermal emfs in the electrical leads, 50 or 60 Hz induced voltages, and rf induced voltages from TV or FM transmitters or from pocket calculators or other digital equipment. A combination of filtering and shielding may be required before reliable low-temperature resistance thermometry is available in the laboratory.

A simple criterion can generally be applied to determine if a resistance thermometer had, in some manner, been heated to a temperature above the temperature of the primary thermometer during calibration. A plot of logarithm of resistance versus logarithm of temperature, i.e., lnR versus lnT, should exhibit a negative slope of ever-increasing magnitude with decreasing temperature. Should the slope decrease with decreasing temperature, the thermometer had probably lost thermal contact with the primary thermometer due to excessive heating and/or a large thermal impedance between the two thermometers.

2.2.4. Thermal Isolation

In most low-temperature calorimetric measurements the sample is isolated in a vacuum. This is readily attained in a leak-tight system by pumping away any He gas which may have been introduced to facilitate cool-down of the cryostat from room temperature. At T < 0.3 K, the cold refrigerator will getter or adsorb any remanent He gas.

The sample must also be protected from electromagnetic radiation. All tubes entering the cryostat from room temperature must be carefully baffled against 300 K infrared radiation. Also, the sample should be shielded against radiation from surfaces at a liquid-helium temperature of 4 K. As an example, a cube of stainless steel 1 cm on a side at a temperature of 0.1 K would change in temperature at a rate of $\approx 1\% \cdot s^{-1}$ if exposed to 4 K radiation. This is an intolerable drift rate and dictates the use of a thermal shield held at a much lower temperature. In a dilution refrigerator, the thermal shield may be attached to the 'still' which maintains a temperature of ≈ 0.7 K. Any holes in the shield should be carefully baffled. The problem of rf radiation should be mentioned again. Not only can rf radiation Joule-heat resistance thermometers, but it has also been observed to Joule-heat electrical heaters used in calorimeters. This heating may be time dependent if metal objects, such as chairs or automobiles, are moving in or near the laboratory. Shielding and/or filtering may be required to suppress this problem [2].

There is also a problem in attempting to provide adequate mechanical support for a sample, yet maintaining the thermal conduction along the mechanical supports small. Often threads or polymer films are used which have small thermal conductivities. But then the sample is not rigidly supported and may vibrate causing the dissipation of heat, some of which enters the sample.

For this reason, the cryostat should be mounted on vibration-isolating supports and other means taken to minimize vibrational levels within the cryostat [2]. The total thermal conductance between a sample and its environment can be further reduced by using electrical leads (for electrical heaters and thermometers) as mechanical supports. The leads should be superconducting to eliminate electronic thermal conduction, and may be stretched taut to help avoid vibrational heating.

2.2.5. Thermal Contact

The general problem of thermal contact has been thoroughly discussed in Ref. [2] and papers cited therein. In brief, electronic conduction should be established if possible. Phonon thermal conduction encounters the Kapitza thermal-boundary resistance, R_B, wherever a dielectric (or a superconductor) is placed in thermal contact with any other material. The boundary resistance varies as T^{-3}, and rapidly becomes more serious with decreasing temperature.

The primary problem in specific heat measurements occurs in placing a sample in contact with the refrigerator so as to cool the sample to low temperature, then breaking this contact so as to thermally isolate the sample for measurements. This problem is sufficiently troublesome that techniques have been devised to completely avoid the necessity of a 'thermal switch.' These will be discussed in Section 2.3; most depend on some time-dependent response of the calorimeter.

It is possible to utilize a mechanical heat switch at low temperatures [7,8], but these tend to be elaborate and to induce vibrational heating of the sample as the contact is opened. An alternative scheme is to use a superconducting foil as a thermal switch [2,9-12]. Application of a magnetic field greater in magnitude than the superconducting critical field transforms the foil to the normal state. The normal state has a large thermal conductivity proportional to T which is contributed by conduction electrons. With the magnetic field off, the thermal conductivity is due only to phonons, is much smaller in magnitude, and is often proportional to T^3. These comments are valid only if the temperature is much less than the superconducting critical temperature since, at the critical temperature, the magnitudes of the thermal conductivities in the normal and superconducting states are equal. The magnetic field needed to operate the switch can be provided by a miniature superconducting solenoid containing the thermal switch.

2.2.6. Application of Heat

The most simple means used to supply heat to the low-temperature calorimeter is by an electrical heater made of resistive wire or foil. Tiny metallic strain gauges may be used as convenient heaters. Generally, four electrical leads are run to the resistive heater, or to two superconducting leads attached to the heater. Two wires carry a measured d.c. current, I; the other two permit a measurement of the voltage, E, across the heater. The power dissipated in the heater is P = IE. It is advisable to also calculate the ratio R = E/I as a check on the I and E readings, as R should be a constant. Voltages are often read with a digital voltmeter because of the speed and accuracy now available in this instrument. However, all digital voltmeters emit rf radiation. It is essential that this radiation not be permitted to enter the cryostat to Joule-heat either the heater or resistance thermometers. As P will be small, it is also necessary to avoid large thermal emfs in the electrical leads and joints. The thermal emfs may exceed the applied voltage and make control of P difficult unless a constant-current source of large internal impedance is used. It is, of course, important to reverse the direction of the current, I, so that thermal emfs in the potential leads may be canceled in the averaged value of E. These problems become more severe for some of the transient heat-capacity measurements discussed in Section 2.3 because, in those measurements, power is applied for very brief periods. Finally, as in any electrical circuit handling very small signals, the heater circuit should be grounded at only one point to avoid ground loops which may induce spurious 50 or 60 Hz currents to flow through the heater.

Most metallic heaters have temperature-dependent heat capacities which can be appreciable at low temperatures and, thus, increase the heat capacity addenda [13]. Also, the electrical leads to the heater are additional sources of heat capacity addenda, thermal conduction heat leaks, and vibrational heat leaks. These problems can be avoided to some extent by using the thermometer as a heater. If a resistance thermometer is used [14,15], the electrical leads must support the larger currents required for heating. Also, the resistance of the thermometer will vary with power level, P, since the temperature of the thermometer varies roughly as $T \propto (P)^{1/4}$. Thus, care must be exercised in determining the total heat, Q, delivered to the calorimeter.

If a magnetic thermometer is used, the paramagnetic material may be heated by applying a large current of high frequency to the mutual-inductance coils [16]. For metallic samples, eddy-current heating is a possibility [17]. Heating by γ-irradiation has also been used [18], but this will also heat other members of the cryostat. These three lead-less techniques do require a preliminary calibration using either an electrical heater or a sample of known heat capacity.

2.2.7. Sample Characteristics

The design of a calorimeter may be dictated by the size of the available samples, by the magnitude of their specific heat, and by their internal thermal time constant. Samples which are very small in size, or of irregular shape, pose special problems. Appropriate calorimeters are discussed in Section 2.3.

The magnitude of the specific heat can itself be troublesome. As an indication of the range of specific heats encountered at 0.1 K, that of Tb is a factor of $\approx 10^9$ larger than MgO. For very large heat capacities, the problem is one of adequate thermal contact to the refrigerator so that the sample can be initially cooled to low temperatures. Here it is important to provide a continuous path for electronic thermal conduction if the sample is metallic. If the heat capacity is very small, as for MgO, the heat capacity of heater, thermometer, and other addenda may exceed that of the sample. It is then imperative that precise measurements of the addenda heat capacity be made and that the calorimeter be compatible with this measurement. Even the heat capacity of small quantities of grease, used to enhance nonelectronic thermal contact, can be troublesome. The specific heats of greases and other amorphous dielectrics are appreciable at low temperatures, roughly $4T(\mathrm{J \cdot K^{-2} \cdot m^{-3}})$ for $T \lesssim 0.3$ K.

Generally the time, τ_{int}, required for relaxation to thermal equilibrium within the sample should be short relative to the time, τ_{ext}, it takes the sample, left to itself, to equilibrate thermally with the refrigerator. The internal time constant, τ_{int}, is proportional to the specific heat divided by the thermal conductivity of the sample. When this time constant is very long, it may be necessary to resort to a diffusive scheme of specific heat measurement. Another internal time constant which can be troublesome is that associated with certain excitations within the sample, excitations which may, for example, be associated with impurities. The excitations may have a long thermal relaxation time to the 'bath' of phonons or electrons in the sample. Examples of appropriate calorimetric techniques are presented in Section 2.3.

The application of a magnetic field during a calorimetric measurement on a superconducting or magnetic sample can cause problems. Superconducting leads, used to reduce conduction heat leak, may be transformed to the normal state having a much larger thermal conductivity. As the magnetic field varies in magnitude, eddy currents will heat any metallic component. Slight vibration of the calorimeter in a magnetic field will also cause eddy-current heating. The magnetic field will also shift the temperature calibration of most thermometers. To avoid this serious problem the thermometer may be placed on a stiff copper wire outside the field [19], the thermometer may be recalibrated at each field magnitude, or a capacitive thermometer may be used which is insensitive to magnetic fields [2].

Additional problems associated with specialized samples, such as radioactivity or thin films, are discussed in the following section.

2.3.   CALORIMETRIC TECHNIQUE

2.3.1.   Adiabatic Calorimetry

We define adiabatic calorimetry in this chapter to mean that the sample and addenda have been thermally isolated as completely as possible from the immediate environment within the cryostat. The heat capacity, C, of the sample plus addenda is given by the ratio $Q/\Delta T$, where Q is the measured quantity of heat added to the calorimeter and ΔT is the change in temperature caused by the addition of Q. The specific heat of the sample deduced from C is generally for constant (zero) pressure.

A very simple example of an adiabatic calorimeter [20] is shown in Fig. 2.1. Both the refrigeration and the thermometry are provided by the cerous magnesium nitrate (CMN) salt. As a magnetic thermometer, the paramagnetic susceptibility of the CMN is monitored by means of a set of mutual inductance coils (not shown in Fig. 2.1) and an a.c. mutual-inductance bridge [2]. This thermometer is calibrated against the vapor pressure of the liquid ^{4}He in the cryogenic bath at temperatures above 1 K. The temperature scale is extrapolated to lower temperatures assuming a paramagnetic T^{-1} temperature dependence of the CMN. Thermal contact to the bath is provided by ^{4}He gas in a vacuum chamber surrounding the calorimeter. Following calibration, the CMN is magnetized in a large magnetic field. The heat of magnetization is removed to the liquid-helium bath by the ^{4}He gas in the vacuum chamber. After the CMN and sample have cooled to the temperature of the bath, roughly 1 K, the ^{4}He gas is pumped away to thermally isolate the calorimeter. Reduction of the magnetic field cools the CMN, and the CMN in turn cools the sample through the copper thermal link, B, of Fig. 2.1. Any residual ^{4}He gas remaining in the vacuum chamber is gettered by the cold surfaces of the CMN pill. To measure a heat capacity, a measured Q is applied via the electrical heater, H, and the temperature change, ΔT, is recorded by means of the CMN thermometer.

There are several problems with this simple design. (i) The heat capacity of the addenda is large since a large quantity of CMN must be used to provide sufficient refrigeration. The specific heat of CMN has a temperature dependence of ≈T^{-2} and, thus, this correction becomes larger at lower temperatures. (ii) It is difficult to remove all the ^{4}He thermal-exchange gas from the vacuum chamber. Hence, measurements are limited to temperatures much lower than 1 K so that the remanent gas remains adsorbed to the CMN. (iii) To provide a large surface area for thermal contact between the CMN and the copper heat link, the CMN crystals were in the form of thin slabs. The resulting shape-correction [21] causes the magnetic thermometer to deviate from a paramagnetic T^{-1} temperature dependence and, hence, the extrapolated temperature scale is in error by ≈5% at 0.05 K. (iv) The calorimeter was utilized for metallic samples, yet only glue was used to provide a phonon thermal contact between sample and copper. This results in a large thermal impedance between sample and copper which varies with temperature as T^{-3}.

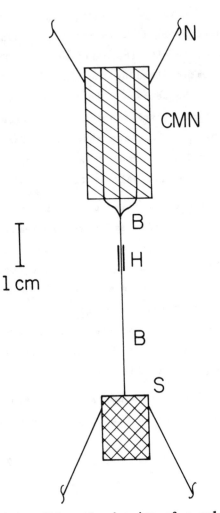

Figure 2.1.   Schematic drawing of a calorimeter for the temperature range 0.06-0.2 K. CMN, cerous magnesium nitrate demagnetization refrigerator <u>and</u> magnetic thermometer, in the form of thin slabs; B, copper thermal connector; H, electrical heater; S, sample; N, nylon-thread mechanical supports. Not shown are vacuum jackets, thermometry coils, the large magnet for demagnetization, or cryogenic-liquid baths.

These several problems have been reduced in the calorimeter design [11,12] of Fig. 2.2. The calorimeter is cooled using a demagnetization refrigerator or a dilution refrigerator. Thermal contact between calorimeter and refrigerator is provided by the superconducting thermal switch, W, of Fig. 2.2. The calorimeter may be recooled by energizing the solenoid, O, to 'close' this thermal switch. No He thermal-exchange gas is needed to cool the calorimeter and so measurements may continue to temperatures as high as ≈1 K.

The pair of carbon resistance thermometers of Fig. 2.2 serve to monitor the behavior of the calorimeter, for example, to be certain that no temperature difference occurs on the two sides of the sample. Accurate thermometry of high resolution is provided by the CMN thermometer which is calibrated against the vapor pressure of liquid ^{3}He. Liquid ^{3}He has the advantages of a higher vapor pressure and no troublesome superfluid film. The long support, B_3, for the CMN thermometer spatially removes the thermometer from other magnetic materials such as sample, S, and the superconducting solenoid, O. The spherical shape of the CMN thermometer was intended to reduce the shape-correction, but in fact does not [21]. Hence, a correction must be applied. A single-crystal sphere of CMN would not require as large a shape correction, but the area of contact to the copper wires in B_3 would be too small to provide adequate thermal contact and a rapid response of the CMN to temperature changes. To further emphasize the seriousness of this problem, it should be noted that the relaxation time, τ_{CMN}, of the CMN thermometer to the temperature of the sample is proportional to the thermal boundary resistance, R_B, at the surface of the CMN multiplied by its heat capacity, C_{CMN}. Since $C_{CMN} \propto T^{-2}$ and $R_B \propto T^{-3}$, $\tau_{CMN} \propto T^{-5}$ and becomes excessive at low temperatures unless the CMN is divided into many layers to increase the area of contact with the copper wires. The CMN has occasionally been crushed to a fine powder to further increase the area of contact [12].

Thermal contact between the metallic sample and the copper mount, B_2, and with the two copper buttons (H in Fig. 2.2 and the contact running to the CMN), were made with an InHg solder which is liquid at room temperature. The oxides were removed from the mating surfaces of copper and sample beneath a protective layer of the

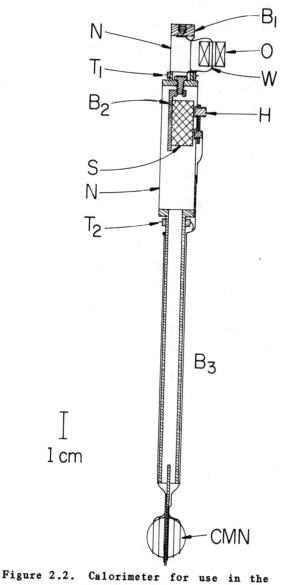

Figure 2.2. Calorimeter for use in the temperature range 0.03–1 K. B_1, threaded copper piece for thermal and mechanical attachment to refrigerator; N, thin-wall nylon tube for thermal isolation and mechanical support; W, tin-foil superconducting thermal switch; O, superconducting solenoid to activate switch; B_2, copper mount for sample S; T_1, T_2, carbon resistance thermometers; H, electrical heater on copper button; B_3, electrically insulated copper wires and epoxy to provide mechanical support and thermal contact to CMN magnetic thermometer.

InHg solder. The copper pieces were then lashed to the sample with nylon thread. The solder extruded allowing the sample and copper to touch, yet prevented the surfaces from oxidizing. This provided excellent <u>electronic</u> thermal contact to the sample. Had glues been used, the rare-earth samples could not have been cooled within a reasonable time limit because of their large heat capacities. For samples having smaller heat capacities, glues or greases have been used successfully to attach samples to the calorimeter [17,22-24].

For the calorimeter of Fig. 2.2, the heat capacity of the addenda alone could be measured. At all temperatures this was less than a 1% correction due to the large heat capacities of the rare-earth samples.

For the discussion thus far, it has been assumed that thermal isolation was complete, that no heat leak occurred. All calorimeters involve some heat leak; there is at least some contact to the refrigerator. An extreme case is depicted in Fig. 2.3. The effect is overly emphasized for the purpose of this discussion. In Fig. 2.3, heat, $Q = P\Delta t$, is applied to the calorimeter during the period Δt. The temperature rises from its initial value of T_i to a maximum value, $T_i + \Delta T_m$, then relaxes back to T_i with time constant, τ_{ext}, because of thermal contact with the refrigerator. A problem occurs because some variable quantity of heat flux, $p(t)$, escapes from the calorimeter even during the period Δt that the heater is turned on. The total quantity of heat added during the heating cycle is [25]

$$Q_{net} = P\Delta t - \int p(t)dt. \tag{1}$$

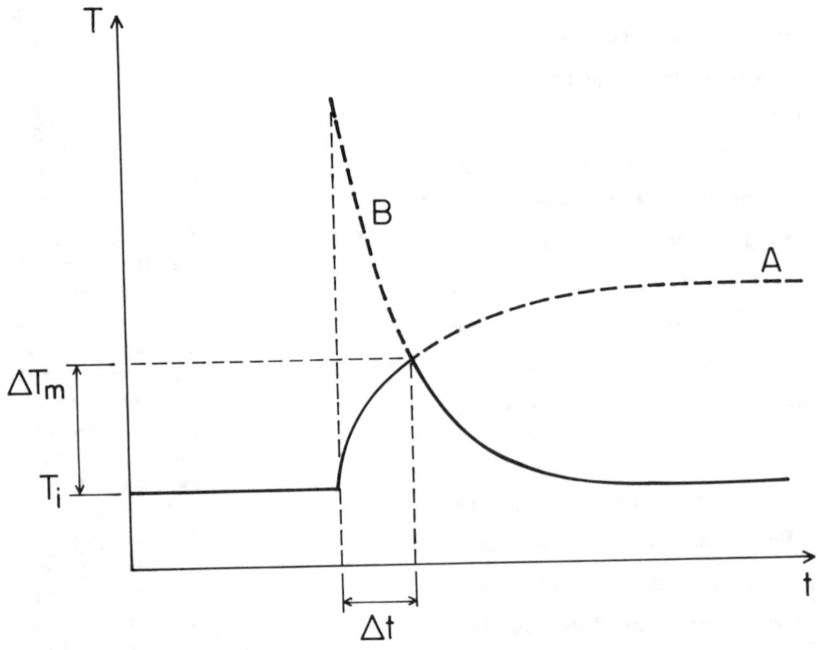

Figure 2.3.   Response of a calorimeter to a heat pulse of duration Δt if thermal isolation from the refrigerator is poor. Solid line, actual temperature excursion; line A, excursion were power left turned on; line B, extrapolation of cooling curve to time heater was turned on.

Since p(t) is a function only of the temperature, $P - p = C(T)dT/dt$. Thus,

$$Q_{net} = P\Delta t - \int_{T_i}^{T_i + \Delta T_m} p\ C(T)\ [P-p]^{-1}dT. \qquad (2)$$

Note that the integration is over the <u>measured</u> temperature excursion ΔT_m. The quantity p may be approximated from the drift curve measured after the heating period. A rough value for $\dot{Q}_{net}$ is calculated using an approximation for the heat capacity C(T) under the integral in Eq. (2). Then $C = Q_{net}/\Delta T_m$, and this approximate result for C(T) obtained over a range of temperatures is reinserted into the integral and the computational process iterated [25] for each temperature. This procedure may be applied even if there is a long-term drift in T both before and after application of heat. If C may be considered to be constant or temperature independent over the excursion ΔT_m, then the computation may be simplified to [26]

$$C = (P\ \tau_{ext}/\Delta T_m)\ (1 - e^{-\Delta t/\tau_{ext}}). \qquad (3)$$

If, in addition, the ratio $\Delta t/\tau_{ext}$ is small [27],

$$C = (Q/\Delta T_m)[1 - \Delta t/2\tau_{ext}]. \qquad (4)$$

Thus, the heating period Δt should be made small, but cannot be made shorter than the thermal-relaxation time for the heater to equilibrate with the sample. An important point in these computational schemes is that, since ΔT_m is the measured temperature excursion, no extrapolation of drift rates to within the heating period is required. However, any time constants internal to the calorimeter must be short relative to τ_{ext} [25]. The external relaxation time τ_{ext} may be calculated from the drift rate of line B in Fig. 2.3.

It is imperative that the net time constant of the electronic circuits used to measure temperature also be short relative to τ_{ext}. This means that, if τ_{ext} is small, a long averaging time cannot be used in a lock-in detector to improve the signal/noise performance of the thermometry.

2.3.2. Continuous-Heating Calorimetry

Having a well-isolated or adiabatic calorimeter, it is possible to use a continuous mode of operation. That is, heat is added at a constant rate P and the corresponding time derivative of the temperature, $\dot{T}$, is observed. Thus, $C = P/\dot{T}$ provided the thermal relaxation times between sample, heater, and thermometer are sufficiently short. If, for example, the relaxation time of the thermometer to the temperature of the sample is τ_T, then $\dot{T}$ should be $\leq 10^{-2}\ T\ \tau_T^{-1}$ for a thermometry error of $\leq 1\%$. A second precaution is that any residual heat leak must be small relative to the applied power P. The continuous-heating mode is especially useful for samples exhibiting very narrow peaks in the specific heat, peaks related to certain phase transitions.

One problem with the continuous heating technique occurs when resistance thermometers are used. A microcomputer must be available to convert the resistance reading to temperature if a continuous readout in real time is needed. This problem has been avoided by the ingenious adaptation [28] of the continuous mode shown in Fig. 2.4. No heater is used on the calorimeter;

rather, heat is extracted at a controlled rate
by means of the metallic thermal impedance Z between the calorimeter and the refrigerator. The
refrigerator is maintained at a low temperature
T_c monitored by thermometer T in Fig. 2.4.
Since the low-temperature thermal conductivity
of a metal is linear in temperature, the heat
flow through the metallic thermal link Z is $P = D(T_h^2 - T_c^2)$ where T_h is the temperature of the
sample as detected by the CMN magnetic thermome-
ter. In operation $T_c \ll T_h$, hence $P = D T_h^2$.
Since $P = C \dot{T}_h$, $D T_h^2 = C \dot{T}_h$ or

$$C = D T_h^2 \dot{T}_h^{-1} = D[d(T_h^{-1})/dt]^{-1}. \qquad (5)$$

But the mutual inductance of the magnetic ther-
mometer is given by the expression $M = M_o + GT^{-1}$. The heat capacity of sample and addenda
is, therefore, simply $C = DG/\dot{M}$, where $\dot{M}$ is the
time derivative of M. The constant G is ob-
tained from the calibration of the CMN thermometer, while D is measured in a preliminary experiment using an electrical heater. Of course,
the thermal impedance of link Z must be much
larger than the thermal impedance between sample
S and support B_2, and any residual heat leak
must be small. Also, the thermal response time
τ_{CMN} of the CMN thermometer must be small, as
for any continuous-heating calorimeter.

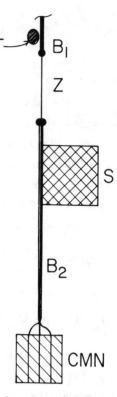

Figure 2.4.   Schematic drawing of a 'continuous-heating' calorimeter
for the temperature range
0.03–1 K. CMN, magnetic
thermometer; B_2, copper
thermal path; S, sample; Z,
brass thermal impedance; B_1,
copper thermal link to refrigerator; T, γ-ray (or
other) thermometer.

The calorimeter of Fig. 2.4 may be raised in temperature so as to repeat a run. To accomplish this the CMN is heated by temporarily increasing the frequency and amplitude of the current
in the mutual-inductance coil.

2.3.3. Transient Calorimetry

The residual heat leaks to the adiabatic calorimeters discussed in Sections 2.3.1 and 2.3.2
have fallen into the range of $\approx$1–10 erg·min^{-1}, or 2–20 nW. The drift rate for the temperature of
an adiabatic calorimeter under the influence of this heat leak is determined by the magnitude of
the sample heat capacity. Assume a heat leak of $\approx 10^{-9}$ W to a sample consisting of 1 cm^3 of Tb
metal at 0.1 K. The drift rate would be $\approx 10^{-8}$ K·s^{-1} and measurements would be both precise and
easy to perform. If MgO were to be measured under the same conditions, the drift rate would be
≈ 6 K·s^{-1}. A measurement of heat capacity under these conditions would be impossible even using
the computational techniques discussed in Section 2.3.1.

Measurement of a small heat capacity at low temperatures, therefore, requires an approach other than adiabatic calorimetry. Several alternative schemes have been used. In each scheme the sample is coupled rather tightly to the refrigerator by means of a thermal link. This has several advantages. (i) Small samples or samples of small heat capacity may be measured. (ii) The sample is rapidly cooled to low temperatures. (iii) There is no need for a thermal switch. (iv) The sample is in good thermal contact with the refrigerator. Hence, the primary thermometer may be placed on the refrigerator, while working thermometers of small thermal mass are placed on the calorimeter and calibrated in situ. (v) The mechanical support is often more rigid, leading to a reduction in vibrational heating. (vi) A measurement is obtained in a short period of time, and, therefore, different signal-averaging techniques become practical for improving the signal/noise in thermometry. (vii) The 'calorimeter' is small and, thus, two or more may be placed in the cryostat.

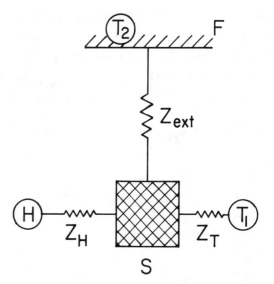

Figure 2.5 shows, schematically, the arrangement for a transient measurement [29-33]. In principle, the arrangement is the same as for an adiabatic calorimeter, except now the external thermal impedance to the refrigerator, Z_{ext}, is made small intentionally. The response to a heat pulse is, therefore, that shown in Fig. 2.3, and the analysis of the data may be done in the same way as discussed in Section 2.3.1. That is, if the heat capacity C is essentially constant over the temperature excursion ΔT_m of Fig. 2.3, then $C = (Q/\Delta T_m)[1-\Delta t/2\tau_{ext}]$ where τ_{ext} is calculated from the drift curve after the heat pulse is completed. Since τ_{ext} may now be rather short, a digital signal averager may be used to greatly improve the signal/noise performance [31]. That is, many pulses of the type shown in Fig. 2.3 are averaged. In this situation it is helpful to have τ_{ext} roughly independent of temperature. Therefore, the thermal link should be selected such that the

Figure 2.5. Schematic arrangement for heat capacity measurements by a transient method. F, refrigerator; S, sample; H, heater; T_1, T_2, thermometers; Z_{ext}, Z_T, Z_H, thermal impedances between sample and refrigerator, thermometer or heater, respectively.

temperature dependence of the product $C\,Z_{ext}$ is small. For metals, as an example, both the specific heat and thermal conductivity are often proportional to T at low temperatures. Thus, the thermal link could be a wire or foil of the same metal.

An alternative method [34] used to compute C is to measure Z_{ext} of Fig. 2.5 by applying constant power P to the sample and measuring the temperature drop between sample and refrigerator using thermometers T_1 and T_2, i.e. $Z_{ext} = (T_1-T_2)/P$. After P is turned off, τ_{ext} is carefully measured from the time dependence of the drift curve B in Fig. 2.3. Then, $C = \tau_{ext}/Z_{ext}$.

The calorimeter might consist only of the sample, with heater and thermometer, suspended by nylon threads (or by the electrical leads) which serve as the thermal link. Or a nonmetallic sample may be pressed against its mechanical support allowing the thermal boundary resistance at the interface to serve as the 'thermal link'. Alternatively, a permanent calorimeter may be constructed to which the sample is attached mechanically and thermally with a small amount of grease. An example of a permanent calorimeter [35] is shown in Fig. 2.6. Note that all members are stretched taut to reduce the vibrational heat leak.

This discussion has assumed that the heat capacity of the addenda is small or can be measured in a separate experiment. The addenda includes the heater, thermometer, greases or glues, and 1/3 of the heat capacity of the thermal link [36]. It has also been assumed that the response times of the heater and the thermometer are short relative to τ_{ext}. The effect of these response times on a determination of the heat capacity of a sample has been discussed in the literature [36].

If the time for the sample itself to equilibrate is long, that is, if τ_{int} is large, then an alternative scheme must be sought. One technique [37] is to thermally connect a slab of the sample material to the refrigerator as shown in Fig. 2.7(a). Heat is applied to the heater H and the thermal conductivity λ calculated from $\lambda = P\,L_1/A(T_2-T_1)$ where A is the cross-sectional area of the sample and L_1 is the separation of thermometers T_1 and T_2 which read temperatures T_1 and T_2. After the power P is turned off, τ_{ext} is calculated from the drift curve which is similar to curve B of Fig. 2.3. Then the heat capacity of the sample is

$$C = \pi^2\,\lambda\,\tau_{ext}\,(A/4L) \tag{7}$$

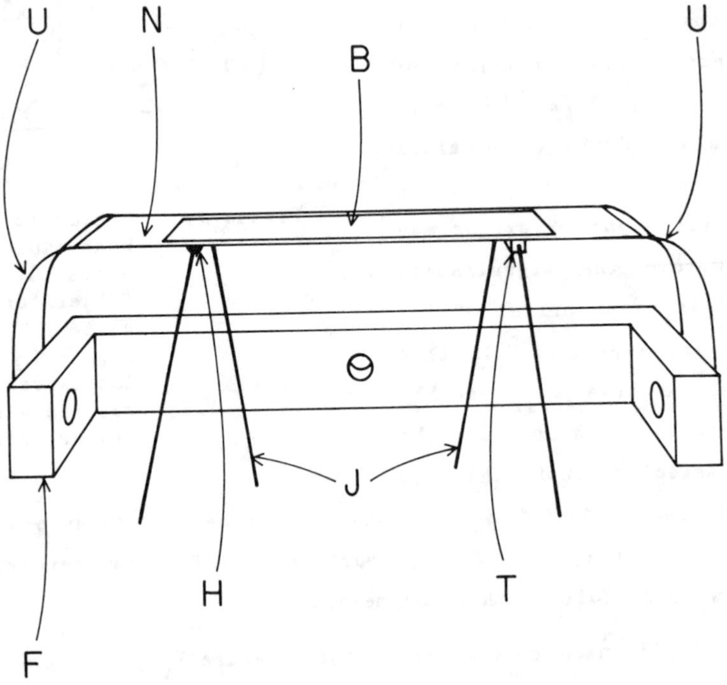

Figure 2.6.   Calorimeter used in a transient mode.   F, copper mount attached mechanically and thermally to refrigerator; U, phosphor bronze springs; N, 10^{-3} cm thick Mylar film to provide mechanical support and a weak thermal link to mount F; B, 10^{-3} cm thick copper foil; H and T, heater and thermometer with NbTi superconducting leads J stretched taut.   Sample is applied to top surface of B.

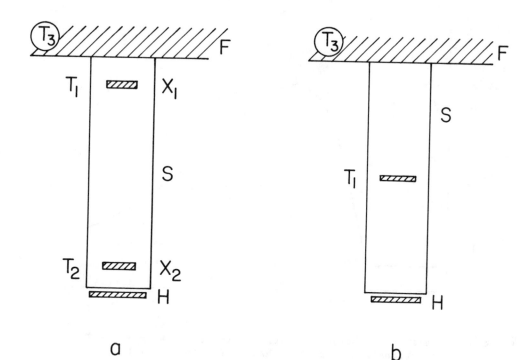

Figure 2.7. Arrangements for transient measurement of heat capacities of samples having small thermal diffusivity. F, refrigerator; S, sample; H, electrical heater; T, thermometers. In (b), thermometer T_1 is placed <u>midway</u> between the ends of the sample.

where L is the <u>total</u> length of the sample. In this technique, the thermal bond to the refrigerator must be of uniform quality over the entire end of the sample. Should part break away, the calculated heat capacity will be too large although the temperature dependence may be nearly correct. An advantage of this technique is that λ is also obtained in the same cryogenic run, should that property be useful to the investigator. If λ is very small, L may have to be small to prevent a residual heat leak from heating the sample. Such heating would prevent a calibration of the thermometers in situ.

Two other techniques are applicable to samples of small diffusivity, although they appear not to have been utilized at temperatures below 1 K. Consider the arrangement of Fig. 2.7(b). If a short heat pulse of total energy Q is applied to heater H, the response of thermometer T_1 located midway between the ends of the sample will be that shown in Fig. 2.8. The heat capacity of the sample may be obtained from [38]

$$C = 0.96 \ Q/\Delta T_m \tag{8}$$

where ΔT_m is the measured excursion of the temperature shown in Fig. 2.8. Alternatively, the heat capacity may be computed from [39,40]

$$C = 1.09 \ Q \ f_1 \ f_0^{-2} \tag{9}$$

where

$$f_n = \int_0^\infty t^n \ \Delta T(t) \ dt \tag{10}$$

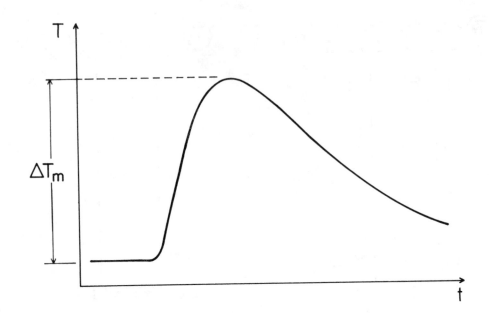

Figure 2.8.   Recording of temperature T versus time t at thermometer T_1 in Fig. 2.7(b)
after application of a short heat pulse to heater H.

and $\Delta T(t)$ is the time-dependent excursion from equilibrium shown in Fig. 2.8. Since this computation is done by a computer, the computer can also fit and extrapolate an exponentially decreasing
ΔT beyond the right edge of Fig. 2.8. Hence, measurements need not be extended to infinite time.
It is argued that this second, integral technique of Eqs. (9) and (10) is superior to using simply
the relation $C = 0.96 \ Q/\Delta T_m$ of Eq. (8) since all of the data are utilized and the signal/noise
ratio is thereby improved. It is possible, using a more involved computation, to obtain C even if
the heat capacity of thermometer and heater are not negligible [40].

      For either scheme (a) or (b) of Fig. 2.7, it is essential that the thermal boundary resist-
ance between sample and refrigerator be small compared to the thermal impedance of the sample,
that is, small relative to $L/(\lambda A)$. It should also be noted that subtle errors may arise if λ is
strongly temperature dependent and, in addition, the temperature of the heater becomes large
compared to the temperature of the refrigerator [41]. In this situation thermal phonons having
different frequencies may diffuse through the sample at different rates, and thus the response of
the thermometers to the heat pulse would be difficult to interpret.

      Radioactive samples produce their own internal heat leak. The relaxation methods described
in this section, which use a link of small thermal impedance attached to the refrigerator, may be
required first to cool the sample to low temperatures and then to permit measurements to be obtained at low temperatures [42].

                              2.3.4.   a.c. Calorimetry

      The transient methods discussed in Section 2.3.3 allow an improvement to be made in signal/
noise through the use of digital signal averaging. Before suitable signal averagers become
readily available, only the lock-in or synchronous-detection method was available for improving

signal/noise. This method utilizes a single a.c. frequency, and it was basically for this reason that low-temperature a.c. calorimetry was developed.

The a.c. calorimeter may be represented by Fig. 2.5. A sinusoidally varying current $I_0 \cdot \cos(\omega t/2)$ is applied to the heater of resistance R, and so the power provided to the sample is

$$P = I_0^2 R \cos^2(\omega t/2) = P_0 \cos^2(\omega t/2).$$ (11)

The temperature of the sample becomes

$$T(t) = T_3 + \frac{1}{2} P_0 Z_{ext} + \Delta T_{ac} \cos(\omega t - \phi)$$ (12)

where T_3 is the temperature of the refrigerator, ϕ is a phase constant of no immediate importance, and ΔT_{ac} is the magnitude of a sinusoidal variation in temperature. This magnitude is given by [43]

$$\Delta T_{ac} = \frac{P_0}{2\omega C} \left[1 + (\omega^2 \tau_{ext}^2)^{-1} + (\omega^2 \tau_{int}^2) \right]^{-1/2}$$ (13)

where τ_{int} is the dominant or longest relaxation time within the calorimeter. The time constant τ_{int} may arise from a small thermal diffusivity of the sample or from the thermal relaxation of thermometer or heater to the temperature of the sample.

It may be seen from Eq. (13) that, in using the a.c. technique, the applied angular frequency should be large relative to τ_{ext}^{-1}, but small relative to τ_{int}^{-1}. Then $C = P/2\tau\Delta T_{ac}$. The magnitude of $\omega\Delta T_{ac}$ may be observed as ω is varied to be certain that there is no frequency dependence, meaning that the correct frequency has been selected for the calorimeter. The a.c. component ΔT_{ac} is much smaller than the d.c. component of the temperature change, namely $(1/2)P_0 Z_{ext}$, but the sensitivity available in lock-in detection permits measurement of the small a.c. portion. Appropriate instrumentation has been discussed in the literature [43,44].

If the thermal diffusivity of the sample is too small, and hence τ_{int} is too large, for the above a.c. technique, an alternative method using an arrangement similar to Fig. 2.7(a) can be applied [45,46]. Again, a current $I_0 \cos(\omega t/2)$ is applied to the heater. The temperature at any point x measured along the sample from the heater is given by

$$T(x,t) = T_3 + \Delta T_{dc} + \Delta T_{ac}$$ (14)

where T_3 is again the temperature of the refrigerator. The time-independent term is

$$\Delta T_{dc} = (1 - \frac{x}{L}) \frac{L}{2\lambda} \frac{P}{A}$$ (15)

with L the length and A the cross-sectional area of the sample. The time-dependent term has the x and t dependence given by

$$\Delta T_{ac} = Y_0 e^{-qx} \cos(\omega t - qx)$$ (16)

where $q = \sqrt{\omega C/2\lambda \ AL}$. By measuring the amplitudes Y_1 and Y_2 of the time-dependent variation at two positions x_1 and x_2 in Fig. 2.7(a), the heat capacity may be obtained from the equation

$$C = \left(\frac{2\lambda AL}{\omega}\right) (x_2 - x_1)^{-2} \left[\ln \frac{Y_1}{Y_2}\right]^2. \tag{17}$$

As in other techniques in which the diffusivity is obtained, the thermal conductivity λ must be measured independently either form the difference in ΔT_{dc} of Eq. (15) measured at the two thermometers, or by actually applying a d.c. power to the heater and measuring λ using thermometers T_1 and T_2 as in the standard manner. Note that this a.c. diffusive measurement of heat capacity provides directly $C/(AL)$, namely the specific heat per unit volume, as does the diffusive measurement of Eq. (7).

Alternatively, the heat capacity of the sample may be calculated from the phases ϕ_1 and ϕ_2 measured at thermometers T_1 and T_2 or, equivalently, by the phase difference $\Delta\phi = \phi_2 - \phi_1$:

$$C = \left(\frac{2\lambda AL}{\omega}\right) (\Delta\phi)^2 (x_2 - x_1)^2. \tag{18}$$

In a.c. measurements utilizing amplitude or phase changes as in Eqs. (17) or (18), the separation of the thermometers must be small relative to the distance L–x between thermometer and the refrigerated edge of the sample. However, computation [46] indicates an error of <1% if $(L-x_1) >$ 1.5 $(\omega C/2\lambda L)^{1/2}$. A separation of this magnitude also reduces the effects of poor thermal bonding between sample and refrigerator. A poor bond would produce an additional, constant term in $T(x,t)$ meaning that the measurements cannot be carried to as low a temperature as in the case of a good thermal contact. If the contact is poor, or if a sample of low thermal conductivity is excessive in length, any extraneous heat leak will raise the temperature of the sample. Hence, the thermometers cannot be calibrated in situ.

Note that both the a.c. diffusive technique discussed here, and the d.c. diffusive technique discussed in Section 2.3.3, provide a measurement of thermal conductivity. Thus, when both C and λ are needed, only one experimental run is necessitated.

2.3.5. Miscellaneous Examples of Low–Temperature Calorimetry

It is difficult to make thermal contact to a sample consisting of many irregular particles. Therefore, a slurry is often used consisting of liquid ^{4}He [47], silicon oil [48], petroleum ether [49], or grease [50]. A large number of small–diameter copper or silver wires are inserted into the slurry to provide thermal contact with the refrigerator.

Helium itself has attracted many experimental investigators. At temperatures above 0.06 K the calorimeter container for the liquid or solid He may be a spherical copper shell [51]. At temperatures to 0.003 K liquid He has been mixed with CMN powder [52]. The CMN serves as a magnetic thermometer; the powdered form provides a large surface area and reduces the thermal relaxation time of this thermometer. In some experiments the CMN also has served as an adiabatic demagnetization refrigerator, having first been cooled to $\lesssim 0.02$ K with another refrigerator. Since the thermal boundary resistance to the wall of the CMN–He cavity varies as T^{-3}, this resistance alone can provide reasonable thermal isolation, eliminating the need for a thermal switch. For temperatures below ≈ 0.003 K, some special means of refrigeration is required such as the direct compressional cooling of liquid ^{3}He [53]. The relatively large thermal conductivity of liquid ^{3}He is helpful in providing a fast τ_{int}.

Films of He are also of interest. To provide a measurable heat capacity, the total area of film must be large. Therefore, the calorimeter is generally filled with sintered copper powder, an expanded graphite (grafoil) [54], or a porous vycor glass [55]. For metallic or superconducting films, such large surface areas are generally not possible. Because of the small mass of sample available in a film of ≈ 1 cm^2 area, one of the transient or a.c. techniques is generally employed to measure the specific heat of a metallic film.

Measuring the heat capacity of a sample under high pressure is difficult because of the large addenda heat capacity associated with the massive calorimeter. The most simple approach, and one used successfully [56], is to apply the pressure at room temperature. The pressurized calorimeter is then cooled to low temperatures.

Some samples have a long internal relaxation time even when the thermal conductivity is large. This long τ_{int} is associated with some internal degree of freedom not coupled closely to the electrons or phonons of the sample. Examples include some nuclear specific heats [57], the quantum-mechanical tunneling modes of impurities [58], or the localized excitations found in amorphous materials [59]. If it is desired to measure the specific heat contributed by these sources, the duration of the heat-capacity measurement must be long compared with the τ_{int}. Thus, a transient scheme may not be possible. On the other hand, a fast or transient measurement may provide the phonon and/or conduction electron contributions unmasked by a (possibly) larger, extraneous contribution.

2.4. SUMMARY

Many techniques have been devised to measure the specific heats of diverse materials under various conditions at temperatures below 0.3 K. All of the needed instrumentation is available commercially, or it may be constructed based on an extensive literature. The experimentalist involved in calorimetry is, therefore, freed to concentrate on the physical or chemical properties of the materials of interest. The low-temperature limit for these 'standard' calorimetric measurements is currently ≈ 0.01 K. General calorimetric measurements to lower temperatures will require different means of refrigeration and thermometry than those described in the present chapter.

2.5. ACKNOWLEDGMENT

Calorimetry in the author's laboratory has been supported in part by the Materials Science Division of the Department of Energy under Contract DE-AC02-76ER01198.

2.6. REFERENCES

1. Hill, R.W., in Experimental Cryophysics (Hoare, F.E., Jackson, L.C., and Kurti, N., Editors), Butterworths, London, England, 264-74, 1961. See also Hill, R.W., in Progress in Cryogenics (Mendelssohn, K., Editor), Vol. 1, Academic, New York, NY, 179-206, 1959.

2. Anderson, A.C., Rev. Sci. Instrum., 51, 1603-13, 1980, and papers cited therein.

3. Metrologia, 15, 65-77, 1979.

4. Soulen, R.J., Jr., J. Phys. (Paris), 39, C6-1166-8, 1978.

5. Schooley, J.F., J. Phys. (Paris), 39, C6-1169-70, 1978.

6. Cotts, E.J. and Anderson, A.C., J. Low Temp. Phys., 43, 437-43, 1981.

7. Gobrecht, K.H., Veyssie, J.J., and Weil, L., Proc. 1966 Low Temp. Calor. Conf. (Lounasmaa, O.V., Editor), Suomalaimen Tiedakatemia, Helsinki, 61-8, 1966.

8. Roach, P.R., et al., Rev. Sci. Instrum., 46, 207-9, 1975.

9. Mueller, R.M., Buchal, C., Oversluizen, T., and Pobell, F., Rev. Sci. Instrum., 49, 515-8, 1978.

10. Krusius, M., Paulson, D.N., and Wheatley, J.C., Rev. Sci. Instrum., 49, 396-8, 1978.

11. Krusius, M., Anderson, A.C., and Holmstrom, B., Phys. Rev., 177, 910-6, 1969.

12. Collan, H.K., Krusius, M., and Pickett, G.R., Phys. Rev., B1, 2888-95, 1970.

13. Cieloszyk, G.S., Cote, P.J., Salinger, G.L., and Williams, J.C., Rev. Sci. Instrum., 46, 1182-5, 1975.

14. Fisher, R.A., Brodale, G.E., Hornung, E.W., and Giauque, W.F., J. Chem. Phys., 68, 169-84, 1978.

15. Fritz, J.J., Goto, T., and Kuntz, R.L., J. Chem. Phys., 70, 2554-9, 1979.

16. Nikitin, L.P., Kogan, A.V., Kulkov, V.D., and Shiryapov, I.P., Sov. Phys.-JETP, 22, 714-6, 1966.

17. Miedema, A.R., Wielinga, R.F., and Huiskamp, W.J., Physica, 31, 1585-90, 1965.

18. Daybell, M.D., Pratt, W.P., and Steyert, W.A., Phys. Rev. Lett., 21, 353-6, 1968.

19. Lambert, M.H., Brock, J.C.F., and Phillips, N.E., Phys. Rev., B3, 1816-25, 1971.

20. Kogan, A.V., Nikulin, E.I., and Patrikeev, Yu.B., Sov. Phys.-JETP, 12, 1565-7, 1971.

21. Anderson, A.C., J. Appl. Phys., 39, 5878-83, 1968.

22. Keyston, J.R.G., Lacaze, A., and Thoulouze, D., Cryogenics, 8, 295-300, 1968.

23. Phillips, N.E., Phys. Rev., 114, 676-85, 1959.

24. Andres, K., Phys. Rev., B2, 3768-71, 1970.

25. Collan, H.K., Heikkila, T., Krusius, M., and Pickett, G.R., Cryogenics, 10, 389-93, 1970.

26. Fagaly, R.L. and Bohn, R.G., Rev. Sci. Instrum., 48, 1502-4, 1977.

27. Durek, D. and Baturic-Rubcic, J., J. Phys., E5, 424-8, 1972.

28. Miedema, A.R. and Haseda, T., Supp. Bull. Instit. Int. Froid, Commission, 1, Londres, 1961, Annexe, 159-65, 1961-5; Physica, 27, 1102-12, 1961.

29. Harrison, J.P., Rev. Sci. Instrum., 39, 145-52, 1968.

30. Gobrecht, K.H. and Saint Paul, M., in Proc. Int. Cryo. Eng. Conf., Berlin, 1970, Iliffe, Guilford, 235-7, 1970.

31. Sellers, G.J. and Anderson, A.C., Rev. Sci. Instrum., 45, 1256-9, 1974.

32. Gerber, J.A., Sample, H.H., and Neuringer, L.J., J. Appl. Phys., 47, 2134-42, 1976.

33. Lasjaunias, L.C., Picot, B., Ravex, A., Thoulouze, D., and Vandorpe, M., Cryogenics, 17, 111-7, 1977.

34. Schutz, R.J., Rev. Sci. Instrum., 45, 548-51, 1974.

35. Anthony, P.J. and Anderson, A.C., Phys. Rev., B16, 3827-8, 1977.

36. Bachmann, R., et al., Rev. Sci. Instrum., 43, 205-14, 1972.

37. Roth, E.P. and Anderson, A.C., J. Appl. Phys., 47, 3644-7, 1976.

38. Bertman, B., Heberlein, D.C., Sandiford, D.J., Shen, L., and Wagner, R.R., Cryogenics, 10, 326-7, 1970.

39. Alterovitz, S., Deutscher, G., and Gershenson, M., J. Appl. Phys., 46, 3637-43, 1975.

40. Gershenson, M. and Alterovitz, S., Appl. Phys., 5, 329-34, 1975.

41. Phillips, W.A., J. Appl. Phys., 51, 3583-5, 1980.

42. Smith, T.L., Anthony, P.J., and Anderson, A.C., Phys. Rev., B17, 4997-5008, 1978.

43. Sullivan, P.F. and Seidel, G., Phys. Rev., 173, 679-85, 1968.

44. Zally, G.D. and Mochel, J.M., Phys. Rev., B6, 4142-50, 1972.

45. Zavoritsky, N.V., in Progress in Cryogenics (Mendelssohn, K., Editor), Vol. 1, Academic, New York, NY, 207-18, 1959.

46. Kogure, Y. and Hiki, Y., Jpn. J. Appl. Phys., 12, 814-22, 1973.

47. Rayl, M., Vilches, O.E., and Wheatley, J.C., Phys. Rev., 165, 692-7, 1968.

48. Conway, M.M., Phillips, N.E., Geballe, T.H., and Kuebler, N.A., J. Phys. Chem. Solids, 31, 2673-8, 1970.

49. Suzuki, H. and Koyama, M., Cryogenics, 15, 38-9, 1975.

50. Azevedo, L.J., et al., J. Phys. (Paris), 39, C6-365-6, 1978.

51. Greywall, D.S., Phys. Rev., B18, 2127-44, 1978.

52. Abel, W.A., Anderson, A.C., Black, W.C., and Wheatley, J.C., Physics, 1, 337-87, 1965.

53. Halperin, W.P., Rasmussen, F.B., Archie, C.N., and Richardson, R.C., J. Low Temp. Phys., 31, 617-98, 1978.

54. Bretz, M., Dash, J.G., Hickernell, D.C., McLean, E.O., and Vilches, O.E., Phys. Rev., A8, 1589-615, 1973.

55. Tait, R.H. and Reppy, J.D., Phys. Rev., B20, 997-1019, 1979.

56. McWhan, D.B., Remeika, J.P., Bader, S.D., Triplett, B.B., and Phillips, N.E., Phys. Rev., B7, 3079-83, 1973.

57. Collan, H.K., Krusius, M., and Pickett, G.R., Phys. Rev. Lett., 23, 11-3, 1969.

58. Sellers, G.J., Anderson, A.C., and Birnbaum, H.K., Phys. Rev., B10, 2771-6, 1974.

59. Laponen, M.T., Dynes, R.C., Narayanamurti, V., and Garno, J.P., Phys. Rev. Lett., 45, 457-60, 1980.

CHAPTER 3

Calorimetry in the Range 0.3–30 K

D. L. MARTIN

3.1. INTRODUCTION

Accurate calorimetry in the 0.3–30 K temperature range has used the same methods as at higher temperatures, i.e., isoperibol and adiabatic calorimeters. However, there are changes in detail for the following reasons:

(1) The very stable platinum resistance secondary thermometer used at immediately higher and overlapping temperatures becomes insensitive as temperature is reduced and, hence, another thermometer is required. The best available today, germanium resistance, can be selected to give good sensitivity over the 0.3–30 K temperature range but may not be reliably stable.

(2) In general, heat capacities are tending to zero as the temperature is reduced. In the lower part of the present range it is found that heat inputs from vibration and radio-frequency generators become significant and may be comparable with those required for the calorimetric measurements. If this stray heat input is approximately constant it may be impossible to cool the calorimeter to the desired temperatures, and there may be thermometry problems owing to calibration errors or self-heating. Intermittent stray heat inputs would clearly degrade the accuracy of measurements. Thus, efforts must be made to reduce these stray heat inputs to negligible levels.

(3) The use of helium exchange gas as a thermal contact medium, either within or around the calorimeter, is fraught with danger. In the former case heats of desorption or evaporation may be comparable with the heat capacity of the calorimeter, and in the latter case there might also be spoiling of the vacuum surrounding the calorimeter, thus leading to uncompensated heat exchange.

(4) Particularly at the lower end of the present temperature range there is a tendency to use isoperibol rather than adiabatic calorimetry. The latter technique is made more difficult by the lack of a sensitive thermocouple for adiabatic shield control, while isoperibol calorimetry becomes more accurate because the different thermometry and heating requirements make the use of thinner (or even superconducting) leads possible and this, together with the very small radiant heat exchange, makes for excellent thermal isolation of the calorimeter from the surroundings.

The temperature range of the present chapter reflects (a) the possible use of fairly conventional techniques, (b) a lower temperature limit attainable with a fairly simple ^{3}He cryostat, (c) an upper temperature limit for accurate germanium thermometry, (d) good overlap with techniques described in the preceding and following chapters, and (e) the fact that a great deal of useful physical information can be obtained from measurements confined to the present temperature range. However, measurements confined to a small part of the present range (e.g., the 'liquid helium range,' ~1–4 K) are liable to misinterpretation because it may then become difficult or impossible to identify correctly all the terms contributing to the specific heat.

Apart from the steady development and improvement of the 'traditional' methods of calorimetry, as discussed above, the last two decades have seen a great deal of work done on 'novel' methods of calorimetry, such as a.c. heating and relaxation methods, particularly directed to the measurement of very small samples. These developments are surveyed in Section 3.8 of this chapter.

In the present chapter is is assumed that the reader is conversant with the general principles of low temperature calorimetry as described in the next chapter. An earlier review of calorimetry [1], while somewhat out of date in places, is still recommended as good background reading. The literature survey for the present chapter was terminated early in 1981, but a few more recent references have been added.

3.2. CRYOSTATS

3.2.1. Location of Cryostat

Some care is advisable here in order to minimize unwanted vibration and radio-frequency heating.

3.2.1.1. Minimization of Vibrational Heating

Heating arises from inelastic dissipation of vibration energy. Vibration amplitudes increase with altitude in a building and, hence, location of equipment on the lowest floor is desirable. If this floor is cast directly onto ground or fill, then vibration should be well damped. Alternatively, the cryostat might be mounted on a pier supported by a large concrete block isolated from the base ground or fill by a thick layer of dry sand. In this case the pier cross-section should be maximized and the height minimized for optimum performance. Another technique uses a large concrete block (~1 ton) set on air springs in order to decouple from local vibrations [2].

Mechanical pumps should be isolated from the cryostat by flexible couplings and some means of damping vibrations (e.g., a sand-box) (Note that single metal bellows may tend to transmit vibration on evacuation and are best used in a counterbalanced or damped configuration [3] to minimize coupling.) Mercury diffusion pumps may be a nuisance owing to 'bumping' [4].

3.2.1.2. Minimization of Radio-Frequency Heating

A basement location on the side of the building away from local commercial transmitters makes the best use of building shielding and may be sufficient even under relatively poor external conditions. Equipment within the building might also be a source of trouble. Typical noise sources are local radio-frequency generators, electric welding sets, and other unsuppressed sparking equipment such as small motors, Tesla coils, and flashers. Ultrasonic equipment might also be a problem, as may computers and computer terminals. A shielded room [5] might be necessary in some cases and is, in some ways, desirable in all cases.

3.2.2. Refrigerants Used in Cryostat

At one time liquid hydrogen was used as a refrigerant at the upper end of the present temperature range and as a shield for liquid helium. However, improvements in cryostat design, such as the 'economizer' [6] which heat exchanges cold evaporated helium gas with the heat conducted down tubes and leads, and the ready availability and relative safety of liquid helium have all but eliminated liquid hydrogen. The low-temperature stage will usually be shielded with liquid nitrogen.

3.2.3. Cryostat Design

Ordinary liquid helium (^{4}He) boils at about 4.2 K at atmospheric pressure and can be pumped to much lower temperatures. However, there is a phase change to a superfluid at about 2.2 K, and the superfluid film present below this temperature may greatly increase the evaporation rate. One method of overcoming this problem is to pump through a small orifice [7] which restricts the film flow. Another problem which often arises is that vibrations develop in the gas in the pumping tube and transmit heat to the helium bath [8]. These have often been overcome with baffles, but a simple and effective method is to hang a piece of darning wool down the pumping tube in the cryostat [9]. Using a cryostat with a single ^{4}He container (surrounded by a liquid nitrogen bath as heat shield), it is usually possible to make calorimetric measurements from about 2.5 K when a small pump is used to pump down the helium. For lower temperatures a second ^{4}He container is often used. This is pumped down while the first ^{4}He container remains at atmospheric pressure, forming a shield at 4.2 K. Alternatively, a single ^{4}He container with a large pumping system may be used. In either of these ways temperatures of about 1 K are available. For still lower temperatures a further container, for ^{3}He, is used. The ^{3}He gas is condensed into the container by contact with the pumped ^{4}He stage at about 1 K. The ^{3}He container is then pumped to reduce the temperature. Only a small quantity (~1 liter at STP) of ^{3}He gas is required, since there is no superfluid transition within the temperature range of interest. Using a mechanical pump alone, temperatures below 0.4 K are available. It has been reported that with careful attention to the size of the pumping tubes and use of a diffusion pump, temperatures close to 0.2 K are attainable [10]. The potential problem of vibration heating from mechanical and diffusion pumps has been discussed in Section 3.2.1.1 above. Thus, the technique of using activated charcoal as a pump is attractive. A description of a typical ^{3}He cryostat of this type, reaching about 0.3 K, has been given [11]. Vibration problems may also be minimized by ensuring a rigid connection between the various stages of the cryostat. (Connecting tubes should be well spaced and it might be advisable to have extra, blank tubes to the lowest stage, which might otherwise have only a single tube supporting it.)

The calorimeter area must be shielded from higher temperature radiation with suitable traps or baffles. If exchange gas is to be used during any stage of the cooling, then attempts must be made to maximize the pumping speed of this system. The various methods of cooling the calorimeter to the lowest temperature will be considered in detail later; these must be allowed for in the cryostat design. The remaining cryostat design problems are the provision for temperature indicators for the various stages, for electrical leads to the calorimeter, and for vapor pressure thermometers for thermometer calibration and/or stability checks. Temperature indicators take many forms. The commonest are (a) vapor pressure thermometers with bulb connected to a dial gauge and permanently filled, and (b) carbon radio resistances, used as thermometers, which are stuck to the can and connected to a Wheatstone bridge circuit with the off-balance indicator a small sensitive meter. The electrical leads to the calorimeter should be chosen to minimize stray thermal voltages and should be well insulated to avoid leakage. The former requirement is often met by using thin copper wire to as low a temperature as possible (4.2 K is quite common) and an alloy such as manganin at lower temperatures. Wires with several layers of insulation are advisable. Quadruple Formel has proved very satisfactory. Bending and kinking wire is to be avoided to minimize thermals which arise from nonuniformities in the wires. It is necessary to thermally anchor the wires, preferably at several temperatures, in order to minimize heat leaks and heat load in the lower temperature stages. One method runs the wires through the helium bath, but good feedthrough seals are then necessary, especially if the helium is to be pumped to the superfluid phase. The other method is to run the leads through the vacuum pumping lines, anchoring (noninductively) a good length (~1 m) at several points. Typically, the anchor points might be the helium-economizer and the 4.2 K stage ^{4}He can. Similarly, the tubes connecting to vapor pressure bulbs are often run through the vacuum pumping lines and thermally anchored at several points. Care is necessary that none of these points is at a lower temperature than that of the bulb.

The cryostat design will also include any special features required (e.g., for measurements to be made in a magnetic field, for condensing in a gaseous sample, etc.).

3.2.4. Cryostat Construction

There are a number of commercial cryostat suppliers, and it is worth considering such a unit. For example, see Ref. [12]. Nevertheless, most cryostats continue to be constructed locally with the aid of resident expertise or one of the well-known texts on low temperature techniques covering both cryostat design and construction [13].

3.3. THERMOMETRY

Ideally, calorimetric measurements should be made on the thermodynamic temperature scale. Various practical scales, thought to approximate to this ideal, have been proposed from time to time. Only in the last few years has there been a coordinated international effort to establish or improve the temperature scale over the whole of the 0.3-30 K region. Hence, calorimetric results in the literature are based on an assortment of scales. This section starts with a brief review. For a recent general review on measurement of temperature, see Hudson [14].

3.3.1. Review of Temperature Scales in the Range 0.3-30 K

The first International Temperature Scales (1927, 1948) had the oxygen point (~90 K) as the lowest calibration temperature. Various scales were established below this temperature using gas thermometry. Probably that most widely used in calorimetry was the National Bureau of Standards 1939 Provisional Scale, which was slightly modified and distributed as the NBS 1955 Provisional

Scale [15]. This scale extended down to 10 K and was carried on platinum resistance thermometers. The International Practical Temperature Scale of 1968 (IPTS-68), also carried on platinum resistance thermometers, was extended down to 13.81 K, the triple point of hydrogen [16]. Simultaneously, with the activity at higher temperatures, various ^{4}He vapor pressure scales were published, the most widely used being that of 1958 (T_{58}) [17]. Soon afterward, a ^{3}He vapor pressure scale was recommended for general use (T_{62}) [18]. Acoustic thermometry at the NBS first showed that the helium vapor pressure scales were inconsistent with temperatures on the NBS 1955 and IPTS68 scales. This acoustic scale has been distributed, via germanium thermometers, as NBS 2-20 [19]. More recently, there have been other acoustic, magnetic, and very precise gas thermometer scales established in the region 2–30 K; for a summary, see Berry [20]. These have confirmed the error in T_{58} and T_{62}, but have also shown that NBS 2-20 is not smooth. A redetermination of both ^{4}He and ^{3}He vapor pressure scales has been reported recently [7], and there has been international agreement on a provisional temperature scale in the 0.5–30 K range (EPT-76) [21]. Apart from these national and international scales, much published calorimetry in the 4–20 K range has been based on 'in-house' scales determined locally using gas thermometers. Careful reading of a published paper should serve to determine which scale has been used. Some of these 'in-house' scales continue in use, one reason being that changes in International Scales are not always for the better [22]. A second reason is to facilitate comparisons of measurements made at different times in the same laboratory [22]. In this connection, it is interesting to note that there was some compromise in deriving the EPT-76 scale, since one objective was that it should be continuous with IPTS-68 at 27.1 K and departure from thermodynamic temperature was permitted in order to achieve this objective [21].

3.3.2. Conversion Between Temperature Scales

The results of accurate calorimetry may be significantly different when evaluated on different temperature scales. There are two basic effects: (a) a difference in the assigned absolute temperature, and (b) a difference in the size of the degree. The problem has been considered in detail by Douglas [23], who showed that, if the differences between the temperature scales are small, the following equation gives the difference in specific heat at the same numerical value of temperature T on each scale:

$$\delta C_p = -\mu \frac{dC_p}{dT} - C_p \frac{d\mu}{dT} \tag{1}$$

where μ is the difference (new minus old scales) at temperature T on the old scale. An independent derivation [24] arrived at the same result and shows the effect of scale changes on results derived from calorimetric measurements in the present temperature range [24,25].

3.3.3. Secondary Thermometers

Early calorimetry in this region used various metallic thermometers, often with some kind of accidental superconducting inclusion. In the early 1950's it was found that some types of carbon radio resistors made very sensitive and fairly stable thermometers for the present temperature

region. However, there were usually significant calibration shifts on each cooling. Such thermometers are now mainly confined to use in high magnetic fields where their low magnetoresistance
is an advantage [26]. The present availability of well-known types of carbon resistors has been
discussed by Rubin [27]. More recently, carbon-glass thermometers have been developed. Their
magnetoresistance is very low [26], but their stability is an order of magnitude inferior to that
of germanium thermometers [28]. Encapsulated germanium resistance thermometers [29] are the most
popular thermometers for the 0.3–30 K range. Early work suggested that they were very reliable
and stable, and led to a false sense of security. Recently, careful thermal cycling experiments
have shown that while many germanium thermometers are very stable others show drifting or irregular behavior on successive coolings, and that a period of stable behavior is no guarantee of continued stability [30]. Hence, careful selection before use should be followed by a calibration
check each time the thermometer is cooled. The calibration shifts appear to correspond roughly to
a temperature-independent proportional change of resistance. Hence, owing to the lower sensitivity here, the calibration shift is greatest at the higher temperatures. Germanium thermometers
should be handled with care. Heating during installation must be avoided; thus, the wires should
be 'heat-sinked' during soldering. It is best to order with a gas filling that will be condensed
out at low temperatures (e.g., N_2), to avoid heat effects associated with helium exchange gas
[31,32]. Adequate thermal contact between the leads and calorimeter is essential. Both carbon
and germanium thermometers are very susceptible to heating by radio-frequency fields. Apart from
the choice of location, discussed in Section 3.2 above, it might be possible to suppress the
effect by the installation of condensers, ferrite beads [31], or filters [33].

Another thermometer, the Rh-Fe resistance thermometer, is being used in standards laboratories [34] and has been used in calorimetry from 6 K up [35]. This thermometer is sensitive over
the whole of the present temperature range and is reported to be very stable. However, the commercial units are much more massive than germanium thermometers, use helium gas for thermal contact, and are less sensitive at the lower temperatures of the present range, necessitating an
unacceptably larger power dissipation. However, the installation of a calibrated thermometer of
this type in the cryostat would serve as an excellent means for checking the stability of other
thermometers. A recent development, the Rh-Fe resistive SQUID, has been used as a thermometer
down to 0.01 K [36]. Thus, new forms of the Rh-Fe thermometer may be useful in very low temperature calorimetry.

3.3.4. Thermometer Calibration Points

Although calibrated thermometers are available from national laboratories and commercial
sources, it is advisable to perform routine calibration checks. These points also form the basis
for 'in-house' calibrations. In the present temperature range both vapor pressure data and superconducting fixed points may be used.

3.3.4.1. Vapor Pressure Thermometry

In the present range neon, hydrogen, and both isotopes of helium are used. The reference
points are strictly the triple and boiling points of neon and hydrogen, a hydrogen point at

(25/76) standard atmosphere, and the boiling point of ^{4}He. For hydrogen a catalyst is used to ensure that the para/ortho ratio is at its low-temperature equilibrium. For ^{3}He a high-purity vapor pressure standard gas is available from the Mound Laboratory, Monsanto Research Corporation, Miamisburg, Ohio 45342, U.S.A. Except for ^{4}He below the λ-point, a small vapor pressure bulb is used. This is connected to a manometer, or other pressure measuring device, via a tube with diameter usually chosen as a compromise between tolerable heat leak, space available, and minimization of the thermomolecular pressure correction. This correction is made using the well-known Weber-Schmidt equation [37]. (For recent discussion of the accuracy of this equation, see Ref. [7] and [20].) There should be a radiation trap in the connecting tube and adequate thermal anchoring, taking care that no point is colder than the bulb. For ^{4}He below the λ-point, the superfluid film flow and evaporation may lead to refluxing and heating of the vapor pressure bulb. It is, therefore, best to measure the pressure of the pumped ^{4}He can rather than use a separate bulb; for an example, see Ref. [7]. Alternatively, all vapor pressure measurements below the λ-point should be made with ^{3}He. The pressure-measuring equipment should be at a stable temperature and a correction made for changes in the density of the manometer liquid with temperature. For mercury, see Ref. [17] and for butyl phthalate, see Ref. [38]. An aerostatic head correction for the weight of gas in the sensing tube must also be considered.

3.3.4.2. Superconducting Transition Points

Since the transition temperature depends on the purity and state of strain of the material, it is recommended that the Superconducting Fixed Point Device SRM767, available from the U.S. National Bureau of Standards, be used [39]. This contains samples of Cd, Zn, Al, In, and Pb and permits the accurate realization of superconducting transition temperatures in the ~0.52 to ~7.2 K range, provided that the magnetic field at the device is close to zero. This requirement may be met using Helmholtz coils [39] or a magnetic shielding material surrounding the device [40]. These devices are individually tested for uniformity of transition temperature within ±1 mK [39]. Other materials in the present temperature range may become available [39]. For recommended values for the transition temperatures, see Ref. [21].

3.3.5. Thermometer Calibration

The calibration points listed in Section 3.3.4 are adequate for calibrating a thermometer in the 0.5–4.5 K range; above 14 K a calibrated platinum thermometer or hydrogen vapor pressure may be used for comparison. However, the only fixed point available between 4 and 14 K is the lead point (~7.2 K), and reliable interpolation with the available secondary thermometers is not possible. The required number of calibration points may be obtained by comparison with a thermometer calibrated on one of the recognized scales or with an 'in-house' interpolation instrument. Barber [41] has pointed out the advantage of calibrating a gas thermometer at both limits of the interpolation interval.

Various relations to represent the calibration of carbon [42] and germanium [43] thermometers have been proposed. The necessity of weighting the observations in a least squares fit is related to the calibration interval [44]. In fitting a calibration over a wide temperature range to a single equation it may be necessary to use a difference table also [44].

3.3.6. Measuring Thermometer Resistance

Whatever method is used, care must be taken that the measuring current does not result in significant self-heating (so that the calibration is altered) and that the thermometry heat input is much less than the heat inputs used for the calorimetric measurements. Direct current measurements are made with conventional or inductive divider-type potentiometers, bridges, or digital voltmeters (DVM). In all cases care must be taken that the appropriate sensitivity is combined with adequate stability and linearity and awareness of thermal voltages, including those generated by the Peltier effect, which reverse direction with change of current direction. The latter have been observed in germanium thermometers but are not of significance in the present temperature range [45]. This problem is avoided with a sufficiently high frequency a.c. measurement, and modern electronics has provided integrating phase sensitive detectors with good noise rejection. Hence, there is an increasing tendency to use a.c. methods. These may be square-wave or sine-wave and include homemade potentiometric and bridge devices and self-balancing, high-precision, commercial potentiometers and bridges [46]. It is an advantage to use inductive dividers as ratio devices since, unlike resistors, the readings do not vary with time or temperature. Care is necessary in using a.c. techniques to measure platinum resistance thermometers at low temperatures. The low resistance necessitates relatively large currents, and effects, possibly due to eddy currents, have been seen [47]. Care is also necessary with the lead wires in the cryostat, twisted pairs being recommended [47,48].

3.3.7. Checking or Extrapolating a Thermometer Calibration Using Specific Heat Measurements

Deviation of the temperature scale from thermodynamic temperature will result in anomalies in specific heat (see Section 3.3.2 of the present chapter). Thus, if the form of the specific heat is known the temperature scale can be adjusted until the correct result is obtained. There are obvious dangers in this process if systematic errors are being made in the specific heat determination. For example, in extrapolating a scale to very low temperatures, it is possible that there is an extraneous supply of heat when the heater is switched on and off (e.g., resulting from a sparking relay) that is negligible at higher temperatures but significant at the lower temperatures. Then an error could be built into the extrapolated scale to compensate this unsuspected perturbation. For an example of the extrapolating technique, see Ref. [49]. This method was also used to suggest that some of the temperatures originally assigned to the superconducting transitions in the NBS SRM 767 Device were incorrect [40]. Later work has shown this to be so [21].

3.3.8. Note on Thermocouples

Thermocouples are not recommended for temperature measurement in this range, but are used differentially in adiabatic calorimeters as a sensor for shield control. While various materials have been used in the past the current recommendation is for gold-iron versus chromel thermocouples [50]. The sensitivity is around 10 $\mu V \cdot K^{-1}$ in the 1–30 K range. Leads from the low temperature area to the outside of the cryostat should be of copper with good thermal anchoring around the junction to the thermocouple wires.

3.4. CALORIMETER DESIGN

Most accurate calorimetry in the 0.3–30 K temperature range has been done with calorimetric apparatus of the 'traditional' type, basically the same as that used at higher temperatures and described in the appropriate chapters of this book. (The theory and design relating to the more recent, 'novel,' methods of calorimetry will be considered later in Section 3.8 of the present chapter.)

In the 'traditional' apparatus the sample is contained in, or is otherwise thermally attached to, the 'calorimeter vessel.' For the sake of brevity the term 'calorimeter' will be taken to mean the 'calorimeter vessel.' The calorimeter, surrounded by a high vacuum, is suspended inside a larger container, the thermal shield. The shield may be held at a constant, or almost constant, temperature; the apparatus is then referred to as an isoperibol calorimeter [51]. Alternatively, the temperature of the shield is always maintained as close as practicable to the calorimeter temperature. This ensures, in theory if not in practice, that there is no heat exchange between the calorimeter and its surroundings, and the apparatus is described as an adiabatic calorimeter. In the literature there are cases where an isoperibol calorimeter with low heat exchange to the environment is described as 'adiabatic.' In the present chapter the term 'adiabatic' will be restricted to apparatus with a controlled shield. Heat is supplied to the calorimeter electrically either in a discontinuous or a continuous mode, and the heat capacity is calculated from the observed temperature increment or rate of change.

The basic principles of calorimeter design may be summarized as follows:

(1) To ensure good thermal equilibrium within the calorimeter. Then the thermometer will correctly indicate the sample temperature, uncertainties in heat exchange will be minimized, and measurements can be made more quickly.

(2) To minimize heat exchange between the calorimeter and its environment. Then corrections are minimized.

(3) To maximize the ratio of sample heat capacity to empty calorimeter heat capacity, thus increasing the accuracy of the sample measurement.

(4) To facilitate easy loading of the sample.

(5) To make the design as universal as possible so that one calorimeter can be used to measure a large variety of samples. This minimizes the time required for installing and measuring calorimeters.

(6) To minimize the number of 'corrections' required for unmeasured material on the calorimeter (such as varnish, solder, etc.).

3.4.1. Thermal Contact Between the Sample and Calorimeter

At higher temperatures this is usually achieved by using an 'exchange gas' within the calorimeter vessel. As detailed in Section 3.1 of the present chapter, such methods are best avoided at the lower end of the present temperature range. For an indication of the errors observed, see Ref. [52]. A second method, also best avoided, is to attach a heater and thermometer directly to the sample with adhesive and estimate the heat capacity of this 'addenda.' The problem here is the difficulty of making an accurate estimate. In order to avoid this problem the thermometer and heater are usually mounted on a piece of copper to which the sample can be attached. This is the 'calorimeter' and the heat capacity can be measured in the absence of a sample. In some cases the thermometer and heater may be carried on separate parts of the calorimeter which can be bolted together in order to determine the heat capacity of the 'empty' calorimeter. Some examples of methods which have been used will now be described. The first seven examples apply to a sample in the form of a single solid block.

Where the method involves the use of copper, it may be convenient or essential to be able to calculate the heat capacity of this component. This is acceptable provided that degassed, high-purity copper is used with great care to avoid contamination (see Section 3.4.7).

3.4.1.1. The 'Tray' Technique

The thermometer and heater are underneath a very light rigid copper surface, preferably machined to be quite smooth. The sample is stuck to the tray with grease [53]. Typically, 0.01 cm^3 of Dow Corning silicone grease may be used. The amount of grease is conveniently dispensed with a hypodermic syringe. Measurements on the 'empty' calorimeter are made with this amount of grease in place. The method works well for a wide variety of samples but fails when the differential thermal contraction becomes too great. Other materials, such as MgO and Al_2O_3 [54], have been used for the tray in order to reduce the heat capacity, and some workers use varnish to attach the sample.

3.4.1.2. The 'Copper Stud' Technique

This is a way of extending the 'tray' method to samples with a larger differential contraction. The sample is drilled and tapped to take a copper stud which sits on the tray calorimeter [55]. Some of the measured amount of silicone grease is used on the stud threads.

3.4.1.3. The 'Differential Contraction' Technique

In the copper stud technique, differential contraction assures thermal contact between the sample and copper stud. It has also been used, for example, in alkali metal calorimetry where the sample is cast, under vacuum, into a copper container with a copper wire running through it [56]. On cooling, the sample contracts onto the wire and makes excellent contact. However, the contraction may also distort the container. This has led to problems when the container itself was being used as the sample in the 'tray' technique [57].

3.4.1.4. The 'Differential Expansion' Technique

With this technique the sample expands into contact on cooling. As an example, a light copper cage was used round an InBi single crystal [58]. This material expands in one direction on cooling.

3.4.1.5. The 'Clamp' Technique

The sample is held in a metal clamp. These have been of many types [59,60], and one or more clamps may encircle the sample or the sample may be gripped between plates. Clearly, the sample may be strained more with this technique than with some of the others. Early methods used massive clamps, but this has proved to be unnecessary.

3.4.1.6. The 'Spring Loaded Foot' Technique

A number of 'feet,' of small contact area, are pushed against the sample [61]. Contact is made with grease and the small contact area minimizes the risk of breakage owing to thermal contraction. If the feet are on springs (beryllium copper), relative movement is possible. This arrangement also minimizes stress on fragile samples.

3.4.1.7. The 'Plug' Technique

A plug containing the thermometer and heater is screwed into a tapped hole in the sample [62]. There may be two plugs, one carrying the thermometer and the other the heater.

3.4.1.8. The 'Loose Powder' Technique

The powder is sealed, without exchange gas, into a container [63] or just put in a 'bucket' [64]. In some cases, which must be determined by experiment, equilibrium is quite fast. The calorimeter should, of course, be equipped with internal fins to reduce the heat conduction path.

3.4.1.9. Compressed Powder Techniques

The powder may be compressed in a copper container which is then attached to the calorimeter [65]. A polymer powder has been compressed with copper powder to improve the thermal diffusivity [66]. Copper vanes compressed with the sample were used for thermal contact. A compressed metallic powder has been used without container or vanes [67].

3.4.1.10. The Heat Exchange Medium Technique

Instead of an exchange gas another material, solid at low temperatures, is used. Glycerol [68,69], silicone oil [70], and petroleum ether [71] have been used in this way. The heat capacity of the contact medium must be measured since the exact composition may depend on supplier, exposure to air, and so on. Unless the sample under investigation is a good thermal conductor it might be necessary to add some copper to promote thermal equilibrium.

It will be noted that in most of these methods the sample is subject to some abuse, either by room temperature compression or by differential contraction as the apparatus is cooled to the measurement temperature. It is possible that in some cases the effect may be significant. For example, a copper band contracted onto a V_3Si crystal altered the specific heat by up to 20% [72].

In some cases it might be necessary to seal the calorimeter containing the sample (e.g., to protect the sample from oxidation or to prevent loss of sample on evacuation of cryostat). If possible a method avoiding the addition of unmeasured material is advisable. For a powder sample or a material which can be poured into the calorimeter (e.g., a molten metal) a small filling tube can be used and, if copper, can be closed with a cold-welding tool [73]. If the tube has been specially made of very high-purity, degassed copper (see Section 3.4.7) a correction for the copper lost can be made. This may be preferable to using solder and flux. An expert can heli-arc seal a flattened tube end with negligible weight change. For larger lids a seal using a gold gasket has been described [74] and appears preferable to another design using Loctite [75]. The sealing is often done with the calorimeter evacuated. Earlier discussion showed that the use of helium gas is best avoided. Care must also be taken to remove air, which can give rise to a large heat capacity anomaly at low temperatures [76].

3.4.2. Calorimeter Thermometer

At the present time this is usually a commercial encapsulated germanium thermometer which relies on heat conduction through the leads for thermal equilibrium. Hence, the lead wires from the thermometer should be copper and should be well stuck to the calorimeter for a length of several cm. The leads between the calorimeter and its surroundings will usually be of a poor thermal conductor. For measurements in magnetic fields a carbon thermometer is preferable and may be calibrated at zero field against an 'in situ' germanium thermometer. For further details on thermometry see Section 3.3 of the present chapter.

3.4.3. Calorimeter Heater

The big problem here is to ensure that all the measured energy is actually supplied to the calorimeter. For this reason the calorimeter heater resistance will usually be high ($\sim$1 kΩ) and the leads between the calorimeter and its surroundings will be a compromise between low electrical resistance and the high thermal resistance required for good calorimeter isolation. Very long leads may give rise to long time effects in lead heat appearing at the calorimeter and are best avoided. In the absence of radiation from the leads, it can be shown that half the heat generated in the leads finds its way to the calorimeter and half to the surroundings. Hence, one potential lead is often connected to one current lead where it leaves the surroundings and the other potential lead is connected to the other current lead where it leaves the calorimeter. Then, ideally, the heat generated in the leads is automatically measured. However, this will not be so if the leads are not well anchored thermally at shield and calorimeter [77] nor if there is significant lead heating leading to radiation from the leads [78]. In the lowest part of the present range some problems can be circumvented by using superconducting leads. These may be of several materials. Niobium has been used, as has lead plated (or solder tinned) constantan or similar material. Niobium cannot be soft soldered but is otherwise more reliable, since the plating on the constantan might develop cracks [79]. However, this limitation is not serious for low-current applications [79]. The heater and thermal tie-down should be noninductive to avoid interaction with stray fields.

Some workers have used an electronic constant-current supply for the heater but others prefer the simplicity of dry, mercury, or lead-acid batteries. Usually the current is fed into a 'dummy' heater, when the heater is not switched on, in order to stabilize the source. The heat supplied to the calorimeter is calculated from a time measurement (usually electronic) and measurement of the voltage across the heater and across a standard resistance in series with the heater. These measurements used to be made with a manual potentiometer, but the use of a digital voltmeter (DVM) is now more common. This must be chosen to have a high input impedance, so that there is no significant shunting of the heater. An integrating instrument is best, since each reading is the result of an integration over several periods of the local a.c. power supply, thus minimizing the effect of a.c. pickup. A number of DVM readings can be made over a heating period in order to take account of small changes of heater resistance, etc., during the period. This problem can also be minimized by choosing a heater material such as Evanohm [80], which has a very low temperature coefficient of resistance in this range. The apparatus should be checked for extraneous heat inputs by operating with zero heater current. There have been cases where relay operation, etc., has caused heat to appear in the calorimeter. It has also been reported that some DVM's will feed heat into the calorimeter (see Section 2.2.6 in the preceding chapter and Ref. [81]). The voltage readings are checked by frequent reference to a standard cell. This standard cell and the current measuring resistor should be checked against standards on a routine basis, and linearity checks of the potentiometer or DVM should be made similarly.

3.4.4. Thermal Isolation of the Calorimeter

The factors contributing to heat exchange between the calorimeter and its surroundings will be considered here.

3.4.4.1. Radiation

The surroundings of the calorimeter should be at as uniform a temperature as possible. Radiation traps should be placed in cryostat tubes so that radiant heat from higher temperature is dissipated in the helium bath, preferably with an earlier trap in the helium 'economizer,' if one is used. The shield immediately surrounding the calorimeter should be a good thermal conductor in order to minimize temperature gradients.

3.4.4.2. High Vacuum

There are two basic ways of obtaining a satisfactory vacuum: (a) To rely on cryogenic pumping. Everything except helium will be condensed out if a tight system is cooled to liquid helium temperatures. Since the concentration of atmospheric helium is not negligible the system should be evacuated to a rough vacuum at room temperature before cooling. (b) To use an active vacuum system. This is essential if an exchange gas is to be used during any stage of the cooling process. If exchange gas is used with any surface below about 6 K it is suggested that measurements should be confined to temperatures below that at which the exchange gas was removed [40,69,82,83]. The hazards of using exchange gas have been detailed earlier, and it is important that an efficient vacuum system be used. It is instructive to use a mass spectrometer type leak detector to monitor the removal of helium [84]. Many hours pumping may be required. Occasional slight heating ('baking') of the surfaces involved will accelerate the removal of gas.

3.4.4.3. Conduction

All wires and tubes leading to the calorimeter should be well anchored thermally (and non-inductively) at several points including, finally, the shield round the calorimeter. The type, length, and thickness of wires and tubes is a compromise between the requirements to minimize lead resistance and maximize thermal isolation. The latter requirement can be calculated from the maximum temperature offsets expected with the heat capacity anticipated. The mass and length of wires and tubes should be minimized to avoid long-time effects (see Section 3.4.3). The calorimeter is usually suspended by threads from above and below with tension applied by damped springs. A taut suspension usually minimizes vibration heating. Various workers prefer nylon, cotton, or blended threads (polyester/cotton). The latter are strong and easy to tie.

3.4.5. Cooling the Calorimeter. Thermal Switches

As detailed in Section 3.4.4.2, exchange gas will not be used in cooling the calorimeter to its lowest temperature. The following methods are used instead.

3.4.5.1. To Rely on Thermal Conduction Through Leads

The calorimeter surroundings are cooled and the calorimeter slowly follows. This might be rather slow but, if the lead length between calorimeter and shield in a well-controlled adiabatic calorimeter has been minimized, the time might not be excessive. The method has the merit of simplicity but calibration checks of the thermometer on the calorimeter are not very certain.

3.4.5.2. To Use a Vapor Pressure 'Pot' on the Calorimeter

The pot might have only one tube or there might be two tubes to allow circulation of helium gas. This method gives rapid cooling and an excellent method for checking the calorimeter thermometer calibration. The disadvantage is that it takes considerable time to obtain a good vacuum in the 'pot' system after cooling. There is always the possibility of small heat effects connected with desorption, but 'flash' heating after boiling off the helium and obtaining a fairly good vacuum serves to remove most of the helium gas. In any case, the surface area is small and the main vacuum surrounding the calorimeter is unaffected. A duplicate set of measurements in the same apparatus using lead conduction for cooling will serve to check for errors.

3.4.5.3. Mechanical Heat Switch

This is probably the most popular method at present. With the switch closed there is fairly rapid cooling and with the switch open there is excellent thermal isolation. The negative factors are (a) complication, (b) heat generated on opening switch, typically 100 ergs though values as low as 10 ergs have been reported, and (c) the thermal contact may not be very satisfactory at the lowest temperatures. Defect (c) is often used to circumvent (b) by making the lowest temperature specific heat measurements with the switch closed.

There are a number of varieties of mechanical heat switch:

(1) The calorimeter is lowered [85] or raised [12] into thermal contact with the surroundings. The method is relatively simple but the thermal contact is less good than that obtainable with other methods.

(2) A wire or strip projecting from the calorimeter is thermally clamped to the surrounding shield. Force is applied via a wire or rod tensioned at room temperature by either a mechanical or pneumatic arrangement. The clamp may be a single lever with mechanical advantage [69,86] or may be a more complicated 'scissors' arrangement of a double jaw clamp. Reference [87] reviews a number of switches of the latter type and discusses the various contact materials used.

(3) Bellows at low temperature. This activates directly a movable jaw which grips a wire or strip projecting from the calorimeter. For an example, see Ref. [59]. Presumably, the installation of a switch of this type is relatively simple, but there must be considerable thermal disturbance when the bellows is filled with liquid helium under pressure. It is claimed [59] that the heat generated on opening the switch is less than that observed with the mechanical type described in the preceding paragraph.

3.4.5.4. Superconducting Heat Switch

Well below the zero-field superconducting transition temperature, the thermal conductivity of a strip or wire of pure metal in its superconducting state is much less than that of the same metal in its normal state (produced by application of a magnetic field). Lead (T_c ~7.2 K) is often used as switch material, with a field of ~0.08 Tesla required to operate the switch. The field can be produced by a superconducting magnet inside the cryostat or a normal magnet outside. The field should be kept away from the calorimeter (possibly by using superconducting screens), so that the thermometer calibration is not affected. The dimensions of the switch material are chosen to give a reasonable cooling rate balanced by reasonable isolation at the higher temperatures, where the ratio of thermal conductivity in the normal and superconducting states approaches unity. Superconducting switches have the advantages of simplicity, but their effectiveness decreases with increasing temperature. However, the advent of DVM measurements enabled heating periods to be made shorter, and good results with a lead heat switch may often be obtained up to about 3 K, provided there is good thermal equilibrium within the calorimeter. If this is not so, then errors may arise owing to differences between the sample and thermometer temperatures [61] and to uncompensated heat exchange [58,88]. Nonlinear drifts may be observed at higher temperatures and must be handled properly in evaluating results [40]. For a further discussion of these points, see Section 3.11.

In summary, the first three methods are available for use over relatively wide temperature ranges, while the last method is restricted to use below about 3 K, where its simplicity and good performance are an attraction.

3.4.6. Checking the Thermometer Calibration 'In Situ'

As discussed earlier, this should be done every time the apparatus is cooled, and serves to check not only the thermometer stability but also the rest of the temperature measuring equipment. The method used depends on the way the calorimeter is cooled (Section 3.4.5). This is most straightforward when there is a vapor pressure pot on the calorimeter. If the vapor pressure pot is connected thermally to the calorimeter through the heat switch, a measurement of the thermal conductivity of the switch together with an estimate of stray heat leaks (vibration, radio-frequency, and thermometer circuit current), which can be made with the switch open, leads to an estimate of any temperature gradients between the thermometers. The thermometer check is much more difficult when lead conduction is used for cooling. It is hazardous to use the reading of a differential thermocouple for transfer because of possible unknown stray voltages in the thermocouple circuit and unknown stray calorimeter heating (vibration, etc.). Thus, a thermocouple setting for zero temperature drift of the calorimeter may be a cold offset to compensate for vibration heating. Similarly, exchange gas may not give temperature equality for various reasons.

The possibility of using a Rh–Fe thermometer as a 'built-in' temperature standard for the present range was discussed in Section 3.3 above.

3.4.7. Checking the Calorimeter Operation. Standard Sample

When the calorimeter has been designed and constructed, or whenever strange results are obtained, it is useful to be able to test the calorimeter in some accepted way. The recommended method is to use a sample of known heat capacity. For the present temperature range, high-purity copper is the accepted material. This was originally supplied by the Argonne National Laboratory [89] but is now available from the U.S. National Bureau of Standards [90]. The results obtained for copper depend, of course, on the temperature scale in use: see Ref. [24].

Copper was chosen because the high-purity material is relatively cheap, a good conductor, and is easily machined. However, care must be taken not to heat the material since it is easily contaminated with transition metals which lead to large specific heat anomalies at low temperatures [91] owing to the Kondo and spin-glass phenomena. It was also found that high-purity copper as received from the supplier may show specific heat anomalies and heating effects [92–94] owing to hydrogen contamination at the 1 ppm level. For this reason the standard material is vacuum heat treated before distribution.

If the calorimeter test yields an anomalous result for copper it is important that the reason be found. The practice of using a 'correcting factor,' based on the known and measured properties of copper, to correct measurements on other materials is unlikely to yield the correct result because of differences in the operating parameters (e.g., ratio of sample and 'empty' calorimeter heat capacities).

3.4.8. Note on Differential Calorimeters

In these calorimeters two or more samples are measured at the same time. They have been used in searching for the small effects of dilute alloying on specific heat [95]. There was an advantage in this method when unstable carbon thermometers were used. Then all the thermometers could be calibrated at the same time at the same points. Even if the calibration had systematic or random errors in temperature, all thermometers were equally affected and any differences in the specific heats of the two or more samples could be identified more readily than if separate calibrations of the thermometers had been made. With the advent of relatively stable germanium thermometers this advantage disappears. Small differences of suspension and wiring could make the effects of stray vibration and radio-frequency pickup different in the different sample assemblies. There appears to be no advantage in using a differential calorimeter in this temperature range. (However, the method is commonly used at much higher temperatures where radiant heat exchange is important.) Rather small differences (<<1%) between samples can be identified with conventional calorimetry; see, for example, Ref. [96].

3.4.9. Note on Measuring Condensed Gases

The problem here is that the calorimeter is usually filled in situ at low temperatures and care is necessary to ensure that the measured amount of gas is, in fact, in the calorimeter and not condensed in the filling tube. The calorimeter is filled through a heated tube and it is arranged that the calorimeter is the coldest surface seen by the gas. When the gas has been liquified in the calorimeter further cooling will produce the desired solid. The filling tube must then be thermally locked to the surface surrounding the calorimeter. Account must be taken of differential contraction between the solidified gas and the calorimeter material to ensure good thermal contact. A good description of a typical apparatus and filling procedure is given in Ref. [97]. An alternative procedure, avoiding some problems, is to fill and seal the calorimeter before insertion in the crystal [98]. It has been reported that oxygen impurity can make a large anomalous contribution to the specific heat, with a maximum around 2 K [99]. Measurements have also been made on solidified gases under pressure. For a typical apparatus, see Ref. [100].

3.4.10. Note on Measurements at High Pressure

For more detail see the chapter entitled 'Calorimetry at Very High Pressures' of the present book. The earliest measurements at low temperature appear to have been on α-uranium contained in a high-purity, beryllium-copper piston and cylinder arrangement. The pressure (10 kbar) was applied at room temperature and retained with a mechanical clamp [101]. Measurements were made down to 0.3 K and the pressure on the sample at the measuring temperature was obtained from the known pressure dependence of the superconducting transition temperature which was clearly revealed by the usual specific heat anomaly. A 25 kbar calorimeter of similar type was used later [102]. For a nonsuperconducting sample the pressure may be deduced from magnetic measurements on a small piece of tin put with the sample in the pressure clamp [103]. This technique suffers from the disadvantage that the calorimeter heat capacity is much greater than that of the sample. However, the effect of these large pressures is often to produce a major change in the sample state which is clearly revealed using this technique for specific heat measurements.

Apart from the work described above, which used the 'traditional' method for specific heat measurement, results have also been obtained by the a.c. heating method (see Section 3.8 of the present chapter). In experiments on tin in the region of the superconducting transition the pressure (up to 10 kbar) was applied using a kerosene-oil mixture [104]. The claimed accuracy was ±3%. A more sophisticated method used low heat capacity and high thermal resistance diamond powder as a pressure transmitting medium in the piston-cylinder pressure cell at measuring temperature [105]. Measurements have been made at pressures up to 22 kbar by this technique which, in large measure, satisfies the optimum requirements for the use of the a.c. method. An accuracy of ±3% was claimed.

3.4.11. Note on Measurements in Magnetic Field

Apart from the necessity of providing the magnetic field, which might be from a conventional electromagnet at room temperature or from a superconducting magnet within the cryostat, the following points might be noted:

(1) Relative vibration of sample and magnet will give rise to eddy current heating. Hence, great care is necessary in this area.

(2) Apart from the sample under investigation, the heat capacity of the 'empty calorimeter' materials may show a magnetic field dependence (e.g., splitting of the levels associated with the magnetic dipole moment of the nucleus).

(3) It has been pointed out that, when using a superconducting solenoid in the persistent mode, a temperature dependent magnetization of the sample may produce a significant change of solenoid current and, hence, of applied field [106].

(4) It may be possible to mount the germanium thermometer on a copper stalk to get it away from the sample and magnetic field [86]. Then the thermometry problems are considerably simplified, since the use of an unstable low magnetoresistance carbon thermometer (see Section 3.3.3) is avoided.

3.4.12. Heat Capacity of Materials Used in Calorimeter Construction

As indicated earlier in this section, it is best if the heat capacity of the 'empty' calorimeter can be measured directly. However, sometimes in sealing a calorimeter a small addition or subtraction of material is necessary and it may be sufficiently accurate to use a published value for the heat capacity of this alteration. Also, some idea of anticipated heat capacity is needed in designing a calorimeter. Care must be taken because the heat capacity of, for example, a varnish is not necessarily the same from batch to batch and will probably depend on heat treatment. Similarly, even 'pure' metals may show a low temperature heat capacity which is drastically affected by small amounts of impurity. (See, for example, the discussion on copper in Section 3.4.7 above.) Also, the published data are not necessarily error free. With these provisos a list of data sources follows. The list is probably incomplete but should serve as a general guide.

The Elements

There are compilations by Phillips [107] and Hultgren et al. [108].

Binary Alloys

The compilation of Hultgren et al. [109].

Resistance Wires

Constantan	0.15–4.2 K [110]
Evanohm	0.15–1.5 K [111]
Manganin	0.2–9.2 K [110,112]
Nichrome	0.05–0.5 K, 2–30 K [113,114]

Wood's Metal — 0.6–30 K [112,115,116]

Soft Solders — 1.3–20 K [117]

Varnishes and Resins

Araldite. Type 1	1.5–20 K [115]
Araldite. 502	1.7–4 K [118]

Araldite. AY111 with HY111 hardener ———————————————————————— 1–2 K [119]
Epibond 121 ——— 1.7–4 K [118]
Epibond 100A —— 1.7–4 K [118]
Epo-Tek. H31 LV —— 4–25 K [120]
Epo-Tek. H62 ——— 4–12.5 K, 20–40 K [120]
General Electric 7031 varnish ———————————————————————————— 0.2–80 K [121–124]

Plastics

Nylon. Polypenco 66 ——————————————————————————————————————— 1–4 K [119]
Nylon. 66 —— 0.3–1.5 K [125]
Polycarbonate. (Lexan) —————————————————————————————————— 0.4–4.2 K [126]
Polyethylene. (Different densities) ———————————————————— 0.3–30 K [125,127]
Polymethyl Methacrylate (Plexiglass, Perspex) —————————— 0.5–4.3 K [128]
Polystyrene ——— 0.5–4.3 K [128]

Glass

Corning 7560 —— 4.5–20 K [129]
Pyrex 7740 —— 0.4–4 K [130]

Carbon Resistors

Allen-Bradley 116Ω (Type TR) material ——————————————————— 0.7–4.2 K [131]
Allen-Bradley 56Ω (1/4 watt) ———————————————————————————— 3–25 K [116]
Allen-Bradley resistor material ————————————————————————— 1–20 K [132]

Greases

Apiezon N ——— 0.4–50 K [133]
Apiezon T ——— 1–350 K [134]
Dow Corning Silicone, High Vacuum ——————————————————————— 1.7–25 K [116,135]

3.5. ISOPERIBOL CALORIMETERS

Most calorimetry in the 0.3–30 K temperature range has been done by this method, in which the calorimeter surroundings are kept at a constant, or approximately constant, temperature while the calorimeter is heated discontinuously. Observation in the 'drift' periods between heatings enables a heat exchange correction to be made. For measurements at the lowest temperatures the surrounding shield is usually attached to a refrigerant bath and will remain at constant temperature or show a slow drift [40,82,94]. Measurements at higher temperatures may be made in the same apparatus but without any refrigerant in the can in contact with the shield. Thus, the shield surrounding the calorimeter will tend to drift up in temperature with the calorimeter and so reduce the heat loss from the calorimeter [94]. Some workers have adjusted and controlled the shield temperature to get low heat exchange [69], while others adjust for each measurement point so that the shield is at the mean temperature for that point [89,136]. The latter procedure minimizes drift corrections but clearly measurements take more time.

The isoperibol calorimeter has the obvious advantage of simplicity but, when the calorimeter temperature departs significantly from that of the surroundings, there may be quite large heat leaks which must be corrected for and which might degrade the accuracy of results, both in view of the magnitude of the correction and also because of resulting temperature gradients in the calorimeter. In contrast, an adiabatic or quasi-adiabatic calorimeter can be operated over a wide temperature range. When the sample under investigation results in long calorimeter equilibrium times, the adiabatic calorimeter is greatly to be preferred, since accurate corrections for heat exchange with the environment may then be difficult or impossible in an isoperibol calorimeter.

3.6. ADIABATIC CALORIMETERS

As indicated in the previous section, the adiabatic method with discontinuous heating is to be preferred when the sample equilibrium is in question. Since time effects associated with unexpected processes (such as annealing [137]) might be missed in an isoperibol calorimeter, it can be argued that adiabatic calorimetry is generally to be preferred despite the added complication. However, at the lowest temperatures (below, say, 2 K), small errors in shield control may begin to degrade results and here isoperibol calorimetry may be best.

The adiabatic shield is controlled using the indication of a differential thermocouple (see Section 3.3.8) between the calorimeter and shield. At one time manual control of the shield heater was necessary, but current practice is to use automatic electronic equipment. The thermocouple output is usually amplified with a chopper-type amplifier and fed to a recorder and controller. The latter usually has three modes of action, i.e., proportional, integrating, and differential [138]. Owing to the large variation in shield heat capacity over the present temperature range, it has often been necessary to alter the control parameters during an experiment. One way of avoiding this is to make the controller output nonlinear so that the control action is much stronger at higher currents (i.e., higher temperatures). A 'homemade' apparatus of this type has been described [48]. It is also possible to build a control system from commercial equipment [139].

A combination of temperature dependent 'stray' thermal voltages and 'stray' heat influx (e.g., vibration) may make the control point for zero calorimeter drift a function of temperature. Rather than readjust the control point, it is common to operate in the 'quasi-adiabatic' mode where there may be small temperature drifts. The usual drift correction methods are applied (Section 3.11). Time effects may be identified as superimposed on steady drifts.

At higher temperatures than the present range there may be uncompensated heat exchange owing to different calorimeter and shield surface temperature gradients during drift and heating periods [140]. Owing to much smaller radiant heat exchange such effects are unlikely in the present range, but they can occur, if, for instance, the differential thermocouple junction and the lead anchoring area are at significantly different temperatures during a heating period. Such effects cannot be detected by varying the heating rate [140]. The adiabatic shield will probably be in at least two parts to allow access to the calorimeter. In the present temperature range it is usually adequate for the differential thermocouple junction to be on the part carrying the lead anchoring area and (most of) the heater. Perfectionists may wish to separately control both parts of the shield to minimize possible temperature gradients between the parts. A light, pure copper or silver shield allows rapid temperature control with minimal temperature gradients.

3.7.   CONTINUOUS HEATING CALORIMETERS

At higher temperatures this method is always used with an adiabatic shield but, in the present temperature range and over small temperature intervals, it is possible to use the isoperibol method when heat exchange is very small and measurable. Some objections have been raised against the continuous heating method as compared with discontinuous-heating adiabatic calorimetry. These relate to possible lack of equilibrium within the sample and between sample and thermometer, and the difficulty of ensuring accurate adiabatic shield control or heat exchange correction in the absence of frequent drift periods. The last mentioned objection is more apparent than real because a similar problem exists during the heating period of a discontinuous heating calorimeter [140]. Some idea of the importance of deviations from equilibrium is obtained by following the calorimeter behavior when the heater is turned off. The continuous heating method is generally faster than the discontinuous method, owing to the absence of frequent drift periods, and transients and associated problems of shield control, etc., are avoided. It is especially useful in situations where a high resolution (small temperature interval ΔT per point) is required together with continuous coverage in the region of a specific heat anomaly.

Cochran et al. [31] have given a careful treatment of continuous heating calorimetry in the present temperature range. They overcome the equilibrium problem, mentioned above, by heating at different rates and extrapolating to zero heating rate. The possible influence of the time constant of the thermometer circuit was also considered. They employed a thermal guard which was always maintained within 0.01 K of the sample temperature. Other, more recent, workers [141] have found a temperature controlled guard to be unnecessary and have used a DVM for measurement. A small sample (~2 g) continuous heating calorimeter, used especially for the investigation of specific heat anomalies related to magnetic phenomena, has been described [142].

3.8.   NOVEL TECHNIQUES

In the last two decades many techniques have been described which differ in some substantial way from the 'traditional' techniques considered previously in this chapter. In some cases the reason behind the method was the simplification of the apparatus by removing the need for a thermal switch, while in other cases the real or supposed problem of vibration heating was minimized by a rigid mounting of the sample, with the inevitable closer coupling to the surroundings. There was also a need to be able to make rather accurate (~1–2%) measurements on small (even 1 mg) samples. The present section attempts a brief survey and classification of methods which have been used within the 0.3–30 K range. A more detailed description of some of these methods is given in the preceding chapter. Reference to the preceding chapter and to the original papers will show that there may be problems with internal equilibrium times and other factors. Thus, these novel methods must be used with care and, where a large enough sample is available, the 'traditional' methods will usually give a more accurate result.

A recent review makes a critical comparison of some of the novel small sample techniques described below and also lists recent developments in measuring still smaller samples [213].

3.8.1. 'Traditional' Methods Adapted for Small Sample

A discontinuous heating apparatus for the 1–20 K range used a ~1 cm diameter button of sample material equipped with miniature heater and thermometer [143]. This was contained in an evacuated can filled with aluminum oxide powder to minimize radiation loss and vibration heating. (The use of a very low heat capacity (silicon) powder as a sample support had been reported previously [144]. The method has not been used recently and was probably an example of 'overkill,' with vibration heating problems less troublesome than was originally feared.) A silicon-tray discontinuous heating calorimeter for ~5 g samples in the liquid helium range was described [145]. An adiabatic calorimeter for ~1 g samples in the 1.5–20 K range, presumably operated in the discontinuous heating mode, yields results of 2% accuracy [146]. A continuous heating small-sample (~2 g) calorimeter has been operated in the liquid helium range [142].

3.8.2. Steady-State a.c. Temperature Calorimetry

The 'traditional' methods of calorimetry are transient and the sample must be recooled to repeat the measurement. In the steady-state a.c. method the sample is thermally coupled, with a relaxation time of several seconds, to a temperature-stabilized heat sink. An a.c. current is passed through a heater attached to the sample. The resultant equilibrium temperature difference consists of two terms, the first a constant depending on the coupling between the sample and the sink, the second a sinusoidal variation at twice the frequency of the current. For a suitable choice of frequency (~20 Hz) the magnitude of the oscillatory temperature variation is inversely proportional to the sample heat capacity [147]. Although this variation is small, it is steady and may be averaged over long periods to get an accurate measurement. For more detail on this technique and possible complications see Section 2.3.4 of the preceding chapter. While not as accurate as the best 'traditional' methods, the method yields fairly high accuracy with very small samples.

The method has been used in a variety of situations. The oscillatory magnetic field dependence of the specific heat of beryllium at 1.5 K was determined [148]. A 1 mm^3 sample of a salt was investigated with very high resolution in the vicinity of the Néel temperature (~2 K) [149]. The method has been used in the 2–10 K range with a laser beam as heat source [150–152]. A detailed discussion of an apparatus for use down to 0.3 K with 1 g samples has been given [153].

3.8.3. Temperature Wave Method

The propagation of a temperature wave down a long, small, cross-section sample is observed using a.c. heating and two thermometers at different positions along the sample. In the latest version [154] the thermal conductivity is obtained from the mean temperature gradient and the diffusivity from the phase difference of the temperature wave at the two thermometer positions. Hence, the specific heat is obtained and has been measured simultaneously with the thermal conductivity on the same sample. This appears to be the attraction of the method, since the samples are rather large and the accuracy not very high (~5%). Results were presented in the 1.4–7.5 K range. For the theory see Section 2.3.4 in the preceding chapter.

3.8.4. Diffuse Temperature Pulse Method (sometimes called Heat Pulse Technique)

The apparatus is similar to that for the immediately preceding methods, but the heating is by a single δ-function rather than steady a.c. After observation of the heat pulse the heating may be repeated and signal averaging techniques used [155]. Results were presented in the 1–10 K temperature range with an apparent accuracy of about 5%. For more detail see Section 2.3.3 of the preceding chapter. A comparison of this method with other methods is given in Ref. [156] and a critical survey of potential errors in Ref. [157].

3.8.5. Temperature Relaxation Methods

The temperature of a small sample, with a short relaxation time to a heat sink, is followed when the sample heater is switched off. In the step method [158] the temperature increment is very small and the heat capacity is determined from the time constant of the exponential decay and measurement of the thermal conductance between the sample and heat sink at that temperature. The sweep method [158] covers a wider temperature range in one cycle. The steady-state temperature is measured as a function of heater power and then, at the highest temperature, the heater is switched off and the temperature is measured as a function of time. The heat capacity, as a function of temperature, can be calculated from the measured (dT/dt). Signal averaging techniques can be applied to both methods. For more information see Section 2.3.3 of the preceding chapter and Ref. [158].

A typical apparatus, measuring small (1–100 mg) samples in the 1–35 K range, uses a silicon chip as a very light 'tray' calorimeter to which the sample is bonded [159]. The accuracy is about ±1%. (By a suitable choice of link between sample and heat sink this apparatus can also be operated as a discontinuous heating isoperibol calorimeter, as in Ref. [143], and also with a.c. heating (Section 3.8.2, above).) Similar equipment using a sapphire 'tray' calorimeter of lower heat capacity has been described [160]. An accuracy of +1% is also obtained in an apparatus where a ~100 mg sample is heated to 10 K and relaxes to 1.5 K in about 30 s in an optional magnetic field [161]. A less accurate (2–5%) apparatus using a 20 mg sample and covering the 4–380 K range by the relaxation method has been described [162]. For a given experiment the relaxation time is fixed by the connections between sample and heat sink. With larger samples (~1 g) the heat leak may be via a mechanical heat switch; the magnitude of the leak can be altered during the experiment by adjusting the switch contact pressure [163]. Laser beam heating has been used with the relaxation method [150] as has an optical flash tube [164].

3.8.6. Resistive SQUID Calorimetry

The resistive SQUID (superconductive quantum interference device) produces an oscillating voltage whose frequency is linearly dependent on the heat current passing through it. Thus, by counting the number of oscillations an integrated heat current is obtained [165]. Two methods for heat capacity measurement with this device have been proposed and preliminary results on a 4 g copper sample with one of these showed a precision of about 0.1% and accuracy better than 1% [165]. Since the equipment for this method is quite different from that normally employed in calorimetry, reference must be made to the original paper for a full description.

3.8.7. Differential Calorimetry with Small Samples

An apparatus [166] covering the 4–370 K range with a claimed accuracy of the order of 1% for 50 mg samples uses a massive copper block heated at a programmed rate. The samples are connected to the block by 'heat columns' constructed from multijunction thermopiles. The power sensitivity at 4 K is 2×10^{-8} W. The heat capacity of high-purity copper is used as a standard.

Another method [167], used over the 1.5–400 K range with a claimed accuracy of 0.5%, couples two samples together with a heat link having a time constant of about one minute. Each sample is equipped with a heater and the apparatus is operated in the discontinuous heating mode with a roughly controlled shield surrounding the sample assembly. If the heat supplied to the two samples is in the ratio of their heat capacities then the temperature difference between them at the end of a heating period is zero. Otherwise, the temperature difference at the end of heating, obtained by extrapolation of the after-heating drift, permits the difference in heat capacity to be obtained.

The laser heating method [151], mentioned earlier in Section 3.8.2 (steady-state a.c. heating method), is often operated as a differential method because the actual heat supplied is not known. One way of overcoming this problem is by alternately illuminating two samples, one of known heat capacity.

3.8.8. Novel Techniques with Larger Samples

These are analogous to some of the small sample techniques considered above. Two examples are discussed.

3.8.8.1. Large Sample Weakly Coupled to Heat Sink

In one example [93] a 10–20 g sample is clamped in a calorimeter connected to the heat sink by a stainless steel tube 3 cm long and 0.3 cm in diameter. The temperature rise caused by discontinuous heating is obtained by extrapolation of the after-heating drift. The heat-sink temperature is increased when a point at a higher temperature is required. The method has been used in the 0.3–5 K range with metallic samples and is claimed to have an accuracy of better than 1%. The apparatus is simple, there being no heat switch while the rigid sample mounting minimizes vibration heating. Furthermore, points at the same temperature can easily be repeated.

A similar method was used in the much more difficult task of measuring alkali halides in the 0.05–2 K range [168]. In this case the sample was rigidly mounted between pointed nylon pins and the heat leak was a piece of fine copper wire.

A long series of results on alloy systems was based on a method where thermal isolation of the sample is sacrificed in favor of rigid support [169]. In this method the calorimeter plus sample is rigidly attached to the cold reservoir with nylon screws. The calorimeter temperature is adjusted to the desired measurement temperature T_M with an auxiliary heater. The calorimeter is then cooled well below this temperature with a mechanical heat-switch and then heated with a second, independent heater. The time to warm through a predetermined temperature interval ΔT, symmetrically disposed about T_M, is measured. The measurement may be repeated any desired number of times at each T_M. This method uses samples of the same mass as 'traditional' methods and appears to produce results, in the 1–4 K range, of similar accuracy. A somewhat similar double heater method with discontinuous heating or cooling has been used for superconductivity studies below 10 K [135].

3.8.8.2. Cooling Curve Method

This is analogous to the sweep method discussed in Section 3.8.5 above. It has been used for a 20 g crystal of CsBr in the 1–6 K range [170].

3.9. SPECIAL SITUATIONS

The ultimate aim of most calorimetry is to determine the heat capacity of a small or large piece of material. Some examples of more complex situations are considered in the present section.

3.9.1. Short Time–Scale Experiments

These are used to search for some process that takes a finite time to occur after the sample is heated. For example, the thermal and transport properties of amorphous materials below 1 K have been rather successfully interpreted within the tunneling model. Further confirmation was sought in a short time–scale measurement. The results [171] were interpreted as showing that the temperature rise produced by a 1 µs heat pulse relaxes over a time period of about 10 ms as the energy in the phonon gas is coupled to the tunneling states. Control experiments were done on a crystalline material. However, later work [172] identified the tunneling states with an observed relaxation time of ~1000 s, the mechanism behind the simultaneously occurring ~10 ms relaxation time not being clear at present. Experiments on gamma–irradiated natural quartz (where the radiation damage results in a two–level system) with control experiments on the same sample after annealing, have also shown a similar transient behavior on the millisecond time–scale. The effect is clearly seen at the lowest measurement temperature (1.4 K) and becomes smaller at higher temperatures, disappearing by 4.2 K [173].

3.9.2. Annealing Experiments

These can sometimes be done in a normal calorimeter. Thus, the supposed annealing of defects, produced by the martensitic transformation which occurs on cooling sodium, was investigated by thermal cycling over the 3–30 K range [137]. In other cases a special apparatus is required, since the defects must be produced at a low temperature. Thus, an experiment has been described where KBr is X–ray irradiated at 10 K and then measured in the 10–50 K range [174]. In another experiment platinum was deuteron irradiated and then measured in the 12–35 K range [175].

3.9.3. Thin Film Experiments

By evaporating a film onto a low–temperature substrate a very thin sample is obtained. The structure is often different from the normal structure of that material (e.g., it is often amorphous) and it may be possible to produce 'alloys' in which the components would segregate at higher temperatures. Clearly, some of the very small sample techniques discussed in Section 3.8 are ideal for measuring such films.

In a typical apparatus [176] the film is evaporated onto a Be foil (very low heat capacity) at a temperature below 20 K. The heat capacity is then determined in the 0.6–4.4 K range by light flash heating and a thermal relaxation method. By cycling above the crystallization temperature and recooling, the difference between the crystalline and amorphous states was determined.

Many of the experiments have been concerned with size effects in superconducting materials. Thus, granular aluminum films have been measured by laser heating with relaxation (Section 3.8.5) or a.c. measurement (Section 3.8.2) techniques [177], by electrical heating with the a.c. technique [178], and by the diffuse temperature pulse (Section 3.8.4) method [179]. The effect of impurities on the superconductivity of indium, including alloys not stable at room temperature, was investigated by the a.c. technique [180].

3.9.4. Small Particle Experiments

These are often rather difficult experiments and, owing to the number of corrections which may be required, the results may not be conclusive. Quite a range of techniques has been used. Loose small particles (~100 Å) of MgO had a much larger specific heat at low temperatures than the bulk [63]. This was ascribed, in part, to external degrees of freedom of the particle. Lead and indium small particle samples (20–60 Å) were prepared by impregnating a porous glass with metal [181]. An enhancement of specific heat owing to low-frequency surface modes was observed and, at the lowest temperatures (<4 K), a rapid decrease of heat capacity thought to be due to a low frequency cut-off of the phonon spectrum caused by the quantum size effect. The quantum size effect in the electronic specific heat was investigated with platinum, which has a very favorable ratio of electronic-to-lattice specific heat at low temperatures [182]. The sample was a granular film prepared by co-sputtering Pt and Si, and measurements in the 1–12 K range were made with a small sample technique. Samples of compressed fine particles have also been used [67].

3.9.5. Experiments on Radioactive Materials

Apart from safety considerations, the chief problem in the present temperature range is the self-heating of the sample. In fact, one technique is to cool the sample with some kind of heat switch and then let it heat continuously, determining the heat output with a subsidiary experiment. One problem here is that the initial rate of heating might be very fast. The other basic method is to use a thermal link to prevent the sample heating out of the range of interest and then apply a relaxation method. An accuracy of +3% is claimed for such an apparatus operating from 2.5 K. However, one sample could not be cooled below 7.2 K owing to the large self-heating [183]. In a more complex apparatus, claimed to have an accuracy of 1% in the 2–60 K range, the sample self-heating is balanced at all times by the leak down the heat link [184]. This is done by automatically controlling the temperature of the sample surroundings. The apparatus is operated in the discontinuous heating mode.

In some cases it might be possible to minimize the self-heating problem by making measurements on a separated isotope which has a much longer half-life than other isotopes of the same element. This was done in Ref. [185].

3.10. AUTOMATED CALORIMETRY IN THE 0.3–30 K TEMPERATURE RANGE

Both the 'traditional' and 'novel' calorimetric techniques described above employ a repetitious series of measurements and lend themselves well to automation. Provided that the equipment is working correctly, the risk of misrecording readings is avoided and the time involved in preparing data for computing is eliminated. Thus, the fatigue associated with long experiments is reduced and, in the more advanced systems, all the measurements and calculations may be performed without the intervention of an operator.

3.10.1. The Development of Automation

The methods used and the degree of automation have gradually evolved over the years. The first step was to fit a duplicate set of contacts to conventional, manually balanced instruments so that the readings were recorded, for example onto paper tape [186], for 'off-line' processing. Another possibility is to transmit these readings directly to an 'on-line' computer for immediate processing, and to return the results to the calorimeter operator so that any unusual features can be followed up immediately [185]. Full automatic data acquisition, using self-balancing instruments with 'off-line' processing (see, for example, Ref. [187]), is not too satisfactory for this temperature range because the rapid variation of heat capacity with temperature necessitates frequent adjustment of the calorimeter heater current in order to maintain an acceptable heating rate. Presumably, this could be done with some fixed preprogramming, but the availability of relatively inexpensive small computers has led to 'on-line' processing with feed-back loop(s) to control the heating rate and possibly other factors (heating period, drift period, etc.) [40,48].

3.10.2. Self-Balancing Instruments

The ready availability of commercial self-balancing instruments has made complete automation possible. The digital voltmeter (DVM) has been discussed earlier and precautions in its use have been detailed (Section 3.4.3). This is the universal choice for heater power determination, and it has also been used for carbon and germanium thermometry. (Some workers [188,189] have even used DVM's for platinum thermometry, but it seems doubtful whether the present sensitivity, linearity, and stability are really adequate for this purpose.)

The ASL 'Cryobridge' [46] is a self-balancing potentiometer designed for thermometry applications in the present temperature region. The ASL automatic Kelvin double bridge [46] was designed for platinum thermometry but has been used successfully with germanium thermometers over the present temperature range [40,44,48]. With the latter instrument the temperature dependence of thermometer sensitivity in the present temperature range means that the integrating band-width (or stepping speed) and the counting logic may both require alteration during even a short heating period. A method of using a microprocessor to make these adjustments, and thus ensure optimum bridge performance at all times, has been described [190].

3.10.3. Some General Considerations

Depending on the degree of automation, the sequence and timing of readings of the self-balancing instruments will be determined by a hard-wired controller [187] or a computer program [191]. Apart from the basic system timer, the readings of an independent timer may also be recorded as a check [48,187]. If a DVM is used to take both heater and thermometer readings, a low thermal switch is necessary to connect the various circuits to the DVM [188,191]. (The thermals in commercial electronic multiplexers are much higher, typically ~1 μV.) Some arrangement for frequent checks of DVM sensitivity and zero during a run is also necessary. If the DVM is used solely for heater power measurements the precautions are less stringent, reed-relays or a commercial multiplexer might be adequate for switching circuits. In any case it is advisable to switch immediately after a reading to allow the thermals to die down before the next reading. In determining heater power a number of readings may be made during a heating period to average the voltage and current. The heater current supply may be obtained from a 'homemade' [40,48,191] or commercial [192] controller. The value of the current is set by the computer after examination of the previous heating interval or heating rate. The thermometer current might also need to be varied during an experiment to balance the requirements of sensitivity with minimum power dissipation [191]. Using a bridge with a constant voltage supply, the current is automatically decreased as the temperature drops and the thermometer resistance rises [40,48]. Some 'on-line' systems make provision to record the actual instrument readings. This is useful if, for example, a variation in temperature scale is to be tested by recomputing the run. It might also be useful in the case of a partial system failure.

Great care must be taken with an automated system to ensure that the data being collected are correct. If the thermometer calibration checks, as recommended in Section 3.4.6, are made, then not only the thermometer but also the automatic measuring instrument and data processing equipment are checked. The DVM is checked using a standard cell and then a shorted input. More exhaustive checks should be made on a routine basis to ensure that no 'bits' are being lost (or added) during instrument readings and that instrument linearity is maintained. Manufacturer's instructions should be consulted for other checks. The computer program for the system should be written so that illegal characters are noted in such a way that their source can be traced. The program must not 'hang-up' or stop on illegal data but must reset in the hope that the problem will not recur. Some legal readings may be in error owing to external noise sources. Methods for detecting and eliminating such readings are discussed in Section 3.11.

3.10.4. Some Typical Systems

The basic problems of calorimetry automation are similar at all temperatures when similar calorimetric techniques are used. Thus, several of the systems already mentioned are designed to cover wide temperature ranges. The following brief summary lists some of the systems which have been used in all or part of the 0.3–30 K range. A potential constructor of such a system will find many useful details in these papers. The accuracy of these systems is generally the same as that obtained with manual measuring equipment and the same calorimetric apparatus. In many cases it may be better, because more time can be spent on the measurements; the equipment runs indefinitely with full attention to the task in hand.

3.10.4.1. 'Off-Line' Systems

These cannot run an experiment without human intervention; see Section 3.10.1.

(1) ASL A7 Kelvin double bridge plus DVM to paper tape [187].

(2) ASL Cryobridge plus DVM to paper tape [193].

(3) Separate DVMs for thermometer and heater to paper tape [194].

3.10.4.2. 'On-Line' Systems

(1) ASL A7 Kelvin double bridge plus DVM. PDP-8 computer [40,48,190].

(2) DVM. Interdata 74 computer [191].

(3) DVM. Sigma 2 computer [188].

(4) DVM. Shared PDP 11/15-PDP 11/20 computer system [189].

(5) DVM. Intel 8080 microprocessor. Continuous heating calorimeter 2-30 K. Numerical filtering of results [192].

3.10.4.3. 'On-Line' Systems for 'Novel' Techniques (Section 3.8)

(1) Automated 'homemade' Wheatstone bridge with Nicholet 1070 signal averaging computer plus PDP-8 computer [159].

(2) DVM plus additional circuitry controlled by an unspecified computer [162].

3.11. ANALYSIS OF DATA. ACCURACY

3.11.1. Reduction of Observations to Molar Specific Heat

These remarks apply to the 'traditional' methods of calorimetry. Similar considerations apply to some of the 'novel' methods but for others reference must be made to the cited papers (Section 3.8).

Consider first discontinuous heating. There is a before-heating drift measurement consisting of pairs of temperature and time (T,t) data and a similar after-heating drift. Immediately after heating the normal drift curve may be perturbed owing to equilibrium time effects between the heater, calorimeter, sample, thermometer, and the wires (and tubes) between the calorimeter and its surroundings. In some cases these perturbations may have a long decay time. If the calorimeter has a controlled adiabatic shield these perturbations do not affect the derived specific heat, and an attempt is made to fit the observed drifts, after equilibrium, to straight lines which are then extrapolated to the mid-point of the heating period to find the temperature rise, mean temperature, and hence heat capacity of sample plus calorimeter from $C_p = \Delta Q/\Delta T$. (It is trivial to calculate ΔQ from the average observed heater potential and current and the heating time.) The situation is more complicated in an isoperibol calorimeter (the price paid for simplicity of experimental equipment!). Firstly, the perturbations owing to equilibrium time effects (even those with short time constant) will affect the heat exchange with the environment during

heating and possibly afterwards. These problems have been considered at length in Chapter 6 of
Ref. [1] and also in Ref. [136]. Secondly, if the sample has a small heat capacity or if there is
close coupling to the surroundings, the drift curves may be exponential. These can often be fitted to an equation of the form

$$T = a + be^{-ct}.$$ (2)

(This involves an iterative, nonlinear, least-squares fit on a computer.) Then the drifts must be
extrapolated into the heating period to find the temperature rise and mean temperature. Two methods have been used (see Chapter 6 of Ref. [1] and Ref. [185]).

For either calorimetric method it might be necessary to apply a 'curvature correction' to the
result to take account of the variation of specific heat over the heating interval (see Chapter 5
of Ref. [1]). Most workers in the present temperature range use so small a heating interval that
this is not necessary. A maximum permissible heating interval can be estimated by assuming a form
for the specific heat (e.g., $C_p = aT + bT^3$) and comparing the integral over ΔT with the mid-point
value.

Finally, consider continuous heating. Under ideal conditions, with good calorimeter equilibrium and adiabatic shield control, the heat capacity is obtained from the slope (dT/dt) or $(\Delta T/\Delta t)$
and measurement of the heat input. It has been suggested that the T versus t curve should be
smoothed before deriving the heat capacity [188]. The situation is more complicated when there
are significant internal equilibrium times and heat leaks. See, for example, Ref. [31]. A method
for samples with good internal equilibrium but small heat capacity and large heat exchange to the
surroundings has been described. This might be termed a semi-continuous heating method since
rather large temperature increments are used [195]. The heat capacity is obtained by an iterative
calculation.

After obtaining the heat capacity of the sample plus calorimeter, the molar specific heat is
obtained by subtracting the previously determined heat capacity of the empty calorimeter and multiplying by the ratio of atomic or molar weight to the sample weight. It might also be necessary
to make small corrections for any materials added to or subtracted from the calorimeter during
sample installation.

3.11.2. The Various Contributions to Specific Heat

Many processes occurring in the sample will contribute to the specific heat. Thus, when the
molar specific heat has been obtained as a function of temperature, the problem is to correctly
identify and quantify the various possible contributions. Some of these are listed below; further
reference should be made to the first chapter in the present book.

3.11.2.1. Lattice Specific Heat

This is always present but might make a negligibly small contribution to the total at low
temperatures. It is conveniently represented by the expression [196]

$$C_p^{\ell} = \sum_1 a_n T^{2n+1}.$$ (3)

The low temperature limiting value of the Debye temperature θ_o^c is given by

$$\theta_o^c = (12\pi^4 R/5a_1)^{1/3} \tag{4}$$

where R is the gas constant.

This Debye temperature should be equal to that (θ_o^{el}) calculated from the low temperature elastic constants, which give a good check on the analysis of results (or even, in some cases, a start for the analysis). Incorrect results may be obtained if the specific heat results do not extend to a sufficiently low temperature, or if they extend only over a small temperature range (e.g., the popular 1–4 K range) when the form of C_p^ℓ may not be well established. For examples see work on a pure metal [197] and on an alloy system [198].

3.11.2.2. Electronic Specific Heat

This is assumed to be linearly dependent on temperature in the present temperature range. However, the electron–phonon enhancement varies with temperature. The effect is small and has not been separated, in the present temperature range, from the other contributions to specific heat [199].

3.11.2.3. Nuclear Specific Heat

The splitting of the energy levels associated with the magnetic dipole and electric quadru-pole moments of the nucleus gives rise to a Schottky anomaly, and the 'high-temperature' tail of this is sometimes seen at the lower end of the present temperature range. It is often sufficient to represent this by the leading term

$$C_N = a_N/T^2. \tag{5}$$

The nuclear term may obscure other contributions to specific heat. It may be possible to make measurements on a separated isotope with zero nuclear spin, for which this term disappears [200]. Conversely, in some cases the nuclear spin–lattice relaxation time is so long that the nuclear heat capacity is not observed. In this case it may be possible to dope the material with an impurity to increase the carrier charge density and thus observe the nuclear contribution [201].

3.11.2.4. Magnetic Specific Heat

With simplifying assumptions, the spin wave contribution to specific heat is expected to vary as $T^{3/2}$ for ferromagnetics and as T^3 for antiferromagnetics. The actual situation may be more complicated. See, for example, Ref. [202]. (Ferroelectrics have been reported to have a $T^{3/2}$ contribution to specific heat [203].)

In some nearly ferromagnetic situations the spins form large super-paramagnetic clusters, and the large specific heat contribution is a multilevel Schottky anomaly. Some assumptions are re-quired in order to separate the various specific heat contributions. For two examples see Refs.

[204] and [205]. Spin fluctuations in exchange-enhanced paramagnetics result in a nearly magnetic field independent specific heat term of the form $T^3 \ln T$ [206]. By making measurements in a magnetic field a distinction can be made between this contribution and the previous, super-paramagnetic cluster contribution [207].

In alloys, especially dilute alloys, when one component carries a magnetic moment, there may be specific heat contributions from the spin-spin interactions (or spin-glass effect [208]) and from the 'single-impurity' spin-compensation (or Kondo) effect [209]. Recent work on spin-glass alloys shows that the form of the specific heat differs from that assumed until recently. See, for example, Ref. [210].

There are many instances where there is a sharp magnetic transition within the present temperature range. See, for example, Ref. [142].

3.11.2.5. Superconducting Transition

Various parameters that can be determined from specific heat measurements serve to distinguish bulk from filamentary effects.

3.11.2.6. Impurities

The effect of impurities on specific heat in this temperature range can be very large. The effect of hydrogen and transition metals on the specific heat of copper was discussed earlier in Section 3.4.7. Good analytical facilities and measurements on two or more samples from different sources may reveal this potential problem.

3.11.3. Analysis and Presentation of Results

The form expected for the specific heat will be known from considerations such as those discussed in the previous section. At one time analysis was done graphically, but a least-squares fit on a computer is now more usual. The basic problem with specific heat data is that it may vary by one or several orders of magnitude over the measuring range. It is important, therefore, that the data be weighted so that fractional and not absolute deviations are minimized. Thus, if the measured and calculated (fitted) data are denoted by superscripts M and C, respectively, the following is minimized (where N is the total number of data points)

$$\sum_{i=1}^{i=N} \left((C_{pi}^C - C_{pi}^M)/C_{pi}^M \right)^2 .$$

The computer program will yield confidence limits on the various coefficients and a plot should be made of percentage deviation of the raw data from the fitted curve. Some deviation may occur because of small temperature-scale errors and should be the same for all samples. It may be necessary to run several fits with different numbers of terms in, for example, the series representing lattice specific heat, until a minimum is observed in the standard deviation. In some cases it might be necessary to break the results into sections to get a good fit. If a reasonable fit cannot be obtained then there must be some contribution to the specific heat which has not been considered. (For example, unexpected structure at the lower end of the fit may indicate a nuclear specific heat contribution.) Experience is a factor in correctly interpreting data. There is also the possibility that one or more of the specific heat points is in error and this should be revealed by the deviation plot or listing. This could be owing to a temporary instrument malfunction, a recording or computing error, or a mechanical (vibration) or electrical noise occurring during a drift or heating period. Inspection of the raw data may serve to identify the problem. (It is very useful here if, when using an automatic data acquisition system, the instrument readings have been recorded as suggested in Section 3.10.3 above.) If the point is found to be in error it is removed from the data and the analysis repeated. It is a great convenience in analyzing data if this rejection (and listing) of suspect data can be done automatically by the computer. There are many ways of identifying such 'outliers' [211]. One method, frowned on by the purists [211], is similar to Chauvenet's criterion [212]. Thus, in a large data set, points with deviations greater than three times the standard deviation may be rejected and the process repeated until all points are acceptable [40,48]. The basic problem, of course, is to identify and reject suspect data without smoothing out a real effect. For this reason all rejected points should be carefully examined in an attempt to find the cause of the deviation and, if necessary, measurements should be repeated. Some experimenters routinely repeat measurements after warming the apparatus to room temperature. This usually serves to confirm results but has, in some cases, led to the study of interesting effects, such as phase changes, resulting in sample instability.

In plotting results it is recommended that fractional deviations from a fitted curve should be shown. In some cases a plot of Debye temperature as a function of temperature is more useful. Owing to the fact that the specific heat may vary by one or several orders of magnitude over the measurement range, a plot of specific heat against temperature is usually of little use or interest. For raw data retrieval purposes (e.g., for possible reanalysis of data by another worker) it is useful if a published paper contains these data. Many authors give their results in tabular form but a clear deviation plot is equally useful and almost as convenient when an automatic digitizer is available.

3.11.4. Accuracy

In the 'traditional' methods the specific heat is derived from the electrical heat input and the observed temperature increment. The former measurement involves voltage, current, and time measurements. (The current is usually determined from the voltage drop across a standard resistor.) Provided the measuring equipment is working correctly and has been recently calibrated, the heat input is easily measured with high accuracy (0.01% or better). However, as discussed in Section 3.4.3 above, great care is necessary with the leads between the calorimeter and its surroundings. Also, as mentioned in Section 3.11.1, if an isoperibol calorimeter 'overheats' while the heater is on then the usual extrapolation of the drift curves may not give the true temperature increment, and thus specific heat, because there has been an uncompensated heat exchange during the heating period.

The correct measurement of the temperature increment depends on an accurate thermometer calibration. Even if this is done, a change in the local radio-frequency 'pollution,' an error in the measuring equipment, or a shift of calibration may all degrade the data. The temperature scale may show small 'wiggles' owing to small errors in the calibration data or small deviations in the fitted calibration curve. These will give rise to corresponding 'wiggles' in the specific heat data but, as a result of curve fitting the data, the final smoothed specific heat might be more accurate than the temperature scale itself.

At the present time the accuracy of a calorimeter is judged from the results of measurements on a standard substance. In the present temperature range copper is used (Section 3.4.7). On this basis, present day techniques may show a precision as good as ±0.1% and an accuracy two or three times worse. Some of the 'novel' small sample methods are significantly less accurate, and there are many instances of poor accuracy with the 'traditional' methods. When using data from the literature the authors' accuracy claims must be judged carefully, especially if the apparatus has not been checked by measurements on a standard sample.

3.12. REFERENCES

1. McCullough, J.P. and Scott, D.W. (Editors), _Experimental Thermodynamics_, Vol. 1, Butterworths, London, England, 1968.

2. Epstein, A.J., Etemad, S., Garito, A.F., and Heeger, A.J., Phys. Rev., B5, 952–77, 1972.

3. Leffert, C.B., Kasner, W.H., and Donahue, T.M., Rev. Sci. Instrum., 27, 1084, 1956; Smith, R.J. and Hegland, D.E., Rev. Sci. Instrum., 36, 709–10, 1965; Kirk, W.P. and Twerdochlib, M., Rev. Sci. Instrum., 49, 765–9, 1978.

4. Darby, J., Hatton, J., Rollin, B.V., Seymour, E.F.W., and Silsbee, H.B., Proc. Phys. Soc., A64, 861–7, 1951.

5. White, D.R.J., _A Handbook Series on Electromagnetic Interference and Compatibility. Vol. 3. EMI Control Methods and Techniques_, Don White Consultants, Inc., Maryland, 1973.

6. Westrum, E.F., Hatcher, J.B., and Osborne, D.W., J. Chem. Phys., 21, 419–23, 1953.

7. Rusby, R.L. and Swenson, C.A., Metrologia, 16, 73–87, 1980.

8. Keesom, W.H., _Helium_, Elsevier, Amsterdam, Holland, 174–5, 1942.

9. Shal'nikov, A.I., Cryogenics, 6, 299, 1966.

10. Walton, D., Rev. Sci. Instrum., 37, 734–6, 1966.

11. Walton, D., Timusk, T., and Sievers, A.J., Rev. Sci. Instrum., 42, 1265–6, 1971.

12. Delhaes, P. and Hishiyama, Y., Carbon, 8, 31–8, 1970.

13. Rose-Innes, A.C., _Low Temperature Laboratory Techniques_, English Univ. Press, England, 1973; White, G.K., _Experimental Techniques in Low Temperature Physics_, 3rd Ed., Clarendon, Oxford, England, 1979; Less useful on cryostat design but good on materials and jointing is Croft, A.J., _Cryogenic Laboratory Equipment_, Plenum, New York, 1970; For a good concise account with particular reference to materials, soldering, and welding see Thornton, F.D., in _Advanced Cryogenics_ (Bailey, C.A., Editor), Plenum, London, England, 493–514, 1971.

14. Hudson, R.P., Rev. Sci. Instrum., 51, 871–81, 1980.

15. Furukawa, G.T., Douglas, T.B., McCoskey, R.E., and Ginnings, D.C., J. Res. Natl. Bur. Stand., 57, 67–82, Footnote 5, 1956; NBS Special Publication 300, Vol. 2, p. 56.

16. Metrologia, 5, 35–44, 1969; Preston-Thomas, H., Metrologia, 12, 7–17, 1976.

17. Brickwedde, F.G., van Dijk, H., Durieux, M., Clement, J.R., and Logan, J.K., J. Res. Natl. Bur. Stand., 64A, 1–17, 1960.

18. Sherman, R.H., Sydoriak, S.G., and Roberts, T.R., J. Res. Natl. Bur. Stand., 68A, 579–88, 1964.

19. Plumb, H.H. and Cataland, G., Metrologia, 2, 127–39, 1966.

20. Berry. K.H., Metrologia, 15, 89–115, 1979.

21. Bureau International des Poids et Mesures, Metrologia, 15, 65–8, 1979; Durieux, M., Astrov, D.N., Kemp, W.R.G., and Swenson, C.A., Metrologia, 15, 57–63, 1979; Durieux, M. and Rusby, R.L., Metrologia, 19, 67–72, 1983.

22. Sostman, H.E., Phys. Today, 30(6), 11–3, 1977.

23. Douglas, T.B., J. Res. Natl. Bur. Stand., 73A, 451–70, 1969.

24. Holste, J.C., Cetas, T.C., and Swenson, C.A., Rev. Sci. Instrum., 43, 670–6, 1972.

25. Holste, J.C., Phys. Rev., B6, 2495–7, 1972.

26. Sample, H.H. and Rubin, L.G., Cryogenics, 17, 597–606, 1977.

27. Rubin, L.G., Rev. Sci. Instrum., 51, 1007, 1980.

28. Besley, L.M., Rev. Sci. Instrum., 50, 1626–8, 1979.

29. Kunzler, J.E., Geballe, T.H., and Hull, G.W., Rev. Sci. Instrum., 28, 96–8, 1957.

30. Besley, L.M., Rev. Sci. Instrum., 51, 972–6, 1980.

31. Cochran, J.F., Shiffman, C.A., and Neighbor, J.E., Rev. Sci. Instrum., 37, 499–512, 1966.

32. Colwell, J.H., J. Low Temp. Phys., 14, 53–71, 1974.

33. Anderson, A.C., Rev. Sci. Instrum., 41, 1446–50, 1970.

34. Rusby, R.J., in Temperature Measurement 1975, Conf. Series No. 26, Inst. Phys., London, England, 125–30, 1975; Besley, L.M., J. Phys. E., 15, 824–6, 1982; Rusby, R.L., in Temperature, Its Measurement and Control in Science and Industry (Schooley, J.F., Editor), Vol. 5, American Institute of Physics, New York, NY, 829–33, 1982.

35. Downie, D.B. and Martin, J.F., J. Chem. Thermodyn., 12, 779–86, 1980.

36. Soulen, R.J., Rusby, R.L., and Van Vechten, D., J. Low Temp. Phys., 40, 553–69, 1980.

37. Roberts, T.R. and Sydoriak, S.G., Phys. Rev., 102, 304–8, 1956.

38. Johnson, D.L., Rev. Sci. Instrum., 39, 399–400, 1968.

39. Schooley, J.F., Evans, G.A., and Soulen, R.J., Cryogenics, 20, 193–9, 1980; Schooley, J.F. and Soulen, R.J., in Temperature, Its Measurement and Control in Science and Industry (Schooley, J.F., Editor), Vol. 5, American Institute of Physics, New York, NY, 251–60, 1982; El Samahy, A.E., Durieux, M., Rusby, R.L., Kemp, R.C., and Kemp, W.R.G., in Temperature, Its Measurement and Control in Science and Industry (Schooley, J.F., Editor), Vol. 5, American Institute of Physics, New York, NY, 261–5, 1982.

40. Martin, D.L., Rev. Sci. Instrum., 46, 1670–5, 1975.

41. Barber, C.R., in Temperature, Its Measurement and Control in Science and Industry (Plumb, H.H., Editor), Vol. 4, Part 1, Instrument Society of America, Pittsburgh, PA, 99–103, 1972.

42. Clement, J.R. and Quinnell, E.H., Rev. Sci. Instrum., 23, 213–6, 1952; Clement, J.R., in discussion to Friedberg, S.A. in Temperature, Its Measurement and Control in Science and Industry (Wolfe, H.C., Editor), Vol. 2, Reinhold, New York, NY, 359–82, 1955.

43. Blakemore, J.S., Winstel, J., and Edwards, R.V., Rev. Sci. Instrum., 41, 835–42, 1970; Collins, J.G. and Kemp, W.R.G., in Temperature, Its Measurement and Control in Science and Industry (Plumb, H.H., Editor), Vol. 4, Instrument Society of America, Pittsburgh, PA, 835–42, 1972; Leung, Y.K. and Kos, J.F., Cryogenics, 19, 531–4, 1979.

44. Martin, D.L., Rev. Sci. Instum., 46, 657–60, 1975.

45. Anderson, M.S. and Swenson, C.A., Rev. Sci. Instrum., 49, 1027–33, 1978.

46. Wolfendale, P.C.F., J. Phys. E., 2, 659–60, 1969. This self-balancing potentiometer is available as the 'Cryobridge' from Automatic Systems Laboratories, Leighton Buzzard, England who also supply automated Kelvin double bridges. For recent developments in this area see several papers in Temperature, Its Measurement and Control in Science and Industry (Schooley, J.F., Editor), Vol. 5, American Institute of Physics, New York, NY, 711–32, 1982.

47. Rusby, R.L., Chattle, M.V., and Gilhen, D.M., J. Phys. E, 5, 1102–5, 1972.

48. Martin, D.L., Bradley, L.L.T., Cazemier, W.J., and Snowdon, R.L., Rev. Sci. Instrum., _44_, 675–84, 1973.

49. Culbert, H.V. and Sungaila, Z., Cryogenics, _8_, 386–8, 1968.

50. Sparks, L.L. and Powell, R.L., J. Res. Natl. Bur. Stand., _76A_, 263–83, 1972.

51. Kubaschewski, O. and Hultgren, R., in _Experimental Thermochemistry_ (Skinner, H.A., Editor), Vol. II, Interscience, New York, NY, 343–84, 1962.

52. Keesom, P.H. and Pearlman, N., Phys. Rev., _91_, 1347–53, 1953; Hornung, E.W., Fisher, R.A., Brodale, G.E., and Giauque, W.F., J. Chem. Phys., _46_, 67–72, 1967; Bystrom, S., Larsson, M., Marklund, K., and Lindqvist, T., Phys. Lett., _44A_, 159–60, 1973; Bloembergen, P. and Miedema, A.R., Physica, _75_, 205–33, 1974.

53. Martin, D.L., Philos. Mag., _46_, 751–8, 1955.

54. Birch, J.A., J. Phys. C, _8_, 2043–7, 1975; Khattak, G.D., Keesom, P.H., and Faile, S.P., Solid State Commun., _26_, 441–4, 1978.

55. Martin, D.L., Phys. Rev., _186_, 642–8, 1969.

56. Rayne, J.A., Phys. Rev., _95_, 1428–34, 1954.

57. Martin, D.L., Can. J. Phys., _48_, 1327–39, 1970.

58. Martin, D.L., Can. J. Phys., _59_, 567–75, 1981.

59. Gmelin, E., Cryogenics, _7_, 225–32, 1967.

60. de Dood, W. and de Chatel, P.F., J. Phys. F, _3_, 1039–53, 1973.

61. Martin, D.L., Proc. Phys. Soc., _83_, 99–108, 1964.

62. Keesom, W.H. and van den Ende, J.N., Leiden Comm., _219b_.

63. Lien, W.H. and Phillips N.E., J. Chem. Phys., _29_, 1415–6, 1958.

64. Webb, F.J. and Wilks, J., Proc. Roy. Soc., _A230_, 549–59, 1955.

65. Flotow, H.E. and Osborne, D.W., Rev. Sci. Instrum., _37_, 1414–5, 1966.

66. Cude, J.L. and Finegold, L., Rev. Sci. Instrum., _42_, 614–5, 1971.

67. Comsa, G.H., Heitkamp, D., and Rade, H.S., Solid State Commun., _20_, 877–80, 1976.

68. Craig, R.S., Massena, C.W., and Mallya, R.M., J. Appl. Phys., _36_, 108–12, 1965.

69. Leadbetter, A.J. and Wycherley, K.E., J. Chem. Thermodyn., _2_, 855–66, 1970.

70. Conway, M.M., Phillips, N.E., Geballe, T.H., and Kuebler, N.A., J. Phys. Chem. Solids, _31_, 2673–8, 1970.

71. Suzuki, H. and Koyama, M., Cryogenics, _15_, 38–9, 1975.

72. Kunzler, J.E., Maita, J.P., Levinstein, H.J., and Ryder, E.J., Phys. Rev., _143_, 390–3, 1966.

73. Filby, J.D. and Martin, D.L., Proc. Roy. Soc., _A276_, 187–203, 1963.

74. Flotow, H.E. and Klocek, E.E., Rev. Sci. Instrum., _39_, 1578–9, 1968.

75. Ashworth, T. and Steeple, H., Cryogenics, _5_, 267–8, 1965.

76. Lien, W.H. and Phillips, N.E., J. Chem. Phys., _34_, 1073–4, 1961.

77. West, E.D. and Ishihara, S., Rev. Sci. Instrum., _40_, 1356–9, 1969.

78. Martin, D.L., Rev. Sci. Instrum., _43_, 1762–5, 1972.

79. Anderson, A.C. and Bloom, D.W., Rev. Sci. Instrum., _40_, 1243–4, 1969.

80. Hust, J.G., Rev. Sci. Instrum., _43_, 1387–8, 1972.

81. Hill, R.W., Cosier, J., and Hukin, D.A., J. Phys. F., _6_, 1731–42, 1976.

82. Martin, D.L., Proc. Roy. Soc., _A263_, 378–86, 1961.

83. Lounasmaa, O.V., Phys. Rev., _133_, A211–8, 1964.

84. Garfunkel, M.P. and Wexler, A., Rev. Sci. Instrum., _25_, 170–2, 1954.

85. Seidel, G. and Keesom, P.H., Rev. Sci. Instrum., _29_, 606–11, 1958.

86. Johnson, D.L., Cluxton, D.H., and Young, R.A., Rev. Sci. Instrum., <u>44</u>, 16–22, 1973.

87. Colwell, J.H., Rev. Sci. Instrum., <u>40</u>, 1182–6, 1969.

88. Lien, W.H. and Phillips, N.E., Phys. Rev., <u>133</u>, A1370–7, 1964.

89. Osborne, D.W., Flotow, H.E., and Schreiner, F., Rev. Sci. Instrum., <u>38</u>, 159–68, 1967.

90. Research Material RM–5 from the Office of Standard Reference Materials, National Bureau of Standards, Washington, D.C. 20234.

91. See, for example, Triplett, B.B. and Phillips, N.E., Phys. Rev. Lett., <u>27</u>, 1001–4, 1971; Martin, D.L., Phys. Rev., <u>B20</u>, 368–75, 1979.

92. Martin, D.L., Rev. Sci. Instrum., <u>38</u>, 1738–40, 1967; Waterhouse N., Can. J. Phys., <u>47</u>, 1485–91, 1969.

93. Gobrecht, K.H. and Saint Paul, M., Proc. 3rd Int. Cryo. Eng. Conf. (Iliffe, Guildford, England), 235–7, 1970.

94. Cetas, T.C., Tilford, C.R., and Swenson, C.A., Phys. Rev., <u>174</u>, 835–44, 1968.

95. Shinozaki, S.S. and Arrott, A., Phys. Rev., <u>152</u>, 611–22, 1966; Montgomery, H., Pells, G.P., and Wray, E.M., Proc. Roy. Soc., <u>A301</u>, 261–84, 1967; Bevk, J. and Massalski, T.B., Phys. Rev., <u>B5</u>, 4678–83, 1972.

96. Martin, D.L., Phys. Rev., <u>B20</u>, 4001–7, 1979.

97. Finegold, L. and Phillips, N.E., Phys. Rev., <u>177</u>, 1383–91, 1969.

98. Colwell, J.H., J. Chem. Phys., <u>51</u>, 3820–32, 1969.

99. Burford, J.C. and Graham, G.M., J. Chem. Phys., <u>49</u>, 763–5, 1968.

100. Fugate, R.Q. and Swenson, C.A., J. Low Temp. Phys., <u>10</u>, 317–43, 1973.

101. Ho, J.C., Phillips, N.E., and Smith, T.F., Phys. Rev. Lett., <u>17</u>, 694–6, 1966; Smith, T.F. and Phillips, N.E., Coll. Int. CNRS, <u>188</u>, 191–5, 1970.

102. McWhan, D.B., Remeika, J.P., Bader, S.D., Triplett, B.B., and Phillips, N.E., Phys. Rev., <u>B7</u>, 3079–83, 1973.

103. Berton, A., Chaussy, J., Cornut, B., Odin, J., Paureau, J., and Peyrard, J., Cryogenics, <u>19</u>, 543–6, 1979.

104. Itskevich, E.S., Kraidenov, V.F., and Syzranov, V.S., Cryogenics, <u>18</u>, 281–4, 1978.

105. Eichler, A. and Gey, W., Rev. Sci. Instrum., <u>50</u>, 1445–52, 1979.

106. Fisher, R.A., Rev. Sci. Instrum., <u>50</u>, 134–5, 1979.

107. Phillips, N.E., CRC Crit. Rev. Solid State Sci., <u>2</u>, 467–553, 1971.

108. Hultgren, R., Desai, P.D., Hawkins, D.T., Gleiser, M., Kelley, K.K., and Wagman, D.D., <u>Selected Values of the Thermodynamic Properties of the Elements</u>, American Society of Metals, Metals Park, OH, 1973.

109. Hultgren, R., Desai, P.D., Hawkins, D.T., Gleiser, M., and Kelley, K.K., <u>Selected Values of the Thermodynamic Properties of Binary Alloys</u>, American Society of Metals, Metals Park, OH, 1973.

110. Ho, J.C., O'Neal, H.R., and Phillips, N.E., Rev. Sci. Instrum., <u>34</u>, 782–3, 1963.

111. Cieloszyk, G.S., Cote, P.J., Salinger, G.L., and Williams, J.C., Rev. Sci. Instrum., <u>46</u>, 1182–5, 1975.

112. Marklund, K., Bystrom, S., Larsson, M., and Lindqvist, T., Cryogenics, <u>13</u>, 671–2, 1973.

113. Stephens, R.B., Cryogenics, <u>15</u>, 481–2, 1975.

114. Lawless, W.N., Cryogenics, <u>20</u>, 527–8, 1980.

115. Parkinson, D.H. and Quarrington, J.E., Br. J. Appl. Phys., <u>5</u>, 219–20, 1954.

116. Waki, S., Denki Shikenjo Iho, <u>32</u>, 1151–6, 1968. This article is in Japanese but the graphical presentation of data is in English.

117. de Nobel, J. and du Chatenier, F.J., Physica, <u>29</u>, 1231–2, 1963.

118. Zoller, P., Decker, P.R., and Dillinger, J.R., Rev. Sci. Instrum., _39_, 1621–2, 1968.

119. Brewer, D.F., Edwards, D.O., Howe, D.R., and Whall, T.E., Cryogenics, _6_, 49–51, 1966.

120. Stewart, G.R., Cryogenics, _18_, 120–1, 1978.

121. Phillips, N.E., Phys. Rev., _114_, 676–85, 1959.

122. Cude, J.L. and Finegold, L., Cryogenics, _11_, 394–5, 1971.

123. Heessels, J.T., Cryogenics, _11_, 483–4, 1971.

124. Stephens, R.B., Cryogenics, _15_, 420–2, 1975.

125. Scott, T.A., de Bruin, J., Giles, M.M., and Terry, C., J. Appl. Phys., _44_, 1212–6, 1973.

126. Cieloszyk, G.S., Cruz, M.T., and Salinger, G.L., Cryogenics, _13_, 718–21, 1973.

127. Tucker, J.E. and Reese, W., J. Chem. Phys., _46_, 1388–97, 1967.

128. Choy, C.L., Hunt, R.G., and Salinger, G.L., J. Chem. Phys., _52_, 3629–33, 1970.

129. Berg, W.T., J. Appl. Phys., _39_, 2154, 1968.

130. Fisher, R.A., Brodale, G.E., Hornung, E.W., and Giauque, W.F., Rev. Sci. Instrum., _39_, 108–14, 1968.

131. Alterovitz, S. and Gershenson, M., Cryogenics, _14_, 618–9, 1974.

132. Jirmanus, M., Sample, H.H., and Neuringer, L.J., J. Low Temp. Phys., _20_, 229–40, 1975.

133. Wun, M. and Phillips, N.E., Cryogenics, _15_, 36–7, 1975; Bevolo, A.J., Cryogenics, _14_, 661–2, 1974.

134. Westrum, E.F., Chou, C., Osborne, D.W., and Flotow, H.E., Cryogenics, _7_, 43–4, 1967.

135. Ehrat, R. and Rinderer, L., Z. Angew. Math. Phys., _24_, 225–47, 1973.

136. Novotny, V. and Meincke, P.P.M., Rev. Sci. Instrum., _44_, 817–20, 1973.

137. Filby, J.D. and Martin, D.L., Phys. Lett., _3_, 244, 1963.

138. Forgan, E.M., Cryogenics, _14_, 207–14, 1974.

139. Gayle, T.M. and Berg, W.T., Rev. Sci. Instrum., _37_, 1740–2, 1966.

140. West, E.D., J. Res. Natl. Bur. Stand., _67A_, 331–41, 1963.

141. Pinel, J. and Lebeau, C., J. Phys. E, _5_, 688–90, 1972.

142. White, J.J., Song, H.I., Rives, J.E., and Landau, D.P., Phys. Rev., _B4_, 4605–10, 1971.

143. Morin, F.J. and Maita, J.P., Phys. Rev., _129_, 1115–20, 1963.

144. Kunzler, J.E., Walker, L.R., and Galt, J.K., Phys. Rev., _119_, 1609–14, 1960.

145. Haemmerle, W.H., Reed, W.A., Juodakis, A., and Kannewurf, C.R., J. Appl. Phys., _44_, 1356–9, 1973.

146. Malyshev, V.M., Topnikov, V.N., and Shchegolev, I.F., Cryogenics, _16_, 50–1, 1976.

147. Sullivan, P.F. and Seidel, G., Phys. Rev., _173_, 679–85, 1968.

148. Sullivan, P.F. and Seidel, G., Phys. Lett., _25A_, 229–30, 1967.

149. Hempstead, R.D. and Mochel, J.M., Phys. Rev., _B7_, 287–99, 1973.

150. Lee, K.N., Bachmann, R., Geballe, T.H., and Maita, J.P., Phys. Rev., _B2_, 4580–5, 1970.

151. Viswanathan, R., J. Appl. Phys., _46_, 4086–7, 1975; Viswanathan, R., in _Analytical Calorimetry_ (Porter, R.S. and Johnson, J.F., Editors), Vol. 3, Plenum, New York, NY, 81–8, 1974.

152. Zubeck, R.B., Barbee, T.W., Geballe, T.H., and Chilton, F., J. Appl. Phys., _50_, 6423–36, 1979.

153. Manuel, P., Niedoba, H., and Veyssie, J.J., Rev. Phys. Appl., _7_, 107–16, 1972.

154. Kogure, Y. and Hiki, Y., Jpn. J. Appl. Phys., _12_, 814–22, 1973; J. Phys. Soc. Jpn., _39_, 698–707, 1975.

155. Bertman, B., Heberlein, D.C., Sandiford, D.J., Shen, L., and Wagner, R.R., Cryogenics, 10, 326-7, 1970. For a correction of Eq. (3) of Bertman et al. and a more detailed description of a similar apparatus, see Sichel, E.K., Miller, R.E., Abrahams, M.S., and Buiocchi, C.J., Phys. Rev., B13, 4607-11, 1976.

156. Alterovitz, S., Deutscher, G., and Gershenson, M., J. Appl. Phys., 46, 3637-43, 1975.

157. Phillips, W.A., J. Appl. Phys., 51, 3583-5, 1980.

158. Bachmann, R., DiSalvo, F.J., Geballe, T.H., Greene, R.L., Howard, R.E., King, C.N., Kirsch, H.C., Lee, K.N., Schwall, R.E., Thomas, H.U., and Zubeck, R.B., Rev. Sci. Instrum., 43, 205-14, 1972.

159. Schwall, R.E., Howard, R.E., and Stewart, G.R., Rev. Sci. Instrum., 46, 1054-9, 1975.

160. Rade, H.S., Feinwerktechnik u. Messtechnik, 83, 230-3, 1975; Stewart, G.R. and Giorgi, A.L., Phys. Rev., B17, 3534-40, 1978.

161. Forgan, E.M. and Nedjat, S., Rev. Sci. Instrum., 51, 411-7, 1980.

162. Griffing , B.F. and Shivashankar, S.A., Rev. Sci. Instrum., 51, 1030-6, 1980.

163. Novotny, V. and Meincke, P.P.M., J. Low Temp. Phys., 18, 147-57, 1975.

164. Kruger, R., Meissner, M., Mimkes, J., and Tausend, A., Phys. Status Solidi, a17, 471-8, 1973.

165. Park, J.G. and Vaidya, A.W., J. Low Temp. Phys., 40, 247-74, 1980.

166. Gavrilov, N.M., Polovov, V.M., and Ponomareva, R.R., Prib. Tekh. Eksp., No. 5, 213-6, 1979; Engl. Transl.: Instrum. Exptl. Tech., 22, 1422-6, 1979.

167. Jones, R.W., Knapp, G.S., and Veal, B.W., Rev. Sci. Instrum., 44, 807-10, 1973.

168. Harrison, J.P., Rev. Sci. Instrum., 39, 145-52, 1968.

169. Green, B.A. and Culbert, H.V., Phys. Rev., 137, A1168-71, 1965.

170. Robbins, R.A. and Marshall, B.J., J. Appl. Phys., 42, 2562-5, 1971.

171. Loponen, M.T., Dynes, R.C., Narayanamurti, V., and Garno, J.P., Phys. Rev. Lett., 45, 457-60, 1980.

172. Loponen, M.T., Dynes, R.C., Narayanamurti, V., and Garno, J.P., Phys. Rev., B25, 1161-73, 1982.

173. Saint-Paul, M., J. Phys. Lett., 41, L169-72, 1980.

174. Schrey, P., Balzer, R., and Peisl, H., J. Phys. C, 10, 2511-21, 1977.

175. Jackson, J.J., Rev. Sci. Instrum., 51, 35-41, 1980.

176. Comberg, A., Ewert, S., and Sander, W., Cryogenics, 18, 79-81, 1978.

177. Greene, R.L., King, C.N., Zubeck, R.B., and Hauser, J.J., Phys. Rev., B6, 3297-305, 1972.

178. Zally, G.D. and Mochel, J.M., Phys. Rev., B6, 4142-50, 1972.

179. Filler, R.L., Lindenfeld, P., and Deutscher, G., Rev. Sci. Instrum., 46, 439-42, 1975.

180. Gibson, B.C., Ginsberg, D.M., and Tai, P.C.L., Phys. Rev., B19, 1409-19, 1979.

181. Novotny, V. and Meincke, P.P.M., Phys. Rev., B8, 4186-99, 1973.

182. Stewart, G.R., Phys. Rev., B15, 1143-50, 1977.

183. Gordon, J.E., Hall, R.O.A., Lee, J.A., and Mortimer, M.J., Proc. R. Soc. London, A351, 179-96, 1976.

184. Trainor, R.J., Knapp, G.S., Brodsky, M.B., Pokorny, G.J., and Snyder, R.B., Rev. Sci. Instrum., 46, 1368-73, 1975.

185. Osborne, D.W., Flotow, H.E., Fried, S.M., and Malm, J.G., J. Chem. Phys., 61, 1463-8, 1974.

186. Blythe, H.J., Harvey, T.J., Hoare, F.E., and Moody, D.E., Cryogenics, 4, 28-35, 1964.

187. Martin, D.L. and Snowdon, R.L., Rev. Sci. Instrum., 41, 1869-76, 1970.

188. Moses, D., Ben-Aroya, O., and Lupu, N., Rev. Sci. Instrum., 48, 1098-103, 1977.

189. Gmelin, E. and Rodhammer, P., J. Phys. E, _14_, 223–8, 1981.

190. Martin, D.L., J. Phys. E, _12_, 478–9, 1979.

191. Joseph, O., Moody, D.E., and Whitehead, J.P., J. Phys. E, _9_, 595–9, 1976.

192. Pinel J., Lebeau, C., and Raboutou, A., Rev. Phys. Appl., _15_, 75–9, 1980.

193. Rindelhardt, U., Herrnkind, U., and Hegenbarth, E., Exp. Tech. Phys., _24_, 541–5, 1976.

194. Bohmhammel, K., Schmidt, H.G., and Wolf, G., Exp. Tech. Phys., _28_, 275–81, 1980.

195. Collan, H.K., Heikkila, T., Krusius, M., and Pickett, G.R., Cryogenics, _10_, 389–93, 1970.

196. Barron, T.H.K. and Morrison, J.A., Can. J. Phys., _35_, 799–810, 1957.

197. Martin, D.L., Phys. Rev. Lett., _12_, 723–4, 1964; Phys. Rev., _B8_, 5357–60, 1973.

198. Padamsee, H., Neighbor, J.E.,and Shiffman, C.A., Phys. Rev., _B13_, 5125–30, 1976.

199. Colet, M., Hermans, F., Van Dorpe, C., and Lambert, M.H., Phys. Lett., _38A_, 321–2, 1972.

200. Lounasmaa, O.V. and Veuro, M.C., Phys. Lett., _40A_, 371–2, 1972.

201. Gregers-Hansen, P.E., Krusius, M., and Pickett, G.R., J. Low Temp. Phys., _12_, 309–17, 1973.

202. Anderson, D.A. and Cracknell, A.P., Phys. Status Solidi, _b56_, 157–61, 1973.

203. Lawless, W.N., Phys. Rev. Lett., _36_, 478–9, 1976; Phys. Rev., _B23_, 2421–4, 1981.

204. Martin, D.L., Phys. Rev., _B6_, 1169–76, 1972.

205. Flotow, H.E., Kuentzler, R., and Osborne, D.W., Phys. Status Solidi, _a34_, 291–5, 1976.

206. Trainor, R.J., Brodsky, M.B., and Culbert, H.V., Phys. Rev. Lett., _34_, 1019–22, 1975.

207. Triplett, B.B. and Phillips, N.E., Phys. Lett., _37A_, 443–4, 1971.

208. Blandin, A., J. Phys. (Paris), _39-C6_, 1499–516, 1978.

209. Rizzuto, C., Rep. Progr. Phys., _37_, 147–229, 1974.

210. Martin, D.L., Phys. Rev., _B21_, 1906–10, 1980.

211. Barnett, V. and Lewis, T., _Outliers in Statistical Data_, John Wiley, Chichester, England, 1978.

212. For a brief account, see Mellor, J.W., _Higher Mathematics for Students of Chemistry and Physics_, Dover, U.S.A., 1955.

213. Stewart, G.R., Rev. Sci. Instrum., _54_, 1–11, 1983.

CHAPTER 4

Calorimetry in the Range 5–300 K

E. F. WESTRUM, JR.

4.1. INTRODUCTION

Heat-capacity data at very low temperatures are clearly useful in delineating the energy spectrum of matter. As one goes into higher – but still cryogenic – temperatures, one enters a thermal chaos in the complexity of the energy spectrum in the thermal properties, such that heat-capacity studies above 10 or 20 K are of considerably less interest to physicists than are those at lower temperatures. On the other hand, the region from 10 to ambient temperatures is of great significance to the chemical thermodynamicists concerned with the thermodynamic function at or above 300 K. This region from 30–300 K is that over which most of the entropy and enthalpy – and consequently the Gibbs energy – are developed. Since the interpretation becomes increasingly statistical in nature, the entropy – particularly that associated with the various transformations of matter – often becomes a more relevant parameter than does the heat capacity. For these reasons, the temperature region of this chapter is of particular importance from the point of view of critical evaluation of ambient temperature chemical thermodynamic data for science and for technology, since the heat-capacity values make such a significant contribution to the thermophysical functions.

Although 10 or even 20 K may represent in many respects a lower practical limit for this region, the utilization of liquid helium as a refrigerant does usually permit the extension of the data to 4 and 5 K, albeit with lower accuracy, so we shall occasionally speak about measurements over this region also. Particularly if platinum thermometry is used to establish a temperature scale through this region, the data may be of relatively low precision, but still useful in establishing the presence or the absence of major anomalies.

Heat capacity data over the cryogenic region, spectroscopic data, and enthalpies of combustion are the basic ingredients for the technological discernment of the enthalpy and the Gibbs energy increment of chemical reactions. In the final utilization of these data for the evaluation of the driving force of chemical reactions, differences in the Gibbs energy increments or entropy increments of – or even between – chemical reactions are important. Since the needed reaction data are small differences between large numbers, errors in excess of more than about 0.1% in the basic numbers are quite intolerable.

Many modern plants rely for their operability upon the clever combination of exothermic and endothermic reaction steps for capital and operating cost minimization. Here the accuracy and the precision of the basic data can be spelled out in large dollar amounts, but will rarely be told because of the proprietary nature of the information. The accuracy and the precision of the Gibbs energy data required for meaningful estimates of equilibrium product distributions is an order of magnitude greater than is that for reaction enthalpies alone. Consequently, the need for more and for precise data, as well as for their critical evaluation, is, therefore, quite overwhelming.

The importance of obtaining the highest practicable accuracy and even higher precision is so great that most – but not all – of the chemical thermophysicists engaged in heat–capacity measurement in the cryogenic region in the United States in recent decades use, or have used, similar resistance thermometers calibrated against the International Temperature Scale above 90 K and against the national standard scale below this temperature [1,2]. They also compare experimental determinations on standard substances [3] for the elimination of systematic errors. As temperatures in technological processes tend toward higher values, the entropy increments – and hence the heat capacities – assume a more dominant role. Although the temperature scale is less important for the underived thermophysical properties, the accuracy of heat capacity is probably limited at the present time by uncertainties in the temperature scale and by temperature determination. An experimental precision of the order of a few hundredth of a percent is, in principle, possible and has been claimed by several laboratories.

Other important constraints are the preparation, identification, and the characterization of the samples themselves. This subject will also be dealt with in the present chapter. A very important consideration – particularly as the temperature approaches zero kelvin – is a general sluggishness and the hysteresis characteristic of phase transition and reorientational motion in substances. Herein arises an important aspect of cryogenic calorimetry and a contrast with the procedures very often used in differential scanning calorimetry over the same or even higher temperature ranges.

Consideration is also given to the important problem of the manner of operating, the scale on which the measurements are made, and the means used in the calculation of the data. With the important trend in the direction of automation of measurements and in the logging of data, calorimetry is not far behind, and consideration of the problems inherent and eliminated by the incorporation of computer–operated calorimeters will also be taken into account.

It should, of course, not be imagined this is the first attempt to summarize the problems of operating in this temperature region. An excellent review – which will be drawn upon whenever feasible in this presentation to keep it terse and brief – is that on adiabatic low–temperature calorimetry [4] as well as the corresponding one on isoperibol calorimetry [5]. A companion chapter which has much of utility applicable to the subambient region is that on adiabatic calorimetry from 3–800 K [6]. Attention is called also to the monograph on low–temperature heat capacities [7] as well as to a chapter dealing with molecular crystals which summarizes many of the problems encountered [8].

4.2. CRYOSTAT DESIGN

Although we sometimes fail to realize that even heat capacities and isothermal enthalpy increments – alias 'specific and latent heats' – had to be discovered also, and that the science of calorimetry has a greater antiquity than is sometimes appreciated [9], in more recent years the concept of evaluating the heat capacity of condensed phases by determining the temperature rise accompanying the input of a measured quantity of electrical energy was first applied by Gaede [10]. In the era of 1910, the concept was developed by Nernst [11] and by his collaborator Eucken [12], which led to the precursor of the modern, adiabatic, vacuum calorimeter. A simple schematic representation of such a calorimeter is presented in Fig. 4.1.

Neglecting for the moment refrigerant and the container for the refrigerant and considering the items 1 through 8 (cf. Fig. 4.1), we note the presence of the crystalline sample, 6, whose heat capacity is to be determined, and which is placed in the sample vessel ('calorimeter'), 5, equipped with a source of electrical energy, 3, and a thermometric device, 7, and all of this suspended within an evacuated enclosure, 8. A suitable means of reducing the above assemblage to the lowest cryogenic temperature of interest is then also a desideratum. However this latter cooling is achieved, it is evident that the interposition of one or more controlled-temperature shields, 4, between the calorimeter and the lowest temperatures allows operation at temperatures considerably higher than that of the refrigerant.

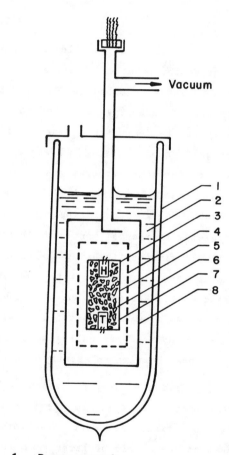

4.2.1. Immersion or Not?

4.2.1.1. Immersion in a Refrigerant Bath

The evacuated vessel surrounding the calorimetric apparatus may – as it was in the earliest model – simply be dipped into a Dewar or other vessel containing a refrigerant with a sufficiently low boiling temperature. For this purpose, the evacuated vessel becomes essentially a 'submarine.' This is a conceptually simple design with the rather significant advantage that leads are brought out through the liquid and hence are 'tempered' in the sense that the heat leak is broken by direct contact with the refrigerant and the cold vapor above the refrigerant surface. This is perhaps the most convenient, the cheapest, and, in many respects, the most expedient way of operation. An alternative arrangement is the aneroid-type cryostat.

1. Dewar vessel
2. Liquified gas refrigerant
3. Electrical heater
4. Adiabatic jacket
5. Calorimeter vessel
6. Crystals
7. Thermometer
8. Evacuated submarine

Figure 4.1. Schematic cryogenic calorimeter with electrical energy input.

4.2.1.2. Aneroid-Type Cryostat

In this mode of operation, the refrigerant(s) are contained in suitable cylindrical vessels, isolated from the surroundings by vacuum. This technique has the advantage that it can be very economical in the amount of the refrigerant required for a series of measurements, and the disadvantage that such vessels are relatively expensive to construct since considerable machine work is required. On the other hand, they are quite rugged devices and, if properly constructed, will serve over long periods of time. They do have the disadvantage that a scheme for tempering the lead wires will need to be found if these are not brought through the liquid refrigerant and the cold gases contained in equilibrium with them. A typical 'aneroid-type' cryostat will be described in Section 4.2.3.1.

4.2.2. Cooling and/or Quenching

The problem in question is essentially that of thermal switching. One must cool the sample prior to the input of electrical energy and the concurrent measurement of heat capacity. A number of techniques have been utilized with good success. Although helium exchange gas provides a convenient and effective means of cooling, the tendency for helium adsorption on surfaces – particularly as the temperature approaches that of liquid helium – necessitates provision of a properly designed pumping system. However, an adequately-long time for degassing may still be involved. As the operating temperatures approach those obtainable only with pumped liquid helium, there is an increased incentive for the complete elimination of exchange gas by a mechanical thermal switch. For an aneroid apparatus, thermal contact between the calorimeter and some portion of the apparatus in thermal contact with refrigerant gas can be achieved. This can be done by the use of a simple windlass to bring the calorimeter successively into contact with the adiabatic shield and the latter into contact with a matching member on the tank containing the low-temperature refrigerant bath. Keesom and Van den Ende [13] obtained thermal contact by lowering the specimen onto a projection from the base of the surrounding vacuum can and lifting the can free of the base by a simple windlass when making measurements. However, according to other authors [14], considerable frictional heating apparently occurred at low temperatures in using this arrangement.

A device already described by Westrum et al. [15] has been used effectively between ambient temperature and 4 K. Little frictional heating results during the process of isolation at 4 K. If the surfaces are kept scrupulously clean, cooling from room temperature to 4 K may be obtained even on the sample of large heat capacity within a period of 10–14 hours.

Should very rapid cooling be desirable, it is possible even with this apparatus to break the vacuum with a 10^{-4} bar of helium gas, to cool directly to about 80 K within a period of 20–30 minutes, and then quickly re-establish the insulating vacuum.

Several aneroid cryostats have been devised [16] which permit the vacuum between the low-temperature cryogen and the calorimeter to be broken only in the region between the calorimeter and the guard shield while maintaining the insulating vacuum surrounding this portion of the cryostat.

A number of other simple heat switches have been described [e.g., 17-19] but because of the complication introduced by the presence of an adiabatic shield, these devices would require some modification were they to be rendered applicable to adiabatic calorimetry.

It is to be noted that an appropriately slow rate of cooling of a sample may be particularly important to prevent the 'freezing-in' of high-temperature phases in regions where they would be metastable. On the other hand, there are occasions in which it is desired to quench a high-temperature phase (and subsequently to measure the heat capacity of that quenched high-temperature phase). For these purposes rapid cooling is a desideratum and can be achieved by some of the techniques already described. It has often been stated that approximately a third of ordinary molecular crystals can be quenched to a vitreous phase by sufficiently rapid cooling. Should the previously described technique not give adequate cooling, it is possible to devise an apparatus in which a jet of liquid refrigerant could be directed at the calorimeter itself for even more rapid quenching [16].

4.2.3. Adiabatic Technique

In many thermophysical calorimetric measurements, determination of the energy lost by heat leak is the factor that most severely limits accuracy; therefore, the decision usually must be made between retaining a significant - but calculable - heat leak or endeavoring to eliminate the heat leak to the point at which it is insignificant - not calculable. For example, if the controlled-temperature shield 4 in Fig. 4.1 is designed to have heat capacity to retain an essentially constant temperature during a measurement, the technique is described as one involving an essentially isothermal jacket. The technique of operation with an essentially isothermal - or better described as isoperibol - jacket was at first used by Nernst and his collaborators. The current cryogenic counterpart, described in detail elsewhere [5] and in a summary fashion herein, still finds many applications today, particularly at temperatures below the range covered in this chapter.

If, on the other hand, the temperature of the shield is so regulated as to be as nearly identical as possible with that of the calorimeter, the technique may be described as one involving an adiabatic jacket or shield. This also has been described in detail elsewhere [4], and will be described in more detail in this presentation. The introduction of the adiabatic method at cryogenic temperatures has been credited to Southard and Andrews [20].

4.2.3.1. Aneroid-Type Adiabatic Cryostat

The first cryostat to be described - one of essentially metallic construction - is depicted in Fig. 4.2 and designed by Westrum et al. [21-26]. The entire cryostat assembly is suspended from the cover plate so that a hoist may be used to remove it from the vacuum tank for leading a sample - or so that alternatively the vacuum can be removed from the rigidly mounted cover plate. The primary function of the cryogenic apparatus is to maintain the calorimeter vessel at any desired temperature between 4 and 350 K in a state of thermal isolation such that virtually no heat is lost or gained by the calorimeter vessel except when introduced by an intermittently operated electrical 'heater' during energy inputs. Two chromium plated, copper refrigerant tanks for

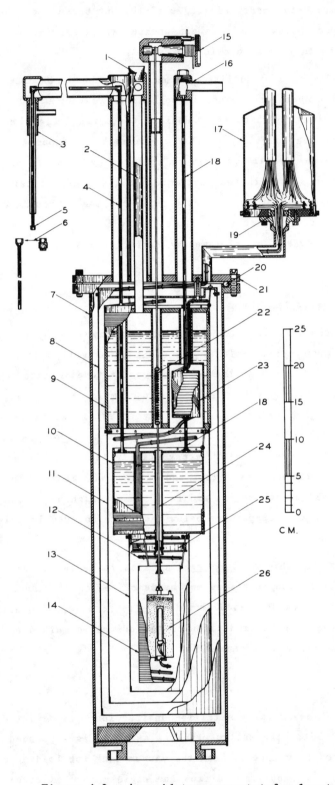

1. Liquid nitrogen inlet and outlet connector
2. Liquid nitrogen filling tube
3. Sleeve fitting to liquid helium transport Dewar
4. Liquid helium transfer tube
5. Screw fitting at inlet of the liquid helium transfer tube
6. Liquid helium transfer tube extender and cap
7. Brass vacuum jacket
8. Outer 'floating' radiation shield
9. Liquid nitrogen tank
10. Liquid helium tank
11. Nitrogen radiation shield
12. Bundle of lead wires
13. Helium radiation shield
14. Adiabatic shield
15. Windlass
16. Helium tank exit
17. Terminal cover
18. Helium tank exit tube
19. Lead seal fitting
20. O-ring cryostat seal
21. Top plate
22. Spring in contact device
23. Economizer
24. Braided silk line
25. Ring
26. Calorimeter

Figure 4.2. Aneroid-type cryostat for low-temperature adiabatic calorimetry [21–25].

liquid nitrogen and for either liquid or solid nitrogen and liquid helium provide low temperature heat sinks. The calorimeter (i.e., the sample container) is suspended from a windlass by a braided silk line. The windlass is used to bring coaxial cones on the calorimeter, adiabatic shield, and the liquid helium tank into direct thermal contact, thereby cooling the calorimeter and the shield. At the desired operating temperature, thermal contact is broken by lowering the calorimeter and the adiabatic shield, and adiabatic conditions are established in preparation for measurements. The adiabatic shield surrounding the calorimeter is suspended from the helium tank by three braided silk cords.

Adiabatic conditions are established at any operating temperature by careful control of all factors that would lead to heat exchange between the calorimeter and its environment. Gas conduction is eliminated by maintaining a high vacuum within the tank. Chromium-plated, copper radiation shields serve not only to conserve refrigerant but also to set up regions of uniform and progressively lower temperatures. Suspensions of low thermal conductivity (such as the stainless steel wire supporting the nitrogen tanks and the silk cords supporting the lower tank, adiabatic shield, and calorimeter) minimize thermal conduction. In a calorimeter of this design, it is important to take special precautions to minimize the thermal conduction along the electrical leads, since these do not pass through refrigerant vapor. The 40 or more leads enter the cryostat through a perforated linen-base Bakelite disk sealed in place with Apiezon-W wax. These leads are thermally anchored to the liquid nitrogen tank and to an 'economizer,' a small cylinder with perforated baffles, which serves as a heat exchanger utilizing the relatively large (compared to the enthalpy of vaporization of liquid helium) heat capacity of the cold, effluent, helium gas to absorb most of the heat conducted down the leads, thereby conserving liquid helium. The leads are anchored to the lower tank from which they pass to a 'floating ring,' which can be maintained very near the temperature of the adiabatic shield and the calorimeter by an electrical heater. Between each of the enumerated stages, the bundle of leads makes loops in the vacuum space to lengthen the conduction path. The final tempering of the leads before they reach the calorimeter is accomplished by varnishing them in a helical groove beneath the heater winding on the cylindrical portion of the adiabatic shield. Upon establishment of adiabatic conditions, the equilibrium temperature of the sample is measured by a platinum-encapsulated, platinum resistance thermometer mounted in an entrant well in the calorimeter vessel. Direct current electrical energy is then supplied to the calorimeter (and sample) by the heating element installed in the same well.

Practical experience has demonstrated that it is possible to provide an equally quick quenching to about 70 K even in the aneroid-type of cryostat without excessive expenditure of liquid nitrogen in the two tanks by simply breaking the insulating vacuum for a few minutes with a small pressure of helium gas and then expeditiously removing it.

The principal advantages of this design of calorimeter are its economy of liquid helium, ruggedness, and ease of reloading with new samples. Calorimeters can be constructed with appropriate size, shape, and features to make them adaptable to samples of different types. The only apparent disadvantage of such a cryostat is its possibly higher initial construction cost.

4.2.3.2. Immersion-Type Adiabatic Cryostat

This particularly simple design of cryostat is advantageous both in respect to the minimal metal fabrication involved and in the facility in which it may be constructed. In its simplest form it may consist of little more than an evacuated container (or 'submarine') in which the calorimeter and/or sample is suspended. The container is then immersed in a suitable refrigerant (e.g., liquid nitrogen, liquid hydrogen, or liquid helium) contained in a suitable glass or metal Dewar for operation over corresponding temperature ranges. The calorimeter devised by Furukawa and coworkers [3,27,28] has been used extensively for measurements on both organic and inorganic substances. The construction consists of dual submarines, the inner one containing a tempering ring, an adiabatic shield, and the calorimeter vessel. This submarine is immersed in a refrigerant — suitable for the lowest temperatures desired — which is contained in a glass or metal Dewar. This Dewar, in turn, is entirely surrounded by a second submarine to facilitate pumping upon the refrigerant and thus achieving lower temperatures. The outer submarine is contained within the liquid nitrogen bath of an outer Dewar. To cool the calorimeter and its contents and to calibrate thermometers (a process which usually requires good thermal contact with the bath) some form of thermal switching is required. In an immersion cryostat, this is usually effected by means of helium exchange gas in the inner submarine. Because helium gas is strongly absorbed on most surfaces at liquid-helium temperatures, considerable time and high-speed pumping may be required to obtain a sufficient degree of degassing to ensure adequate thermal isolation for the measurement. However, as in the previously described aneroid cryostat, it is possible to use a mechanical heat switch between the calorimeter, adiabatic shield, and tempering ring to achieve thermal contact.

4.2.3.3. Combination Immersion-Aneroid-Type Adiabatic Cryostat

A combination of the two preceding designs has been created by Sterrett et al. [29] especially for measurements on materials in the vitreous phase. It has the advantages of versatility in producing a desired thermal history of the glass sample and reliability over long periods of continuous operation necessary for studying 'kinetic effects' associated with the glass transition. This cryostat employs a single submarine housing, the aneroid-type of helium tank, and heat exchanger similar to that involved in the cryostat of Westrum et al. [30]. As with the immersion-type cryostat, cooling at higher rates than provided by the thermal switch may be obtained by breaking the vacuum with helium gas and utilizing the cooling power of the refrigerant contained in the Dewar vessel surrounding the assembly.

4.2.3.4. Filling-Tube-Type Cryostat

For making heat-capacity measurements on materials which are gaseous below — or not far above the ambient (room temperature) — it is convenient and sometimes highly desirable to distill the sample directly into a calorimeter already located within the cryostat at reduced temperatures. To this end a calorimetric cryostat provided with a filling tube (typically of thin-wall, stainless-steel tubing) is useful.

Although in most other respects such a cryostat may be similar to any of the three previously described designs, some of the problems imposed by inclusion of the filling tube require attention. During operation, it is essential that the temperature of the filling tube continuously be maintained slightly above that of the calorimeter so that material will not be distilled from calorimeter into the filling tube. However, in order to maintain adiabatic conditions, it is necessary that this excess temperature be carefully regulated. It is customary to provide a heater winding about the tube and, in addition, suitably-placed thermocouples to survey the temperature of the tube during measurements (and during loading). Another useful feature of this design is the possibility of distilling material directly from the calorimeter into a suitable receiver (a weighing vessel) and thereby determining directly the enthalpy of vaporization of the sample. The technique utilized here is rather similar to that of the corresponding operation in a calorimeter with an isoperibol jacket. Designs of successful adiabatic filling-tube-type cryostats [15,28,31] are numerous; an example is the calorimeter used by Furukawa [31] especially for enthalpy of vaporization measurements. Here the provision of a throttling valve to control the rate of vaporization is a valuable asset in determining the enthalpy of vaporization of substances. Moreover, even in the course of heat-capacity measurements it is also a convenience because it minimizes the required exactness of the temperature control of the filling tube, since the valve may be closed during the course of ordinary heat capacity and enthalpy of transition measurements.

4.2.3.5. Control of Heat Exchange with the Calorimeter

Adiabatic calorimetry in practice: Although in principle, adiabatic conditions may simply be described as the absence to heat transfer between system and surroundings, in practice, the problem of maintenance of adiabatic conditions is more difficult. This stems from the problem of achieving uniform exterior temperatures on the calorimeter and on the surroundings, the determination of a zero difference in temperature between these two entities, and in the maintenance itself of the desired temperature for the surroundings. As a matter of fact, most of what passes for 'adiabatic calorimetry' in the literature could better be described as 'quasi-adiabatic calorimetry.' This is indeed convenient and suggests that as a test for experimental adiabaticity, the absence of a temperature drift of the calorimeter replace the convention of a zero difference in temperature between calorimeter and surroundings. The thermometer current in a calorimeter does generate a certain amount of electrical energy and consequently a temperature drift; in a difficult equilibrium situation it is perhaps better to cancel this by the imposition of slight loss of energy by radiation from the calorimeter in order to maintain a constant temperature of the calorimeter. Although measurements are sometimes made by dynamic – continuous energy – input we here prefer to discuss discontinuous, i.e., intermittent, heating. This is the type of measurement which will generally be implied throughout the remaining discussion in this chapter unless the contrary is indicated. Measurements of heat capacity are usually made with increasing temperature and by the intermittent methods. In some instances – for example, at a magnetic transition – where the study of the hysteresis involved is important, one sometimes makes measurements (interpreted as heat capacity values) with decreasing temperatures by setting the temperature of the adiabatic shield at a fixed temperature, resistance, or emf (thermocouple) increment below that of the calorimeter and utilizing the continuous cooling method. When interpreted by an appropriately devised computer program, reliable values of heat capacity can be interpolated even in transition regions especially if appropriate adjustment is made for the effect of heat loss by radiation. An analysis of other types of calorimetry including continuous heating are made in the chapter of this volume entitled 'Adiabatic Calorimetry Above 300 K.'

The adiabatic jacket: The calorimeter vessel (defined as that part of the system in which energy input and corresponding temperature changes are accurately measured) and the adiabatic jacket are designed to have relative heat-leak coefficients as small as feasible. This permits reduction both in the uniformity of temperature control necessary over the various portions of the jacket and in the precision of control required, without excessive total heat transfer. The radiative heat transfer is kept small by gold-plating the outer surface of the calorimeter vessel and that of the jacket to achieve high reflectivity. The convective heat-leak coefficient is virtually eliminated by the maintenance of high vacuum; the heat-transfer coefficient of conduction is reduced by minimizing the cross-sections and thermal conductivities of the lead wires and filling tubes insofar as possible.

The relative temperature between the adiabatic jacket and the calorimeter vessel is determined by differential thermocouples. These thermocouples are designed to give as high a sensitivity (i.e., as large dE/dT as feasible) and to have themselves a low thermal conductivity for minimizing transfer. Chromel-P vs. constantan, Chromel-P vs. gold-cobalt (2.1 atomic percent Co) or other gold/silver alloy wires may be used to ensure high sensitivity over the entire working range. Often four different wires are utilized with two-junction thermels to achieve this purpose. The location of the thermocouples on the calorimeter vessel for the adiabatic jacket is important in order to obtain as nearly as possible the same temperature distribution on both.

The adiabatic jacket – like the calorimeter – usually approximates a cylinder, which may need to be at least partially dismantled to permit loading and unloading of the calorimeter. Thermocouples are normally provided on the top end, the middle cylindrical portion, and the bottom end even though but a single thermocouple or thermel is provided on the more integral surface of the calorimeter vessel. As a matter of expediency, the ends of the adiabatic jacket are often referenced to the middle of the cylindrical portion to minimize thermal conduction to the calorimeter vessel. During the equilibration period, the outer surface of the calorimeter vessel is nearly isothermal, so the temperature of the adiabatic jacket must also be isothermal. During the energy input, the calorimeter vessel is more likely to have a small temperature gradient although the placement of the calorimeter heater and the design of the calorimeter should endeavor to minimize this. The distribution of heaters on the jacket is as important as that of the thermocouples, and if the three sections of the adiabatic shield are not controlled independently it is especially important that the heater winding be designed with resistance in proportion to the mass of the part of the jacket which they control. The best practice is usually to control separately the three portions of the jacket. However, since the ideal temperature control of the adiabatic jacket cannot be achieved, the calorimeter should be designed to have the smallest practicable heat transfer coefficient. Note that adiabatic jacket control is described in Section 4.7.2.

The conductive leak along the lead wires: The leads may be wound in good thermal contact with the shield over the cylindrical portion. However, since the temperature difference between the refrigerant tanks and the calorimeter difference is of the order of several hundred kelvins, it is usually convenient to provide a tempering ring to reduce the gradient in the lead bundle preceding its junction with the adiabatic jacket. A differential thermocouple between the ring and the adiabatic jacket together with a suitable electronically-controlled heater on the ring is also provided.

The heater lead wires: Over most temperatures, an important heat leak between calorimeter and adiabatic shield involves the conduction along the lead bundle. Although the size of most leads involved may be minimized to practical mechanical, thermal and strain emf, and electrical lower limits, that of the heater current leads requires consideration. The Ohmic energy generated in these leads between the adiabatic jacket and the calorimeter vessel must be accounted for even though it is kept to probably less than 0.1% of the total energy. But even when the temperatures of thermal contact at both ends of the current leads are different the energy generated in these leads is dissipated equally to the jacket and to the calorimeter vessel [32]. Therefore, by attaching one potential heater lead at the jacket end of one current lead and the other potential lead at the calorimeter end of the second current lead, the energy generated in the current leads will be correctly accounted for in the power measurements. This avoids the corrections which would be required from most other configurations of the potential leads (cf, [4] for further discussion).

Guard shield: To minimize the adjustments required for automated operation of the cryostat whether the temperature control is in an automated mode or in manual operation, it is convenient to provide another (cylindrical) shield surrounding the adiabatic jacket. The temperature of this is controlled at a constant decrement from that of the adiabatic jacket itself. This cannot be introduced without some sacrifice, inconvenience, and other constraints because it also must be cooled by the refrigerant and imposition of additional surfaces in the contact link between the calorietmer and the refrigerant tank. This restricts the rate of the cooling process and perhaps even the ultimate minimum achieved temperature, and requires expenditure of somewhat more helium than otherwise would be involved. For this reason, the helium tank and the radiation shield to it are sometimes operated as a guard shield. This compromise works reasonably well, since the only temperatures at which such guarding is important are those temperatures in which radiation becomes increasingly important, i.e., above pumped liquid nitrogen temperatures.

4.2.4. Isoperibol Technique

In the earliest heat-capacity calorimeters developed during the first and second decades of the present century, the outer wall of the vacuum space surrounding the calorimeter was usually in thermal contact with the liquid bath of boiling refrigerant. When the temperature of the calorimeter was far from that of the bath, the heat leak became large and introduced serious errors in the measurement of heat capacity. These errors were eventually reduced as shown in Fig. 4.1 by the imposition of a metallic thermal shield intermediate between the calorimeter and the surrounding bath whose temperature could be controlled. Giauque and his collaborators [33,34] developed apparatus for the measurement of heat capacities and of enthalpies of transition, fusion, and vaporization over the 5-300 K range. In their apparatus, the calorimeter was surrounded by a

massive thermal jacket made of copper and lead in the insulating vacuum space. The jacket was
provided with an electrical heater and with thermocouples to measure its temperature. If the temperature of this isoperibol jacket were adjusted to be approximately midway between the initial
and final temperatures of the determination, the temperature of the calorimeter would never be far
(say ±5 K) from that of the jacket. By utilizing such an apparatus, the accuracy of heat capacity
measurements were reduced to several tenths of a percent between about 30 and 300 K. Measurements
made with this type of apparatus have often been loosely described by the term 'Isothermal Calorimetry.' Here the word isothermal refers not to the behavior of the calorimeter – whose temperature necessarily changes during heat capacity measurement – but to that of the massive block whose
temperature remains approximately constant during the course of a single determination. Stout [5]
mentions many other modifications of calorimetric apparatus of isoperibol design. No attempt at a
comprehensive survey of the literature will be made here.

4.2.4.1. Typical Isoperibol Cryostats

An isoperibol cryostat for low-temperature heat capacity measurements on solids is that of
Cole et al. [35]. Many of the features were copied from earlier designs of Giauque and his co-
workers at the University of California. In this version, the inner and outer tubes of the case
are soldered to flanges at the top which form part of an O-ring seal. Most other joints are soldered. The space between the tubes is filled with Santocel insulation and evacuated at the ambient temperature to a pressure of about 10 μm of mercury. The Pyrex glass Dewar vessel of 8.5
liters capacity is contained in a vacuum-type outer can of monel metal with a stainless steel
ball-vee mechanical joint at the top. The inner vacuum can is suspended by a monel tube about 2
cm outer diameter and through it the thermally insulating vacuum is pumped. The various thermocouples and leads to the resistance thermometer and heater also pass through it. The submarine is
made from monel tubing of about 9 cm outer diameter and the vacuum seal at the top of the can is
made with low-melting solder contained in a small annular groove on the bottom section of the can.

The massive thermal block is made in two sections, the top containing 3.2 kg of lead and 1.0
kg of copper and the bottom section a heavy-walled cylinder containing 2.4 kg of copper. The
joint between the two sections is a cone of 8° included angle and the weight of the shield is carried on screws fastened into the bottom section extending through two clearance holes in the top
section. Hence, the push brass rods threaded into the top of the shield and pushing against the
flat surface on the bottom are used to dismantle the conical joint for removal of the bottom portion of the shield. A heater of constantan wire is wound in helical grooves on the outside of the
top and bottom sections of the shield. Radial holes drilled through the base of the bottom section provide access ports for vacuum but prevent direct access of radiation from the outside to
the calorimeter. Thin sheets of German silver and of aluminum foil to minimize radiation exchange
between the shield and the vacuum can are wrapped around the shield. The inside of the shield is
gold-plated or lined with aluminum foil. All wires leading to the calorimeter are brought into
good thermal contact with the top of the shield but they are electrically insulated from it.
Copper-constantan thermocouples are used to measure the temperature of the top and bottom of the
shield and of the vacuum can.

Other calorimeters for solids and liquids as well as for gases have been described in the literature for the isoperibol mode of operation.

4.3. CALORIMETER DESIGN

4.3.1. General Considerations

As has already been indicated, the design of the calorimeter itself is crucial to the successful operation of an adiabatic calorimeter. Apart from the obviously important considerations of chemical, thermal, and mechanical stability are the problems of ensuring uniformity of temperature on the outer surface even during the energy input, of convenience in loading and unloading the samples, and of ensuring good dynamics for the thermal diffusivity and thermal equilibration processes. Mechanical strengths are occasionally important, particularly in dealing with the freezing of materials which expand upon solidification (e.g., water and salts of organic acids, some metals) or, for example, in dealing with materials of high vapor pressures. It is a desideratum to have the heat capacity of the calorimeter vessel small compared to that of the sample to give as favorable a set of circumstances as possible for the accurate determination of the heat capacity of the latter. In this exposition we shall be content to describe a single calorimeter in detail and then mention variants of this and the problems encountered thereby. Such a calorimeter is useful for typical samples of small crystals of ionic or molecular solids or even samples of zone-melted or flame-fused crystalline substances. By simple modification it can be utilized for liquids, for larger crystalline or shaped samples of various sorts.

4.3.2. Calorimeter with Internal Thermometer and Heater

Most of the details of the calorimeter may be seen by reference to Fig. 4.3 and the accompanying legend. The heater and thermometer are inserted within an entrant well and the bundle of leads passes from them, around the spool to ensure that the leads will attain the temperature of the calorimeter surface prior to their departure in the bundle. Calorimeters of this type have been widely used in many laboratories. This particular one has been described by Westrum et al. [4]. In this calorimeter, Apiezon-T was used to obtain thermal contact between the heater-thermometer assembly. It will be noted that the heater is wound bifilarly on a grooved, copper mandrel which is closely fitted within the reamed well of the calorimeter. The thermometer fits snugly within the heater sleeve. The differential thermocouple (or thermel) in the adiabatic jacket control system is mounted with the same type of grease for thermal contact within a sleeve (or sleeves) attached to the side of the vessel. Vertical vanes of thin copper foil aid in the distribution of heat from the electrical heater and in the subsequent re-establishment of thermal equilibration.

When a specimen to be measured is obtained in the form of a suitably-shaped (machined, cast, vitreous, or a grown single crystal block or cylinder), it is possible to use a similar style of calorimeter if the heater thermometer well is displaced from the cylindrical axis of the calorimeter and the sample(s) encased by sprung vanes of foil attached to the entrant well. If necessary, the entire top of the calorimeter may be made removable in order to insert the specimen by

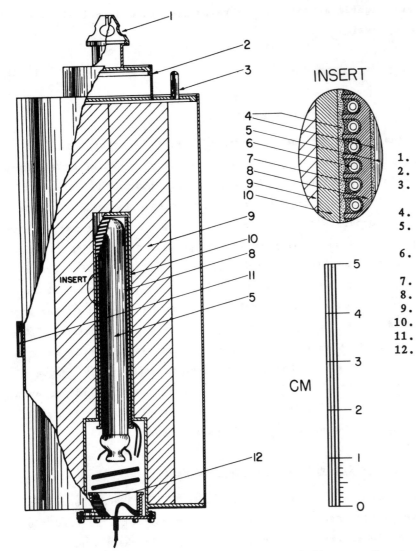

INSERT

1. Thermal contact cone
2. Monel cupola
3. Solder-capped monel tube for helium seal-off
4. Grease for thermal contact
5. Capsule-type platinum resistance thermometer
6. Fiberglas-insulated No. 40 Advance wire
7. Formvar enamel
8. Gold-plated copper heater core
9. Copper vanes
10. Gold-plated copper heater sleeve
11. Differential thermocouple sleeve
12. Spool to thermally equilibrate leads with calorimeter [37]

Figure 4.3. Cross-sectional diagram of calorimeter for solids.

increasing the diameter of the monel cupola to that of the calorimeter. Low-melting solder is convenient to seal the sample suitably within the calorimeter. For this purpose Cerroseal-50 (50% In/50% Sn) is convenient.

Because certain samples readily decompose in the inevitable temperature rise of the calorimeter during the soldering operation, if a thermal dam is not incorporated, it is desirable to provide a cover or cap such as the ones to which the thermal contact cone is attached, with a small neck or cupola of a poorly conducting material such as stainless steel or monel. This does not – as it may appear to do – imply that the temperature of the cover and the thermal conduction cone lag significantly behind the calorimeter. Helium gas in the sample space at 2 torr pressure provides good thermal diffusivity. With a thermal dam incorporated into the calorimeter, low-melting solder may be affixed without significant change in the temperature of the calorimeter or its contents. It is convenient to make the soldered seal and then to evacuate the calorimeter in a suitable vacuum vessel and to cover a pinhole in the tip of a monel tube with solder to seal the small helium pressure required for thermal diffusivity. A device for this purpose has been described [4].

An alternative closure which has proven very practicable and which avoids the adjustment of mass required in soldering is that of using a threaded monel neck with a sharp, circular, knife edge on its uppermost edge. A gold gasket may then be clamped into place by a threaded cover as has already been described elsewhere [6]. This type of seal has proven very adaptable between 1 and 800 K and avoids the necessity of the adjustment of the amounts of solder between the full and the empty calorimeter. In loading such a calorimeter, the sample is placed within the calorimeter, the gasket clamped into place by means of a wrench (remotely operated through an O-ring seal in a metal vacuum chamber) after the pressure of helium in the chamber has been adjusted to that desired within the sample space. After completion of the seal, its integrity can be tested with a helium leak detector. Many modifications of this technique and indeed of other modifications of the calorimeter have been described elsewhere [36,37].

A careful account is kept of the weights of the calorimeter, of the sample, and of the amounts of solder and helium on the filled calorimeter as well as on the empty calorimeter. The heat capacity of the empty calorimeter is determined in separate series of experiments.

The modification required to adapt such a calorimeter for use with condensed gases is essentially the replacement of the soldered or gold-gasketed seal with a filling tube.

4.3.3.   Calorimeter with Thermometer-Heater on Outer Surface

A good representative of this type of calorimeter is the gold calorimeter used in measurements for condensed gases described by Giauque and his collaborators [38-42]. This vessel, also, has a resistance thermometer mounted on a cylindrical axis in an entrant well and eight (vertical) gold vanes to improve the thermal contact between the calorimeter and the sample. A filling tube is provided for the operation. A later version of this type of calorimeter was described by Murch and Giauque [43]. This calorimeter was designed for samples which are solid or liquid at 300 K and consequently has a small filling tube over which a cap is soldered after sample and helium gas have been introduced into the calorimeter. A thermometer-heater made of 0.12 mm diameter wire of gold containing 0.1% silver was wound on the external cylindrical surface of the calorimeter over two layers of China silk, each 0.12 mm thick and the entire assembly cemented down with Formvar varnish. The completed thermometer is covered with two additional layers of silk floss cemented with Formvar varnish. In this instance, the gold thermometer-heater was compared with a platinum thermometer at the beginning and end of each heat-capacity measurement. Hence, during the drift period and the energy input the gold thermometer gave a continuous measure of the surface temperature of the calorimeter.

4.3.4.   Calorimeter for Liquids and Condensable Gases

In measurements of heat capacities of condensable vapors, liquids, or low-melting solids it is often expedient to use circular (horizontal) vanes perpendicular to the axis of the heater well [44] to prevent settling of the solid phase during fractional-melting studies and possible temperature homogeneities. Such vanes may be conveniently stamped from thin noble-metal foil. Alternatively, such horizontal vanes (spaced somewhat further apart), or vertical (radial), or even spiral or screw-shaped vanes may be machined from the same metal cylinder from which the heater-thermometer well is produced. Although many closures have been used to seal liquids into the calorimeter, one that has been found very convenient is the demountable valve with a small diameter aperture and gold-gasketed seal. A suitable valve-like mechanism for closing this demountable valve directly from the vacuum line may be provided with a suitable spring wire snap ring and groove to permit removal and reinsertion of the entire threaded plug carrying a gold gasket, so that a reasonably-sized aperture is available for the distillation process [45].

4.3.5. Special Calorimeters

Occasionally, the reactivity of samples will require the use of special metals or materials for calorimeter construction. An example is in the use of a nickel calorimeter for transition-metal hexafluorides. Here difficult brazing and welding operations were eliminated by the use of circular V-shaped flanges projecting into sharper angled grooves which proved to be vacuum tight over the entire temperature range [45]. A glass-lined copper calorimeter was employed for measurements on perchloric acid hydrate [46]. In instances where the strength or chemical resistance of poor thermal conductors such as stainless steel or other alloys is a desideratum, it is possible to electroform a copper or a silver thermal conducting layer on the exterior to provide for suitable temperature homogenization over the calorimetric vessel. The provision of a reflective gold plate (usually over successive copper and silver plates to minimize diffusion) provides for better reflectivity of heat.

4.3.6. Calculations Involved in Calorimetric Operation

Excellent mathematical analysis of the corrections for heat leaks on the basis of Newton's law of cooling, the thermal relaxation times, and of the calculation of heat capacity neglecting thermal gradients in the calorimetric system have been presented by Stout [5]. His discussion also includes the very important corrections for thermal gradients in the calorimetric system. When a calorimeter, such as that described in Section 4.3.3 is employed with an external thermometer-heater, the $(A/E)B_N$ correction first described by Giauque and Wiebe [38] must be made. In this mode of correction the temperature rise in a heat-capacity measurement is calculated from the resistance of the thermometer wound on the external surface of the calorimeter. During the heating period, the resistance of the thermometer-heater is also measured and thus the surface temperature of the calorimeter is known. The trend of the surface temperature and the mean temperature of the calorimetric system are both taken into account. Contributions from this important term approximate 1% near 300 K.

The related γB_N correction for calorimeters with internal thermometer-heaters is also explained [5], as is the effect of a temperature gradient between calorimeter and sample. The attention of the reader is called to the corrections discussed by Stout [5] and to other calculational adjustments for heating by thermometer current, and for addenda (solder and helium) to the heat capacity of the empty calorimeter. The calculation of the heat capacity and the derived thermodynamic properties is also treated. Note that a further discussion concerning the calculation of the heat capacity as well as of the derived thermal physical function is contained in Section 4.10 of this chapter.

4.4. THERMOMETRY AND TEMPERATURE SCALES

4.4.1. Thermometric Sensors

If a special thermometer – e.g., a gold thermometer-heater, a non-strain-free thermometer, a thermistor thermometer for the determination of minute temperature increases with high precision, etc. -- is used, this thermometer must be calibrated in terms of a standard thermometer. For many purposes this is done during the fore and after drift periods of each heat-capacity measurement against either a strain-free standard platinum-resistance thermometer or against a standard copper-constantan thermocouple which in turn has been calibrated by comparison with a gas thermometer. In many countries, the use of the Leeds and Northrup platinum-encased, capsule-type, platinum resistance thermometer is routine for the calorimeter thermometer. The temperature sensitivity of this thermometer in the region near 10 K and below is rather low and there has been increasing interest in the utilization of an iron-rhodium thermometer for heat-capacity measurements involving the 5-30 K region. Some measurements also have been made with miniature platinum resistance thermometers, which often prove not to be strain-free and are consequently somewhat irreproducible on cycling. They are not often bifilarly wound.

4.4.2. Thermometer Calibration and Thermometric Scales

Whatever mode of thermometry is used, calibration is important and needs to be not only ascertained carefully initially, but occasionally checked at certain fixed points such as the boiling point of hydrogen, of oxygen, of nitrogen, and/or the triple point of water. Cells either commercially available or the apparatus described by Stimson are convenient [2]. Occasionally, even a platinum resistance thermometer is calibrated by comparison with a standard thermometer of the same sort calibrated by the National Bureau of Standards or other standardizing laboratory, or against other standards.

Although initially smooth tables of the temperature/resistance ratio were prepared for manual calculations, the preparation of a temperature scale by computer using, for example, Lagrangian interpolation coefficients greatly simplifies the task. Until such a time as the thermodynamic temperature scale can be realized directly, thermometer calibrations are usually referred to the International Temperature Scale of 1948 (IPTS-48) for the region between the boiling point of sulfur and of oxygen or against the more recent revision – that of 1968. (Unfortunately, more recent tests of the IPTS-68 temperature scale have shown it to be somewhat flawed and less accurate than its 1948 predecessor.) Below the oxygen boiling point the scale of Hoge and Brickwedde [1] is used.

Thermometers calibrated by the U.S. National Bureau of Standards have an unfortunate discontinuity in the derivative of the resistance vs. temperature at the junction of the IPTS with that of Hoge and Brickwedde. The artificial kinks and curves of measured heat-capacities occasioned by this discontinuity have seldom been successfully smoothed out in the construction of the smoothing of the resistance-temperature table.

4.4.3. Temperature Calibration and Thermometric Scales

4.4.3.1. Special Thermometers

For some purposes, such as the more precise determination of the very small temperature increments, a less reproducible but higher-sensitivity thermometer is utilized, and is calibrated during the fore and after drift period of each heat capacity measurement against a more reliable absolute thermometer. The special thermometer is then used as an interpolation device of higher sensitivity than that of a platinum resistance thermometer. An example of this is in the work of Suga and his collaborators [47,48], who have used this approach with a thermistor in the delineation of small peaks on a normal curve where ordinary heat capacity measurements show little more than scatter in the experimental points.

4.4.3.2. Thermocouples and Thermels

For determination of temperature differences for provision of adiabatic control, etc., for rough determinations of absolute temperature, and finally as a check of thermometric calibrations, single or multi-junction thermocouples are often employed. For this purpose, appropriate junctions are assembled from appropriate alloy wires to give maximum sensitivity over large temperature ranges. A wide assortment of different kinds of wires have been used by different experimentalists.

4.5. THE TREND TOWARD AUTOMATION IN DATA ACQUISITION

Equilibrium adiabatic calorimetry can be a tedious operation, particularly in the vicinity of phase transformations where equilibrium times often approach the better part of a day. One of the earliest successful attempts to provide reasonably automated acquisition of data prior to the general availability of computers in the laboratory was the achievement by Stull [49], who used a rather complex arrangement of shields. His device achieved an overall accuracy of approximately 0.3%. This was thus significantly less than possible by manual operation, but it does not reflect any basic inferiority of ultimate automated operation. Other pioneering endeavors in this direction were those done by Furukawa [31] at the National Bureau of Standards. A similar computer operation devised by Grønvold [50] represented a sophisticated early computer endeavor which proved to be exceedingly successful in the region above ambient temperature, and which resulted in the production of much data on important chalcogenides and pnictides [51]. Since that time, a number of devices for operation and data-logging as well as for the processing of the resultant data has been described in the literature. These are of varying levels of complexity, and insufficient data have been presented to totally judge the overall reliability of the system's construction.   Although no attempt will be made to be exhaustive in this treatment, the reader is urged to consult the individual references and the authors' claims for the success of the operation.   One very innovative one was constructed by Chang at the National Bureau of Standards; this was based on a motor activated double potentiometer circuit and resulted essentially in the automated operation of an already well-established calorimeter [52].   Another based on the use of an automatic six decade A-C bridge (Automatic Systems Laboratory model ASL-6) for recording the measured property of the temperature sensing device has been described by Schaake and his collaborators [53].

The use of a digital computer operating on a real-time time-sharing mode in the operation of a super-ambient calorimeter has also been described [54].   Here the sample temperature and the power to the sample/calorimeter heater are logged <u>continuously</u> and the heat capacity calculated for specified temperature intervals.

Yet another, using a dedicated microcomputer for the super-ambient region has been developed by Andrews et al. [55].   This unit also uses an ASL bridge as the temperature sensor.   A rather similar application of an ASL bridge and a dedicated microcomputer over the cryogenic region has also been constructed [56,57].   This device has been utilized for around-the-clock operation and provides acquisition of high-precision, high-accuracy data over most cryogenic regions of temperature without human oversight.   It has proven equally convenient also over the super-ambient region.   This system provides a continual print-out of the data with enough detail so that the quality of the operation can be carefully monitored, subsequently or on stream, by the heat capacities provided after each energy input.

Another automated low-adiabatic calorimeter operating over the range of 5–300 K has been described by Beyer [58].   The calorimeter instrumentation consists of automated adiabatic shield control and digitally controlled data acquisition.   Temperature measurements are made potentiometrically using a programmable d.c. voltage source and a digital nanovoltmeter.   Energy measurements are made with a 6.5 digit voltmeter.   The program also provides immediately evaluated heat capacities.

Yet another automated calorimeter is almost unique in that it is reported to operate without benefit of a computer [59].   The continuous heating adiabatic method presented yields a graphical representation of the inverse heat capacity as a function of temperature without the use of a computer.   Its principal advantages are its simple construction, high-level electrical signals, sample size approximating one gram, and (reported) exemplary behavior through first-order transitions.   Its operation is said to be that of complete automation from 80–300 K.

It is anticipated that a significant improvement in this mode of operation will appear in many areas of calorimetry and provide data of greatly improved quality over that taken manually.

4.6.   THE TREND TOWARD SMALLER SAMPLES

In the early days of thermophysical calorimetry, massive samples of hundreds of grams were customarily employed.   Although it is evident that this does result in considerable convenience in the procurement of highly precise data, as one gets beyond the readily available elements and compounds into the region of prohibitively-expensive, explosively-dangerous, difficult-to-prepare molecular crystals, hydrothermally-prepared mineralogical-type specimens, etc., the calorimetrist is increasingly obliged to operate on much smaller samples. Staveley [60] and others [15,61,62] pioneered in devising calorimeters for studies on samples smaller than customarily used. Another small calorimeter utilizing a small non-bifilarly wound, platinum-resistance thermometer has been described by Falk et al. [63] and utilized as a sensor together with an ASL-6 a.c. bridge. An even more significant advance, however, has been achieved in the USSR [64] in which reasonably precise data have been obtained in calorimeters employing between one and three grams of sample. The authors report, moreover, the achievement of relatively rapid thermal equilibration. In yet another report [65], a similar calorimeter is described in which small capsules, approximately 4 mm in diameter and 30 mm in length, are wound with a Karma heater and provided with a Leeds and Northrup capsule-type platinum-resistance thermometer mounted on the adiabatic shield. A two-junction thermel is employed to monitor the difference between calorimeter and shield. In this way, the lead wire bundle to the calorimeter is greatly reduced and the small mass of sample is confronted only with the relatively even smaller mass of the calorimeter in the absence of even a microscale platinum-resistance thermometer.

This device is proven to be practical on samples between 1 and 4 grams in mass. Under most circumstances, the minute temperature difference noted by the differential thermocouple is not significant even though the provision of an integration of the imbalance of the ASL-6 bridge is used to provide approximately 1.5 additional digits to that normally provided by the bridge itself. With only slight improvement in measurement sensitivity, results of quality entirely comparable to that obtained with more massive samples can be provided.

A further development relative to the measurement of the heat capacity of small samples by the ordinary adiabatic measurement is that of Paukov and his collaborators [66,67]. Their calorimeter with a capacity of 0.3 cm^3 was used to determine the heat capacity of benzoic acid with an experimental error between 0.2 and 0.3% over the range from 12–300 K.

4.7. MEASUREMENT TECHNIQUES AND ADJUVANT CIRCUITRY

4.7.1. Energy Circuitry

Adiabatic calorimetry involves a number of electrical measurements and control features. In addition to regulating electrical power input to the calorimeter, determining the energy input and the resultant temperature (under adiabatic or quasi-adiabatic conditions) by resistance thermometry, one must also control the temperatures of the adiabatic jacket surfaces, the guard shield, and the electrical lead bundle relative to the temperature of the surface of the calorimeter. This is done by heaters which operate by analog controllers in conjunction with differential thermocouples. Such circuits are widely described in review articles [4–6] and indeed in many of the original papers describing the cryostats and measurements. Certainly, the provision of high-sensitivity microvolt meters has greatly simplified measurement problems.

Inasmuch as the measurement of the temperature increment is usually the factor limiting the accuracy of determination of heat capacity, energy circuitry problems do not usually assume an important aspect in the accuracy of the measurements. For example, in the use of the potentiometric methods, the calibration of the standard cell is of small consequence to the measurement of resistance, which involves only the comparison of the thermometer resistance with that of the standard resistor. However, the calculation of energy from the measured current (i.e., the potential drop across the standard resistor) and the potential across the heater involves this calibration twice. The fractional error in the energy from the standard cell is, therefore, twice the error in the standard cell's electromotive force. Minimization of the error requires frequent check of the standard cell voltage against that of a recently calibrated absolute voltage standard.

4.7.2. Resistance Thermometry Circuitry

Experimental thermodynamics has been said to be as much a study of thermometry as it is a study of the fundamental properties of macroscopic systems. Pursuant to that statement and the already emphasized importance of the determination both of temperature increment and absolute temperature in the heat capacity measurements, it might seem obligatory upon the author of this chapter to discuss in greater detail both the thermometric sensor and the circuitry used to ascertain the temperature indicated by the sensor. However, in view of the detailed presentations which have already been made on these subjects, this will be avoided. These include Refs. [4–6] and the very recent appearance of three articles dealing, respectively, with platinum resistance thermometry by Furukawa [68], with practical thermometers and temperature scales (including NBS-IPTS-68, EPT-76, etc., and discussions of magnetic, electronic paramagnetic, nuclear paramagnetic, rhodium-iron resistance thermometers, germanium resistance thermometers, thermistor thermometers, etc.) by Mangum [69], and with quartz-resonator thermometry by Walls [70]. Suffice it to say that manual or automated potentiometric methods, bridge methods, and to an increasing extent, digital microvolt meters are being used to realize the temperature scale in resistance thermometry. Some of the special thermometers do indeed require special techniques which are referred to in the references already given.

The attention of the reader is called also to the excellent discussion in the chapter of this volume entitled 'Calorimetry in the Range 0.3–30 K' on temperature scales (3.3.2), on conversion between them, on their effect on heat-capacity measurements, on secondary thermometers (3.3.3), on thermometry calibration points (3.3.4), on calibration (3.3.5), and on measurement of resistance (3.3.6).

Particularly with potentiometric resistance thermometry the constant current or potential source of the thermometer current may also influence the choice of alternative modes of instrumentation.

4.7.3. Adiabatic Jacket Control Circuitry

In isoperibol calorimetry, the maintenance of an approximately constant temperature is achieved simply by the massiveness of the jacket itself. In adiabatic calorimetry, or for that matter, in quasi-adiabatic calorimetry, the circuitry used for sensing the temperature differential between the calorimeter and the various portions of the surroundings to be maintained at appropriate temperatures will vary considerably from one cryostat to another. Under the tutelage of the Calorimetry Conference in the United States, investigators developed conventional shield control by the use of several channels of commercial components to replace manual control by a skilled operator. Certainly the full advantage of the adiabatic method – particularly the computerized adiabatic method – can be conveniently realized only with automatic, electronic controls so that a true equilibrium state may be approached even over a period of many hours, without undue strain on the operators. As already indicated, thermocouples are commonly used to detect temperature differences; the main requirements for manual control are those of adequate sensitivity and stability with relatively small thermal lag for the thermocouples and the jacket heaters. To minimize the heat leaks to the calorimeter, it is often convenient to refer to a selected point on the main adiabatic jacket other than to the calorimeter. The choice of whether a.c. or d.c. is to be employed for jacket temperature control may depend on the nature of the thermometry utilized in the calorimeter. The block diagrams, idealized circuitry, and details have already been discussed adequately and the reader is referred to this source [4].

In updating the practical part of Ref. [4], it should be noted that in a recent calorimeter [71] five parallel channels of feedback circuitry have been used to control the temperature of the three parts of the adiabatic jacket, of the tempering ring, and of the guard jacket. Each channel is referenced to a thermocouple between the controlled surface and a reference point. For the cylindrical portion of the adiabatic jacket the reference point is the calorimeter itself, but for the other three channels the reference point is the middle of the adiabatic jacket.

Each channel is referenced to a thermocouple between the controlled surface and a reference point. The signals developed by the thermocouple—minute analogs of the temperature differences—are greatly amplified using Keithley 150 B microvolt-ammeters. Their high-level outputs are directly connected to Leeds and Northrup CAT Model 80 controllers. The amplifier also includes a zero-offset capability which is used for 'bucking' adventitious emf's in the thermocouples and for facilitating the achievement of zero-drift conditions. The controllers compute the power needed by the control surface from approach, proportional, integral, and differential terms and control programmable power supplies (Kepco Model PCX-MAT) which heat the surfaces. A three-channel potentiometric recorder is used to record simultaneously any three of the five channels. In this instance, the control of the shield heater is done by d.c. so as to prevent the introduction of spurious a.c. signals into the thermometric circuits and possible saturation of the a.c. bridge input.

4.8. SPECIAL TECHNIQUES

Although the great bulk of data reported over the 5–300 K range has been determined by the techniques either of equilibrium isoperibol or adiabatic calorimetry, other special techniques have assumed an increasingly important role with the passage of time. A few of these techniques will be rather tersely described here.

4.8.1. Inverse Temperature Drop-Calorimetry

Although calorimetry by the method of mixtures is usually done over the super-ambient temperature range or even the high temperature range and is described in the chapter of this volume entitled 'Adiabatic Calorimetry above 300 K,' the technique has been used occasionally by an inverse-temperature drop — that is, a drop from a low temperature to a higher temperature. This avoids the difficulty of slow equilibration and hysteresis at low temperatures and permits the determination of enthalpies over suitable ranges of temperature. This is often useful in the confirmation of heat-capacity values determined by other means or enthalpies of transition for metastable phases. An example is provided by studies on stishovite [72].

4.8.2. Calorimetry at High Pressures

Relatively little equilibrium calorimetry has been done at high pressures by direct measurement. These studies would be particularly significant for the determination of enthalpies of transitions as a function of pressure. Much of this work at present has been done by dynamic scanning calorimetry, and little can be said about the variation in the enthalpies of transition measured at equilibrium as a function of pressure. A case in point is that of ammonium chloride, which was studied initially by Dutch investigators [73] and subsequently re-examined by Paukov et al. [74].

4.8.3. Calorimetry in Magnetic Fields

Relatively little work has been done over the 5–300 K range since this technique is usually employed at temperatures much closer to absolute zero. It is interesting, however, to note that direct methods for the evaluation of magnetic heat capacity have been devised and tested [75]. Such methods do provide an interesting comparison with calorimetric schemes where applicable.

4.8.4. Calorimetry by the Thermal Relaxation Method

In research on the properties of new materials, one is often obliged to work with small samples. A method for determining relative heat capacities of samples as small as 10 mg has been reported by Hatta [76]. This method is no stranger to the very low temperature region (cf. chapter entitled 'Calorimetry in the Range 0.3–30 K' in this volume) but its utilization over the 5–300 K range is rather novel. Either the continuous, (dynamic) warming methods or the intermittent or pulse method require reasonably good thermal isolation of the sample. This is increasingly difficult to achieve as the thermal conductivity of the lead wires of the thermometer, for example, is not negligible and the heat capacity of the addenda is significant. Consequently, a method for measuring heat capacities under the existence of a thermal leak rather than under adiabatic conditions has been achieved by suspending a small sample from fine (thermocouple) wires in a medium of thermal exchange gas surrounded by a thermal bath. The junction of fine thermocouple wires is bonded to the sample and each wire is thermally terminated at the wall of the thermal bath. If a constant heat flux Q is supplied to the sample by thermal radiation, the temperature T of the sample increases with the relaxation time τ, and the thermal difference between the sample and the thermal bath reaches a saturation value, ΔT_m. If T and ΔT_m are measured, one may determine a relative heat capacity of the sample by the method utilized by Bachmann et al. [77] over the range 1–35 K utilizing a thermal leak provided by the wires instead of by thermal exchange gas. Somewhat later, Schutz [78] extended the measurements to temperatures below 1 K. Djurek and Baturic-Rubcic [79] have also extended the method toward high temperatures. In the application of this method, over the 5–350 K region, a disc-like sample and the thermocouple wires are attached to the rear surface of the absorber with Aquadag (made of graphite and water). A constant heat flux is supplied by visible radiation from a stable halogen lamp. The emf generated at the thermocouple corresponds to the temperature by which the sample exceeds that of the thermal bath. If heat is applied to the sample in a step-wise fashion, the temperature difference increases with the relaxation time. The saturation value of the temperature difference is $\Delta T_m = QR$, and $T = CR$, where C is the heat capacity of the sample, R is the thermal resistance between the sample and the thermal bath, and Q is the intensity of the heat flux. Hence, we obtain $C = TQ/\Delta T_m$. At best, only relative heat capacities can be ascertained and the accuracy approximates about 3%. Reasons for such errors include the relatively large contribution of varnish to the total heat capacity, and a certain sluggishness in the response of the amplifier. The necessity of comparison between an experimental sample and a reference material is a contributory aspect of the lower accuracy.

The method does, however, have the advantage that a very small sample (10–500 mg) is all that is required, and it does moreover make possible the measurement of the heat capacity, even under high gas pressure, in various atmospheres, even in the range in which radiation loss cannot be avoided.

4.8.5. Laser-Flash Calorimetry

A.c. calorimetry developed by Sullivan and Seidel [80] is a method which affords very high resolution in heat capacity and has proven particularly useful in the study of critical phenomena as well as in the magnetic and electric transitions of metals and salts. Its promise for extension into the magnetic field and to high pressure effects on heat capacity makes it of special interest. The high-speed measurement of heat capacity of metallic materials at very high temperatures as developed by Cezairliyan et al. [81] is one of the most successful developments of the pulse-heating method, whether a d.c. current of less than one microsecond duration or pulse heating by electron beam is used as the energy source. However, this application is beyond the temperature range of the present chapter. Laser-flash calorimetry is also a pulse-heating method, with a flash of a ruby (or other type) laser as the heat source.

The laser-flash method differs from other methods of pulse-heating in being capable of measuring samples of both electrically conducting and insulating materials. Initially developed by Takahashi [82], it is not necessarily limited to the super-ambient range, and satisfactory results have been obtained at temperatures ranging from 80–1100 K [83]. Laser energy has also been used as a source of energy in the 2–20 K region [84]. One of the advantages of the laser-flash method is that the mass of samples required for measurement is small. The normal size sample is from 30–500 mg. In the early application of the method, the advantages—such as simultaneous determination of heat capacity and thermal diffusivity data—had been offset by the unreliability of the resulting values as compared with those obtained by more conventional methods. The discrepancies of data obtained by the laser-flash method are associated with adjustments for the differences in the reflectivity of the surfaces of the samples and the problem of obtaining an accurate measurement of the energy of the flash itself. Many of these problems have been subsequently solved by Takahashi and his co-workers [85], who review much of the recent progress on the methods.

Their apparatus—except for the optical path—is similar to those already described. Samples are in the form of small (sometimes pelleted) discs 8–12 mm diameter and 0.5–5 mm thick. Powdered samples may be encapsulated in an aluminum container of the pellet size described above. A thin, absorbing plate attached to the front surface of the sample is made of glassy carbon, of the same diameter as the sample and approximately 0.2 mm thick. It is attached to the sample surface with about a milligram of silver paste or silicone grease and absorbs the impinging laser energy. It provides the advantage of a uniform absorption efficiency and enables one to determine the absolute heat capacity of the sample. Moreover, in measurements just below a phase transition, the presence of the disc prevents an excessive temperature rise of the front surface of the sample through the transition and back down again with its deleterious effect on the achievement of equilibrium. This improved method has proved experimentally capable of determining heat capacity anomalies quite precisely, with temperature increments less than 0.5 K.

The most important components in this type of heat capacity measurement are the precise measurement of the sample temperature and its increments as well as of the energy applied to the sample. The temperature rise is usually detected with a thermocouple attached to the back surface of the sample with silver paste. A high-precision digital microvoltmeter may be used to detect this and to record it as often as twice a second. The temperature rise of the sample is determined by the graphical plot of the emf as a function of time. To minimize heat loss from the sample very fine thermocouple leads are used.

Obviously, adjustments must be made for the heat capacity of the addenda such as the absorbing disc, the silver paste, the silicone grease, the thermocouple junction, and – if used – the aluminum container for powdered materials. In a typical experiment, these amount to approximately 5% of the total for pelleted samples whereas they may aggregate 50% of the total for encapsulated samples. The overall imprecision in the measurement of heat capacity of a pellet sample by this method is about 0.5% over the range from 80–800 K. When an aluminum container is used, the precision is lower but the estimated experimental error is less than 2%. A number of applications have been described [85].

4.8.6. High-Resolution Heat Capacity Calorimeter

Experimental measurement of thermodynamic quantities with high resolution both with respect to the independent variable – usually the temperature – and in the quantity measured is a desideratum. The development of an unusually successful calorimeter of this type at Osaka by Seki, Suga, Matsuo, and others is described in a series of papers of which only the most significant are mentioned here [47,48,86]. The most noteworthy feature of the apparatus is the use of a thermistor thermometer to enable precise measurement of small temperature increments in the calorimeter. The platinum resistance thermometer also stationed on the calorimeter is used for the absolute temperature scale and for the calibration of the thermistor in the fore and after drift periods. Obviously, for high-resolution measurements either in transition regions or in heat capacity regions showing small thermal effects, it was necessary to maintain strict adiabatic regulation — sometimes for long periods of time. The superb measurements of the symmetrical shape of the mostly second-order anomaly in stannous chloride dihydrate [86] is a good example of the power of the calorimeter. The further detection of the lower Curie point in Rochelle salt between 230 and 310 K (which amounts to only 0.2% of the total heat capacity) is yet another example [48]. The results presented in these works clearly show the utility of the sophisticated, high-level calorimeter with precision approximating 0.05%.

4.8.7. Flow Calorimetry

Without detailed presentation, it should be noted that the heat capacity of fluids—both liquid and gaseous—are being increasingly determined by flow calorimetry. Although the utilization of flow calorimetry for both gaseous thermophysics and thermochemistry has been traditional for many years, use of the capillary-scale tubes for flow-calorimetric systems for liquids is increasingly yielding more precise determinations of heat capacities as well as thermomechanical measurements on mixing, solution, and reaction.

4.9. TRANSITIONS – PHASE AND OTHERWISE

Although the occurrence of transitions of many sorts in the course of heat capacity measurement provide much of the thrill and excitement of these measurements, they do represent special problems. After detection and mapping out of a transition it is good practice to ascertain, first of all, that one can cool through the transition in reproducible fashion, before making heat capacity measurements at the lower temperatures on samples which may contain — to a greater or less extent — a metastable phase. One may make direct enthalpy-type measurements through the transition repeatably after different cooling rates (and in some instances after different amounts of undercooling) and thereby establish whether or not reproducibility has been achieved. Subsequently, if the hysteresis involved permits, one may wish to map out — using increasingly small temperature increments — the actual shape of the heat capacity through the transition region. The curve of the heat capacity drawn through this region may then be integrated by appropriate methods and compared with the enthalpy increments made by direct measurement. When these are in satisfactory accord, an appropriate integration over the curve gives the entropy increment associated with the transition.

4.10. REFERENCE STANDARDS FOR THERMOPHYSICAL CALORIMETRY

The importance of the utilization of standard reference materials for testing the calibration of the thermophysical calorimeter over the 5–300 K range will be apparent to any practicing experimentalist. Fortunately, the need has been clearly recognized by--among other bodies--the (U.S.) Calorimetry Conference, who have encouraged the production of a series of standard materials for use over this range. These materials, available from the National Bureau of Standards as certified reference materials for heat capacity, include such substances as copper (Cu), synthetic sapphire (Al_2O_3), normal-heptane (C_7H_{16}), and benzoic acid ($C_7H_7O_2$). Experimentalists are urged to use whichever of these materials are most appropriate for the calibration and utilization of the particular design of calorimeter, the substances and any special needs of the work, in order to ensure the accuracy of the reported measurements.

References to the publication of articles dealing with these calibration materials are readily available. For example, the heat capacity of a National Bureau of Standards' sample of benzoic acid [87] has been measured by Busey [88], by Cole et al. [35], by Furukawa et al. [89], and by Osborne et al. [90]. The first two authors use an isoperibol-type of calorimeter, the latter two an equilibrium adiabatic type. The comparison of the results of the various investigators shows that there is agreement within the mutual indexes of precision except at the lowest temperatures (near 10 K). Inasmuch as the National Bureau of Standards' temperature scale [1,91] was used by all investigators, systematic errors would not be disclosed by comparison of data from these laboratories.

4.11. PROCEDURES FOR DATA HANDLING AND PRESENTATION

4.11.1. Calculation of $C_p{}^o$ and Other Thermophysical Properties

4.11.1.1. General Principles

It may readily be shown that for a calorimeter the increment in the internal energy, ΔE, of the entire calorimetric system, over a temperature increment ΔT, can readily be interpreted in terms of the equation $\Delta E/\Delta T = (\Delta H/\Delta T)_p + B$, where H is enthalpy and the correction term B is occasioned only by the presence of the helium gas present for conduction under a pressure of the order of 2 torr over the entire temperature range. Additional corrections are, however, required when the sample being measured has a vapor pressure of larger magnitude, particularly when the vapor space in the calorimeter is relatively large.

Consequently, utilizing appropriately devised computer programs, measured energy increments and temperature increments are converted to values of $(\Delta H/\Delta T)_p$. Such programs for example have been provided to many laboratories from the original 'MICHITHERM' and 'FITAB' programs [92,93]. Such programs include a correction for the nonlinearity of C_p vs. T by expansion in a Taylor's series around the mean temperature of the measurement [5]. Provided good judgment has been made in the temperature increments of the run, this correction should be less than 0.1% over most of the temperature range. By means of this 'curvature correction' the values of $(\Delta H/\Delta T)_p$ are converted to the limit

$$\lim_{\Delta T \to 0} (\Delta H/\Delta T)_p \text{ or } (\partial H/\partial T)_p$$

and hence to the defined C_p values. Provided the pressures involved are substantially below an atmosphere the values are readily interpreted as C_p^o values. (It should, however, be noted that when the sample has a significant vapor pressure the sublimation or vaporization of the sample as well as the heat capacity of the gaseous phase will need to be taken into account, and the heat capacity at saturation, C_s, will be determined.)

4.11.1.2. Derived Quantities

Although the heat capacity is a very sensitive and useful parameter detailing the energy spectrum of matter, it is particularly powerful at very low temperatures. At higher temperatures, and for other purposes, the entropy or the Gibbs energy may be more useful, or perhaps other derived thermodynamic quantities are desired. These properties are also readily provided by the programs mentioned or by equivalent programs. Usually, minor problems are encountered in the division of the heat capacity vs. temperature curve into appropriate segments to make the fitting convenient, in view of the several-thousand-fold variation in C_p even over the range 5–350 K. (An inflection point is a convenient place to make such a break.) In addition, the appropriate thermodynamic functions integrated over the region of anomalies and/or transitions must also be incorporated and, finally, the calculation of the derived thermodynamic quantities requires an extrapolation of the heat capacity from temperatures in the region 5–10 K to zero kelvin.

4.11.2.   The Extrapolation to Zero Kelvin

In order for the extrapolation to zero kelvin to be valid it is necessary that only small contributions to the thermodynamic quantities arising from the lattice vibration and/or the low-temperature premonitory effect of a magnetic ordering transition be involved. In cases involving paramagnetic substances, special consideration of the magnetic contribution to thermodynamic functions must be made, particularly if these occur below the temperature from which extrapolation is made.

It is usually found convenient to utilize the Debye limiting law: $C = \alpha T^3$. This result has been shown to be valid for many substances and is often adequate for chemical thermodynamic purposes for simple solids, e.g., monatomic elements, as well as for the heat capacity of complex solids, which often tend toward zero in a fashion rather similar to that for monatomic elements. However, when electronic complications are involved, additional considerations may be required in extrapolation of the heat capacity.

4.11.2.1. Conduction Electron and Debye Contributions

For solids possessing conduction electrons (e.g., metals) the conduction electron contribution may be treated as that due to a degenerate Fermi–Dirac gas. The electronic heat capacity at low temperatures varies with the first power of the absolute temperature [94], i.e.:

$$C_{el} = \gamma T. \tag{1}$$

The electronic heat capacity may greatly exceed that due to the lattice vibrations at temperatures below 3 K. Consequently, extrapolations for a substance with such electronic contributions might well be made in terms of the equation:

$$C_p = \gamma T + \alpha T^3 \hspace{6cm} (2)$$

in which α and γ are the constants representing, respectively, the Debye term for the lattice vibration and the term due to the conduction electrons. It is convenient to plot C_p/RT vs. T^2, since on this basis Eq. (2) becomes linear with slope α/R and intercept γ/R at $T^2 = 0$.

Such a plot is also of considerable utility in extrapolations to zero kelvin even for substances which may not be expected to follow the equation, particularly if data for comparison dealing with a family of rather similar substances are available. Heat capacity plots in this format for a series of organic and inorganic molecular crystals are presented in Figs. 4.4 and 4.5 and, although significant deviations from linearity are observed, the trend within families is obvious.

4.11.2.2. The Zero Point Entropy

In the presentation of thermophysical properties, assuming that the experimenter has either used reliable data taken near zero kelvin or has determined from the nature of the material that magnetic, electronic, or other transitions are unlikely to be relevant, he is still confronted with the question of zero point entropy and the disorder present in the solid below the lowest temperature measurement.

The question as to the extent to which the third law of thermodynamics holds is a matter beyond the scope of the present chapter, but attention is called to the fact that very often naturally-occurring mineral samples do exhibit an ordered state whereas those prepared synthetically by hydrothermal methods on a laboratory time scale often show a lack of order. Further discussion may be found elsewhere [95,96]. Amorphous materials, vitreous substances, etc., and undercooled (metastable) solids and liquids may likewise be anticipated to show interesting behavior. Moreover, metastable phenomena such as that involved in the pyrite/marcasite system [97], should also be noted.

4.11.3. Other Matters of Calculational Relevance

4.11.3.1. Routine Details

The reader is urged to compare the detailed discussion of these points already presented in the several chapters of Experimental Thermodynamics, Vol. 1 [4–6].

4.11.3.2. Errors Associated with Thermophysical Data

Two volumes entitled 'The Critical Evaluation of Thermodynamic Data' [98] are in production. There an extensive discussion of this matter will be given by several authors with expertise in the various branches of thermophysical measurement. For this reason only a few comments will be made here.

As has already been emphasized, the accuracy of any calorimetric system operating over the 5–350 K range may be tested against one of the several standard samples for heat capacity use.

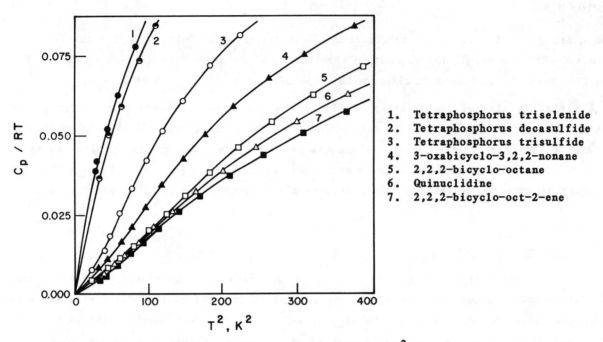

1. Tetraphosphorus triselenide
2. Tetraphosphorus decasulfide
3. Tetraphosphorus trisulfide
4. 3-oxabicyclo-3,2,2-nonane
5. 2,2,2-bicyclo-octane
6. Quinuclidine
7. 2,2,2-bicyclo-oct-2-ene

Figure 4.4.   Heat capacity plots in the form C/T vs T^2 for organic and inorganic molecular crystals.

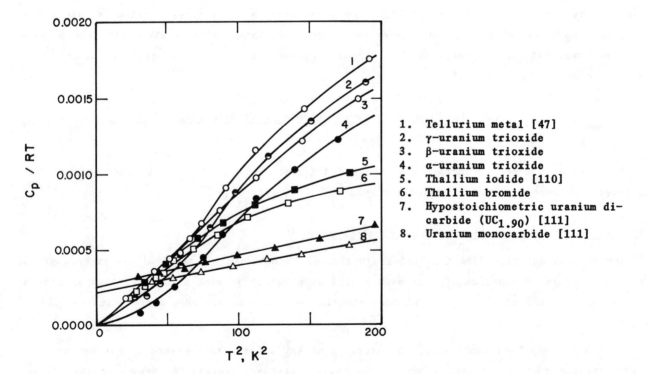

1. Tellurium metal [47]
2. γ-uranium trioxide
3. β-uranium trioxide
4. α-uranium trioxide
5. Thallium iodide [110]
6. Thallium bromide
7. Hypostoichiometric uranium di-carbide ($UC_{1.90}$) [111]
8. Uranium monocarbide [111]

Figure 4.5.   Heat capacity plots in the form C/T vs T^2 for metals and inorganic solids.

This should be a mandatory prerequisite to the publication of data obtained in a new calorimetric system, since the accuracy of the absolute thermometry, the temperature increments, and the overall working of the system can readily be established. Apart from this, the importance of a careful and thorough definition, analysis, and characterization of the sample is obvious. If the test is on a standard sample of established purity, and if the measuring equipment is working correctly and has been recently calibrated, one may reasonably expect to find accuracy and precision better than 0.1%, except where hysteresis, failure to establish a stable equilibrium phase, etc., are involved.

4.11.3.3. The Lattice Heat Capacity

Very often the difference between the measured heat capacity and the lattice heat capacity is used to establish an excess heat capacity of some sort, be it Schottky contribution, a phase-transitional contribution, or one of the many other contributions (such as the magnetic, ferro-electric, order-disorder, glass transition, etc.) or one of the many specialized 'named' transitions which often occur. Although in the very low temperature range the lattice-heat-capacity contribution is often vanishingly small, this is seldom the case in the 5-350 K range, and an integral part of the evaluation of excess heat capacities involves the resolution of the lattice heat capacity. This besetting problem has confronted thermophysicists for decades. Corresponding state theories have been less than satisfactory; Lindemann schemes are suggestive. For many years the Latimer scheme [99,100] has been widely used. This scheme is based essentially on the mass dependence of the cation with a predetermined contribution for the anion. Unfortunately, in many publications the lattice heat capacity – and consequently the resolution of the excess contribution – is made by good(?) judgment only.

Since the lattice contribution is typically 80-100% of the total measured heat capacity, and if it is going to be estimated, the procedure must be one of relatively high reliability and accuracy. Related problems have been discussed by Saxena [101], by Cantor [102], and by Kieffer [103]. Over the 5-350 K range where entropy development largely occurs, we have found in several instances that a volume-weighted scheme is more relevant than one based on mass dependence [104–109]. Evidence of this predominance of volume over mass – in the dynamic region – is presented by Fig. 4.6, where the heat capacities of a series of four isostructural ionically-crystalline disulfides are presented as a function of the molar volume. The smooth trend contrasts greatly with the irregularities of cationic masses given digitally underneath the points of Fig. 4.6. In other instances, the general trends of cationic mass with molar volume in isoanionic series are often parallel. But, where a difference occurs, as it does for example over the lanthanide contraction, the volume effect predominates. This is reasonable in view of the optical type of vibrations involved in this region, although it is clear that as one goes to a more acoustical type of vibration at lower temperatures the significance of the mass will be enhanced.

4.12. COMPARISON OF DSC AND ADIABATIC APPROACHES

With the development of computer-driven differential scanning calorimetry (DSC) this has

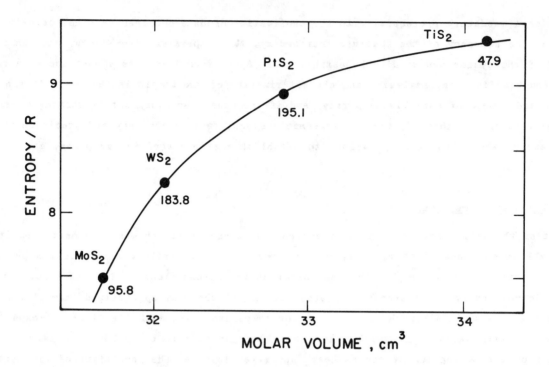

Figure 4.6. Entropy versus molar volume for an isostructural series. The numbers beneath the
points are the cationic (atomic) masses.

become an increasingly important technique, and 'heat-capacity data' are often reported on the
basis of DSC measurements. Some of the values reported are quite reasonable, others are not, and
it is often quite difficult to decide which are to be trusted. It should be noted that the DSC
approach involves dynamic measurements under constraints of continuous heating. Usually the rate
of heating is quite rapid and, in the presence of significant hysteresis, significant errors can
occur. Transitions – particularly small transitions with marked hysteresis – may not be visible
in the DSC trace even when they have been detected by direct adiabatic calorimetry and their presence is known. Moreover, in using the very minute samples for DSC purposes, great care must be
taken to avoid contamination, to obtain accurately determined masses of sample, and to know with
sufficiently high precision the heat capacity of the adjuvant capsule, etc. Perhaps the greatest
hazard is the failure to recognize the important effect of hysteresis and the difference between
the dynamic methods and one involving equilibration at each state in the measurement.

Certainly for many endeavors the DSC determinations will prove to be quite adequate. Certainly they require less effort and less total time. However, the automation of both endeavors
has made the operator-time requirement more comparable for the two methods, and with some further
development (particularly on smaller samples) of the equilibrium adiabatic determinations, the
increment of time required to do these measurements may often be worth the additional certitude
provided by equilibrium measurements over those taken dynamically.

Although it was earlier asserted that emphasis would be placed on intermittent (discontinuous) heating calorimetry, since this is clearly preferable whenever hysteresis or long thermal
equilibration times are involved, there are many situations where rapid equilibration obtains.

For example, two recent papers on copper metal [112,113] are in excellent agreement with earlier results [114]. Both dynamic (continuous) heating and intermittent heating were involved. Moreover, the work on α–Al$_2$O$_3$ [115] should be consulted as further evidence. However, the reader should not imagine that high-precision, continuous-heating, adiabatic calorimetry is nothing more than DSC. Here again, care must be taken as sluggish transformations can often be totally missed whenever the scanning rate is too high for thermal equilibration.

Another application of continuous adiabatic calorimetry is in the measurement of heat capacity with descending temperatures in studying transitions or hysteresis loops [45]. This should, however, not be confused with the ingenious method of intermittent equilibrium calorimetry with descending temperature [116] utilizing quantitative introduction of heat from the calorimeter to a cold reservoir under otherwise adiabatic constraints.

4.13.   CONCLUDING REMARKS

Although the tendency to be terse and to reference more detailed sources rather than incorporate full details here has kept the chapter within bounds, several comments are still appropriate despite these constraints. McGlashan described clearly in thermodynamic parlance the nature of the calorimetric process [117]. Moreover, the Workshop on Techniques for Thermodynamic Properties [118] provides a current detailed treatment by experts of the topic of this (and other) chapters.

No attempt has been made to discuss the source of critical evaluated thermophysical and chemical thermodynamic values, but attention is called to such a list in a compilation by Freeman in a forthcoming publication in the CODATA Directory on Chemical Thermodynamics [119] which not only provides such sources but also incorporates on-line computer data bases and data centers.

Chemical thermodynamics may be an almost unique discipline in the availability of a Bulletin of Chemical Thermodynamics [120] which endeavors to list and to index not only relevant progress over the calendar year previous, but to summarize the general research programs of laboratories throughout the world and to indicate their completed – but yet unpublished – endeavors in the same indexing scheme.

4.14.   ACKNOWLEDGMENT

The author expresses his continuing appreciation to the National Science Foundation, Structural Chemistry and Chemical Thermodynamics Program within the Chemistry Division for their continuing support of his research endeavors on which his recent experience has been largely predicated.

4.15.   REFERENCES

1. Hoge, H.F.J. and Brickwedde, F.G., J. Res. Natl. Bur. Stand., 22, 351–73, 1939.
2. Stimson, H.F., J. Res. Natl. Bur. Stand., 42, 209–17, 1949.

3. Douglas, T.B., Furukawa, G.T., McCoskey, R.E., and Ball, A.F., J. Res. Natl. Bur. Stand., 53, 139–53, 1954.

4. Westrum, E.F., Jr., Furukawa, G.T., and McCullough, J.P., in Experimental Thermodynamics (McCullough, J.P. and Scott, D.W., Editors), Vol. 1, Chap. 5, Butterworth's and Co., Ltd., London, England, 133–214, 1968.

5. Stout, J.W., in Experimental Thermodynamics (McCullough, J.P. and Scott, D.W., Editors), Vol. 1, Chap. 6, Butterworth's and Co., Ltd., London, England, 133–214, 1968.

6. West, E.D. and Westrum, E.F., Jr., in Experimental Thermodynamics (McCullough, J.P. and Scott, D.W., Editors), Vol. 1, Chap. 9, Butterworth's and Co., Ltd., London, England, 333–67, 1968.

7. Gopal, E.S.R., Specific Heats at Low Temperatures, Plenum Press, New York, NY, 1966.

8. Westrum, E.F., Jr. and McCullough, J.P., in Physics and Chemistry of the Organic Solid State (Fox, D., Editor), Vol. 1, Interscience Publishers, New York, NY, 1–178, 1963.

9. McKie, D. and Heathcote, N.S., The Discovery of Specific and Latent Heats, Edward Arnolds and Co., London, England, 1935.

10. Gaede, W., Phys. Z., 4, 105–8, 1902.

11. Nernst, W., Sitzber. Kgl. Preuss. Akad. Wiss., 12,13, 261–82, 1910; Chem. Abstr., 4, 2397, 1910.

12. Eucken, A., Phys. Z., 10, 586–9, 1909.

13. Keesom, W.H. and Van den Ende, J.N., Commun. Kamerlingh Onnes Lab. Univ. Leiden, 219b, 1932.

14. Kostryukova, M.O. and Strelkov, P.G., Comp. Rend. (Dokl.) Akad. Nauk USSR, 90, 525–8, 1953.

15. Westrum, E.F., Jr., Hatcher, J.B., and Osborne, D.W., J. Chem. Phys., 21, 419–23, 1953.

16. Westrum, E.F., Jr., personal communication.

17. Rayne, J.A., Aust. J. Phys., 9, 189–97, 1956.

18. Ramanathan, K.G. and Srinivasan, T.M., Philos. Mag., 46, 338–40, 1955.

19. Webb, F.J. and Wilks, J., Proc. R. Soc. London, A230, 549–59, 1955.

20. Southard, J.C. and Andrews, D.H., J. Franklin Inst., 209, 349–60, 1930.

21. Westrum, E.F., Jr. and Chou, C., Annexe 1955–3, Supplement au Bulletin de l'Institut International du Froid, Paris, 308–10, 1956.

22. Westrum, E.F., Jr. and Greenberg, E., J. Am. Chem. Soc., 78, 4526–8, 1956.

23. Grenier, G. and Westrum, E.F., Jr., J. Am. Chem. Soc., 78, 6226–7, 1956.

24. Westrum, E.F., Jr., 'The Low Temperature Heat Capacity of Neutron Irradiated Quartz,' in Proc. IVe Congres International du Verre, Paris, 396–9, 1956.

25. Westrum, E.F., Jr., Chou, C., Machol, R.E., and Grønvold, F., J. Chem. Phys., 28, 497–503, 1958.

26. Westrum, E.F., Jr., in Advances in Cryogenic Engineering (Timmerhaus, K.D., Editor), Vol. 7, Plenum Press, New York, NY, 1962.

27. Furukawa, G.T., Douglas, T.B., McCoskey, R.E., and Ginnings, D.C., J. Res. Natl. Bur. Stand., 57, 67–82, 1956.

28. Scott, R.B., Meyers, C.H., Rands, R.D., Jr., Brickwedde, F.G., and Bekkedahl, N., J. Res. Natl. Bur. Stand., 35, 39–85, 1945.

29. Sterrett, K.F., Blackburn, D.H., Bestul, A.B., Chang, S.S., and Horman, J., J. Res. Natl. Bur. Stand., 69C, 19, 1965.

30. Furukawa, G.T. and Piccirelli, J.H., 'Calorimetric Determination of the Purity of Benzene (IUPAC–59),' International Union of Pure and Applied Chemistry, Commission on Physicochemical Data and Standards, Cooperative Determination of Purity by Thermal Methods, Report of the Organizing Committee, July 14, 1961.

31. Furukawa, G.T., private communication.

32. Din, F. and Cockett, A.H. (Editors), Low–Temperature Techniques, George Newnes Ltd., London, England, 1960.

33. Giauque, W.F. and Stout, J.W., J. Am. Chem. Soc., _58_, 1144–50, 1936.

34. Hildenbrand, D.L. and Giauque, W.F., J. Am. Chem. Soc., _75_, 2811–8, 1953.

35. Cole, A.G., Hutchens, J.O., Robie, R.A., and Stout, J.W., J. Am. Chem. Soc., _82_, 4807–13, 1960.

36. Westrum, E.F., Jr., 'Developments in Low-Temperature Calorimetry,' in _Thermochemie_, International Colloquium No. 201 of CNRS, Paris, France, 103–18, 1972.

37. Westrum, E.F., Jr., Hatcher, J.B., and Osborne, D.W., J. Chem. Phys., _21_, 419–23, 1953.

38. Giauque, W.F. and Wiebe, R., J. Am. Chem. Soc., _50_, 101–22, 1928.

39. Giauque, W.F. and Johnston, H.L., J. Am. Chem. Soc., _51_, 2300–21, 1929.

40. Blue, R.W. and Giauque, W.F., J. Am. Chem. Soc., _57_, 991–9, 1935.

41. Giauque, W.F. and Egan, C.J., J. Chem. Phys., _5_, 45–54, 1937.

42. Kemp, J.D. and Giauque, W.F., J. Am. Chem. Soc., _59_, 79–84, 1937.

43. Murch, L.E. and Giauque, W.F., J. Phys. Chem., _66_, 2052–9, 1962.

44. Huffman, H.M., Chem. Rev., _40_, 1–14, 1947.

45. Westrum, E.F., Jr., unpublished data.

46. Trowbridge, J.C. and Westrum, E.F., Jr., J. Phys. Chem., _68_, 42–9, 1964.

47. Matsuo, T., Tatsumi, M., Suga, H., and Seki, S., Solid State Commun., _13_, 1829, 1973.

48. Tatsumi, M., Matsuo, T., Suga, H., and Seki, S., Proc. Jpn. Acad., _50_, 476, 1974.

49. Stull, D.R., Anal. Chim. Acta, _17_, 133–43, 1957.

50. Grønvold, F., Acta Chem. Scand., _21_, 1695, 1975.

51. Grønvold, F. and Westrum, E.F., Jr., Inorg. Chem., _1_, 36–48, 1962.

52. Chang, S.S., in _Proceedings of the 7th Symposium on Thermophysical Properties_, ASME, New York, NY, pp. 75 and 83, 1977.

53. Schaake, R.C.F., Offringa, J.C.A., Vanderberg, G.J.K., and van Miltenburg, J.C., Recl. Trav. Chim. Pays-Bas, _98_, 408–12, 1979.

54. Cash, W.M., Stansbury, E.E., Moore, C.F., and Brooks, C.R., Rev. Sci. Instrum., _52_, 895–901, 1981.

55. Andrews, J.T.S., personal communication.

56. Falk, B.G., Evans, B.J., and Westrum, E.F., Jr., J. Chem. Thermodyn., _11_, 367, 1979.

57. Westrum, E.F., Jr., 'Computerized Adiabatic Thermophysical Calorimetry,' in _Proceedings NATO Advanced Study Institute on Thermochemistry_ (Ribeiro da Silva, M.A.V., Editor), Reidel, Boston, MA, 745–67, 1984..

58. Beyer, R.P., 'Automation of a Low-Temperature Calorimeter,' in _Proceedings of the Workshop on Techniques for Measurement of Thermodynamic Properties_, IC-8853, U.S. Bureau of Mines, 113–23, 1981.

59. Junod, A., J. Phys. E., _12_, 945–52, 1979.

60. Staveley, L.A.K. and Gupta, A.K., Trans. Faraday Soc., _45_, 50–61, 1949.

61. Westrum, E.F., Jr., Hatcher, J.B., and Osborne, D.W., J. Chem. Phys., _21_, 419–23, 1953.

62. Westrum, E.F., Jr. and Feick, G., J. Chem. Thermodyn., _9_, 293–9, 1977.

63. Arvidson, K., Falk, B., and Sunner, S., Chemica Scripta, _10_, 193, 1976.

64. Gorbunov, V.E., Gavrichev, K.S., Gurevich, V.M., and Sharpataya, G.A., in _Abstracts of Poster Papers, 6th International Conference on Thermodynamics_, Merseburg, 1980. See also prototype in Gorbunov, V.E. and Palkin, V.A., Zh. Fiz. Khim., _46_, 1625, 1972.

65. Gorbunov, V.E., Gurevich, V.M., and Gavrichev, K.S., in _Proceedings of the VIII All-Union Conference on Calorimetry and Chemical Thermodynamics 1979_, Ivanovski Chemical Technological Inst., Ivanov, 113, 1979.

66. Paukov, I.E., Anishin, V.F., and Anishimov, M.P., Zh. Fiz. Khim., _46_, 778–81, 1972.

67. Sukhovei, K.S., Anishin, V.F., and Paukov, I.E., Zh. Fiz. Khim., 48, 1589–93, 1974.

68. Furukawa, G.T., 'Platinum Resistance Thermometry in Thermodynamic Measurements,' in Proceedings of the Workshop on Techniques for Measurement of Thermodynamic Properties, IC-8853, U.S. Bureau of Mines, 7–26, 1981.

69. Mangum, B.W., 'Practical Thermometers and Temperature Scales,' in Proceedings of the Workshop on Techniques for Measurement of Thermodynamic Properties, IC-8853, U.S. Bureau of Mines, 27–50, 1981.

70. Walls, F.L., 'Future of Quartz Resonator Thermometry,' in Proceedings of the Workshop on Techniques for Measurement of Thermodynamic Properties, IC-8853, U.S. Bureau of Mines, 51–62, 1981.

71. Andrews, J.T.S., Norton, P.A., and Westrum, E.F., Jr., J. Chem. Thermodyn., 10, 949–58, 1978.

72. Holm, J.L., Kleppa, O.J., and Westrum, E.F., Jr., J. Geochim. Cosmochim. Acta, 37, 2289–307, 1967.

73. Trappeniers, N.J., Ber. Bunsenges. Phys. Chem., 70, 1080, 1966.

74. Amitin, E.B., Kovalevskaya, Ya.A., Lebedeva, E.G., and Paukov, I.E., Zh. Fiz. Khim., 50, 2996, 1976.

75. Wolf, W.P., Meissner, H.E., Catanese, C.A., and Scott, P.D., Colloq. Int. Cent. Nat. Rech. Sci., 180, 93, 1970.

76. Hatta, I., Rev. Sci. Instrum., 50, 292–5, 1979.

77. Bachmann, R., DiSalvo, F.J., Jr., Geballe, T.H., Greene, R.L., Howard, R.E., King, C.N., Kirsch, H.C., Lee, K.N., Schwall, R.E., Thomas, H.U., and Zubeck, R.B., Rev. Sci. Instrum., 43, 205, 1972.

78. Schutz, R.J., Rev. Sci. Isntrum., 45, 548, 1974.

79. Djurek, D. and Baturic-Rubcic, J., J. Phys. E, 5, 424, 1972.

80. Sullivan, P. and Seidel, G., Phys. Rev., 173, 679–85, 1968.

81. Cezairliyan, A., Morse, M.S., Berman, H.A., and Beckett, C.W., J. Res. Natl. Bur. Stand., A74, 65–92, 1970; Cezairliyan, ibid, C75, 7–18, 1972.

82. Takahashi, Y., J. Nucl. Mater., 51, 17, 1974.

83. Yokokawa, H., Takahashi, Y., and Mukaibo, T., in Thermodynamics of Nuclear Materials, 1974, Vol. 2, IAEA, Vienna, 419–30, 1975.

84. Lee, K.N., Bachmann, R., Geballe, T.H., and Maita, J.P., Phys. Rev. B, 2, 4580–5, 1970.

85. Takahashi, Y., 'Recent Developments in Experimental Methods for Heat-Capacity Measurements,' in Chemical Thermodynamics – 4 (Rouquerol, J. and Sabbah, R., Editors), Pergamon Press, Oxford, England, 323–31, 1976.

86. Tatsumi, M., Matsuo, T., Suga, H., and Seki, S., Bull. Chem. Soc. Jpn., 48, 3060–6, 1975.

87. Ginnings, D.C. and Furukawa, G.T., J. Am. Chem. Soc., 75, 522–7, 1953.

88. Busey, R.J., J. Am. Chem. Soc., 78, 3263–6, 1956; ORNL-1828, Office of Technical Services, Dept. of Commerce, Washington, D.C.; Cole, A.G., Huchens, J.O., Robie, R.A., and Stout, J.W., J. Am. Chem. Soc., 82, 4807, 1960.

89. Furukawa, G.T., McCoskey, R.E., and King, G.J., J. Res. Natl. Bur. Stand., 47, 256–61, 1951.

90. Osborne, D.W., Westrum, E.F., Jr., and Lohr, H.R., J. Am. Chem. Soc., 77, 2737–9, 1955.

91. National Bureau of Standards, (publicly undocumented) informal modification of temperature scales.

92. Westrum, E.F., Jr. and Justice, B.H., 'Electronic Levels and Thermodynamic Properties of Cerium(III), Thulium(III), and Luretium(III) Oxides,' in Rare Earth Research Conference Proceedings, Coronado, CA, 821–32, 1968.

93. Justice, B.H., Thermal Data Fitting with Orthogonal Functions and Combined Table Generation. The FITAB Program, 1969.

94. Sommerfeld, A., Z. Phys., 47, 1–32, 1928.

95. Westrum, E.F., Jr., 'Thermophysics of Mineral and Rock Systems,' in Proceedings NATO Advanced Study Institute on Thermochemistry (Ribeiro da Silva, M.A.V., Editor), Reidel, Boston, MA, 719-44, 1984.

96. Westrum, E.F., Jr., 'Calorimetry of Phase, Schottky, and Other Transitions,' in Proceedings NATO Advanced Study Institute on Thermochemistry (Ribeiro da Silva, M.A.V., Editor), Reidel, Boston, MA, 695-718, 1984.

97. Grønvold, F. and Westrum, E.F., Jr., J. Chem. Thermodyn., 8, 1039-48, 1976.

98. Westrum, E.F., Jr. and Armstrong, G.T., 'The Critical Evaluation of Thermodynamic Data,' to be published under the auspices of CODATA and IUPAC Commission I.2 (in preparation).

99. Latimer, W.M., J. Am. Chem. Soc., 43, 818-26, 1921.

100. Latimer, W.M., J. Am. Chem. Soc., 73, 1480-2, 1951.

101. Saxena, S.K., Science, 193, 1241-2, 1976.

102. Cantor, S., Science, 198, 206-7, 1977.

103. Kieffer, S.W., Rev. Geophys. Space Phys., 17, 1-34, 1979; ibid., 20-34, 1979; ibid., 35-59, 1979; ibid., 18, 862-6, 1980; ibid., 20, 827-49, 1982.

104. Westrum, E.F., Jr., Chirico, R.D., and Gruber, J.B., J. Chem. Thermodyn., 12, 717-36, 1980.

105. Chirico, R.D., Westrum, E.F., Jr., Gruber, J.B., and Warmkessel, J., J. Chem. Thermodyn., 11, 835-50, 1979.

106. Chirico, R.D. and Westrum, E.F., Jr., J. Chem. Thermodyn., 12, 71-85, 1980; ibid., 311-27; Ibid., 13, 519-25, 1981.

107. Westrum, E.F., Jr., 'Morphology of Heat-Capacity Curves. Resolution of Transitional, of Schottky, and of Lattice Contributions,' in Proceedings of the 6th International Conference on Thermodynamics, Leipzig, German Democratic Republic, 1-14, 1981.

108. Westrum, E.F., Jr., Chirico, R.D., and Gruber, J.B., in The Rare Earths in Modern Science and Technology (McCarthy, G.J., Rhyne, J.J., and Silber, H.B., Editors), Vol. 2, Plenum Press, New York, NY, 387-92, 1980.

109. Westrum, E.F., Jr., J. Chem. Thermodyn., 15, in press.

110. Takahashi, Y. and Westrum, E.F., Jr., J. Chem. Eng. Data, 10, 244-6, 1965.

111. Westrum, E.F., Jr., Suits, E., and Lonsdale, H.K., in Advances in Thermophysical Properties at Extreme Temperatures and Pressures (Gratch, S., Editor), American Society of Mechanical Engineers, New York, NY, 156-61, 1965.

112. Downie, D.B. and Martin, J.F., J. Chem. Thermodyn., 12, 779, 1980.

113. Robie, R.A., Hemingway, B.S., and Wilson, W.H., J. Res. U.S. Geol. Survey, 4, 631, 1976.

114. Martin, D.L., Can. J. Phys., 38, 17, 1960.

115. Martin, D.L. and Snowden, R.L., Rev. Sci. Instrum., 41, 1869, 1970.

116. Chihara, H., personal communication, 1982.

117. McGlashan, M.L., J. Chem. Educ., 43, 226, 1966.

118. Gokcen, N.A., Mrazek, R.V., and Pankrats, L.B. (Compilers), Proceedings of the Workshop on Techniques for Measurement of Thermodynamic Properties, held at Albany, OR, Aug. 21-23, 1979, U.S. Bureau of Mines Information Circular 8853.

119. Freeman, R.D., CODATA Bulletin No. 55, CODATA Directory Chapter on Chemical Thermodynamics (Westrum, E.F., Jr., Editor), Pergamon, London, 134 pp., 1984.

120. IUPAC: Thermochemistry, Inc., Stillwater, OK, Bulletin of Chemical Thermodynamics, 23, 1980.

CHAPTER 5

Adiabatic Calorimetry Above 300 K

C. R. BROOKS and E. E. STANSBURY

5.1. INTRODUCTION

This review deals with the design, construction, and operation of adiabatic calorimeters for use above 300 K, and discusses the accuracy of such calorimeters in terms of the measurement of the specific heat of solids. Adiabatic calorimeters are used to measure the specific heat of solids and liquids, heats of solution and formation, and heat effects associated with structural changes. However, we will restrict our review to the use of adiabatic calorimeters to measure the specific heat of solids, as this is the subject of this volume. Specific heat of solids can be measured by several methods. Above 1300 K the accuracy of the measurements by adiabatic calorimetry degrades to a level at which specific heat values of better accuracy can be obtained by drop calorimetry or pulse calorimeter. However, accurate values of the specific heat cannot be derived from drop calorimetry if nonequilibrium states are produced in the sample. Pulse calorimetry employing resistive (Joule) self-heating of the sample is limited to electrical conductors.

Terminology used in calorimetry has been discussed by Ginnings [1]. A calorimeter is used in the present review to mean a device or instrument to measure heat effects associated with changes in state. The term 'calorimeter' is frequently used to mean the whole experimental apparatus, but more precisely it means that part of the system in which the heat effects occur. This raises a problem in distinguishing between the 'sample,' the 'cell,' and the 'calorimeter.' Where we feel it is necessary, we will distinguish between the cases where the calorimeter is the sample itself and where it is a cell (a capsule which may contain a sample).

Our review is limited to calorimeters designed to measure the specific heat directly and we will not include twin calorimeters or differential scanning calorimeters nor calorimeters specifically designed to measure heats of solution or heats of formation. However, reference will be made to calorimeters designed to measure the specific heat as well as these heat effects.

As in all adiabatic calorimetry, the main problem in high-temperature calorimetry is minimizing heat exchange between the calorimeter and its surroundings. In high-temperature calorimetry, two factors cause heat exchange to become more difficult to control as the temperature of measurement increases. One is that the temperature difference between the calorimeter and ambient temperature (300 K) leads to increasing heat exchange by conduction. The other factor is that heat exchange by radiation becomes proportionately greater due to its fourth power dependence on temperature. Another important problem in high-temperature calorimetry is the severely-restricted choices of materials of construction.

In the material which follows, a detailed section is presented on the design and construction of the sample, cell, and adiabatic shields in order to minimize heat exchange. (Materials of construction for high-temperature calorimetry are reviewed in an appendix to this volume.) A separate section is devoted to the temperature measurement which is a critical factor in determining the accuracy of specific heat measurement. Corrections which must be applied to the measured specific heat and especially those designed to correct for heat exchange are described in detail. Errors in the measurement of the specific heat and the accuracy of the measurements are analyzed. There is a brief section describing current use of computers in adiabatic calorimetry. Finally, there is a brief survey of high-temperature, adiabatic calorimeters which updates the 1968 list by West and Westrum [2].

In concluding this introduction, we acknowledge drawing extensively on our own research experience in writing this review. After struggling with problems in high-temperature adiabatic calorimetry for some years, it is difficult to do otherwise. We hope, however, that this experience has given us the insight to appreciate the work of all who participate in this area of investigation and that we have accomplished a reasonably-balanced and thorough review.

5.2. THE PRINCIPLE OF ADIABATIC CALORIMETRY

The principle of adiabatic calorimetry is quite straightforward, but its implementation can be difficult because of the nonadiabaticity of real calorimeters. Basic to accurate adiabatic calorimetry is design in order to minimize heat exchange between the calorimeter sample or cell and the surroundings, and an understanding of how to correct for the heat leakage which does occur. The heat exchange and design aspects of the calorimeter sample or cell and of the adiabatic shields are described in detail in Section 5.4. Corrections for heat leakage are covered in Section 5.6.

There are generally two different modes of operation of an adiabatic calorimeter. These are continuous heating (sometimes called dynamic adiabatic calorimetry) and discontinuous heating (sometimes called step-wise or intermittent heating). They both involve the same basic relation to calculate the specific heat, but different approaches are taken to deal with nonadiabatic conditions (i.e., to correct for heat exchange).

5.2.1. Continuous Heating

Consider the case where the sample is the calorimeter (i.e., the sample is not encapsulated in a cell). It has mass, m, and energy and is supplied to the sample at a constant rate, P. The mass of peripheral material, such as the heater or thermometer, is assumed to be negligible. While energy is supplied to the sample, its temperature, T, is measured as a function of time, t. The specific heat is given by

$$c_p = \frac{P}{m} (dT/dt)^{-1}. \tag{1}$$

The specific heat at a given temperature is thus obtained from the derivative of temperature as a function of time at that temperature, as illustrated in Fig. 5.1a, if adiabatic conditions prevail

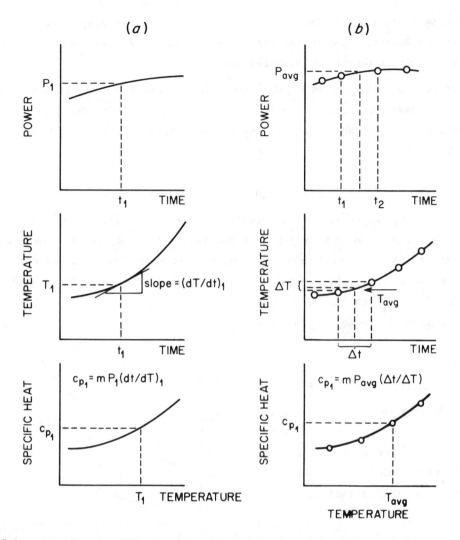

Figure 5.1. Schematic diagram illustrating how c_p is determined from the power and temperature data. (a) depicts the relationships in an ideal continuous-heating experiment, and (b) data measured in a real continuous-heating experiment.

and if the sample has a uniform temperature at any time. The power P can be time-dependent; only its value at the temperature of interest need be known. However, usually calorimeters operate so that the power is approximately constant or changes slowly with time.

In actual operation, the time and temperature usually are measured in finite increments from which approximate values for the time derivative can be calculated. For example, the time increment for the temperature to change 10 K may be measured. (A typical value might be 200 s.) In this case, the specific heat is calculated directly from these increments. Here the average power in the time increment is used, and again it is usually approximately constant. The specific heat is then given by

$$c_p = \frac{P}{m} \, (\Delta t / \Delta T).$$

(2)

The value of c_p is assigned to the time-averaged temperature over the increment ΔT (see Fig. 5.1b).

Actually, the c_p calculated by Eqs. (1) or (2) is that of the sample and the interactive components such as the heater. The correction for the heater and other equipment (should these not be negligible as assumed above) is described in Section 5.6.1.

If the sample is contained within a cell, then the procedure just described must be followed for an empty cell and then the identical cell with the sample in it. The heat capacity of the sample is then obtained by subtraction of the two measured heat capacities. This method is illustrated in Fig. 5.2.

5.2.2. Discontinuous Heating

Discontinuous heating can be used with encapsulated or unencapsulated sample. This procedure is illustrated in Fig. 5.3. Power is applied to the sample for only a measured time Δt. If the system is adiabatic, which is assumed here, the temperature will rise to a new equilibrium value, and ΔT is used to calculate c_p. The average specific heat over the temperature increment ΔT is again given by Eq. (2). Also, if a cell is used, measurements must be made on the cell empty and full.

5.2.3. Nonadiabatic Conditions

The previous descriptions refer to calorimeters operating under adiabatic conditions. Such conditions do not exist in reality. Equations (1) and (2) must then be modified to include the heat exchange, Q_L, as follows:

$$P \ dt = m \ c_p \ dT + Q_L. \qquad (3)$$

The main goal of calorimeter design is to minimize Q_L by surrounding the calorimeter (sample or cell) with thermal shields (referred to as adiabatic shields or guards), whose temperatures are generally controlled as closely as possible to the temperature of the calorimeter. Design can only minimize heat exchange between the calorimeter and the surroundings; practically, the heat exchange must be estimated or measured or the experiments performed such that Q_L, though finite, is cancelled. Methods for correcting for Q_L are described in Section 5.6.

5.3. GENERAL DESIGN AND OPERATION

The design and operation of the calorimeter

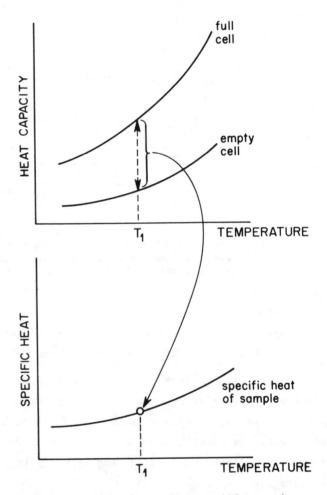

Figure 5.2. Schematic diagram illustrating determination of c_p of a sample from the heat capacity of a full and empty cell.

system determine the accuracy with which the specific heat and other heat effects can be measured. Important factors are the geometric design, the materials used, the method of taking and correcting the data, and the temperature control of the adiabatic shields. The design aspects are described in detail in Section 5.4. Corrections to the data are the subject of Section 5.6, and temperature control is discussed in Section 5.8 in conjunction with computer operation of the calorimeter. In this section, we briefly outline the important features of the design and operation of calorimeters.

5.3.1. Design

The geometric design of the calorimeter and the selection of materials for the components are based on many factors. The important ones are listed here. Note that the criteria for some factors may be contradictory to those of others.

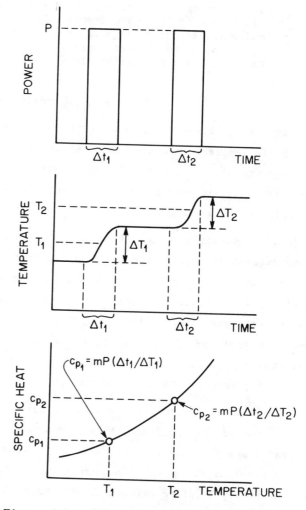

(1) The calorimeter sample or cell must be surrounded with radiation shields to minimize heat exchange with the surroundings. The shields and the calorimeter geometry are chosen to minimize the radiating surface area. The materials should be chosen to have low emittance for the sample or cell and high reflectivity for the shields.

(2) The shields should be made of materials with the highest thermal conductivity in order to minimize temperature differences on their surfaces. Also, the geometry should be chosen to minimize these differences.

Figure 5.3. Schematic diagram illustrating how c_p is determined from the power and temperature data for discontinuous heating. Here the drift rate both before and after the application of power is assumed to be zero.

(3) The calorimeter sample and cells are directly in contact with wires (e.g., thermocouple wires or heater wires) which allow exchange of heat by conduction with the surroundings. These wires should contact a conduction shield whose purpose is to maintain these leads at the temperature of the cell or sample in order to minimize heat exchange by conduction. This requires good thermal contact between the shield and the wires, yet high electrical resistance. These wires should have a low thermal conduction area (e.g., small diameter) and should have a low thermal conductivity.

(4) Temperature differences in the calorimeter sample or cell should be minimized through heater design and the geometry of the sample or cell.

(5) Shield design depends upon geometrical factors to minimize heat exchange, on the type of measurement control, and on the method of heating the shields. Thus, massive shields

with poor heat transfer may make temperature control difficult and require more energy to heat them.

(6) Although the thermal conductivity and emittance are main factors to consider in the selection of the materials of construction, other properties are important. A prime consideration is the thermodynamic stability of all materials with the atmosphere. Metallic materials generally require the use of a vacuum or an inert atmosphere. Such properties as strength and thermal expansion may be important, and in choosing electrical insulating materials, the decreasing electrical resistivity at high temperature must be considered.

5.3.2. Operation – Continuous Heating

If the calorimeter operation is initiated at 300 K, the procedure is quite straightforward. The shield temperature controllers and shield power supplies are activated and the sample or cell heater power set to give the desired heating rate. Reliable data cannot be taken until at least the main adiabatic shield comes under control, which depends on the particular calorimeter system and also on the heating rate.

In continuous heating, it is most common to employ a constant (or nearly constant) energy input rate to the cell (empty or full) or to the sample. For example, if an electrical heater is used and the power supply operated with either a constant current or constant voltage, the power supplied to the heater depends upon the resistance of the heater, which is temperature-dependent. However, for practical heater materials, the range of the resistance increase from 300 K to the upper limit of operation does not exceed about 10%. The heating rate depends as well upon the heat capacity of the calorimeter and heat leakage, both of which are temperature-dependent. Thus, the heating rate is usually not constant.

Temperature control of the shields surrounding the calorimeter is critical to maintaining adiabatic conditions. Temperature differences between shield and calorimeter must be continually monitored and the magnitude of these differences used to control power to the shields. Although this control may be manual, commercial controllers with proportional, integral, and derivative action greatly simplify this task, and under computer operation control algorithms within computer programs perform the function. Optimum controller parameters do change with increasing temperature due to the increased radiative heat exchange rate and temperature-dependent changes in the thermal diffusivity of the calorimeter and shields, but changes in these parameters are gradual, and readjustment is usually not a problem. Thermocouple emfs need to be determined with sensitivities of one microvolt or less, which requires amplification of these signals for control purposes.

The acquisition of data using a computer will be described in Section 5.8. Here the description will apply to acquisition of data using manual methods and is typical of the procedures followed. If the thermometer is a thermocouple, its emf voltage is measured with a precision potentiometer or microvoltmeter. If the thermometer is a resistance thermometer, the associated voltage drops are measured likewise with a precision potentiometer or a bridge.

The temperature must be measured as a function of time to relate instantaneous energy input to the associated temperature rise. When a potentiometer is used for thermocouple voltage measurement, this instrument is set to a voltage which is slightly greater than that of the present calorimeter thermocouple. The imbalance of these two voltages is detected (e.g., visually with a null detector). When the imbalance becomes zero as a result of heating the sample, the operator activates the second of the two timers employed to time successive temperature intervals. The first timer is then reset to zero. This process is repeated giving sets of successive temperature (e.g., 10 K) and time intervals. The sample is heated by Joule heating in a coil of wire in close thermal contact with it. The sample-heater power is calculated as the product of the voltage drop E across the coil and the current I through the coil. This current is measured by the voltage drop across a standard resistance in series in the heater circuit but external to the system. The voltage drops may be measured with potentiometers or, more conveniently, with digital voltmeters. The total energy input over the time interval Δt is given by

$$\int_0^{\Delta t} EI dt.$$

The specific heat is calculated using Eq. (2). If it is established that the voltage E and current I vary smoothly and monotonically, they are obtained from an average of the corresponding readings at the beginning and at the end of the temperature increment.

The maximum heating rate is usually determined by the ability of the operator to take and record the necessary readings, to set the potentiometer and timers, and to monitor and adjust the temperature control of the shields. Our experience limits the minimum time increment to about 100 s, which corresponds to an upper limit on the heating rate of approximately $0.2 \text{ K} \cdot \text{s}^{-1}$ (but this depends on the heat capacity of the calorimeter).

5.3.3. Operation – Discontinuous Heating

The operation of the calorimeter using discontinuous (step-wise) heating is similar to that just described for continuous heating. The same type of system will be used here to illustrate the operation.

The system is brought to steady state with no power to the specimen. The adiabatic shield temperature controllers are adjusted to give very small 'drift,' as evidenced by the change in temperature of the calorimeter. Once an acceptably small, linear temperature change is observed over several minutes, power is applied to the specimen heater and the interval timer started. The voltage drops across this heater and across the standard resistor in series with it are measured. The temperature rise of the calorimeter is monitored and when the desired increase (e.g., 10 K) is achieved, the power and the timer are turned off simultaneously. Just before turning off the power, the voltage drops across the heater and the standard resistor are measured again. The temperature of the calorimeter is monitored for several minutes until a linear temperature change (drift) is again observed. The adiabatic shield temperature offset remains the same as at the beginning of the heating period. Once a linear temperature change of the calorimeter is again established, the shield controller may be readjusted to minimize the calorimeter temperature drift for the next measurement.

The temperature increment must be sufficient to allow about 100 s for the necessary measurements during the application of power. The temperature drifting before and after the heating period is usually observed for several minutes. The specific heat is calculated using Eq. (2). The ΔT is obtained by extrapolation of the linear temperature-versus-time curves before and after the heating period to the time midpoint (see Section 5.6.4).

5.4. ADIABATIC SHIELD AND CALORIMETER CELL AND SAMPLE DESIGN

In the design of adiabatic shields and calorimeter samples and cells, the guiding principle is the minimization of heat leakage between the calorimeter and the shield. Design objectives include minimization of temperature differences on the surface of the shields and the sample or cells, minimization of heat leakage between the shields and the sample or cell by radiation, and minimization of heat leakage between the sample or cell and the surroundings by conduction along the necessary wires. Heat transfer is governed by the geometry (size and shape) of the components and the heat transfer properties (thermal conductivity and emittance) of the materials used. In principle, the heat transfer equations can be solved to determine the geometries and materials which minimize heat exchange. Although this rarely can be done rigorously, simplified heat transfer calculations are essential to effective calorimeter design.

The heat transfer equations and their relations to the design of calorimeters have been examined in detail in excellent papers by West [3] and by Ginnings and West [4]. We will not describe the relations in any detail here, but rather draw on information from their papers in reviewing the importance of critical evaluation of heat transfer in calorimeter design. Heat transfer aspects of other types of calorimeters have been treated in detail by West and Churney [5] and by Zielenkiewicz [6].

5.4.1. Radiation Shields

First, we consider the shield design in which the physically isolated shield is heated to maintain its temperature identical to that of the calorimeter. That is, it is designed so its temperature will follow the temperature of the calorimeter and, hence, minimize heat leakage by radiation. Several desirable characteristics of a radiation shield are listed in Table 5.1. The most important problems in design are temperature gradients along the surfaces of the shield and temperature control of the shield.

To emphasize the importance of radiation in contributing to heat exchange, we will consider an example. The heat exchange by radiation between two surfaces, one at absolute temperature T and the other at (T+ΔT), is given by

$$Q = AF\sigma[T^4 - (T+\Delta T)^4] = AF\sigma[4T^3 \Delta T + \ldots] \simeq 4AF\sigma T^3 \Delta T \tag{4}$$

where we have assumed that the two surfaces are of equal area A. σ is the Stefan-Boltzmann constant (5.67×10^{-8} watts$\cdot$m$^{-2}\cdot$K^{-4}). F is a factor which depends upon the geometry of the two surfaces and includes the emittance, assumed to be the same for the two surfaces.

Table 5.1.   Some Desirable Characteristics of Radiation Shields

---

a. High thermal conductivity to minimize temperature differences on the surface.

b. Low emittance to minimize absorption and emission of radiation.

c. Surface stability to maintain reproducible emittance.

d. Size and shape should be such as to make temperature response rapid for good temperature control.

e. Accurate temperature control to minimize non-reproducible heat exchange.

f. Size and shape should be such as to minimize power required to heat shield.

g. Thermocouples can be attached to it, preferably by welding.

h. Available in shapes and sizes for ease of fabrication.

i. Material should be machinable, deformable, and weldable for ease of fabrication.

Figure 5.4 shows the quantity Q/AF as a function of temperature T for different values of ΔT [7]. Note that increasing the temperature difference from 0.1 K to 1.0 K increases the heat exchange by a factor of approximately 10.

To further illustrate the problem, consider the following estimation of heat loss from a calorimeter sample which is 2.0 cm in diameter, 7 cm long, and weighs 200 g. The surface area is 44 cm^2. Assume that the radiation shield has approximately the same surface area and emittance as the sample and that it is 0.1 K lower in temperature. From Fig. 5.4, $Q/AF = 2 \times 10^{-3}$. Thus, the rate of temperature decrease dT/dt of the sample is given by $Q = (44)(0.1)(2 \times 10^{-3}) = (200)(c_p)(dT/dt)$. If the specific heat c_p is taken to be 0.4 $J \cdot gm^{-1} \cdot K^{-1}$ (the value for Cu at 1000 K), then $dT/dt = 10^{-4}$ $K \cdot s^{-1}$, or 0.4 $K \cdot hr^{-1}$. If ΔT is 1.0 K, then $dT/dt = 4$ $K \cdot hr^{-1}$. In our calorimetric technique [13], we can determine the heat exchange by measuring the temperature of the sample as a function of time with no power to the sample. If conditions were adiabatic, then $dT/dt = 0$. However, we frequently encounter 'drift' rates near 1000 K similar to the above calculated value; they increase to as high as 10 $K \cdot hr^{-1}$ at 1300 K.

This example shows the importance of minimizing temperature differences on the radiation shields and calorimeter surfaces and of minimizing temperature differences between shields and calorimeter.

Minimization of temperature differences on the shields requires the use of high thermal conductivity material. The most desirable material is Ag. Many calorimeters have been designed using Ag, but its use is limited to about 1100 K. Cu can be used to about 1300 K, and it has a thermal conductivity only slightly less than Ag. For the higher temperature ranges, Mo, W, or Ta are usable. They have a reasonably low emittance, can be obtained in sheet form, and are weldable.

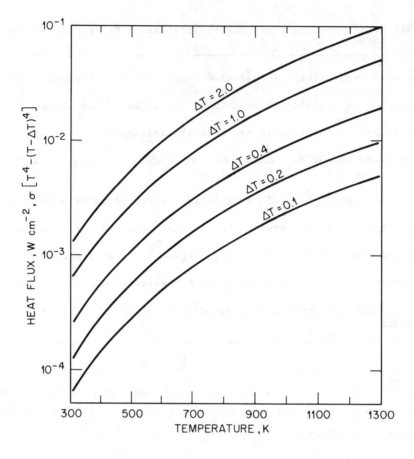

Figure 5.4. Heat flux for radiative heat transfer as a function of temperature and temperature difference between two surfaces [7].

That temperature differences exist on shield surfaces has been documented by Naoki et al. [8]. They measured the temperature differences between three separate points on the surface of a spherical Ag shield as a function of temperature. Figure 5.5 shows their results. These measurements were made in vacuum, reduced pressure Ar, and Ar at atmospheric pressure. Although the presence of a gas should increase the response of the shields due to gaseous conduction and, hence, improve temperature control, convection currents cause severe temperature differences on the surface of even high conductivity Ag. Thus, a vacuum is desired instead of a gaseous environment.

We have observed similar temperature differences on the surfaces of shields. For example, in our earlier calorimeters, a thick-walled cylindrical shield also served as the vacuum container. A Ni or Ni-30% Cu alloy (Monel) tube, 1 cm thick with spiral grooves on the outside, containing nichrome heater wire threaded through ceramic beads, was used. The inside surface was mechanically or chemically polished. In one design, a 2 cm thick layer of Ag was cast around the outside of the tube and grooves for the heater machined in the Ag layer. The purpose was to improve heat conduction longitudinally along the tube. However, we still measured temperature differences of several degrees (e.g., 5 K) between the point where the control thermocouple was welded, at a level opposite the sample, and locations 20 cm distance. Part of this may have been due to eventual separation of the Ag and the tube at their interface after repeated cycling to 1300 K.

Most radiation shield designs employ embedded heaters, although they have the disadvantage of requiring sufficient size to contain the heater and, hence, are relatively massive. This requires considerable power for heating and results in poor response to temperature control. To avoid these problems, we use thin (e.g., 0.03 cm thick) sheets for some shields and heat by radiation from a bare-wire nichrome heater located about 0.5 cm away, as shown in Fig. 5.6. The response of these shields is rapid, allowing excellent temperature tracking. The shields should be uniform in thickness, as variations will affect the uniformity of surface temperature. For temperatures approaching 1300 K and above, W and Mo are satisfactory shield materials. Below 1300 K, we use Pt and Ni for most of our thin shields. They are easy to fabricate into desired shapes and readily permit attachment of support wires and thermocouples by welding. Ni can be chemically polished to increase the reflectivity.

It is worth noting that the relative areas of the shield and the sample or cell are important in setting the magnitude of heat leakage by radiation. For the case of concentric cylinders, each of finite area, the heat exchange rate is given by

$$\dot{Q} = \frac{\varepsilon_1 \varepsilon_2}{\varepsilon_2 + \varepsilon_1 (1-\varepsilon_2)(r_1/r_2)} \ \sigma(T_1^4 - T_2^4) \qquad (5)$$

where the subscript 1 refers to the inner cylinder and 2 to the outer cylinder. It is seen that heat exchange is minimized if the ratio r_1/r_2 is as close to unity as possible. Thus, the radiation shields should be located close to the cell or sample. Also note that under conditions of $r_1 \ll r_2$, heat exchange is governed by the radiation properties of the surface of the smaller cylinder, and not the larger.

For thermal symmetry, it is desirable to make the calorimeter sample or cell and the

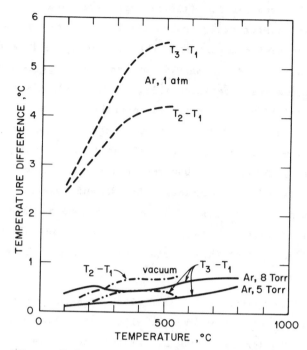

Figure 5.5. The temperature distribution on the surface of an adiabatic shield as a function of temperature [8]. T_1, T_2, and T_3 refer to the temperature at three different positions on the surface of the shield.

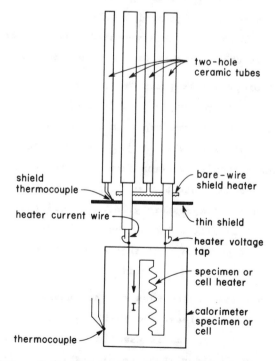

Figure 5.6. Schematic diagram illustrating the geometry for a thin shield heated by radiation from a bare-wire heater.

shields spherical. Calorimeters have been constructed with spherical shields [9,10], but the sample heater design makes a spherical cell or sample awkward. It is more convenient to use a cylindrical geometry with the samples or cell centered concentrically inside a cylindrical shield which is closed by shields on each end. It is common to heat each shield independently. In our calorimeters, the three shields are physically separate with temperature control of the cylindrical shield most critical in controlling heat exchange by radiation. We have tested this sensitivity of heat exchange by heating the calorimeter system to a given temperature (e.g., 1000 K), then, with no power to the calorimeter, measuring its 'drift' rate dT/dt. If adiabatic conditions obtain, $dT/dt = 0$. However, usually a finite rate is observed. We then introduce temperature offsets to the individual shields and observe the change in drift rate. The bottom and top shields may be offset 1 K or more before a measurable effect on the drift rate is observed. However, changes as low as 0.1 K in the offset of the cylindrical shield may affect the drift rate, confirming that this shield, which presents the greatest area to the calorimeter, is most critical with respect to control.

The cylindrical shield which we have used very successfully is a thin-walled Ni tube which is directly heated by current within it. The tube is suspended from the top of the calorimeter frame, which is at 300 K. At this point, one power lead connects to Cu rods which extend down parallel to the tube and on the outside of it. At the end, these rods are connected to the bottom of the tube. The upper 15 cm section of the tube is Cu having a wall thickness of 0.32 cm. The next 15 cm is monel (Ni-30% Cu) to minimize heat conduction from the hotter Ni part. Several Ag fins are brazed to the Cu and monel to serve as radiation fins to assist in cooling this section. The lower section (55 cm) is Ni with a wall thickness of 0.025 cm, and this section serves as the cylinder shield. Centered inside this Ni section is a suspension frame supporting the calorimeter sample. Both the directly-heated tube and the calorimeter suspension unit are located concentrically inside a thick-walled Ni tube which serves as an auxiliary shield and the vacuum container.

A programmable power supply is used to obtain current up to 100 A at 10 V. For temperature control, a thermocouple is welded directly to the outside of the tube, and its signal is compared to that of the thermocouple on the sample, the difference amplified, and this signal used to activate a controller which adjusts the current. One problem with this system is voltage pickup by the thermocouple junction from current flowing through the tube. However, we find that any error present is reproducible as long as the thermocouple is not moved. Another problem arises from longitudinal temperature gradients, and these (and, hence, the heat loss) are quite sensitive to the position of the thermocouple. The main advantage of this shield design is the extremely quick thermal response of this shield and the excellent temperature control achieved. We have used this shield to 1300 K with a control of ±0.1 K.

To minimize the mass of the shields and the power required to heat them, most calorimeter designs have at least one auxiliary temperature-controlled shield. This sometimes serves as the vacuum container and is usually concentric with the rest of the system. It usually is operated from 1-10 K colder than the main shields. If it is not, its temperature control unduly affects the temperature control of the main shields.

The use of unheated (passive) radiation shields is highly recommended to reduce heat exchange. Placed between the outside of the temperature-controlled shields and the inside of the calorimeter vacuum container, these shields reduce considerably the heat lost from the outside surface of the shields and, hence, reduce their required power. Ginnings and West [4] have discussed the importance of radiation shields and have illustrated their effectiveness. Figure 5.7 shows the results of one of their examples. Here is shown the reduction in heat transfer with the number of shields inserted between two parallel plates 5 cm apart. The emittance of all surfaces was assumed to be 0.5. One surface was at 400 K and the other at 500 K. In this example, the use of four shields reduced the heat transfer by about a factor of four.

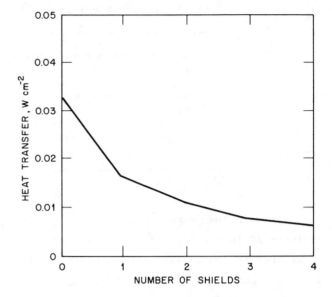

Figure 5.7. Illustration of the effect of the number of shields between two surfaces 5 cm apart, each having emittance of 0.5, one at 400 K and the other at 500 K (adapted from Ginnings and West [4]).

Passive radiation shields are seldom used between the calorimeter sample or cell and the main adiabatic shields as they will affect the response of the shields and may make temperature control more difficult.

5.4.2. Conduction Shields

As will be illustrated, heat conduction along thermocouple wires and heater leads from the sample or cell to regions of lower temperature constituents a main source of heat leakage. The purpose of a conduction shield is to bring the temperature of these wires to the temperature of the sample or cell before they exit this region. The conduction shield is usually a block with its own heater, and its temperature is maintained by the use of a controller to that of the sample or cell. (However, it can just as well be one of the radiation shields.) This type of conduction shield is often called a thermal block, tempering block, or thermal contact.

Some desirable characteristics of a conduction shield are itemized in Table 5.2. The most important factor is good thermal contact of wires with the block. For low-temperature (below 400 K) calorimetry, this is usually achieved by the use of fine wires which are wrapped repeatedly around the block and held in place with lacquer. The thin insulation on the wire and the thin layer of lacquer have sufficient electrical resistance to adequately insulate the wire from the block, and yet give good thermal contact, and the thin insulating coating and fine wire provide a short thermal conduction path. However, in high-temperature calorimetry, this approach is not feasible as the electrically insulating material is ceramic (although mica can be used to about 1000 K) and it is difficult to obtain a good thermal contact with the shield.

Table 5.2.   Some Desirable Characteristics of Conduction Shields

---

a.   Good thermal contact between shield and wires from sample and cell.

b.   Good electrical insulation between shield and wires from sample and cell.

c.   Other characteristics same as those for radiation shields (Table 5.1).

Also, in low-temperature calorimetry, the heater leads are usually very fine Cu wires to minimize Joule heating. However, in high-temperature calorimetry, Cu wires are usable only to about 1300 K. Materials which will go to higher temperatures have higher electrical resistivities, and increasing the cross-sectional area to minimize heating increases the heat conduction.

Most high-temperature calorimeters do not use a physical conduction shield, but instead accept heating of the wires by radiation from adjacent radiation shields as they exit the system. For the calorimeter sample or cell heater wires, another approach has been to choose the heater lead wires so that the heat generated by the current flow in them will approximately compensate for the heat loss. This approach is valid only for continuous heating.

The heat conduction along the lead wires is proportional to the temperature gradient, which is approximately proportional to the difference between the temperature of the calorimeter and that of the ambient environment, usually near 300 K. Thus, in low-temperature calorimetry, as the calorimeter is heated from a low temperature to obtain the data, this difference becomes smaller, and heat conduction along the leads towards the calorimeter becomes less. In addition, the thermal conductivitiy decreases as the temperature of the wires increases, so heat exchange decreases. In high-temperature calorimetry, however, this temperature difference continually increases as the calorimetry is heated, so that heat loss becomes an increasingly severe problem.

The heat transfer relations related to this conduction problem have been discussed in detail by Ginnings and West [4]. To illustrate how serious this heat exchange is, consider the following experiment. In one of our calorimeters, using a copper specimen, with the arrangement shown in Fig. 5.8, the drift rate dT/dt of the specimen was $-0.28°C \cdot hr^{-1}$ at 595 K. The calorimeter was then cooled to 300 K and the voltage taps and heater leads were disconnected so that the conduction area was reduced by about 75%. The calorimeter was then reheated, and the drift was measured to be $-0.18°C \cdot hr^{-1}$ at 595 K.

In this calorimeter, the thermocouple leads and ground leads are brought up to the top radiation shield; then the wires in ceramic insulation are laid in grooves in the Cu shield, on top of which is a Cu lid. The wires in ceramic insulation exit the radiation shield, pass down along the outer radiation shield, then up to the top of the calorimeter to a junction block (at 300 K). The total path is approximately 50 cm. We estimate the heat loss by assuming the two thermocouple wires and the ground wire are each 0.013 cm in diameter and each 50 cm long. Two of these are Pt

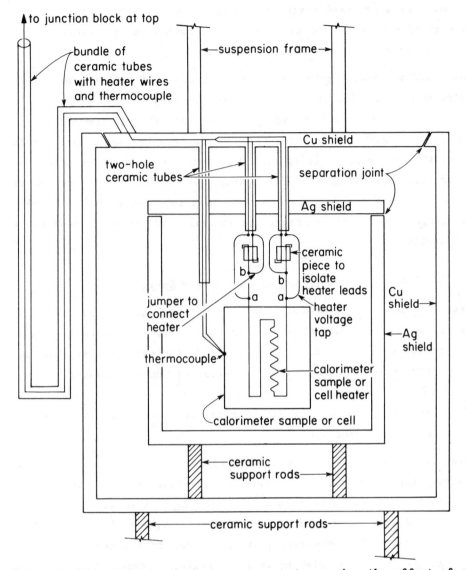

Figure 5.8. Schematic illustration of the arrangement to examine the effect of conduction along the heater wires on the heat loss.

and one is Pt–13% Rh alloy. However, we use for the thermal conductivity of all three that of Pt at 595 K, about 0.8 W·cm^{-2}·K^{-1}. Now assuming that no heat is transferred radially from the wires to the surroundings, only longitudinally along the wires to the junction block at 300 K, then the heat loss is

$$Q = (0.8 \text{ W·cm}^{-2}\text{·K}^{-1}(0.00013 \text{ cm}^2) \ (300 \text{ K})/50 \text{ cm} = 0.00062 \text{ W}$$

for each wire, or for all three wires this is about 0.0019 W. From the measured drift rate at this temperature under this geometry, we get the actual heat loss to be

$$Q = (160 \text{ g}) \ (0.425 \text{ J·g}^{-1}\text{·K}^{-1})(0.18 \text{ K·hr}^{-1})/(3600 \text{ s·hr}^{-1}) = 0.003 \text{ W}.$$

In spite of the uncertainties in the calculation, this clearly indicates that heat loss by conduction along the lead wires is significant. Thus, even if suitable radiation shields are used, heat loss by conduction must be considered seriously. Note also that the thermal tempering to the shield may not be very effective due to the necessity of heat transfer through the electrically-insulating, low thermal conductivity ceramic insulation.

The measurements and calculations in the preceding paragraphs indicate that passing the thermocouple wires in ceramic insulation through the region between radiation shields (in this case for about 50 cm length) does not transfer sufficient heat radially to compensate for heat loss. Thus, the use of a conduction shield is important. Up to 1100 K, some designs combine the main radiation shield and the conduction shield, with the lead wires simply attached to the shield before exiting. For electrical insulation, thin (e.g., 0.004 cm thick) mica sheets are used. However, such thin sheets of high-temperature ceramics are too fragile for use. The use of larger, and hence less fragile, ceramics increases the thermal conduction path and makes sufficient surface contact difficult.

5.4.3. Sample and Cell Heater Design

The heat leakage of the calorimeter sample or cell depends not only on the shield design, but also on the design of the heater. But, there is a basic difference between the use of a calorimeter sample with or without a cell. Usually the use of a calorimeter sample alone means that the measurements of the apparent specific heat must be corrected for heat leakage to obtain the true specific heat of the sample. However, with the use of a cell, it is sufficient that the design feature a low and reproducible heat leakage for both the cell empty and the cell with the sample in it. These two cases are discussed below.

There are three commonly-used methods of introducing a measurable energy into the sample or cell. The most common, emphasized in the following discussion, is to use an internal electrical resistance heater. Another method is to insert into the sample a small mass of another material of known specific heat and at a known initial temperature. This is the method of mixing, which Krauss [11] has referred to as 'reverse calorimetry,' since, if the inserted material is at a lower temperature than that of the sample, cooling is obtained instead of heating. High measurement accuracy has not been realized with this method, so it is rarely used. Bell and Hultgren [12] used this method quite successfully to measure the heat capacity of liquid Bi. The third method used is electron bombardment, which has the advantage that no connecting wires to the sample or cell are required and, hence, one source of conduction heat leakage is eliminated. This method will be described briefly at the end of this section.

If the calorimeter is heated by a heater on its surface, then the problem of proportioning the energy between the calorimeter and the shields facing the calorimeter must be considered. This is avoided by using an internal heater in the calorimeter. The problem here is that the heater temperature will be above the sample temperature, so that some of the measured electrical energy may be conducted along the heater leads from the sample towards the shield. To prevent this requires that the heater leads attain the temperature of the sample before they leave the sample. The customary heater design is a spiral of resistance wire (e.g., nichrome) wound on a ceramic tube. If such a heater is inserted directly into a hole in the sample, the losses due to conduction can be unacceptable. To minimize this problem, we use the design shown in Fig. 5.9 [13]. The heater leads carrying the current pass almost through the sample before they are attached to the heater. Thus, as heat is conducted from the hot heater along the heater current leads, this heat is partially transferred out to the sample. The voltage drop required to calculate the electrical power is measured just as the current leads exit the sample.

This heater design is relatively simple to construct, and we have used this successfully on solid samples and on thin-walled, welded containers in which pure metals are sealed for measurement into the liquid region [14]. Of course, the mass and the heat capacity must be known. As the measured electrical power is used to calculate the heat capacity of the sample plus the heater, it is desirable to keep the mass of the heater small to minimize its heat capacity. As will be discussed in Section 5.6.1, correction for the heater is usually negligible if the mass of the heater is less than 0.01 that of the sample. We usually use samples from 100 to 400 g, so the heater must be from 1 to 4 g, which is not difficult to achieve.

It would be desirable to pass the heater current leads through the sample several times, but this complicates the construction and insertion of the heater and increases its mass. It is also desirable to have minimum Joule heating in the current lead wires outside the sample. However, the use of low-resistivity wire (e.g., Cu) is not advisable due to its high thermal conductivity. Thus, with the design in Fig. 5.9, the heater wire is chosen to generate some heat to compensate for heat conduction.

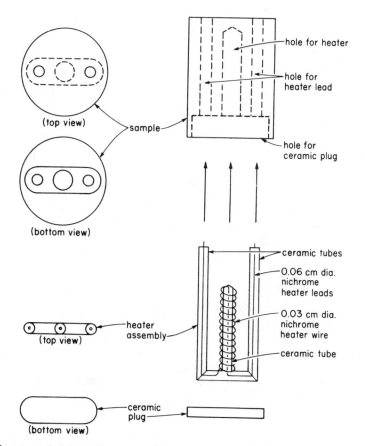

Figure 5.9. Sample heater design in which the heater current leads pass up through the sample to assist in thermally equilibrating the leads [13].

The problem of temperature differences in the sample must be considered. For metallic materials, we have estimated the temperature difference between the surface of the inside of the heater
hole and the outside surface of the sample to be typically about 1 K. Such a gradient does not
significantly affect the specific heat measurements (except in the vicinity of sharp phase transitions), but there is a difference between the actual temperature measured on the surface with a
thermocouple and the average temperature. This difference depends upon the thermal conductivity
of the sample. However, we have measured [13] the specific heat of a 250 g solid α-Al$_2$O$_3$ sample
of the geometry shown in Fig. 5.9, and within the scatter of our data (approximately $\pm 1\%$), we
found the data to agree with those of the National Bureau of Standards.

When using a calorimeter cell an internal or an external heater can be used. The design depends somewhat upon the type of thermometer. Platinum resistance thermometers are sometimes used
for measurements up to 1000 K, in which case the cell heater and the thermometer are incorporated
directly into the cell. West and Westrum [2] have described such a configuration in some detail.
If a thermocouple is used for the thermometer, it can be welded on the side of the cell or inserted in a receptacle hole which is inside the cell and, hence, surrounded by the sample. The same
configuration can be used if the cell is heated externally. In this case, the cell heater is
intimately in contact with the cell surface, usually inserted in the walls of the cell.

Naito et al. [8] have made a detailed study of the heat leakage from a cell with an internal
and with an external heater. The configuration they used is shown in Fig. 5.10. They measured
α-Al$_2$O$_3$, whose heat capacity is well known, so that they could estimate the heat leakage. Their
results depended upon the heating rate and they found that the heat leakage was less for an external heater (Fig. 5.11). Also note that in both cases the heat leakage was not identical for the
full and the empty cell, but the difference was less for an external heater. Since in their design the external heater was placed just on the inside wall of the cell, the heat transfer between

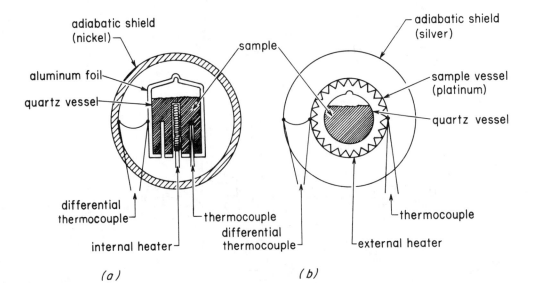

Figure 5.10.   The configuration used by Naito et al. [8] to study the effect of cell heater type
on heat leakage.

this heater and the cell was only slightly influenced by the presence of the sample. Also note that there is less heat leakage if the heating rate is lower. The influence of heating rate is discussed further in Section 5.6.2.

The problem of heat leakage due to heater leads has been considered in detail by Wollenberger [15] and others [16–20]. Wollenberger points out that the systematic error arising from heat leakage along the heater leads can best be reduced by a thermal short circuit. Thus, either a conduction shield should be used or a combined radiation-conduction shield. He points out the importance of a high heat conduction path where the heater leads are to be brought to thermal equilibrium and cites the use of mica sheets for this purpose. He also suggests the use of small diameter, sheathed wires. He also points out that BeO would be a desirable insulation for such sheathed wires as its thermal conductivity is similar to that of metals.

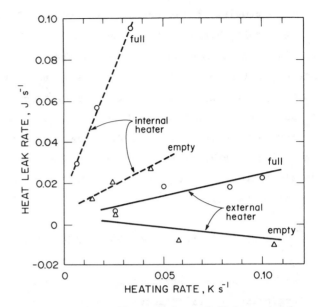

Figure 5.11.   The results of Naito et al. [8] on the heat leakage of a full and an empty cell with an internal and external heater (see also Fig. 5.10).

The problem of heat generated in the heater leads has been examined in detail [16–20]. If the thermal resistance between the heater wire and the sample or cell, as well as between the wire and some external conduction (or radiation) shield, is low, then when the heater sustains a current the wire attains a maximum temperature in its middle. Thus, heat is generated which flows both ways, towards the shield and towards the sample or cell. A common practical solution to this is to locate one heater voltage tap at the sample or cell and the other at the shield, assuming that half the heat generated in each lead flows into the sample or cell. Another approach, using the same assumption, is to place the voltage taps midway along each heater lead wire.

However, when using a cell, a different power is required to heat the cell at the same rate when full and when empty. Thus, the heater leads will attain a different temperature, and radiation losses will not be the same for the two cases. Martin [20] has discussed this problem and has pointed out that there are two ways to minimize it. One way is to surround each heater lead with a tubular radiation shield and anchor thermally one tube at the cell and the other at the shield. The other way is to use a heater of high electrical resistance relative to the resistance of the leads.

In addition to the heater, the other important consideration in designing the calorimeter cell is the material and geometry to minimize temperature gradients in the cell and sample. Here the material should be of high thermal conductivity and high temperature stability. Ag or Cu are suitable to about 1000 K, and for higher temperatures Mo has been used [21]. The geometry frequently includes internally-connected material to transfer the heat more evenly from the heater to the regions of the sample.

To completely circumvent the problem of heat leakage associated with the sample or cell heater, Wollenberger and his co-workers [22,23] have designed and used a calorimeter in which the sample or cell are heated by an electron beam. Thus, there are no physical connections from a heater to the surroundings. The design uses an electron beam accelerated by about 5,000 V. The beam input energy is controlled by adjustment of the bombardment time. The beam can be deflected to allow scanning of the surface of the sample. Steffen and Wollenberger have used this method of heating a twin calorimeter [24] operating from 300 to about 800 K. They report the accuracy of the specific heat measurements to be about $\pm 4\%$. This method of heating is promising, and there should be continued improvement in the development in its use to heat not only the sample or cell, but also shields.

5.5. TEMPERATURE MEASUREMENTS IN HIGH-TEMPERATURE CALORIMETRY

Techniques and problems in the measurement of temperature in adiabatic calorimetry relate to determination of calorimeter temperature, to the calculation of the specific heat, and to the determination of calorimeter and shield temperatures required to maintain adiabatic conditions. Although measurement of a temperature interval is implied by the form of the specific heat expression used for calculation purposes (Eq. 2), in fact the slope of the curve of temperature with respect to the temperature-sensitive physical variable (e.g., resistance or thermal emf) is effectively involved rather than absolute values of the temperature at the limits of the interval. Recognition of this fact can significantly reduce errors in temperature interval measurement since the slopes of such functions are much less sensitive to variations in individual thermal elements of a given type and to changes in calibration over time. Frequently, the accuracy with which the temperature interval can be determined is controlled more by the sensitivity of the instrumentation than by the absolute calibration of the sensor. Accuracy of determination of the temperature interval is also improved by the use of statistical methods. In the discontinuous-heating technique for measuring the specific heat, the process of measuring the temperature drift before and after an energy addition to the calorimeter and extrapolation of the drifts to a mean temperature is essentially a statistical treatment of the temperature determinations. The majority of the experimental techniques used in the past have depended on the measurement of a single value for the temperature at the limits of an interval. Introduction of computer techniques has made it possible for the almost continuous measurement of temperature and the use of curve fitting to increase considerably the accuracy of determination of a temperature interval. In contrast, for adiabatic shield control the temperature of the calorimeter and the shield need to be monitored at the same instant of time and corrective changes of power to the shield made based on the differential output of sensors attached to the shield and to the calorimeter. In general, statistical techniques have not been used for shield control; therefore, for sensor pairs used for shield control, retention of matched output at any temperature over the time-temperature history of use of the calorimeter system is important for maintaining adiabatic conditions. Lack of temperature matching can be compensated to various extents by empirically determined offsets of the shield temperature to give zero drift rate of the calorimeter. However, this technique is more difficult to implement for continuous heating measurements than for discontinuous heating methods.

For adiabatic calorimetry above 300 K, three types of temperature sensors may be considered: resistance thermometers, thermocouples, and radiation pyrometers. Depending upon the material being measured, the accuracy of the measurement and the temperature range of the measurement, some generalizations can be drawn. First, with precision measurements on thermally reversible substances using interrupted heating techniques to 800 K, specimen temperatures are usually determined by resistance thermometers and shield control is maintained by thermocouples. Second, thermally irreversible substances require measurements using continuous heating techniques, which usually necessitate choosing thermocouples for measuring both the sample temperature and for shield control. Regardless of sensor type, adiabatic calorimetry has not been extended much above 1300 K since temperature differences and uniformity of temperature over the surface of both the calorimeter and shields must be held to 0.1 K in order to realize an accuracy in the specific heat of 1%. At higher temperatures, the accuracy of these measurements is generally insufficient for practical use in calorimetry and will not be considered further here.

Optical pyrometers that might be considered for use above 1100 K have found limited use and will not be considered in this review. Before reviewing the characteristics of resistance thermometers and thermocouples, it is useful to examine the important factors governing choice of these two sensors. These are listed with brief comments in Table 5.3, and reference will be made to these factors in the following discussion.

Table 5.3. Important Factors Governing the Choice of a Thermometer

---

1. Temperature range of reliable operation.

2. Accuracy and reproducibility: related to the sensitivity of the sensor, susceptibility to physical and chemical changes and to the magnitude of the output of the signal.

3. Response time: inherent response time of the sensor and response time of the thermal coupling of the sensor and the object being measured.

4. Instrumentation: accuracy and reproducibility, response time, and adaptability to computer interfacing.

5. Thermal contact: ability to place the sensor in direct contact with sample, calorimeter, and shields.

6. Thermal conductivity: a factor in the heat transfer between sensor and object to be measured.

5.5.1. Resistance Thermometry

Resistance thermometers are used extensively in adiabatic calorimetry to about 600 K and to a limited extent to 800 K. They are used almost exclusively for measuring sample temperature with the discontinuous heating technique. This limitation results from the difficulty of making direct thermal contact between the resistance element and the object whose temperature is being monitored. The discontinuous heating method allows time for thermal equilibrium between the resistance thermometer and its surroundings and extrapolation procedures used in determining the specific heat are reasonably tolerant of the thermal lags. In contrast, these thermal lags may be prohibitive when using continuous heating. The thermal lag is a consequence of the size of the thermometer, particularly if it is in the usual sheath, and the difficulty, if not inability, to provide direct thermal contact with the object whose temperature is being measured. The problem becomes more difficult as the heating rate is increased. Since the thermal lag is more critical in introducing error in matching the shield and the calorimeter temperature (i.e., maintaining adiabatic conditions), thermocouples (individually or as thermopiles) are required for shield temperature measurement.

Resistance thermometry depends upon establishing a relationship between the resistance of the sensor (or the ratio of this resistance to the resistance at some reference temperature) and temperature. The major factors governing their use are allowable physical configurations (size and shape of the resistance element), ability to establish reproducible functional relationships between resistance and temperature coefficient of the resistivity, and sensitivity of the resistivity to physical and chemical inhomogeneities. For most applications where measurements extend from 300–773 K, pure metals are usually selected for the resistance element because of their almost linear temperature dependence of resistivity and their larger temperature coefficient of resistance as compared to alloys. The three most commonly-used metals are Pt, Cu, and Ni. Ni has the largest coefficient over the range 300–600 K, but the temperature dependence is complicated by the ferromagnetic transition that occurs just above this range. Cu has a slightly larger coefficient than Pt, but the disadvantages are lower strength, lower melting point, and greater susceptibility to contamination. For these reasons, Pt is used almost exclusively for precision resistance thermometry.

Pt resistance thermometry is usually limited to a maximum temperature of 903 K (630°C) where glass and mica construction materials are satisfactory [25]; to this temperature uncertainties of interpolation of temperature are 0.001 K. Quartz sheaths are structurally sound to about 1400 K, but conventional mica used to support the resistance element deteriorates objectionably above 900 K. Also, above this temperature the choice of support materials is limited because of progressive electrical shunting effects resulting from decreasing resistivity with increasing temperature. To offset this shunting, the Pt element is redesigned to decrease its resistance by a factor of about 100 (25.5–0.25 ohms), but this necessitates an increase in the precision of the instrumentation in order for the lower resistance elements to yield comparable accuracy of measurement of temperature. With appropriate precautions, uncertainties (25) in interpolation of temperature become ~0.001 K from 273–773 K, 0.001–0.03 K from 773–903 K, and 0.002–0.02 K from 903–1337 K.

The major problems encountered in using resistance thermometers for temperature measurements in adiabatic calorimetry relate to the size of the thermometer with the accompanying problem of adequate thermal contact, to the heat capacity of the thermometer itself, and to thermal conduction from the calorimeter along the leads and supporting structure of the thermometer. There is also the question of maintenance of calibration, which is a function of the susceptibility of the resistance element to physical and chemical inhomogeneities resulting from handling and contamination. Strain can be introduced by mechanical shock and excessively rapid heating and cooling. Effects of these factors can usually be detected by redetermination of the resistance of the thermometer at the ice point. These characteristics and precautions have been reviewed in detail by Anderson and Kollie [25].

5.5.2. Thermocouple Thermometry

Thermocouples are used extensively in high-temperature adiabatic calorimeters for shield control. They are frequently used for temperature measurement of the sample from room temperature and are used almost exclusively for both sample and shield temperature measurements above 800 K. Two obvious advantages are small size of the sensing element and ability to maintain direct physical contact with the object being measured. Important factors which should be taken into consideration in these applications include the magnitude of the thermal emf, which relates to the sensitivity required in the instrumentation; increased thermal emf provides the most direct way of increasing the magnitude of the signal-to-noise ratio. The sensitivity of the thermocouple (usually expressed in $\mu V \cdot K^{-1}$) is particularly important when the sensor is used to measure the temperature interval entering directly into the specific heat calculation. Further, consideration must be given to the temperature range over which the thermocouple is operable and to its sensitivity to physical inhomogeneities (elastic and plastic strain) and to chemical inhomogeneities. The latter is particularly important where chemical inhomogeneities may exist in the initial thermoelement wire or where there is transfer of chemical species to or from the environment. In some applications, strength, ductility, and fabricability may be major factors for consideration. Important thermocouple combinations, their range of use, and sensitivity are given in Table 5.4.

Table 5.4. Temperature Range and Sensitivity of Selected Thermocouples, 300–1300 K [25]

Thermocouple	Maximum Temperature, K[a]	Sensitivity, $(\mu V \cdot K^{-1})$	
		at 300 K	at Limit
Pt: Pt-10% Rh	1480	5	12
Pt: Pt-13% Rh	1480	6	13
70% Pt-13% Rh:94% Pt-6% Rh	1975		12
50% Ir-50% Rh:Ir	2273	6(Avg.)	
74% W-26% Re:W	3200	15(Avg.)	
Iron-Constantan	650	50	55
Copper-Constantan	475	40	53
Chromel-Alumel	1150	40	40
Chromel-Constantan	750	60	75

[a]Approximate upper limit for No. 8 wire (0.033 cm dia). Smaller diameter wires are frequently selected for use in calorimetry: also conditions in calorimetry may alter these limits.

5.5.3. Thermocouple Errors

The theory of the thermal emf, characteristics of the more important thermocouples, precautions in use, and sources of error have been examined extensively in the literature [25–30]. The following discussion draws significantly on the review by Anderson and Kollie [25], but with emphasis on applications to adiabatic calorimetry at elevated temperatures. The sources of errors which may be encountered in thermocouple thermometry are listed in Table 5.5.

Table 5.5. Sources of Error in Thermocouple Thermometry

a.   Thermal shunting
b.   Electrical shunting
c.   Calibration
d.   Decalibration
e.   Extension wire
f.   Reference junction
g.   Data acquisition

a. Errors due to thermal shunting: A thermocouple or individual thermoelement attached directly to a shield or to a calorimeter wall results in temperature changes at the point of contact leading to temperature measurement error called thermal shunting. The magnitude of the error is a function of the physical properties of the thermocouple and contacted materials, geometrical factors, and the heat sources or sinks governing the heat flow. With respect to the latter, recognition must be given to the fact that in the usual vacuum environments, the thermocouple wire is heated either by conductive transfer from the contacted material or by radiation. The effect in both cases is a thermal lag of the thermoelement wire at distances from the contacted surfaces and, therefore, heat flow from these surfaces. The mass of the thermocouple is frequently a controlling factor leading to use of the smallest size compatible with other factors; a particular consideration is that small wires have large surface-to-volume ratios, and any contamination can lead to large errors. Low thermal conductivity thermoelements are desirable, but selection based on this criterion is not generally possible. A number of techniques are available to reduce temperature gradients along the wires at the point of contact (see Section 5.4.2), including deep insertion into the calorimeter and shield wall, extending thermal contact of the shield on the thermoelement with the object to be measured, and, with differential thermocouples or thermopiles, confinement of the 'jumper' thermoelement to the heated zone between calorimeter and shield, thus avoiding the thermal gradients that would otherwise exist on carrying these wires outside the system. Problems of thermal shunting are significantly different with discontinuous (interrupted) versus continuous heating techniques, and it may be difficult to assess which technique is more affected. Continuous heating generally leads to larger lag of temperature of the thermoelement wires relative to the contacted surface and, therefore, steeper temperature gradients with accompanying greater heat transfer. Although the discontinuous heating method allows for temperature equilibration, each energy addition does introduce transients which lead to temperature measurement error. Intrinsic thermocouples (thermoelements attached individually to a surface) show faster response time and generally result in less thermal shunting, conduction from only two points of contact having less effect on the local temperature. Rapid heating, however, may lead to temperature gradients between the contacting thermoelements which can generate significant error. Sheathed or compacted thermocouple assemblies are particularly prone to thermal shunting errors.

b. Electrical shunting errors: Electrical shunting errors are traceable to electrical conduction through insulating materials, usually ceramics in high-temperature calorimetry, and to surface conduction resulting from deposition of conductive layers on the ceramics, frequently metals or oxide species of higher conductivity than the underlying ceramic insulating material. The problem increases with increasing temperature due to the decrease in bulk resistivity with temperature and the increase vapor transport leading to surface contamination. The resulting electrical shunting or leakage is encountered between thermoelements in multihole insulators and between thermoelements and the calorimeter or shields, both of which may be grounded. The resulting errors may become quite significant where leakage can occur from the sample heater circuit through insulation material to thermoelements to ground, resulting in measurable potential drops superimposed on the thermal emfs. Electrical shunting can frequently be reduced by insulating the individual thermoelements; however, the use of uninsulated, free-standing wires may lead to significant error at the higher temperatures due to electron emission. This emission from exposed specimen heaters may be quite significant; heaters may operate at several hundred degrees above the sample and introduce dual errors of leakage of heater current (heater current incorrectly measured) plus the thermocouple error resulting from flow of this current along thermoelements. Anderson and Kollie [24] conclude that electrical shunting is minimized by using dense, high-resistivity insulators in which wires loosely fit and the insulator is loosely fitted inside a metallic sheath. This provides less metal-to-ceramic contact, and the metal sheath acts as an electrical shield. Unfortunately, this construction may be objectionably bulky in a number of calorimeter systems. It follows from these criteria that compacted thermocouple assemblies (sheathed thermocouples) may exhibit maximum electrical shunting. This is an obvious disadvantage to their use.

c. Calibration errors: Intrinsic thermocouples (thermoelements individually attached to a metallic surface), fused thermocouple junctions, and sheathed thermocouples may require calibration for adiabatic calorimetry depending on use and required accuracy. Reporting of specific heats, temperatures and enthalpies of phase transformations, or assignment of specific temperature to other measured properties requires calibration of thermocouples for high precision in order to give the temperature dependence of the property. This calibration may be conducted by the user or as part of a purchase specification. For the calculations of the specific heat, determination of temperature intervals of a few degrees are required; the accuracy of this determination is generally more a function of the sensitivity of the measuring instruments than the accuracy of the thermocouple. Although calibration may show deviations of several degrees between reference temperatures and apparent temperatures obtained by relating measured thermal emfs to published tables of emf versus temperature, such deviations are almost always uniform over the temperature range. That is, the slope of the actual emf-temperature function remains sufficiently close to values derived from standard tables (or accepted functional representation) to have negligible effect on the accuracy of the specific heat measurement. Also, the change in specific heat with temperature is usually small enough that errors in the specific heat-temperature relationship are small.

Thermocouples, conventionally constructed by fusing and sheathed or unsheathed, may be calibrated directly. With intrinsic thermocouples, calibration must be conducted in place with the thermoelements contacting the metal to be measured or the elements joined so that calibration is conducted as for a conventional couple. The junction is then cut and the intrinsic couple made with the contacting surface. Calibrations are conducted by reference to standard thermocouples, standard resistance thermometers, or to the assigned freezing temperatures of pure samples. Current manufacturing practice produces thermoelements which, as uncalibrated thermocouples, reproduce the International Practical Temperature Scale (IPTS) [31] within a few degrees. Table 5.6 shows limits of error to be expected from uncalibrated couples according to ANSI Standard C96.1 [32]. Also shown are uncertainties in calibrated values using the indicated methods, both at the calibration points and when interpolation between these points is required.

Uncertainties in calibration are due to both external and internal factors. External factors include differences in precision of the instrumentation used for calibration and in subsequent application and to physical factors such as thermal and electrical shunting as described in the previous sections. Internal factors causing uncertainties in calibration can be traced to chemical, microstructural, and physical inhomogeneities in the thermoelements which generate thermal

Table 5.6. Calibration Characteristics of Commonly Used Thermocouples

| Type | Temp. Range (K) | Limits of Error[a] | | Calibration Uncertainty[b] | | | |
| | | | | Freezing Point Method | | Comparison Method[c] | |
		Standard	Special	At Calib. Points	Interpolated	At Comparison Points	Interpolated
Pt–Pt:10% Rh or Pt–Pt:13% Rh	273–800 800–1750	±2.8 K ±1/2%	±1.4 K ±1/4%	±0.2 K	±0.3 K	±0.3 K	±0.5
Copper– Constantan	214–365 365–650	±0.8 ±3/4%	±0.4 ±3/8%				
Iron– Constantan	273–550 550–1030	±2.2 ±3/4%	±1.1 ±3/8%	±0.2	±1.0	±0.5	±1.0
Chromel P– Alumel	273–550 550–1530	±2.2 ±3/4%	±1.1 ±3/8%	±0.2	±1.0	±0.5	±1.0
Chromel P– Constantan	273–590 590–1150	±1.7 ±1/2%	±1.2 ±3/8%	±0.2	±0.5	±0.5	±0.5
Pt:30% Rh–Pt: 6% Rh	1150–1975	±1/2%					

[a]Values in ANSI Standard C96.1 [32] have been converted from degrees F to K.

[b]From ASTM, STP 470 [28].

[c]Pt–Pt: 13% Rh Reference Standard.

emfs when present in thermal gradients; if gradients are significantly different (as is frequently the case) during calibration and in use, temperature measurement error will result. The problems are summarized critically for noble metal thermocouples by McLaren and Murdock [33]:

> 'This study has shown that serious problems involving quenched-in vacancy strain, surface and internal oxiding and deoxiding processes in thermoelements, immersion characteristics at fixed points of both individual and combined thermoelements, variations in thermopower with heat treatment, thermal conduction losses in thermoelements at low (0 to 700°C) temperatures, selection of the most suitable materials for the thermocouple and preparatory procedures, etc., exist even at this late date in the understanding of the basic properties, calibration and use of the standard thermocouple in thermometry; uncontrolled, these factors undoubtedly lead to the propagation of large unknown systematic uncertainties in the accuracy of subsequent temperatures scaled in the 630 to 1000°C range.

> 'Unfortunately, the recently revised, IPTS (1968), international temperature scale, in the range 0 to 1064°C, is itself based on measurements that were determined in experimental arrangements for the metal freezing point realizations with both thermocouple and gas thermometry that could not fail to introduce immersion and radiation-loss uncertainties of large but unknown magnitude.'

d. Decalibration errors: Chemical, microstructural, or physical inhomogeneities introduced into thermoelements subsequent to calibration cause error if the inhomogeneity is in a temperature gradient. Chemical inhomogeneities result predominately from reactions with the gaseous environment of the thermocouple and can involve transport of species either to or from the thermoelements. Vacuum and inert atmospheres may lead to an environment of low partial pressure of oxygen capable of producing volatile suboxides, which provide a transfer mechanism from ceramic and other materials to the thermoelements. Several of these contaminants are low-vapor-pressure metals which cannot be removed from noble metal thermocouples by annealing in air. Vapor phase transport from materials such as Al_2O_3, ZrO_2, and ThO_2, and by impurities such as Fe and Si in these materials, has been observed [33]. Because of the sensitivity of noble metal thermoelements to impurity pick-up, particular problems have been encountered in sheathed thermocouple construction. Due to the confinement of the elements in a sheath, with limited access of oxygen, impurity pick-up has been observed from both the granular ceramic and sheath material.

In order to minimize calibration change in these sheathed thermocouples it is necessary to specify careful cleaning and drying of the sheath and use of dry, high-purity ceramic as the embedding material. Best stability seems to result from the use of high-purity MgO, presumably due to its thermodynamic stability, and to the high vapor pressure of magnesium which at elevated temperatures prevents significant retention of Mg by the noble metal thermoelements.

e. and f. Extension lead and reference junction errors: Extension wires selected to match acceptably the thermoelectric characteristics of the thermoelements of a thermocouple are rarely used in calorimetry. Their general use is to decrease substantially the cost of noble metal wire between the vicinity of the couple and measuring instruments. Errors introduced into the measurement depend on the mismatch in thermoelectric properties of the thermoelement and their extension lead wires and failure to maintain the pairs of a junction at the same temperature. The precision sought in using thermocouple thermometry in calorimeter systems usually precludes use of extension lead wires, and the usual proximity of a reference junction to the calorimeter allows restricting lengths of noble metal thermoelements until the cost is reasonable.

Analytical expressions or tabular values of thermal emf versus temperature are based on a reference (or cold) junction temperature. Junctions held at the reference temperature are usually between each thermoelement and copper wires leading to measuring instruments. The reference junction is usually designed for operation at the ice point, which is defined as 273.15 K (0°C) and is the temperature at which ice is in equilibrium with air-saturated water at one atmosphere pressure. An alternative is the use of a triple-point cell in which water, ice, and water vapor are in equilibrium; the temperature of this cell is 273.16 K. With reasonable care, the ice-point cell is accurate to 0.01 K; a triple-point cell can be constructed which is accurate to 10^{-4} K, but this accuracy is generally not required in thermocouple thermometry. Conventionally-constructed, ice-point reference junctions contain the junctions in individual, closed-end tubes immersed in the ice-water mixture. These tubes usually contain a small amount of oil or mercury for increased thermal contact between the junction and the ice-water environment. These tubes should be sealed at the point of entry of the wires to prevent condensation of moisture from the air. Many of the problems associated with ice-point reference junctions are circumvented by using reference junction blocks or cavities which are held at a fixed temperature (usually slightly above ambient). The temperature of these units can be held easily to ±0.01 K using resistance or thermistor thermometers and relatively simple temperature control schemes.

Major sources of error in measurement of temperature by thermocouples can be associated with improper precautions in the use of ice-water reference junctions. Following are the more important sources of error:

(1) Errors due to inhomogeneities in thermoelement and lead wires. Rather large temperature gradients may exist along wires passing from ambient to ice-bath temperatures. If these gradients span chemical or physical inhomogeneities in the wires, extraneous thermal emfs are generated. The major source of error is failure to recognize that careless handling of wires, usually bending and straightening, can produce local amounts of cold work and, hence, physical inhomogeneities leading to error if these wires are in the temperature gradient. Careful handling and possible annealing of wires are precautions against this source of error. Most copper wires will have negligible chemical inhomogeneities and these inhomogeneities in the thermoelement wires are usually small enough to not cause serious error. In comparison, the temperature gradients in the calorimeter system are much larger and, therefore, chemical and physical inhomogeneities can be a significant source of error.

(2) Errors due to thermal shunting. Wires and their accompanying isolating tubes provide paths for heat conduction to the thermoelement junction from the surroundings at ambient temperature. Because of the high thermal conductivity of copper lead wires, the fact that thermal conduction can raise the junction temperature above 273.15 K (0°C) may be overlooked. The resulting error is increased when the magnitude of the shunting differs for the individual thermoelement junctions of a couple. Precautions are, therefore, necessary to provide sufficient immersion of tubes into the ice-water mixture and to use small diameter copper lead wires.

(3) Errors due to moisture condensation. Tubes containing thermoelement-lead-wire junctions should be sealed external to the ice-bath container. Otherwise, condensation of moisture from the air having access to the tube contributes to thermal shunting and may lead to error in the thermal emf resulting from voltages generated by electrochemical corrosion.

(4) Errors due to improper maintenance of reference junctions. These errors result from failure to recognize that intimate contact of the tubes containing the junctions must be maintained with the ice-water mixture to assure the reference junction temperature of 273.15 K. Even though the ice-water mixture is usually contained in insulated (vacuum) containers, melting will occur, resulting in ice floating at the top of the bath and

exposure of the tubes to water only. Temperature gradients in this ice-free water may be great enough to heat the thermoelement junction above the reference temperature. These problems may be reduced by using temperature-controlled junctions as described previously.

g. Data acquisition errors: Errors in addition to those in the previous four categories may be broadly classed as data-acquisition errors, in that they arise from sources external to the immediate calorimeter system. These sources include (1) electrical noise from electromagnetic radiation and electrostatic coupling, (2) noise due to lack of adequate common-mode rejection of interconnected components, (3) excessive response time of instruments and communication line filters, (4) transients and thermal emf errors generated at terminal and switch interfaces, and (5) errors due to the precision and maintenance of calibration of the measuring systems. Although these errors may be significant in measuring calorimeter heater voltage drops and voltage drops across the standard resistance to measure heater current (usually millivolts to volts), they are of greater significance in the measurement of thermocouple emfs, where signal integrity in the submicrovolt range is required.

The sources of these errors and their magnitude are generally difficult to assess. Therefore, a thorough knowledge of where the errors can originate and of the use of best practices to minimize them is essential to the consistent measurement of the specific heat with an accuracy of better than 1%. Different sets of problems arise when a calorimeter system is based on use of batteries for heater current and instrumentation versus the use of electronic power supplies and instruments or the use of completely computerized systems. In each arrangement, it is necessary to recognize those sources of error which arise in the immediate calorimeter system (calorimeter and shields), those which can be related to instrumentation, and those which are associated with interconnecting communication lines and peripheral equipment such as reference junctions.

Electrical noise from electromagnetic radiation and electrostatic coupling results from 60 Hz sources such as shield heaters, vacuum pump motors, and nearby equipment; in extreme cases severe interference may be traced to surrounding induction heating equipment and radio transmission. These can significantly affect measurements with low-level electronic amplifiers, voltmeters and null indicators used with potentiometers, and long-line transmission to digital computers. Common-mode noise usually arises from 'ground loops' associated with interconnection of two or more electronic instruments to the same 60 Hz power source or to independently grounded inputs and outputs to instruments with insufficient input/output isolation. These types of noise errors can be minimized by careful attention to grounding and shielding [34]. Generally, thermocouples should be grounded at the couple and a grounded shield should surround the leads from as near the source as possible to the instrumentation. Where preamplification of the thermocouple signals is used, low-level differential (isolating) amplifiers are generally desired. Since leads and their shields from such amplifiers may also require grounding for stability, high input-to-output impedance is required to prevent a ground loop error.

Data acquisition errors due to lack of adequate common-mode rejection and transient and thermal errors are generally more significant in rapid data collection and in calorimeter operation by computers. An exception arises if there is insufficient precaution in minimizing temperature gradients across terminal pairs passing low-level signals. These terminals may have such different thermoelectric properties relative to the connecting wires that thermal emfs of a few microvolts are generated; care in matching materials and enclosing junction pairs to maintain them at the same temperature are obvious precautions. Error may still arise due to transient temperature changes on closing switches. However, switching errors become more serious with high-speed-computer data acquisition. Switch contact 'bounce,' contact potentials, and thermal emfs due to frictional heating can give large transient errors when one is taking readings at subsecond intervals.

When calorimeter systems are operated manually, using potentiometers and null indicators for balance, the accuracy of the measurement of power to the calorimeter and of thermocouple emfs is clearly related to the accuracy of these instruments. Acceptance of the manufacturers calibration or periodic calibration against standards is common practice. However, systems employing complex switching, electronic instruments, and computer data acquisition require calibration, and there is no substitute for high-quality potentiometers as calibration sources. By this means, high-level and low-level signals can be injected into the measuring system and the resulting signal detected. Signal integrity with respect to noise and total amplifier gain for the system is thereby established relative to the accuracy of the potentiometer calibration source. This capability is particularly important with the increase in use of preamplifiers and isolation amplifiers in thermocouple measuring systems. Using techniques for computer operation of calorimeter systems described in Section 5.8 and careful calibration, integrity of thermocouple emf of ± 0.25 µV can be realized. This represents an improvement over values reported by Anderson and Kollie [25] in their discussion of data acquisition errors.

5.5.4. Factors Governing the Use of Thermocouples in High-Temperature Adiabatic Calorimetry

Table 5.4 shows that several thermocouple combinations exhibit thermoelectric powers which are from four to seven times that of the Pt:Pt-19%Rh or Pt:Pt-13%Rh thermocouples. In spite of lower sensitivity, these noble metal thermocouples exhibit calibration uncertainties, expressed in degrees kelvin, which are only about one-half as large as those realized for the higher sensitivity thermocouples. This significantly better performance leads to almost exclusive use of Pt:Pt-(10 or 13% Rh) couples in high-temperature calorimetry. Characteristics contributing to this performance include ease with which high-purity platinum and homogeneous Pt-Rh alloys can be produced, high melting temperatures, general freedom from oxidation, simplicity of annealing in air by resistance heating and good structural stability with respect to internal ordering. Current instrumentation permits determining the thermal emf to about 0.1 µV (0.01 K), which reduces the measurement error below the calibration errors due to thermal and electrical shunting. Since the Pt:Pt-10%Rh thermocouple defines the IPTS [31] from 903.89-1337.58 K and is used extensively below and above this temperature range, a major effort has been made in establishing tables and analytical expressions for the emf-temperature relationships for this couple. The procedures are discussed in detail in NBS Monograph 125 [31]. The thermoelectric voltage is given as a power series expansion over successive temperature ranges from 223-2041 K, the ranges most relevant to adiabatic calorimetry being 223-904 K and 904-1337 K. This analytical representation is an improvement over previous functions in that the second derivative changes smoothly over the entire temperature range except at 903.89 K and 1337.58 K, the antimony and gold points. This character of the thermoelectric voltage-temperature curve is essential in specific heat determinations when temperature intervals are derived from emf measurements by reference to specific values in the table (a few degrees apart) or by using the slope of the power series representation. Previous tables in NBS Circular 561 based on the interpolation equation of Roeser and Wensel [35] led to serious error in calculation of temperature intervals when the interval spanned the calibration point temperature used in establishing their interpolation functions.

Noble-metal thermocouples used in adiabatic calorimetry are subject to thermal and electrical shunting errors and decalibration errors of the types discussed previously. Sheathed, open junction, and intrinsic thermocouples are used. Direct attachment to the calorimeter or to shield wall leads to thermal conduction from the point of attachment to a 'cold spot,' leading to the measured temperature lagging the bulk temperature of the material to which the couple is attached. Since the effect depends on the mass of the thermoelement wires and the area of contact, errors will be different on shield and calorimeter, thus contributing to temperature differences introducing heat leakage and deviations from adiabatic conditions. Thermal shunting can be decreased by embedding the sensor in the calorimeter, but this is usually not feasible on shields. Since this requires insulated or sheathed thermoelements to prevent contact with the calorimeter, the increased mass may contribute to thermal shunting. Use of very small thermoelement wires (0.002 cm) will reduce this thermal shunting, but smaller wire is more rapidly contaminated, leading to calibration errors. It is also impractical to carry these fine wires external to the calorimeter system. Welding short lengths to the larger diameter wires is easily accomplished, but these welds in steep temperature gradients will generate spurious emfs.

5.6. CORRECTIONS TO THE MEASURED SPECIFIC HEAT

The measured specific heat is subject to and must be corrected for two major errors. One is a contribution from the heater heat capacity. This error is avoided if a cell is used; if not, it must be minimized by heater design and usually can be calculated with acceptable accuracy. The other error is due to heat leakage between the calorimeter and the surroundings, the correction for which is more uncertain.

Corrections for heat leakage are usually made for the following contributions. (i) A separable contribution to leakage results from inability to control the shield to the temperature of the sample or cell. (ii) Even if the temperature of the shields is controlled properly, there is still heat leakage by radiation due to shield-and-sample or cell-surface temperature differences, and by conduction along lead wires. (iii) When a resistance thermometer is used to measure the sample temperature, then the heat generated in it must be included in the energy supplied to the sample. With proper design and measurement methods, all three of these corrections can be calculated.

The section which follows describes the correction to the measured specific heat for the heater. The remaining sections deal with methods for correcting for heat leakage.

5.6.1. Correction for the Sample Heater

In measuring the specific heat where the unencapsulated sample is the calorimeter, the measured c_p, calculated from Eq. (2), is first corrected for heat leakage, as described in Section 5.6.2. Then the correction for the specific heat of the heater is made. However, we will examine the heater correction first. Thus, in this section, it is assumed that the specific heat which is to be corrected for the heater is already corrected for heat leakage.

Following Pawel [36], and using the continuous heating technique, a simple energy balance gives the necessary starting relation

$$c_p'(m_s + m_h)\, \Delta T_s \simeq m_s c_{p_s} \Delta T_s + m_h c_{p_h} \Delta T_h. \tag{6}$$

Here c_p' is the measured specific heat corrected for heat leakage and m_s and m_h are the mass of the sample and the heater, respectively. c_{p_s} is the corrected or 'true' specific heat of the sample, and c_{p_h} is the specific heat of the heater. c_p' is calculated based on the temperature difference, ΔT_s, determined by the sample thermometer; it is assumed that this ΔT_s also applies to the sample alone. The temperature response of the heater and sample is shown schematically in Fig. 5.12. After an initial transient, the rate of temperature rise of the heater is less than that of the sample. Nauman [37] determined the relation between ΔT_s and ΔT_h (for the same Δt) in the following way. The electrical resistance for nichrome (Ni–20% Cr) rises with temperature, then decreases to a minimum at about 1000 K, then increases again. Since the voltage drop across the

heater and the current are measured during a run, the resistance of the heater as a function of sample temperature can be obtained. If the heater is nichrome, then the degree of mismatch on the temperature axis between the curve of heater resistance versus sample temperature and that of the resistance–temperature of nichrome gives the temperature excess of the heater. From this, it was found that $\Delta T_h \simeq 0.55\, \Delta T_s$. By assuming that the heat from the heater is transferred to the inner surface of the sample heater hole by radiation, Pawel [36] estimated about this same value. Then, Eq. (6) becomes

$$c_{p_s} = c_p' + \frac{m_h}{m_s} (c_p' - 0.55\, c_{p_h}). \tag{7}$$

For typical heater and sample materials, this correction is <0.1% if $m_h/h_s < 0.01$. Thus, for a 100 g sample, the heater should be less than 1 g, which is of sufficient size to be amenable to construction.

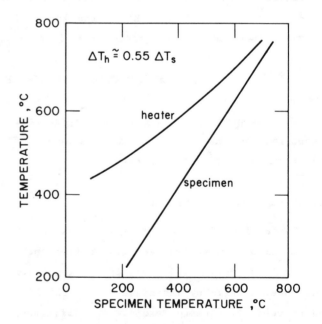

Figure 5.12.   Schematic illustration of the relation of the specimen temperature to that of the internal heater.

5.6.2. Correction for Heat Leakage from the Sample--Continuous Heating

The best method of correcting for heat leakage is to heat the sample discontinuously. This method is described in the next section (5.6.3). It is quite applicable where the heat absorbing processes (e.g., increased lattice vibrations, electron excitations, magnetic disordering, etc.) are reversible. However, if there are structural processes which are to be examined which have a rate that is dependent upon the heating rate, then continuous heating must be used. Correction for the heat leakage under these conditions is the subject of this section.

When data must be taken continuously, it becomes unclear exactly how to proceed to correct for heat leakage. The usual approach is to employ heating rate as a variable and to then examine ways to use these data and the heating rate to obtain the correct c_p. West [3] has pointed out that this is a 'time-honored test' for adiabatic calorimeters. However, West [38] and others [8,15] have questioned whether such a test is valid.

Their argument is as follows. Let a calorimeter be heated continuously at a constant rate β. Further, let the temperature control point on the sample (or cell) and on the shield (or shields) be maintained truly identical. There still exist temperature gradients on the sample or cell and shield surface which lead to heat leakage and give rise to an apparent specific heat c_{pa}. As measured by the thermometer of the sample, the temperature of the sample rises ΔT during a time increment Δt. Thus,

$$m\, c_{pa}\, \Delta T = m\, c_p\, \Delta T + \dot{Q}_L\, \Delta t \tag{8}$$

where c_p is the correct specific heat, and $\dot{Q}_L$ is the rate of heat leakage. The time increment is clearly inversely proportional to β, so $\Delta t = a/\beta$, where a is a constant.

The heat leakage is determined by the nonuniform temperature on the shield and sample (or cell), even though the points on each for temperature control are identical. West [38] argues that these differences are greater the higher the heating rate (we have observed this effect), and he assumed that the rate of heat leakage $\dot{Q}_L$ is proportional to β. Thus, $\dot{Q}_L = b\beta$, where b is a constant. The total heat leakage during the temperature increment ΔT is then

$$Q_L = \dot{Q}_L\, \Delta t = (b\beta)(\Delta t) = (b\, a/\Delta t)(\Delta t) = ab.$$

Thus, the heat leakage for a given ΔT is independent of the heating rate and cannot be detected by measurements in which the heating rate is varied.

But Naito et al. [8] have measured heat leakage which is dependent upon the heating rate (see Fig. 5.11). They found the leakage to be linearly related to β, so that

$$Q_L = c + d\beta \tag{9}$$

where c and d are constants. However, c was quite small, so that $\dot{Q}_L \simeq d\beta$. Using Eq. (9) yields

$$c_{pa} = c_p + \frac{c}{m} + \frac{d}{m\beta} \tag{10}$$

in which c_p depends inversely on the heating rate. Such a dependence is expected if one argues that for an infinite heating rate there is no time for heat loss. Under such a condition

$$c_{pa} = c_p + \frac{c}{m} \ . \tag{11}$$

The work of Naito et al. [8] indicates that the relations obtained by West [38] overlook a heating rate dependent factor in the heat transfer relations. This effect is negligible if the constant d is small. If the temperature on the surface of the shields and sample or cell is uniform, then the constant c is low. If both constants are low, then the measured (calculated) specific heat c_{pa} closely approximates the true specific heat c_p in Eq. (10). Apparently the magnitudes of these constants depends upon the design of the calorimeter but they are difficult to predict.

In our high-temperature calorimetry, we have found the measured specific heat c_{pa} to be quite dependent on the heating rate. We use the following method to obtain the corrected specific heat. Consider an energy balance on a sample and heater (which we will refer to here as the sample) during a time increment Δt. The temperature rise of the sample is measured as ΔT_a, where the subscript refers to the fact that this ΔT_a is used to calculate the apparent specific heat c_{pa}. The rise in temperature if there were no heat leakage is $\Delta T'$, and the corresponding correct specific heat is c_p'. Due to heat leakage, energy $\dot{Q}_L$ is unaccounted for and corresponds to a contribution ΔT_L to the measured temperature rise ΔT_a. Thus,

$$\dot{Q}' = \dot{Q}_T + \dot{Q}_L$$

$$\frac{m \, c_p' \Delta T'}{\Delta t} = \frac{m \, c_{pa} \Delta T_a}{\Delta t} + \frac{m \, c_p' \Delta T_L}{\Delta t} \ . \tag{12}$$

We designate the contribution to the rate of temperature rise from heat leakage $\Delta T_L / \Delta t$ as R. Then Eq. (12) becomes

$$c_p' = \frac{c_{pa}}{1 - \dfrac{R}{(\Delta T_a / \Delta t)}} \tag{13}$$

where it has been assumed that $\Delta T' \simeq \Delta T_a$. Now we can write

$$\frac{\dot{Q}_T}{m} = c_{pa} \frac{\Delta T_a}{\Delta t} \tag{14}$$

where $\dot{Q}_T / m = k$ is the energy input rate. Then Eq. (13) can be written as

$$c_p' = \frac{c_{pa}}{1 - \dfrac{R \, c_{pa}}{k}} \ . \tag{15}$$

One method of making measurements so that Eq. (15) can be utilized to correct the specific heat is as follows. If there is no energy input to the sample, then R becomes the rate of temperature change during these conditions. If the system is adiabatic, R = 0; if not, it is finite. We refer to R as the 'drift rate.' To apply Eq. (15), drift data are obtained at several temperatures, then a curve is fitted to these data to obtain values of R at each temperature for which values of c_{pa} have been obtained. An indication of the validity of this equation is to take c_{pa}

ata at several input rates k, then apply the correction using the same R-temperature curve. Such n application after Pawel [36] is shown in Fig. 5.13.

There are two situations which will make this approach inapplicable. One is the determination of drift data while the sample is undergoing structural changes which evolve or absorb heat, o that the drift rate does not properly reflect the heat leakage. This creates difficulties in btaining drift in some alloys on which we have measured the specific heat. The more fundamental roblem is whether the assumption that the drift rate, determined with the system static, gives he same heat leakage at the same temperature as is occurring during heating. We have measured he temperature at several points on shields and have found that the temperature gradients are not dentical during drift and during dynamic operation. In some cases, we have found that the appli- ation of Eq. (13) using a R-temperature curve will not give corrected data on a common curve, and e have interpreted this to mean that these gradients are sufficiently different to make this type f correction inapplicable.

In such a case, we have taken two approaches to correct the data. One is to take data at ifferent heating rates and plot these against c_{pa}/k. At an infinite heating rate (infinite k), he heat leakage is zero, so that a curve through these points can be extrapolated to $c_{pa}/k = 0$ to btain the corrected c_p. This must be done for each temperature for which c_p values are desired. his method of correction is illustrated in Fig. 5.14. To make clear how the correction is ap- lied, consider the correction at 440°C. Five sets of c_{pa} data were obtained, each at a different nergy input rate (k value). For each k value, a smooth curve was drawn through the data points. hen, c_{pa} values at 400°C were obtained from these curves. These five values were plotted against pa/k, fitted by a straight line, and then extrapolated to $c_{pa}/k = 0$ to obtain the corrected c_p alue of 0.4507. Also shown are two extrapolated values obtained by varying the straight line. his leads to an uncertainty in the extrapolated value of about ±0.5%. For this correction pro- dure to be absolutely valid there should be no temperature gradients across the sample, heater, nd thermometer. As the heating rate increases, this is approximated less, and the procedure comes more questionable.

The other method of obtaining the corrected specific heat c_p when the static drift rate is plicable is to take data at two different energy input rates k, then at a given temperature use e two c_{pa} values with the two k values to solve Eq. (15) for the corrected specific heat c_p. e success of this approach is illustrated in Fig. 5.15.

The approaches to correct the measured specific heat described in the paragraphs above raise o concerns. One is associated with the fact that the higher the heating rate (energy input te) the hotter the heater and, hence, the higher the heat leakage by conduction along the heater ads. Whether the corrections above account for this properly is not clear. The other concern how sensitive the temperature distribution on the surface of the shields and the sample is to e heating rate. The correct application of the relations above depends upon these temperature adients being identical at a given temperature for all heating rates.

The use of heating rate as a test of the suitability of the calorimeter design is common. wever, the insensitivity of the measured data to heating rate is no guarantee that heat leakage

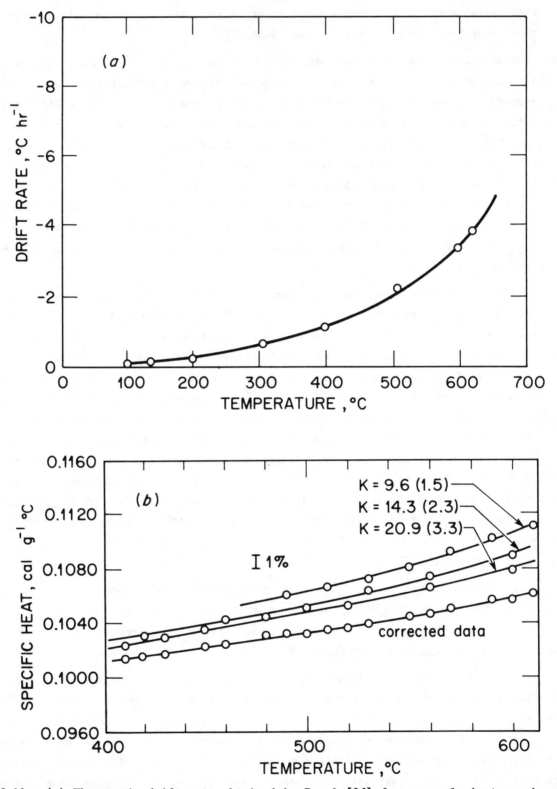

Figure 5.13. (a) The static drift rate obtained by Pawel [36] for one calorimeter using a Cu sample. (b) The top three curves are the apparent heat capacity of Cu measured using continuous heating. Each curve is for a fixed energy input rate k (in $cal \cdot g^{-1} \cdot hr^{-1}$); the approximate heating rate ($K \cdot min^{-1}$) is in parenthesis. These three sets of data were corrected by Eq. (15) using the drift data in (a) to obtain the corrected data curve.

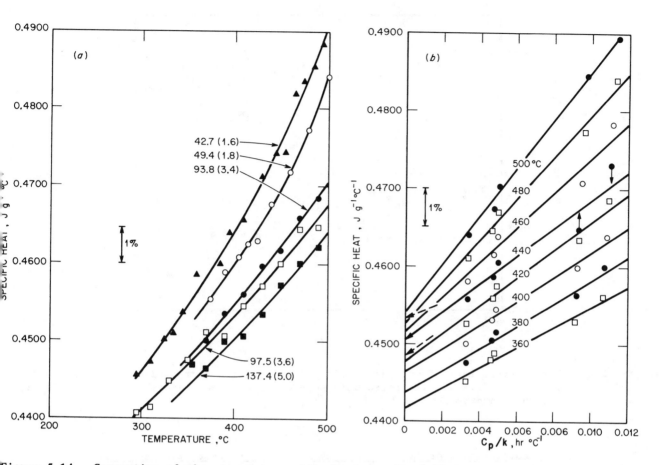

Figure 5.14. Correction of the apparent specific heat c_{pa} [in (a)] for a Cu–17.1 at.% Ga alloy by plotting c_{pa} versus c_{pa}/k and extrapolating [in (b)] to infinite heating rate (infinite k value) to obtain the corrected c_p. Here k is the energy input rate $(J \cdot g^{-1} \cdot hr^{-1})$ and the approximate corresponding heating rate (in $K \cdot min^{-1}$) is in parenthesis. See text for details. The data are from Brooks [39].

is not present. West [3] has shown that heat exchange due to nonisothermal conditions on the surface of the calorimeter and the shields cannot necessarily be revealed by varying the heating rate. Also, sensitivity of the data to heating rate does not necessarily reflect heat leakage. For example, if the temperature difference between the sample and the heater is sensitive to the heating rate, then c_p will depend upon heating rate, even if the heater design is such that heat leakage is negligible. To summarize, the detection of a dependence of c_p on heating rate is an indication of significant heat leakage or other design problems. The absence of the dependence of c_p on heating rate is no guarantee that the data are correct.

In the case being considered in this section, where it is required that the sample be heated continuously, we know of no methods other than those described above to correct the data. The accuracy of the corrections can be estimated by varying the heating rate, trying the three methods described above correcting the data, and using samples and heaters of different sizes. This approach is the best, and perhaps only, method to determine the correct specific heat under this restriction. The best method, however, is to obtain data by heating discontinuously and to use a cell. The correction methods involved under these circumstances are described in the next two sections.

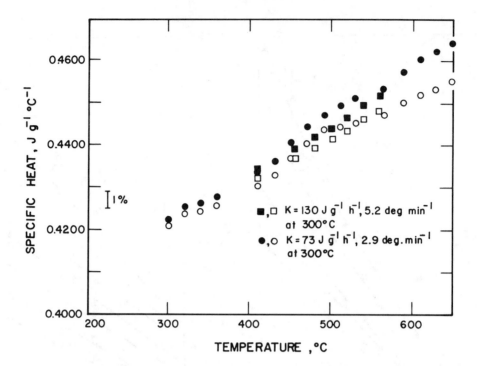

Figure 5.15. The specific heat of Cu. The solid points are data corrected only for the heater. The open points are the solid data points corrected as described in the text (from Stansbury and Brooks [13]).

5.6.3. Corrections for Heat Leakage from the Sample--Discontinuous Heating

As mentioned previously, there are three sources of heat leakage: (i) leakage to the surroundings, (ii) leakage due to heat generated in a resistance thermometer, and (iii) leakage due to shield temperature control limitations. Leakage due to heat generated when passing a current through the resistance thermometer is usually quite negligible. Of course, the use of a thermocouple eliminates this error source. Leakage to the surroundings is assumed to be corrected by drift before and after the step heating. The procedure is to adjust the shield control points (relative to the sample) to give a low drift rate. The sample temperature is monitored with time to assure that steady state is attained. The sample is then heated for a known time and the temperature monitored until steady state is again attained. These temperature data during drift are then extrapolated to the midpoint of the time interval, and the ΔT obtained from these extrapolated curves (Fig. 5.16) is used in Eq. (3) to calculate c_p. (Note that in this case only a heater correction is applied to the data.) This method of correcting the specific heat is subject to uncertainty about the heat leakage during the heating period. This is minimized by using a low heating rate. Leakage due to shield temperature control fluctuations can be corrected for if the amount of leakage is known as a function of temperature control error signal (i.e., the difference in temperature between the sample and the shield). This can be determined by offsetting the controller set point different known amounts and measuring the drift rate using a sample of known specific heat. Application of the correction, however, requires the temperature difference between the shields and the sample as a function of time. This can be obtained if the temperature control process is suitably slow, but at high temperature the temperature difference fluctuations have a high frequency, so that this information is difficult to obtain. Thus, this correction is usually neglected. (However, see the next section.)

5.6.4. Corrections Using a Cell

If the specific heat is measured using a cell, then there is no correction for heater and none for heat leakage if the leakage is identical for the cell empty and full. Section 5.5.3 shows that this condition is not necessarily met. Thus, to obtain the most accurate specific heat data using a cell, it is best to measure on the cell full and empty using discontinuous heating, obtaining the specific heat for each case by the method described in the previous section (5.6.3), and then subtract these data to obtain the corrected c_p. To optimize the accuracy of the specific heat, the data on the full and on the empty cell should be corrected for heat generated in the resistance thermometer (if one is used) and for leakage due to shield temperature control limitations, as described in the previous section (5.6.3). For a calorimeter operating to 1050 K, Grønvold [40] has described these corrections in detail and discussed their contribution to the uncertainty in the specific heat.

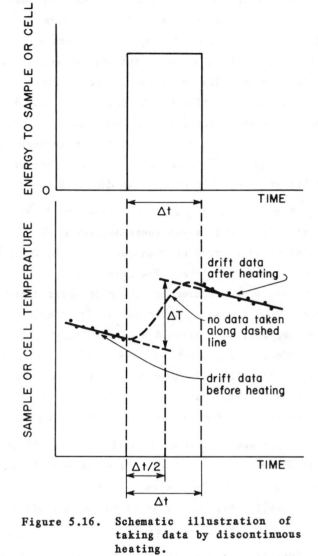

Figure 5.16.   Schematic illustration of taking data by discontinuous heating.

5.7. ERRORS AND ACCURACY OF SPECIFIC HEAT MEASUREMENTS

In this section we summarize those aspects of the measurements which allow estimates of the accuracy of the specific heat. First, we define several terms related to the concept of accuracy. The accuracy is the best estimate of how close the average measured data are to the true or accepted data. It should properly be reported as a percentage. The inaccuracy, or error, would then be 100 minus the accuracy. However, it is common to report the error (on a percentage basis) and yet refer to it as the accuracy. Generally, there is no difficulty understanding how different writers are using these terms.

Reporting the accuracy as defined above can be misleading if, as usual, it is based on the average of a group of data. Rather, information about the variation in the group of data is more significant. This can be reported in several ways [41,42]. The range can be reported as the difference from the average of the maximum and minimum values in the group, again reported on a percentage basis. The variation can be based on statistical treatments where it is common to report the standard deviation or the variance. This is more meaningful than reporting the range, as an indication is obtained of the distribution of the data about the average or mean value. For example, a value of the specific heat of a certain material at a certain temperature can be compared to a new group of data, say 15 values, obtained at this temperature, by reporting them as all within ±1.5% of this value, and 85% of them within 1.0% of this value.

When measuring quantities where the true or accepted value is not known, the accuracy is usually indicated by the repeatability or by the precision. We consider repeatability to be the variation in a group of measured data when the measurements are repeated under as nearly the same conditions as possible. The precision (or reproducibility) is the variation in a group of measured data when measurements are made under different operating conditions – heating rate, sample size, etc. The two quantities are frequently reported as the maximum percent variation from the average, but, as mentioned above, it is useful to state statistical quantities, such as standard deviation. The precision is frequently reported as the ratio of the data to the average, based on percentage. Thus, in this scheme, the reproducibility is related to precision; if the reproducibility is ±0.8%, then the precision is 99.2%. Note that repeatability and precision do not allow comparison directly to the true value of the measured quantity.

Since measured specific heat data are obtained as a function of temperature, the precision is best indicated by reporting the variation of the data from a curve fitted to the data (say, by least squares). The accuracy is indicated conveniently by reporting the variation of the data from accepted data by a plot of percentage difference versus temperature.

Errors can be classified as accountable (determinate) or unaccountable (indeterminate). We consider accountable errors those associated with the measurement of the quantities in Eq. (2) used to calculate the specific heat. Errors associated with these quantities are established by independent methods, such as calibration of timers or potentiometers. Most accountable errors are caused by systematic errors, which always tend to have the same algebraic sign, and which cannot be normally detected by statistical treatment of the data. Unaccountable errors are associated with errors of relatively fixed magnitude, but which are unrecognized, and with random errors. Random errors are usually considered to be reflected in the precision of the measurements and, hence, are examined by statistical treatment of the data. Thus, for measurements (such as specific heat) being made on a material of unknown property, or on material whose property is not well established, the minimum error is taken to be that given by the systematic error, and this value is then expanded to include the error based on the precision.

5.7.1. Systematic Errors

The systematic errors in the quantities in Eq. (2) depend upon the method of measurements and the instrumentation used. The errors in measurements of the mass, time interval, and power to the cell (if a resistance heater is used) can all be specified accurately. Table 5.7 gives estimated values of the errors associated with the quantities in Eq. (2) used to calculate c_p. These are usually dictated by the sensitivity and accuracy of the instruments used. Values are presented from the work of three different groups, representing calorimeters covering three different temperature ranges.

The limiting quantity in the systematic errors is ΔT. This was discussed in Section 5.6. For example, using thermocouples and precision potentiometers (Stansbury and Brooks [13]), a resolution of about 0.03 K is available. However, this error may exist at both extremes of ΔT, so that the error if $\Delta T = 30$ K is about 0.2%. The total systematic error in c_p for the example in Table 5.7 is about 0.2%, limited by the error in ΔT.

The error in the actual temperature measurements, used to represent the temperature of the sample, depends upon the type of thermometer and its calibration. In the case of thermocouples, parasitic voltages affect the accuracy of this quantity greatly. Typically, uncalibrated Pt–Rh type thermocouples have an accuracy of about 1–2 K, depending on the temperature range. Thus, the c_p–T curve may be shifted this amount on the temperature axis from its correct position. Frequent calibration of the thermocouple should make it accurate to 0.1–2 K, depending on the temperature, in the range 300–1800 K. The Pt resistance thermometer is more accurate. Its resistance at the triple-point of water should be measured occasionally to check stability. Grønvold [40] has calibrated his thermometer in situ by observing the melting of metals encapsulated in his quartz container.

An error which is usually systematic is the heat capacity of the heater, corrections for which were described in Section 5.6. The error due to the heater can be kept small by using a small mass heater, and in some designs this is sufficient to make this error negligible.

Table 5.7. Approximate Values (%) of Systematic (Determinate) Errors, Usually Based on Instrument Resolution, Obtained by Three Groups. These Errors (on a % basis) were Added to Obtain the Error in Calculating c_p by Eq. (2). The Reported Error (last line) Includes Systematic and Random Errors.

	West and Ginnings [43] (300–780 K)	Grønvold [40] (300–1050 K)	Stansbury and Brooks [13] (300–1300 K)
I	0.01	——	0.02
E	0.01	——	0.01
Δt	0.001	——	0.03
$EI\Delta t$	——	0.02	——
m	0.001	0.001	0.0006
T	0.003	0.0005	0.1
ΔT	0.01	0.01	0.15
c_p (calculated from Eq. 2)	0.03	0.01 ($\leq$0.2 including corrections)	0.2
c_p (reported)	$\leq$0.2	$\leq$0.2	0.7

5.7.2. Random Errors

The main source of random errors is that associated with heat leakage. Even if the adiabatic shields maintain the temperature of their control points exactly at that of the calorimeter, heat leakage may still occur. As described in Section 5.4, design minimizes this, and this error can be tested for by varying the experimental parameters, such as heating rate and calorimeter mass. The other main source of random errors is that associated with temperature control. Problems associated with this were discussed in Section 5.4. The magnitude of this error is also tested for by varying parameters such as heating rate and size or mass. The random error usually is determined by multiple experiments in which experimental parameters are varied, and it is this error that partly determines the precision (see Section 5.7.1).

5.7.3. Accuracy of Specific Heat Measurements

The uncertainty in evaluating indeterminate errors requires an experimental approach to evaluating its magnitude. Measurements should be made using different sample masses, different heating rates, different heater masses, and different sample and shield geometries. This approach gives the reproducibility of the measurements, and it gives the lower limit on the error. Of course, it usually is greater than that based on the determinate error.

It is common practice now to measure the specific heat of α-Al_2O_3 as a test of the accuracy of the measurements. Over many years, several laboratories have made measurements of the specific heat of samples from one source of this material (referred to as the Calorimeter Conferences sample), and this has allowed an intercomparison of various techniques. Measurements of high-purity α-Al_2O_3 at the National Bureau of Standards have established reference data on the specific heat of this material over a wide temperature range. Indeed, samples of high-purity α-Al_2O_3 can be purchased now from the National Bureau of Standards (Standard Reference Material 720) with a certified specific heat accurate from 0.1% at 300 K to $\pm$0.3% at 1200 K. (Research on which these data are based is given with the certificate.) However, it is available only in small rods (about 0.2 cm dia. and 0.9 cm long). It can be purchased from other sources in high purity, but it is difficult to fabricate into specific geometries, and it has a low thermal conductivity. Measurements on α-Al_2O_3 are highly recommended to help evaluate the accuracy of the measurements from a calorimeter.

Molybdenum metal is also available from the National Bureau of Standards as a special reference material, NBS SRM 781, with certified specific heat from 273-2800 K. (Research on which these data are based is given with the certificates.) Thus, this material can be used to extend the temperature range of applicability of the specific heat of Al_2O_3 .

From the information in the literature, it is obvious that as the upper temperature limit of operation of calorimeters increases, the assessed accuracy at this limit decreases. There is sufficient information to allow stating the approximate minimum error expected in the measured specific heat if the calorimeter is carefully designed, operated properly, and the measured data corrected properly. Around 300 K, measurement can be made to $\pm$0.1%, around 500 K to $\pm$0.3%, and 1000 K to $\pm$0.5%. Above 1000 K, the difficulties in reproducing the heat leakage because of the increasing importance of leakage by radiation increases the error significantly. Further, in this range there have not been sufficient measurements on identical materials from a common source to allow detailed intercomparison. At 1300 K, an error of ±1.0% is not unexpected, and at 1500 K it is probably ±2%. Platinum could also serve as a useful specific heat standard material from 300–2000 K. It can be obtained in high purity, it is quite inert, and it has a high thermal conductivity; of course, it is expensive. However, the data in Fig. 5.17 clearly show that before Pt can be effectively used as a reference material, systematic measurements of the specific heat, similar to those carried out on α–Al$_2$O$_3$, need to be undertaken. Here the variance possible among measured specific heat values above 300 K is illustrated by results as measured by eight investigations, representing four different techniques. Note that the data scatter about 4% from 300 K to about 1700 K. Table 5.8 lists the method used to obtain each set of data and the approximate error given by the investigators (or estimated by us from their error analyses). Clearly, the disagreement in Fig. 5.17 is outside the bounds of the reported error. The magnitude of the differences cannot be associated with the purity of the sample, except near the melting point (2042 K) where vacancy formation may be influenced by the solute content. Thus, the differences are due to unknown errors, perhaps partially consisting of unknown or uncorrected heat leakage.

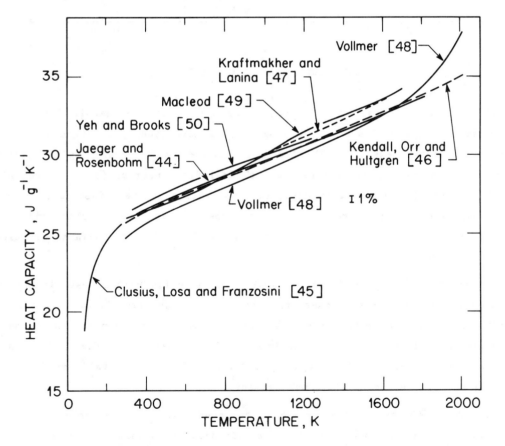

Figure 5.17. The specific heat of Pt. See Table 5.8 for details.

Table 5.8. Source of the Specific Heat Data of Pt in Fig. 5.17, the Method Used to Measure the Specific Heat and the Estimated Error in the Measurements.

Investigators and Year Data Reported	Temperature Range (K)	Calorimetric Method	Estimated Error in c_p (%)
Jaeger and Rosenbohm [44] (1939)	217–1870	Drop	±0.1
Clusius, Losa, and Franzosini [45] (1957)	10–273	Adiabatic	$\sim\pm1$
Kendall, Orr, and Hultgren [46] (1962)	340–1440	Drop	$\sim\pm1$
Kraftmakher and Lanina [47] (1965)	1000–2000	Pulse	±1–3
Vollmer [48] (1971)	300–2000	Adiabatic (see Ref. 57)	±2
MacLeod [49] (1972)	400–1700	Drop	$\sim\pm1$
Yeh and Brooks [50] (1973)	350–1200	Adiabatic	$\sim\pm1$
Seville [51] (1974)	1200–1900	Temperature Modulation	$\sim\pm1$

5.8. COMPUTER OPERATION OF CALORIMETERS

The availability of high-quality mini- and microcomputers, programmable power supplies, programmable amplifiers, and high-speed recording devices makes the application of computers in high-temperature calorimetry feasible. The two major problems are: realizing the same accuracy in the measurement of the variables entering into the calculation of c_p (e.g., by Eq. 2) and designing algorithms which will permit control of the temperature of the adiabatic shields equal to that of the commonly-used analog controllers. It now appears that these problems can be resolved, with enhanced calorimeter performance. In the remainder of this section, data acquisition aspects will be discussed [52,53].

Whether heating continuously or discontinuously, the specific heat is usually obtained from Eq. (2). The voltage drop across a standard resistance in series with the calorimeter heater and the voltage drop across the heater itself must be measured to obtain the power to the calorimeter. The temperature rise ΔT and the time increment Δt must also be measured. The latter can be determined easily to better than 0.001 s, although a precision of 0.01 s is usually sufficient, so that the error in c_p is not limited by the time measurement.

The criterion for the design and selection of components for computer-based control and data acquisition is retention of the integrity of the analog signals as integer counts in the computer code. Signals to be converted include voltage drops across the specimen heater and standard resistance measuring the specimen current, and thermocouple emfs. Since the latter signals are the smallest and, therefore, present the major problem, the following discussion emphasizes computer-based measurements of thermocouple emfs. With a Pt:Pt-13% Rh thermocouple, the output must be detected within ± 1.0 µV to hold the average error in determining a 20 K interval to $\pm 0.5\%$. This interval is usually determined by taking absolute readings of thermocouple emf at the limits of the interval. To accomplish this, the analog-to-digital interface must be selected to convert ± 1 µV of input to ± 1 digital count. Considering the necessity of obtaining both absolute and differential temperature values on multiple thermocouples, the interface will generally involve a multiplexer to select any of several thermocouples for input, variable gain amplifiers, filtering circuits, and analog-to-digital (A/D) converters. If a multiplexer is not used, an interface must be provided for each input; this is expensive and may create an error when, for example, signals from two thermocouples are being compared. Accuracy would then involve the ability to duplicate interface systems.

Interface components may be selected individually or purchased integrated into versatile digital voltmeters (DVMs) capable of having operating parameters set by and digital output transmitted to the computer. Although these DVMs have a wide range of characteristics, careful attention must be given to specifications to ensure that they meet user needs. Typically, these instruments have a low level range of 10^5 µV with a resolution of ± 1 µV. Accuracy of the DVM output depends on selection of operating parameters (range, internal samples per output value, and filtering) and may range from ± 10 µV for a single 1 ms reading to ± 3 µV by averaging over a 64 ms interval or by filtering the analog signal. Some instruments permit selective averaging of very large numbers of samples taken over times extending to minutes. A few instruments are available with a 10^4 µV lower range span, 0.1 µV resolution, and accuracy of ± 0.4 µV; the response time to 0.01% of final value, however, is 1-3 s. These representative performance values are given since they must be considered in writing computer code. Of particular importance is matching response time, accuracy of reading, and heating rate; also, considerations may be significantly different for data acquisition for calculating the specific heat as compared to thermocouple emf measurements required to calculate power adjustments for adiabatic shield control.

The alternative to the commercial DVM is selection of the individual components of the analog-to-digital interface. This approach is usually less expensive and allows greater flexibility in setting specifications for each component, but may require design and construction of components and usually requires more complex computer programs. The following discussion is a brief analysis of factors involved in selecting individual components of an analog-to-digital interface. It also provides insight into the design considerations of DVMs used as interface devices. A typical order of components is multiplexer, amplifier, filter, and A/D converter. A 12-bit A/D converter has a resolution of one count in 2048 (with one bit used to designate polarity) or $\pm 0.05\%$ corresponding to ± 0.005 V with a maximum input of 10 V. A common recommendation is to select an A/D converter with a maximum input of 10 V and select preamplification at the source of the signal such that the input voltage is as near 10 V as reasonable. Since preamplifiers having linearity and gain accuracy of 0.01% or better are readily available, it is reasonable to expect an integrity of voltage drops across heaters and standard resistances of about $\pm 0.05\%$ in the computer code for a single reading. Amplifiers with programmable gain are particularly useful in allowing selection of maximum gain to provide inputs as near to 10 V as possible when, as usual, heater and thermocouple outputs are otherwise unacceptably small.

Computer-based measurements of thermocouple emfs usually present the critical problem because of their small magnitude and the accuracy required for determination of temperature intervals. As indicated previously, with Pt:Pt-13% Rh thermocouples, the output must be detected within ± 1.0 μV in order to hold the average error of determining a 20 K interval to $\pm 0.5\%$. To convert this limiting ± 1 μV resolution to the limiting ± 1 digital count output from the A/D converter (corresponding to ± 0.005 V input) a preamplifier with gain of $3 \cdot 10^3$ is required. This gain, however, limits the thermocouple output to 2000 μV, because otherwise the allowed $+10$ V input limit to the A/D converter is exceeded; also, most preamplifiers have a maximum output of 10 V. These limitations are met by using programmable voltage suppression units, in this example, five increments of 2000 μV. These suppression units can be built with step accuracy approaching 0.1 μV with computer code activating the steps on either increase or decrease of input signal. To retain sub-microvolt integrity of signals from several thermocouples, the multiplexer used for selecting the thermocouple and for incrementing the suppression steps must be constructed with low-thermal, usually gold-plated, microswitches enclosed in well-insulated (isothermal) enclosures.

To realize the full capability of DVMs, preamplifiers, A/D converters, and low-level multiplexer systems, it is essential to pay strict attention to the best principles of shielding and grounding [34] and to place the interface components as close as reasonable to the signal source. It is generally desirable to ground the thermocouple, usually by direct attachment, to the calorimeter and shields which are grounded, and then to use shielded leads with this shield grounded as close as possible to the point of grounding of the thermocouple element itself. With careful grounding and shielding, using voltage suppression in increments of 1000 μV and internally averaging three A/D converter outputs within 60 ms, we have been able to hold the integrity of a thermocouple emf to better than ± 0.5 μV corresponding to about ± 0.05 K. To accomplish the observed temperature resolution, it is essential to make this thermal emf measurement when power to the shield is turned off. We include in our computer code a command which operates relays to turn off power to this shield heater for 0.1 s, during which time interval the thermocouple emfs are measured. During this period of time the shield temperature changes negligibly, and adiabatic conditions are satisfactorily maintained.

The capability of the computer to store large quantities of data permits faster statistical treatment of measured quantities. However, these data are treated somewhat differently when using continuous versus discontinuous heating. In either case the energy input rate can be measured at intervals of a few seconds and averaged over a given period of time to allow a statistically evaluated input of energy during the time interval. Thermal emfs measured at similar intervals permit establishing statistically the temperature as a function of time by least squares fitting. This significantly improves correlating temperature-time intervals during continuous heating. With the discontinuous heating technique, it improves the extrapolation of drifts in temperature (before and after an energy input) to a selected reference temperature at which the specific heat is calculated.

Additional advantages of computer operation are apparent. All parameters required in the specific heat calculation can be introduced at the computer terminal, including specimen mass, drift rate, and heater corrections together with all conversion factors. It is particularly convenient to include with the code analytical expressions or tabular values for conversion of thermal emfs to temperatures, including calibration factors for a particular thermocouple. Finally, the measured and corrected values are immediately available for fitting to analytical expressions relating specific heat to temperature.

5.9. SURVEY OF ADIABATIC CALORIMETERS

In 1968, West and Westrum [2] listed several high-temperature adiabatic calorimeters with some of their important characteristics. In Table 5.9 we have brought their compilation up to date. There is no attempt to list all of the details of the construction or operation. The temperature range of operation of each calorimeter is listed, along with the approximate precision.

5.10. CLOSURE

A major objective of this review has been a detailed presentation of the procedures and problems encountered in the design and construction of high-temperature, adiabatic calorimeters, since this information has not been available in one source. It has been emphasized that if the error in the specific heat measured by adiabatic calorimetry is to be kept near 1% approaching 1300 K and above, then critical attention must be given to heat transfer aspects of the design in order to minimize heat leakage. This involves not only a careful analysis of the origins of heat leakage but recognition of those options that may be available through the selection of materials that would minimize the leakage. Finally, as regards accuracy assessment, there is no substitute to making careful and detailed measurements of reference materials, such as α-Al_2O_3 and Mo whose values are well established, in evaluating the capabilities of a calorimeter system.

5.11. ACKNOWLEDGMENTS

The writing of this chapter was supported partially by the Department of Energy under contract DE-05-79-5951. The calorimeter work in our laboratory was supported from 1950 to 1978 by the United States Atomic Energy Commission and later the Energy Research and Development Administration. Mr. Alejo Garcia made some of the measurements referred to in this paper. We especially thank Mr. D. A. Ditmars and Dr. D. L. Martin for careful reading of the manuscript and for suggestions to improve it.

Table 5.9.   Survey of Adiabatic Calorimeters Operating Above 300 K. This List Covers the
Time Since the Compilation of West and Westrum [2] in 1963.

Author, Date Published (Reference), Temperature Range, and Approximate Accuracy (Error) in Specific Heat	Construction, Nature of Sample and Operating Technique
Martin and Snowdon, 1966 [54] 300–475 K ±0.2%	Stainless steel cell with Ag radiation shield as outer surface; Ag plated, Cu shields; Pt resistance thermometer; continuous heating.
Karasz and O'Reilly, 1966 [55] 15–600 K ±0.3%	Ag cell, Rh plated on outside; Pt resistance thermometer; discontinuous heating.
Leadbetter, 1968 [56] 300–773 K ±0.2%	Ag cell; discontinuous heating; Ag shield; Pt resistance thermometer.
Braun, Kohlhaas, and Vollmer, 1968 [57] 300–1400 K ±2%	Solid sample, or container for liquids; continuous heating.
Sale, 1970 [9] 650–1750 K ±1.5%	Solid sample; spherical shields of Mo and Ta; discontinuous heating; thermocouple thermometer.
Malinsky and Claisse, 1973 [21] 900–1800 K ±1.5%	Solid sample; Mo shields; Pt–10% Rh thermocouple thermometer; continuous heating.
Steffen and Wollenberger, 1973 [24] 300–700 K ±4%	Electron beam heated, twin calorimeter; thermocouple thermometer.
Naito, Inaga, Ishida, Saito, and Arima, 1974 [61] 400–1000 K ±(1–5%)	Quartz sample container in spherical Pt cell; spherical Ag shield; thermocouple thermometer; continuous heating.
Andrews, Norton, and Westrum, 1978 [58] 300–550 K ±0.1%	Ag cell; Cr-plated Cu shields; Pt resistance thermometer; discontinuous heating.
Schmidt and Maksimov, 1979 [59] 300–800 K ±0.2%	Quartz sample container in an Ag cell; Ag shield; Pt resistance thermometer; discontinuous heating.
Oetting and West, 1981 [60] 300–700 K ±0.6%	Adapted calorimeter by West and Ginnings [41] for small samples (e.g., 10 cm^3).

5.12.   BIBLIOGRAPHY

In this section, review articles pertinent to high-temperature, adiabatic calorimetry are listed which are not given in the References.

Stansbury, E.E., in Proceedings of 3rd Conference on Thermal Conductivity, Gatlinburg, TN, limited distribution, 639–60, 1961.

Krauss, F., Z. Metallkd., 49, 386-92, 1958.

Wittig, F.E., Pure Appl. Chem., 2, 183-204, 1961.

Grønvold, F., in Thermodynamics, Vol. I, International Atomic Energy Agency, Vienna, Austria, 35-52, 1966.

Westrum, E.F., Jr., in Advances in High Temperature Chemistry (Eyring, L., Editor), Vol. I, Academic Press, New York, NY, 239-57, 1967.

Komarek, K.L., Z. Metallkd., 64, 325-41, 1973.

Takahashi, Y., Pure Appl. Chem., 46, 323-31, 1976.

5.13. REFERENCES

1. Ginnings, D.C., in Experimental Thermodynamics. Volume I. Calorimetry of Non-Reacting Systems (McCullough, J.P. and Scott, D.W., Editors), Plenum Press, New York, NY, 1-13, 1968.

2. West, E.D. and Westrum, E.F., in Experimental Thermodynamics. Volume I. Calorimetry of Non-Reacting Systems (McCullough, J.P. and Scott, D.W., Editors), Plenum Press, New York, NY, 333-67, 1968.

3. West, E.D., J. Res. Natl. Bur. Stand., 67A, 331-41, 1963.

4. Ginnings, D.C. and West, E.D., in Experimental Thermodynamics. Volume I. Calorimetry of Non-Reacting Systems (McCullough, J.P. and Scott, D.W., Editors), Plenum Press, New York, NY, 85-131, 1968.

5. West, D.D. and Churney, K.L., J. Appl. Phys., 39, 4206-15, 1965.

6. Zielenkiewicz, W., Bull. Acad. Polon. Sci., Ser. Sci. Chim., 14, 583-7, 1966, and other papers in this journal both before and after 1966.

7. Cash, W.M., 'The Data Acquisition from and the Theoretical Modeling of a Dynamic Adiabatic Calorimeter,' Univ. of Tennessee, M.S. Thesis, 1976.

8. Naito, K., Kamagashira, N., Yamada, N., and Kitagawa, J., J. Phys. E, 6, 836-40, 1973.

9. Sale, F.R., J. Phys. E, 3, 646-50, 1970.

10. Sale, F.R. and Normanton, A.S., in Metallurgical Chemistry (Kubaschewski, O., Editor), HMSO, London, England, 19-28, 1972.

11. Krauss, R., Z. Metallkd., 49, 386-92, 1958.

12. Bell, H. and Hultgren, R., Metall. Trans., 2, 3230-1, 1971.

13. Stansbury, E.E. and Brooks, C.R., High Temp.-High Pressures, 1, 289-307, 1969.

14. Corboba, G. and Brooks, C.R., Phys. Status Solidi A, 7, 503-8, 1971.

15. Wollenberger, H., Phys. Status Solidi A, 2, 511-20, 1970.

16. Ginnings, D.C. and West, E.D., Rev. Sci. Instrum., 35, 965-7, 1964.

17. Neighbor, J.E., Rev. Sci. Instrum., 37, 497-9, 1966.

18. West, E.D. and Ishihara, S., Rev. Sci. Instrum., 40, 1356-9, 1969.

19. Bellarby, P.W., J. Phys. E, 4, 153-6, 1971.

20. Martin, D.L., Rev. Sci. Instrum., 43, 1762-5, 1972.

21. Malinsky, I. and Claisse, F., J. Chem. Thermodyn., 5, 615-22, 1973.

22. Wollenberger, H., Z. Metallkd., 49, 629-33, 1958.

23. Wollenberger, H. and Wuttig, M., Z. Metallkd., 51, 503-9, 1960.

24. Steffen, H. and Wollenberger, H., Rev. Sci. Instrum., 44, 937-43, 1973.

25. Anderson, R.L. and Kollie, T.G., CRC Crit. Rev. Anal. Chem., 6, 171-221, 1976.

26. Pollock, D.D., The Theory and Properties of Thermocouple Elements, STP 492, American Society for Testing and Materials, Philadelphia, PA, 1971.

27. Plumb, H.H., Editor, <u>Temperature – Its Measurement and Control in Science and Industry</u>, Vol. 4, Part 3, Instrument Society of America, Pittsburgh, PA, 1972.

28. ASTM, <u>Manual on the Use of Thermocouples in Temperature Measurement</u>, STP 470, American Society for Testing and Materials, Philadelphia, PA, 1970.

29. Kinzie, P.A., <u>Thermocouple Temperature Measurement</u>, Wiley, New York, NY, 1973.

30. Darling, A.S. and Selman, G.L., in <u>Temperature – Its Measurement and Control in Science and Industry</u> (Plumb, H.H., Editor), Vol. 4, Part 3, Instrument Society of American, Pittsburgh, PA, 1633-44, 1972.

31. NBS, <u>Thermocouple Reference Tables Based on the IPTS-68</u>, NBS Monograph 125, U.S. Department of Commerce, Bureau of Standards, Washington, DC, 1974.

32. ASA, <u>Temperature Measurements with Thermocouples</u>, ASA-C96, American Standards Association, 1964.

33. McLaren, E.H. and Murdock, E.G., in <u>Temperature – Its Measurement and Control in Science and Industry</u>, Vol. 4, Part 3, Instrument Society of America, Pittsburgh, PA, 1543-60, 1972.

34. Morrison, R., <u>Grounding and Shielding Techniques in Instrumentation</u>, Wiley, New York, NY, 1967.

35. Roeser, W.F. and Wansel, H.T., J. Res. Natl. Bur. Stand., $\underline{10}$, 275-87, 1933.

36. Pawel, R.E., 'The Application of Dynamic Adiabatic Calorimetry to the Copper-Nickel System from 50 to 620°C,' Univ. of Tennessee, Dissertation, 1956.

37. Nauman, E.B., 'Analysis of a High Temperature Adiabatic Calorimeter,' Univ. of Tennessee, M.S. Thesis, 1961.

38. West, E.D., Trans. Faraday Soc., $\underline{59}$, 2200-3, 1963.

39. Brooks, C.R., 'A Calorimetric Investigation of Anomalies in the Specific Heat of Binary Solid Solutions of Copper Containing Aluminum and Gallium,' Univ. of Tennessee, Thesis, 1962.

40. Grønvold, F., Acta Chem. Scand., $\underline{21}$, 1695-713, 1967.

41. Parratt, L.G., <u>Probability and Experimental Errors in Science</u>, Dover, New York, NY, 1961.

42. Brinkworth, B.J., <u>An Introduction to Experimentation</u>, American Elsevier, New York, NY, 1968.

43. West, E.D. and Ginnings, D.C., J. Res. Natl. Bur. Stand., $\underline{60}$, 309-16, 1958.

44. Jaeger, F.M. and Rosenbohm, E., Physica, $\underline{6}$, 1123-5, 1939.

45. Clusius, K., Losa, C.G., and Franzosini, P., Z. Naturforsch., $\underline{12}$, 34-8, 1957.

46. Kendall, W.E., Orr, R.L., and Hultgren, R., J. Chem. Eng. Data, $\underline{7}$, 516-8, 1962.

47. Kraftmakher, Y.A. and Lanina, E.G., Sov. Phys.-Solid State, $\underline{7}$, 92-5, 1965.

48. Vollmer, O., 'Die Spezifische Warme von Rhodium, Palladium und Platinim Bereich hoher Temperaturen,' Universitat zu Koln, Dissertation, 1971.

49. MacLeod, A.C., J. Chem. Thermodyn., $\underline{4}$, 391-9, 1972.

50. Yeh, C.C. and Brooks, C.R., High Temp. Sci., $\underline{5}$, 403-13, 1973.

51. Seville, A.H., Phys. Status Solidi A, $\underline{21}$, 649-58, 1974.

52. Cash, W.M., 'The Direct Digital Control of a Dynamic Adiabatic Calorimeter,' Univ. of Tennessee, Dissertation, 1979.

53. Cash, W.M., Stansbury, E.E., Moore, C.F., and Brooks, C.R., Rev. Sci. Instrum., $\underline{52}$, 895-901, 1981.

54. Martin, D.L. and Snowdon, R.L., Can. J. Phys., $\underline{44}$, 1449-65, 1966.

55. Karasz, F.E. and O'Reilly, J.M., Rev. Sci. Instrum., $\underline{37}$, 255-60, 1966.

56. Leadbetter, A.J., J. Phys. C, $\underline{1}$, 1481-8, 1968.

57. Braun, M., Kohlhass, R., and Vollmer, O., Z. Angew. Phys., $\underline{25}$, 365-72, 1968.

58. Andrews, J.T.S., Norton, P.A., and Westrum, E.F., J. Chem. Thermodyn., $\underline{10}$, 949-58, 1978.

59. Schmidt, N.E. and Maksimov, D.N., Russ. J. Phys. Chem., <u>53</u>, 1084-7, 1979.

60. Oetting, F.L. and West, E.D., J. Chem. Thermodyn., <u>14</u>, 107-14, 1982.

61. Naito, K., Inaba, H., Ishida, M., Saito, Y., and Arima, H., J. Phys. E, <u>7</u>, 464-8, 1974.

CHAPTER 6

Drop Calorimetry Above 300 K

D. A. DITMARS

6.1. INTRODUCTION

Calorimetry, in common with all experimental measurement science, has developed historically within bounds imposed by available measurement technology and by the spectrum of construction materials. The first calorimetric measurements of heat content were made in the easily-accessible temperature range between the ice and steam points and were usually based on some variation of the 'method of mixtures.' In this method, two systems at known thermodynamic equilibrium states – one, a specimen under investigation and one, a carefully characterized reference system ('calorimeter') – are brought together ('mixed'), cf. Fig. 6.1*. Normally, the specimen temperature is greater than the calorimeter temperature. The heat content of the specimen is inferred from the measured change in state of the calorimeter through the independent application of some energy-calibration procedure to the calorimeter, either before or after the specimen experiment. Early specific heat data were obtained in this way using only the simplest modes of construction and instrumentation.

The introduction of improved construction materials and fabrication techniques, along with electronic power supply, temperature control, and data recording, has greatly expanded the temperature range attainable in drop calorimetry and simultaneously lowered the levels of imprecision and inaccuracy of this method. It is today possible (though, admittedly, with great effort) to employ drop calorimetry well into the temperature range of molten refractory materials.**

Drop calorimetry is a slower technique due to its inherent dependence on thermal equilibration throughout fairly massive systems. It is still a technique most efficiently applied with significant operator attention to critical phases of each datum measurement. Nevertheless, if one

---

*For practical reasons, this is usually effected by physically <u>dropping</u> a specimen into a calorimeter; hence, the oft-used terms 'drop calorimetry,' 'drop calorimeter,' etc. We adhere to this usage throughout this chapter.

**At the same time, these developments have created a fertile field for the extension of adiabatic calorimetry to high temperatures and for the introduction of temperature-modulation techniques and of scanning and other transient techniques for measuring heat capacity. These newer techniques excel drop calorimetry chiefly in their susceptibility to automation and in their higher rate of data production. In carefully designed and executed experiments over restricted temperature ranges, some of them can be competitive with drop calorimetry in accuracy, if not in initial or maintenance cost.

243

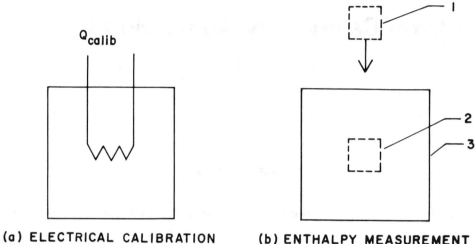

$$Q_{calib}$$

(a) ELECTRICAL CALIBRATION   (b) ENTHALPY MEASUREMENT

1 - Specimen initially at thermodynamic equilibrium outside calorimeter
2 - Specimen at new equilibrium state inside calorimeter
3 - Calorimeter whose own equilibrium state is changed by addition of specimen

Figure 6.1. Ideal drop-calorimetric experiment.

desires a single apparatus capable of covering a very wide temperature range, it is still undoubt-
edly the most accurate technique available above 300 K. In addition, it is applicable to a very
wide range of specimen types and can employ relatively unsophisticated instrumentation.

This chapter on current drop-calorimetric technique and apparatus in the temperature range
above 300 K is intended primarily for the evaluator of thermodynamic data derived from drop-
calorimetric experiments. The aim is to present, in the context of the basic thermodynamic
principles underlying drop calorimetry, sufficient detail regarding apparatus design, operating
technique, measurement instrumentation, and data treatment to enable the reader to judge the
extent to which reported experimental drop-calorimetric data may be subject to error. Those con-
templating the addition of a drop calorimeter to their facilities should be aware that drop calorimeters cannot normally be purchased 'off the shelf' and that this chapter is not intended as a
handbook of design. Creating such a facility can be a major undertaking of, conservatively, one
to two year's duration. Would-be designers and builders are encouraged to consult more detailed
calorimeter descriptions appearing in the literature (in part, cited in the References of this
chapter), and especially to cultivate personal contact with experimentalists in the field.

6.2. FUNDAMENTAL CONCEPTS AND RELATIONSHIPS IN DROP CALORIMETRY

6.2.1. Ideal Drop-Calorimetric Experiment

The basic concepts required for performing a drop-calorimetric experiment or for interpreting
data derived from this technique are described in this section in the context of an 'ideal experiment.' Thereafter, real apparatus and real procedures are introduced, and the way in which these
contribute to inaccuracy in the end result (an enthalpy datum) is outlined.

In all drop calorimetry the specimen, first brought to an initial thermodynamic equilibrium state [state while outside the calorimeter (see Fig. 6.1)], is translated rapidly to be within the calorimeter boundary. As a result of this change, the calorimeter proceeds from its own initial thermodynamic equilibrium state prior to adding the specimen to some final state while containing the specimen. The specimen proceeds at the same time to its final equilibrium state within the calorimeter. In the ideal experiment, the specimen and the calorimeter each have the same temperature throughout in their final states. The physical quantity which one aims to obtain through this measurement is the enthalpy change of the specimen upon entering the calorimeter. To achieve this, at least one additional experiment ('calibration') is necessary. In this calibration experiment, one measures the amount of heat, Q_{calib}, required to produce the identical change in state of the calorimeter as the specimen. Generally, electrical energy is the most practical source for Q_{calib}.

Consider now the specimen in the calorimeter and the thermodynamic definition of its enthalpy:

$$H = U + PV. \tag{1}$$

For an infinitesimal change in the equilibrium state of the specimen,

$$dH = dU + PdV + VdP. \tag{2}$$

However, the heat dQ transferred between the specimen and calorimeter in this change,

$$dQ = dU + PdV. \tag{3}$$

Therefore,

$$\frac{dH}{dT} = \frac{dQ}{dT} + V\frac{dP}{dT}, \tag{4}$$

where dT is the infinitesimal change in the thermodynamic temperature of the specimen that is associated with its change of state.

Under isobaric conditions (which apply in all drop calorimetry, excepting measurements on volatile liquids),

$$\left(\frac{dH}{dT}\right)_p = \left(\frac{dQ}{dT}\right)_p \quad C_p, \tag{5}$$

where C_p denotes the specific heat.

Thus, the change in specimen enthalpy under these conditions is equal to the heat transferred between the specimen and the calorimeter:

$$\int_{T_c}^{T_i} dH = H_{T_i} - H_{T_c} = Q = \int_{T_c}^{T_i} C_p dT, \tag{6}$$

where T_i denotes the initial specimen temperature, and T_c denotes the final temperature common to both specimen and calorimeter. Further, the entropy and Gibbs energy functions can be calculated:

$$S_{T_i} = S_{T_c} + \int_{T_c}^{T_i} \frac{C_p}{T} dT, \tag{7}$$

and

$$-\frac{(G_{T_i} - H_{T_c})}{T_i} = S_{T_i} - \frac{(H_{T_i} - H_{T_c})}{T_i} \ . \tag{8}$$

6.2.2. Real Drop-Calorimetric Experiment; Some Error Sources

Drop calorimeters for measuring the relative enthalpy of substances above 300 K have historically been classified according to some significant descriptor which characterizes the heat-measuring part of the calorimeter. Thus, one has as principal drop-calorimeter types, liquid bath calorimeters, metal-block calorimeters, adiabatic receiving calorimeters, and phase-change or true isothermal calorimeters. The measurement error associated with any particular datum produced with one of these calorimeters depends on several factors which are common to all these apparatus types as well as on factors which are type-specific. We hope that the present and following sections heighten the reader's awareness of potential sources of error in drop-calorimetric experiments. Naturally, this array of error sources will in general not apply entirely to any specific experiment, nor can the reader expect to have provided for him in each instance from the literature all the information he needs to properly assess each error source.

6.2.2.1. Sample

Although drop calorimetry is an appropriate technique for a broad range of substances, it does have important limitations. It is in general not applicable to samples which react chemically or dissociate in the temperature range of interest. It is desirable that the sample have low or negligible vapor pressure. Hermetic encapsulation will ensure its physical integrity and prevent reaction with the environment. The working sample pressures due to gases within such containment will not ordinarily reach values at which the pressure coefficient of heat capacity is quantitatively important. However, if a substantial portion of the sample is in the vapor state at the initial temperature, then the measured heat must be corrected for the heat of vaporization. The method of calculating this correction has been discussed in length [1,2].

If the sample experiences solid-solid phase transformation in the temperature range of interest, it is essential that the experimenter demonstrate, e.g., by X-ray structural analysis, that despite the rapid sample-cooling rates experienced during drop-calorimetric measurements (these may approach 1000–2000 $K \cdot s^{-1}$) the sample attains each time in the calorimeter the same thermodynamic state.

Should the desired result of the enthalpy measurements be the thermodynamic properties of a pure phase, as opposed to those of some undifferentiated mixture, then the determination of sample purity and correction of the measured enthalpy for the contributions of identified impurity phases must be addressed. In the absence of reliable enthalpy data for the impurity phases, a first approximation to this correction may be calculated under the assumption that each phase present contributes to the heat capacity of the total sample in proportion to its molar concentration. Meriting special attention in this regard are impurity phases of low atomic weight such as hydrogen.

6.2.2.2. Sample Temperature

The two essential temperature measurements in a drop-calorimetric experiment are that of the sample in the furnace prior to its fall into the calorimeter and that of the calorimeter heat-measuring element itself during the course of the experiment. The usual instrumentation for measurement of sample and calorimeter temperature in drop calorimetry includes thermocouples, resistance thermometers, quartz crystal thermometers, and in the highest temperature range, optical pyrometers. Aside from the questions of calibration and correct employment of temperature instrumentation, the data evaluator must judge the extent to which the _measured_ temperature represents the _sample_ temperature.

Generally, in drop calorimetry at temperatures below those at which optical pyrometry can be applied (roughly, 1300 K), the sample temperature is not measured directly. Rather, the furnace temperature is measured and one relies on some temperature zone within the furnace, designed to be closely isothermal, to ensure attainment of a uniform sample temperature. Here, one must consider whether the residence time of the sample in the furnace has been sufficient to allow full temperature equilibration. This is especially true in the temperature vicinity of melting transformations, in which considerable amounts of heat must sometimes be transferred to the sample over rather small temperature gradients.

The magnitude of reported values for enthalpy and heat capacity are slightly dependent upon the temperature scale on which they are expressed. While current practice stresses the importance of specifying the temperature scale employed and such information as fixed calibrating points along with the reported drop-calorimetric data, this has not always been so in the past. The data evaluator must determine details of the temperature scale for reported data as best he can and make the required corrections to the reported enthalpy or heat-capacity data if the accuracy of the data warrants. Fortunately, these corrections are often small or at most of the same order of magnitude as the measurement precision. Details of correcting thermodynamic data expressed on the temperature scale T-48 to the scale T-68 are given in [3,4].

6.2.2.3. Sample Heat Loss; Encapsulation

The transfer of the sample from the furnace to the calorimeter unavoidably involves a loss of heat to the surroundings which is not detected by the calorimeter. For unencapsulated samples, this loss can be a significant fraction (2-5%) of the total sample heat at very high temperatures, and a correction for it must be included in the reported enthalpy data. The amount of heat lost from the sample will be much smaller if the sample is encapsulated during measurement. (Encapsulation may also be required for reasons unrelated to heat loss, as when a sample is finely divided or in the liquid state, or has significant vapor pressure.)

The heat loss from an unencapsulated sample to its surroundings has a radiative component and if not dropped in a vacuum, conduction and convection components. The accuracy with which one can calculate the heat loss from such a sample will depend on successful modeling of the surroundings through which the sample travels on its path to the calorimeter. Calculation of the radiative loss also requires knowledge of the total hemispherical emissivity for the sample. The convection loss has been estimated by some authors [5,6] from a relation representing the heat loss by a sphere cooling in a gas stream [7].

If a sample remains encapsulated during measurement, it is sometimes assumed that the enthalpy increment due to the sample alone can be obtained by subtracting from the relative enthalpy determined for the sample plus its container, the corresponding relative enthalpy for the container alone (a 'blank,' measured in a second experiment). This procedure implicitly assumes that no heat is lost by the sample itself in dropping into the calorimeter. Practically, this could be achieved by designing the sample container such that the temperature of its inner surface does not change during the short dropping time. Any heat loss by the sample itself prior to entering the calorimeter constitutes a contribution to systematic error in the measurement. Ginnings [8] has estimated that at 1200 K, this systematic error in enthalpy is not likely to exceed 0.03%.

The enumeration of contributions to imprecision and inaccuracy from the heat-measuring part of the calorimeter is deferred till the next section, as these contributions vary with the type and design of drop calorimeter.

6.3. CLASSIFICATION AND DESCRIPTION OF DROP-CALORIMETRIC APPARATUS

6.3.1. Furnaces

The role of the furnace in conventional drop calorimetry is to provide a temperature-controlled region in close proximity to the calorimeter within which the sample can be retained prior to its transfer to the calorimeter. Frequently, the furnace and calorimeter are joined together as a hermetic unit to permit continuous control of the pressure and composition of the atmosphere surrounding the sample. Energy can be supplied to furnaces by resistive heating using Nichrome or platinum alloy heating elements [1]. These have an upper practical temperature limit of about 1500–1700 K. At higher temperatures, resistance furnaces using refractory metal [9,10] or graphite [11–13] heating elements have been successful. Induction heating furnaces* have been employed up to 2800 K [14–16]. An innovative but less-frequently applied design variation is electron-beam heating [17,18].

Situating a furnace which operates at 1200 K or above this temperature close to a calorimeter such that there is an open path for radiation from one to the other can introduce an additional source of systematic error in heat measurement. In such cases, it is necessary that the furnace or calorimeter or both of these be provided with radiation-blocking shutters which are opened for the brief period required to admit a sample to the calorimeter. These have appeared in both manual [1] and automatically-operated [19] versions. In any case, it is necessary to measure the heat contribution, if any, of the furnace to the measured enthalpy.

---

*Induction heating is also the usual heating method in Levitation Calorimetry to attain steady temperatures for extended time periods in excess of 3600 K. However, in this technique, one is concerned with small, unencapsulated samples and no furnace as such is used. See the discussion of Levitation Calorimetry elsewhere in this volume.

6.3.2. Calorimeters

Calorimeters used in drop calorimetry fall into three major classes: liquid-bath calorimeters, metal-block calorimeters, and phase-change calorimeters. All these calorimeters consist of a central (heat-measuring) part which is contained within some surrounding thermal shielding. It is necessary, in order to perform an accurate heat measurement with such a system, to have an accurate value for the total heat capacity of the central part*. In addition, it is essential that account be taken of heat exchange between the central part and its surrounding shielding both before, during, and after transfer to it of a sample from the furnace. The major contribution of the calorimeter to the overall inaccuracy of enthalpy data arises ultimately from the experimenter's inability to accurately calculate or measure this heat exchange. The shielding is conventionally held either at some unvarying temperature chosen with the aim of minimizing heat exchange (isoperibol mode of operation) or its temperature is controlled through heat addition to equal the temperature of the central part (adiabatic mode of operation). In this section, the essential structure and instrumentation is presented for the above-mentioned three major classes of calorimeter.

6.3.2.1. Liquid-Bath Type

In this type of drop calorimeter, illustrated in Fig. 6.2 and described in much more detail by Eitel [23] and Barton [24], the central measuring part consists of a liquid bath of a heat capacity yielding a conveniently measurable temperature rise for the least amount of heat anticipated for a single measurement. Water is commonly chosen as the calorimetric working substance; however, on occasion, other liquids have been used [20]. The sample can also be immersed directly in the working substance [21]. This, however, may lead to splashing or vaporization which would lower the calorimeter heat capacity. In Fig. 6.2, a version is shown with a closed-end tube preventing contact of sample and liquid [22]. Thermocouples, resistance thermometers, or thermistors would be appropriate as temperature-sensing elements and are preferable to strictly analog thermometers such as the mercury-in-glass type used by early experimenters. An electrical calibration is possible for this type of calorimeter, but is perhaps unnecessary. At the relatively high inaccuracy level of this type of calorimeter (typically, one to several percent) it would be possible to calibrate through heat measurements on some reliable standard reference material, such as α-Al_2O_3 [25]. This apparatus description is given largely out of historical interest, since some heat-capacity data which are still frequently cited (e.g. [21]) were obtained with just such apparatus. The experimental measurements required for calibration of this type of calorimeter or for performing enthalpy measurements with it are essentially the same as those required for operation of a metal-block isoperibol calorimeter. These are discussed below.

6.3.2.2. Metal-Block Type

The isoperibol, metal-block calorimeter, schematically illustrated in Fig. 6.3, is the most common type of drop calorimeter. In this type, a metal of high thermal diffusivity serves as the

---

*Phase-change or true isothermal calorimeters have, by definition, an infinite heat capacity. In this case, one is concerned not with measuring temperature changes, but volume or mass changes; see Section 6.3.2.4.

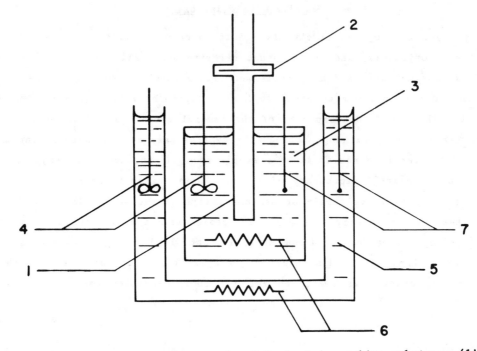

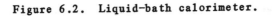

1 – Sample tube; 2 – Radiation shutter; 3 – Calorimetric working substance (liquid);
4 – Stirrers; 5 – Thermal shield; 6 – Calibration and control heaters; 7 – Measuring
and control thermometers

Figure 6.2.   Liquid–bath calorimeter.

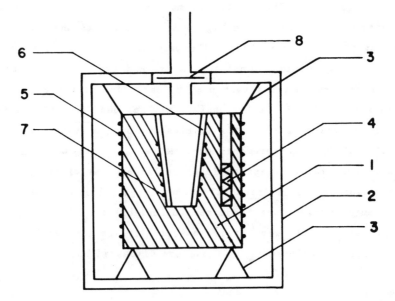

1 – Calorimetric working substance (solid metal); 2 – Constant–temperature enclosure;
3 – Supports; 4 – Capsule thermometer; 5 – Distributed (resistance) thermometer;
6 – Heater insert; 7 – Calibration heater; 8 – Radiation shutter

Figure 6.3.   Isoperibol metal–block calorimeter.

calorimetric working substance. Resort to a solid working substance eliminates the energy of (liquid) stirring as well as errors which arise from splashing and vaporization in the liquid–bath calorimeter. At the same time, the high thermal diffusivity of the metal reduces temperature gradients throughout the working substance, and promotes more rapid temperature equilibration. Copper, with a thermal diffusivity 1000 times that of water, is the metal usually used due to its low cost, ready availability, and easy machinability. Aluminum blocks have been used occasionally; silver, with a thermal diffusivity 50% greater than that of copper, might be technically feasible, but it is doubtful if its higher cost would be justified by the reduction in temperature gradients. Typically, tens of kilograms of metal are required for the block.

The block is positioned by a support of high thermal resistance within a shield which can be set and held at a constant temperature. Two alternate types of support are indicated in Fig. 6.3; a hanging–type support [19] and a resting–type support [26]. The shield is usually maintained at some temperature in the 20–30°C range to within one to several millikelvins. This has been accomplished by use of a hollow shield filled with a circulating, temperature–regulated fluid [17], by use of a shield bearing a fluid circulation coil [10], or – most usually – by direct immersion into a temperature–regulated bath [27]. Heat transfer between block and shield has been reduced by polishing and gold–plating the surfaces of these parts [28] as well as through control of the pressure and composition of the gas in the space between block and shield [29].

Block temperatures have been measured with spatially–concentrated thermometers (encapsulated platinum resistance thermometers or quartz crystal oscillator thermometers) embedded in the metal block [17,27], as well as with spatially–distributed resistance thermometers wound onto the outer block surface [30]. Calibration of this type of calorimeter (i.e., measurement of its total heat capacity) should take place in a way that reproduces as closely as possible the heat–flow patterns encountered in a typical enthalpy measurement. Therefore, the calibrating heater is generally wound onto an insert fitting closely within the sample well (Fig. 6.3). One or more radiation shields are installed between the furnace and calorimeter in order to intercept radiation. The literature contains an abundance of references to the instrumentation and operation of metal–block calorimeters. References [12,14,31–39] describe other variations of the instrumentation discussed above.

During both calibration and enthalpy–measurement experiments with either metal–block or liquid–bath calorimeters operated in an isoperibol mode, the indicated temperature of the calorimetric working substance and its surroundings is recorded as a function of time during each of three time intervals. These intervals (Fig. 6.4) comprise an equilibrating fore period, 1, prior to heating the working substance, the heating period itself, 2, and lastly, an equilibrating after–period, 3. During electrical calibration, in which current from some highly–stable source is passed through the calibration heater, the potential differences across this heater and an external standard resistor connected in series with it are also measured as a function of time.

The total sample heat content is calculated as the product of the total calibrated heat capacity of the calorimetric working substance and its _effective_ temperature increase. This effective temperature increase is the temperature increase of the calorimetric working substance which would have been obtained under strictly <u>adiabatic</u> conditions. In reality, heat loss from the working substance to its surroundings takes place through all available heat radiation and conduction paths as soon as the working–substance temperature rises above that of its surroundings. This heat loss can be well in excess of 10% of the total sample heat; hence, accurate calculation of the loss becomes a central problem of isoperibol calorimetry. In Fig. 6.4, the temperature behavior of the working substance under real isoperibol operating conditions and under ideal adiabatic conditions are indicated. The <u>measured</u> temperature increase will always be less than the effective temperature increase. The heat loss is calculated as the product of the maximum difference between these increases at the end of a heating period and the total heat capacity of the working substance. Detailed treatment of this calculation

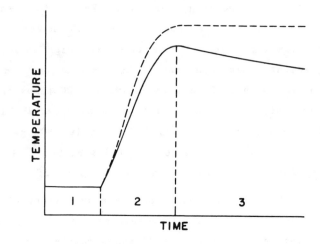

1 – fore–period; 2 – heating period;
3 – after–period
—— measured temperature (real isoperibol conditions)
--- effective temperature (effective adiabatic conditions)

Figure 6.4.   Temperature of isoperibol calorimeter working substance.

requires consideration of heat transfer in the fore and after periods to determine the effective end of the heating period. There exist several excellent treatments of this calculation [29, 40–42] which, in addition to examining the consequences of alternate conceptual models for the calorimeter, take into account the effect of such extraneous heat sources as self–heating or radioactive decay.

The overall inaccuracy of an enthalpy measurement with an isoperibol calorimeter depends on the inaccuracy of the measured sample temperature as well as on the experimenter's success in accounting for heat loss during the interval of translation into the calorimeter and for heat exchange between the calorimetric working substance and its enclosure. As a rough guide, with a carefully designed and instrumented isoperibol calorimeter, overall enthalpy inaccuracy not exceeding 0.2–0.4% should be attainable up to 1200 K. From 1200 K to about 3000 K, this inaccuracy may be expected to increase to 2–3% at the highest temperature.

6.3.2.3. Adiabatic–Receiving Type

In adiabatic–receiving calorimeters (Fig. 6.5), the massive isothermal enclosure of the isoperibol calorimeter has been replaced by a less–massive, heated enclosure whose average temperature is at all times during an experiment kept as close as possible to that of the calorimetric working substance within. The working substance is similar in many respects to that of the block isoperibol calorimeter described in the previous section, but is usually less massive. On occasion, a liquid bath has been employed as a working substance for this type [43]. The success one may expect in measuring good average temperatures for the calorimeter elements depends strongly on

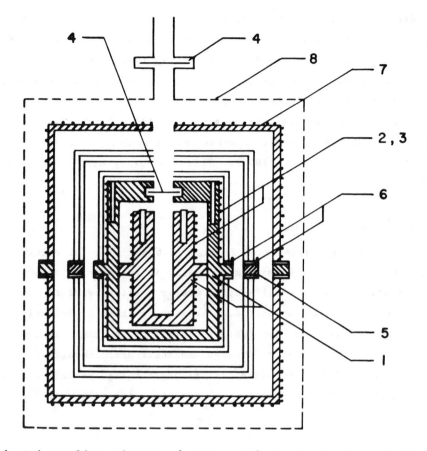

1 - Calorimetric working substance (solid metal); 2, 3 - Heaters and thermometers;
4 - Radiation shutters; 5 - Thermal shield; 6 - Differential temperature measured
between elements shown; 7 - Thermal guard with heater, thermometers; 8 - Additional
thermal shielding, vacuum containment

Figure 6.5. Adiabatic receiving calorimeter.

the calorimeter thermal design and on the spatial distribution of the temperature sensors. Note
that the design [14] illustrated in Fig. 6.5 employs multiple thin radiation shields to reduce
surface thermal gradients. Other examples of adiabatic-receiving calorimeters with further discussion of construction and operating features of specific apparatus can be found in [16,44-50].

The temperature and electric-energy-measurement instrumentation for this type is identical to
that employed in calorimeters of the isoperibol type and should also have the capabilities of precision emf, resistance, time, and frequency measurements. The differential thermometer elements
which sense the temperature difference between the working substance and its surrounding shield
are usually distributed wire windings connected in a Maier bridge configuration [51]. It is not
practical to attempt manual control of the thermal shield temperature. This control function can
be handled more efficiently using some version of an electronic servo-control circuit which incorporates proportional, derivative (rate), and integral (reset) control elements [29].

The temperature and time data required for operation of this type of calorimeter are similar
to those required for the operation of isoperibol calorimeters. A continuous record is also made

of the temperature difference between the working substance and the thermal shield. Since this difference is maintained close to zero at all times, even during the heating period, the relatively large heat-transfer correction of isoperibol calorimetry (see above) is not encountered in adiabatic calorimetry. For the most accurate work, though, the heat transfer caused by even small deviations of the shield temperature from the working substance temperature should be estimated. One method for calculating this heat transfer is given in [52]. A well-designed adiabatic-receiving calorimeter should be capable of enthalpy measurements with an inaccuracy not exceeding 0.3% in the temperature range 300 K to about 1300 K. This inaccuracy can be expected to increase to about 0.6% near 2300 K.

6.3.2.4. Phase-Change (Isothermal) Type

In this type of calorimeter, illustrated in Figs. 6.6 and 6.7, the working substance exists in two phases, either solid/liquid or liquid/gas, ideally in thermodynamic equilibrium. Heat from some source — either a hot sample or a calibration heater — transferred to the working substance through the sample tube, brings about an isothermal change in phase distribution (melting or boiling) within the working substance. The temperature of this phase change at essentially atmospheric pressure having been well-established beforehand, no additional temperature measurements within the calorimeter are required. The temperature of this phase change becomes the reference temperature for enthalpy measurement, and the resultant changes in total volume of the working substance

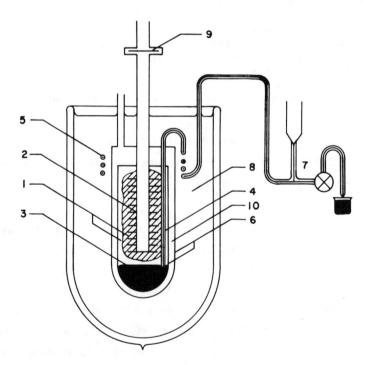

1 – Calorimetric working substance (solid and liquid phases shown); 2 – Sample tube;
3 – Mercury; 4 – Egress/ingress capillary; 5 – Tempering coil; 6 – Calorimeter envelopes;
7 – Mass-accounting system; 8 – Isothermal bath; 9 – Radiation shutter; 10 – Gas gap

Figure 6.6. Phase-change calorimeter (solid-liquid).

are employed as the measure of energy transferred. Figure 6.6 illustrates a solid/liquid phase-change calorimeter and Fig. 6.7, one of the liquid/gas type. Since both types of calorimeter illustrated in Figs. 6.6 and 6.7 rely ultimately for their accuracy on the unchanging thermodynamic properties of both phases of their working substances, a necessary condition is that the phases in equilibrium have the same composition. The conventional way to achieve this is by using highly-purified, solute-free working substances.

Solid-Liquid Systems: The two most commonly-used working substances for solid-liquid phase-change calorimeters (Fig. 6.6) have been water [53] and diphenyl ether (DPE) [54] with solid-liquid equilibrium temperatures at atmospheric pressure of 273.15 K and 299.99 K, respectively. Other substances have been proposed and used [55]. In this type, the liquid working substance is confined within a rigid, constant-volume container by a pool of mercury at the bottom of the container. This mercury can flow through a coiled tempering reservoir to an exterior mass-accounting system. In this manner, volume changes in the working substance are detected as mass changes at the mass-accounting system.

The parameter of practical interest for energy-measuring purposes is the 'calibration constant' of the calorimeter, i.e., the correspondence between energy absorbed from some heat source within the calorimeter sample tube and the change in some reference mass of mercury recorded at the mass-accounting system. This calibration constant depends only on the properties of the working substance, is the same for all phase-change calorimeters using the same working substance, and can in principle be calculated from the formula,

$$\Gamma = H_f [(v_s - v_\ell) \rho_{Hg}]. \tag{9}$$

Here, Γ is the calibration constant expressed in heat units per unit mass of mercury, H_f is the heat of fusion of the calorimeter working substance, v_s and v_ℓ are the specific volumes of the solid and liquid phases, respectively, and ρ_{Hg} is the density of mercury at the calorimeter equilibrium temperature. In practice, it is more usual to measure Γ by a direct electrical calibration using a heater constructed and operated to closely reproduce the same temperatures and heat flows within the calorimeter as are found during an enthalpy measurement. Widely accepted values are:

$$\Gamma_{ice} = 270.48 \pm 0.03 \text{ J} \cdot g_{Hg}^{-1} \text{ [1] and}$$

$$\Gamma_{DPE} = 79.109 \pm 0.023 \text{ J} \cdot g_{Hg}^{-1} \text{ [56]}.$$

Preparation of the calorimeter for a heat measurement is preceded by creating an isothermal bath for the inner calorimeter elements. This bath usually consists of an ice-water mixture for the ice calorimeter and a stirred, thermostatted liquid bath for the DPE calorimeter. Then, a fraction of the working substance is solidified by introducing some refrigerant into the sample tube. After this, the calorimeter is allowed to reach a steady thermal state. For a well-designed calorimeter the working substance will undergo steady-state melting or freezing at a rate equivalent to a net energy interchange rate between the inner calorimeter elements and the isothermal bath on the order of 100 µW or less. The magnitude and direction of this energy interchange ('heat leak') are determined by the temperature difference between the working substance and the isothermal bath and by heat flow through the calorimeter supports as well as by the pressure exerted on the working substance by the hydrostatic head of the mass-accounting system.

A single heat measurement requires only five primary calorimeter data: the steady-state 'heat leak' before and after a heating period (additional heat comes from either a sample or calibrating heater during heating period), the mercury reference mass at the accounting system before and after the heating period, and time. The heating period is defined as the minimum time interval from the onset of heating by sample or by calibrating heater to re-establishment of a steady-state 'heat leak.' The measured heat can then be calculated from the expression,

$$Q_{meas} = \frac{\Delta m \pm q(\Delta \tau)}{\Gamma} \tag{10}$$

where Δm is the change in mercury reference mass, q is the observed calorimeter heat leak expressed in the units mass of mercury (measured at the accounting system) per unit time, $\Delta \tau$ is the length of the heating period, and Γ is the calorimeter calibration constant. One may see [50] for a detailed discussion of the measurement of q. The main contribution to the measured heat is contained in the Δm term and can be anywhere from 100-10000 J. The heat-leak correction is small, typically less than 0.25 J, and its sign depends on the direction of net heat exchange between the working substance and its environment in the steady state.

Many design variations of phase-change calorimeters have appeared in the literature. For further material on ice calorimeters, the interested reader is referred to [57-69] in addition to the sources already cited above. DPE calorimeters are treated in [70-72].

The contribution of the calorimeter proper to the overall inaccuracy of a measured enthalpy datum contains contributions from inaccuracy in the calibration constant, in the mercury mass measurements, and in the estimation of the heat-leak correction. It is not likely to exceed 0.05-0.10% and below ~1300 K should be comparable with the combined error associated with sample temperature measurement and compensation for heat lost during translation of the specimen into the calorimeter. The smallest obtainable overall inaccuracy of enthalpy data in the range 273 K to ~1300 K, measured with a solid-liquid phase change calorimeter on a pure substance having no phase transitions, should be about 0.1-0.2%. Above 1300 K, errors in temperature measurement can be expected to increase and predominate.

Liquid-Vapor Systems: The liquid-vapor phase-change calorimeter is illustrated schematically in Fig. 6.7. It depends for its measuring principle on the vaporization of the working substance by a heat source (either a sample for enthalpy measurement or a calibrating heater). the mass of vaporized material is obtained either by direct weighing after condensation [73] or by calculating it from volumetric flow, pressure, and temperature data on the vaporized material which must be continuously recorded during the course of an experiment [74]. The calorimeter calibration (i.e., the energy required to vaporize unit mass of the working substance) is measured using the same type of stable current source to energize the calibrating heater as in other calorimeter types discussed above, and the same precision instrumentation is used to measure heater voltage and current. The total heat transferred during an enthalpy measurement is equal to the product of this calorimeter calibration and the total mass of working substance vaporized during the enthalpy measurement. If there is a significant background flow due to steady-state heat leak into the calorimeter, this must be independently determined and applied as a subtractive correction to the total measured heat.

The liquid-vapor calorimeter has appeared rather infrequently in the recent literature. This is perhaps attributable to the complications of the instrumentation for volume or mass measurement. Two features are in its favor: (1) construction of the calorimeter can be relatively simple and may consist of two concentric Dewars to contain the working substance and its isothermal bath, and (2) use of an inert working substance such as argon provides a desirable environment for high-temperature systems which might otherwise react with the calorimeter atmosphere. Most often, low-temperature refrigerants such as argon [74], nitrogen [75], or one of the chloro-fluoro-methane series ('Freons') [76,77] have been employed. One example of a boiling-water drop calorimeter has appeared [73].

The limited amount of data reported from liquid-vapor phase-change calorimeters show them to have apparently an inaccuracy about twice that of isoperibol block or solid-liquid phase-change calorimeters used in the same temperature range.

6.4. ANALYSIS AND PRESENTATION OF DROP-CALORIMETRIC DATA

The utility of any series of relative enthalpy data obtained through drop calorimetry, aside from its practical engineering value, lies in ones ability to extract from it the magnitude and temperature dependence of certain thermodynamic functions for the substance of interest.

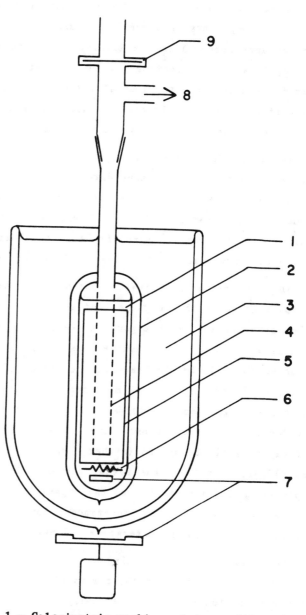

1 - Calorimetric working substance (liquid)
2 - Insulating Dewar
3 - Isothermal bath
4 - Screen sample retainer
5 - Radiation shield
6 - Calibrating heater
7 - Magnetic stirring
8 - Vapor to flow/mass measuring apparatus
9 - Radiation shutter

Figure 6.7. Phase-change calorimeter (liquid-gas).

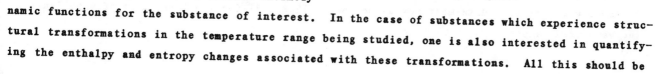

In the case of substances which experience structural transformations in the temperature range being studied, one is also interested in quantifying the enthalpy and entropy changes associated with these transformations. All this should be carried out in a way relating the smoothed data to the measured data with the aid of statistical and random components of overall inaccuracy. If related thermodynamic data for the substance of interest have appeared in the literature, it is reasonable to expect that these be considered, where applicable, in treatment of the present data. At any rate, few data exist in isolation and relating or comparing experimental enthalpy or heat-capacity data to existing literature data is an important part of their presentation. This section discusses some, though by no means all, of the considerations and methods which play a part in the treatment of drop-calorimetric data.

The starting point of all drop-calorimetric data treatment lies in the calculation from the raw, measured relative enthalpy data, of the enthalpy for the substance under investigation relative to its enthalpy at the selected reference temperature. The usual reference temperature assumed in the treatment of drop-calorimetric data is either 273.15 K or 298.15 K. It is assumed in the following that repeatability at the calorimeter terminal temperature has been established for the thermodynamic state of the sample being measured.

The raw, measured data will consist of relative enthalpy values for the sample of interest, perhaps together with a capsule. If the sample is encapsulated, there will also be relative enthalpy data for the empty capsule corresponding to the same or nearly the same furnace temperatures as were used for the sample measurements. All these enthalpy data will have already been corrected for calorimeter heat leak or other effects specific to the type of calorimeter used. A correction is applied to each measured enthalpy datum for which the terminal temperature during the equilibration period after a heat measurement is different from the assumed reference temperature (as in isoperibol calorimetry). This correction is calculated as the product of the total heat capacity of the system (either sample with capsule or empty capsule alone) and the temperature difference between the actual terminal temperature and the reference temperature. If the sample heat capacity or that of the capsule components is not known, estimated heat-capacity values must be used. For many substances, heat-capacity data may be found in one of the reliable data compilations in the literature [78-81].

Next, all relative enthalpy data on the sample are corrected for heat lost during the translation from the furnace to the calorimeter or, if a capsule was used, for the contribution of the empty capsule to the relative enthalpy. For unencapsulated samples, a direct calculation of heat loss must be made for each furnace temperature and this loss added to each enthalpy datum measured at that furnace temperature. If encapsulated samples are used, subtraction of the measured empty-capsule enthalpy from the measured enthalpy of the sample plus capsule will yield the desired sample enthalpy, corrected for heat loss and capsule contribution. The empty-capsule enthalpy data can be either measured directly - one enthalpy datum at each temperature for which sample-plus-capsule enthalpy data exist - or smoothed enthalpy data for the capsule can be obtained from a function fitted by the method of least squares to all existing empty-capsule enthalpy data. If the encapsulated sample has a significant vapor pressure at the temperatures of enthalpy measurement, the sample enthalpy data must now be corrected [2,82,83] for the total heat of vaporization of the vaporized sample.

At this point, one has a discrete set of enthalpy/temperature pairs (H,T) extending over some temperature range. It is assumed now that one is not dealing with a system in which chemical reaction or heat of radioactive decay contribute significant energy; also, that in the temperature range covered, the system does not experience change of phase, premelting, solid-solid reaction, or vacancy formation. In principle, it is possible to calculate the average heat capacity at the average temperature of two such pairs (identified, respectively, by the subscripts '1' and '2') by evaluating* $(H_2-H_1)/(T_2-T_1)$. In practice, this may produce a misleading large variance of the calculated heat-capacity values due to the difference in size of the temperature intervals and the computation of (H_2-H_1) as a relatively small difference of two much larger numbers.

A more satisfactory representation of the heat capacity can usually be obtained by first fitting some suitable function to the enthalpy data by the method of least squares and then differentiating this function to obtain the heat-capacity function. The algebraic form chosen for the enthalpy function should in general contain a minimum numbers of terms consistent with the requirement that this enthalpy function fit the data within the known precision of measurement. Usually, a low-order linear polynomial with temperature as the independent variable will suffice. At this point, experimental factors which might justify assigning relative statistical weights to the data or rejecting grossly deviating data should be considered.

When great difficulty is encountered in finding a suitable single function to represent the enthalpy over an entire temperature range an alternate approach may be useful. This consists of subdividing the larger temperature range and fitting the data within each sub-range with a suitable low-order polynomial, under the constraint that all pairs of polynomials defined over adjacent sub-ranges together with their low-order derivatives shall constitute continuous functions. This fitting technique is known as the 'Method of Splines' [85]. It provides great flexibility insofar as one can arbitrarily choose the sub-range boundaries and the form of the function within each sub-range. There exists a wide variety of software in available software libraries for performing least-squares regression and spline-fitting. As one instance, one may cite Ref. [86].

In order to derive absolute entropy values [cf. Eq. (7)] it is necessary to combine heat-capacity data derived from drop calorimetry above room temperature with data obtained on the same substance through heat-capacity calorimetry at low-temperatures. Only in exceptional cases will the two sets of heat-capacity data merge smoothly. Since the inaccuracy of drop-calorimetric data generally increases somewhat near to its reference temperature and is usually greater than the inaccuracy associated with heat-capacity calorimetry, some procedure for assigning relative statistical weights to the data of the two sets should be adopted. Fitting the combined data sets simultaneously with a spline function is the most satisfactory way to derive a composite heat-capacity function.

Enthalpy and heat-capacity data for materials containing low levels of impurity phases, having structural transformations in the temperature range of the drop-calorimetric data, exhibiting

---

*If this method is followed, it is necessary that account be taken of the fact that the <u>true</u> heat capacity at the mid-temperature of a temperature interval is likely to differ from the calculated <u>average</u> heat capacity due to curvature in the enthalpy function H(T). The calculation of this curvature correction is given in detail in [84].

premonitory phenomena (as premelting or vacancy formation), or in which heat of radioactive decay is present, require special treatment. Methods of treatment appropriate to data subject to these effects have been outlined in [1]. Reference [41] discusses at length the treatment of drop-calorimetric data involving heat of radioactive decay.

Each measured drop-calorimetric datum has associated with it, by the very nature of the measurement process, an unknown error. The data as finally presented are of limited value without some statement by the experimenter containing his beliefs about this error. However, the interpretation of statements related to the error in drop-calorimetric data requires particular care. One encounters 'accuracy' statements in the literature having justification covering the entire range from pure, unsubstantiated guesswork to detailed analyses which statistically combine all known and knowledgeably-estimated sources of error. Semantically speaking, one is concerned not with 'accuracy,' but with 'inaccuracy,' i.e., the small range of values (expressed, say, as a percent of the whole) centered about a measured datum which is considered to contain with some high probability the 'true' value. In the following, we indicate what we consider to be a desirable way to arrive at a value for inaccuracy of drop-calorimetric data. All of the concepts are drawn from Refs. [87-89].

Inaccuracy is a term we use for the combination of systematic and random components of error. A source of systematic error influences equally (in either an absolute or a relative sense) all data and arises, for instance, from gross error in the measured mass of sample or capsule, unidentified impurity phases in the sample, error in the calibration of measuring instruments (as pyrometers, voltmeters, standard resistors, analytical balances), etc. The experimenter should review painstakingly the entire enthalpy-measurement process and assign to each suspected error source a credible maximum percentage effect upon the enthalpy data. In the search for systematic error, it will be of special value to have enthalpy data on recognized standard reference materials such as Al_2O_3 [25] or Mo [50] measured at the same time as the substance under investigation. These estimates are then combined in quadrature, i.e., as the square root of the sum of squared errors. Random error varies from datum to datum. It arises from the individual processes involved in the measurement of temperature, mass, or electric potential as well as from the inexact reproduction from experiment to experiment of temperature fields within the calorimeter and the resulting differences in 'heat leak.' In drop calorimetry, this class of error is observed in the variance within repeated measurements of enthalpy data at a single temperature or in the random deviation of measured enthalpy data from a fitted curve in a temperature interval within which no phase transitions are to be expected.

The conventional statistical measure of these deviations is the computed standard deviation, s [90], of data points from the fitted curve. Now in conjunction with a determinate number of data and a fitting function with a fixed number of disposable constants, there is associated a certain probability that any of the measured data lie within a range $\pm s$ around the 'predicted value' calculated from the fitted function. The probability that the measured data lie within some larger range $\pm t \cdot s$, where $t > 1$, will be correspondingly greater. In order that the predicted values given by the fitted function be correct with a high degree of probability, one chooses a value of t in the range 2-3 [91]. The (overall) inaccuracy is then calculated as the arithmetic sum of the root-squared systematic error and the random error, t·s, and is stated together with the probability level one has chosen.

6.5. REFERENCES

1. Douglas, T.B. and King, E.G., 'High-Temperature Drop Calorimetry,' in Experimental Thermodynamics (McCullough, J.P. and Scott, D.W., Editors), Vol. I, Plenum Press, New York, NY, 1968.

2. Osborne, N.S., J. Res. Natl. Bur. Stand., 4, 609-29, 1930.

3. Douglas, T.B., J. Res. Natl. Bur. Stand., 73A(5), 451-70, 1969.

4. Rossini, F.D., J. Chem. Thermodyn., 2, 447-59, 1970.

5. Bonnell, D.W., Rice Univ., Ph.D. Thesis, 1972.

6. Stephens, H.P., High Temp. Sci., 6, 156-66, 1974.

7. McAdams, W.H., Heat Transmission, McGraw-Hill, New York, NY, 265, 1954.

8. Ginnings, D.C., Douglas, T.B., and Ball, A.F., J. Res. Natl. Bur. Stand., RP2110, 45, 23-33, 1950.

9. Braun, M., Kohlhaas, R., and Vollmer, O., Z. Angew. Phys., 25(6), 365-72, 1968.

10. Stout, N.D., Mar, R.W., and Boo, W.O.J., High Temp. Sci., 5, 241-51, 1973.

11. Chekhovskoi, V.Ya. and Sheindlin, A.E., Prib. Tekh. Eksp., 1, 197-9, 1963.

12. Guseva, E.A., Bolgar, A.S., Gordienko, S.P., Gorbatyuk, V.A., and Fesenko, V.V., High Temp. (Engl. Transl.), 4(5), 609-12, 1966.

13. Sheindlin, A.E., Chekhovskoi, V.Ya., and Reshetov, L.A., Prib. Tekh. Eksp., 2, 153-6, 1963.

14. West, E.D. and Ishihara, S., 'A Calorimetric Determination of the Enthalpy of Graphite from 1200 to 2600 K,' in Advances in Thermophysical Properties at Extreme Temperatures and Pressures (Gratch, S., Editor), ASME, New York, NY, 146-51, 1965.

15. Morizur, G., Radenac, A., and Cretenet, J.S., High Temp.-High Pressures, 8, 113-20, 1976.

16. Dennison, D.H., Gschneider, K.A., Jr., and Daane, A.H., J. Chem. Phys., 44(11), 4273-82, 1966.

17. Fredrickson, D.R., Kleb, R., Nuttall, R.L., and Hubbard, W.N., Rev. Sci. Instrum., 40(8), 1022-5, 1969.

18. Steffen, H. and Wollenberger, H., Rev. Sci. Instrum., 44(8), 937-43, 1973.

19. Proks, I., Eliasova, M., Zlatovsky, I., and Zauska, J., Silikaty (Prague), 21(3), 253-64, 1977.

20. Andrews, D.H., Lynn, G., and Johnston, J., J. Am. Chem. Soc., 48, 1274-87, 1926.

21. Umino, S., Sci. Rep. Tohoku Imp. Univ., 1(15), 597-617, 1926.

22. Conway, J.B. and Hein, R.A., 'Enthalpy Measurements of Solid Materials to 2400 C by Means of a Drop Technique,' in Advances in Thermophysical Properties at Extreme Temperatures and Pressures (Gratch, S., Editor), ASME, New York, NY, 131-7, 1965.

23. Eitel, W., in Thermochemical Methods in Silicate Investigation, par. 111 ffg., Rutgers Univ. Press, New Jersey, 1952.

24. Barton, A., in A Textbook on Heat, Chapt. II, Longmans, Green and Co., New York, NY, 1938.

25. NBS Certificate, Standard Reference Material 720, Aug. 26, 1970.

26. Marchidan, D.I. and Ciopec, M., Rev. Roum. Chim., 16(8), 1145-54, 1971.

27. Chekhovskoi, V.Ya., Tarasov, V.D., and Zhukova, I.A., High Temp. (Engl. Transl.), 12(6), 1088-91, 1974.

28. Dworkin, A.S. and Bredig, M.A., J. Phys. Chem., 64, 269-72, 1960.

29. MacLeod, A.C., Trans. Faraday Soc., No. 530, 63 (Pt. 2), 300-10, 1967.

30. Blachnik, R., Igel, R., and Wallbrecht, P., Z. Naturforsch., 29a, 1198-201, 1974.

31. Southard, J.C., J. Am. Chem. Soc., 63, 3142-6, 1941.

32. Kirillin, V.A., Sheindlin, A.E., and Chekhovskoi, V.Ya., Inzh.-Fiz. Zh., 4(2), 3-17, 1961.

33. Fomichev, E.N., Kandyba, V.V., and Kantor, P.B., Izmer. Tekh., 5, 15-8, 1962.

34. Marchidan, D.I. and Ciopec, M., Acad. Repub. Rom. Stud. Cercet. Chim., 17(9), 737-62, 1969.

35. Sheindlin, A.E., Belevich, I.S., and Kozhevnikov, I.G., High Temp. (Engl. Transl.), 8(3), 563-5, 1970.

36. Lindroth, D.P. and Krawza, W.G., U.S. Bur. Mines Rept. Invest. 7503, 25 pp., 1971.

37. Buchnev, L.M., Volga, V.I., Dymov, B.K., and Markelov, N.V., High Temp. (Engl. Transl.), 11(6), 1072-6, 1974.

38. Chemykin, V.I., Zedina, I.N., and Vaisburd, S.E., Inzh.-Fiz. Zh., 34(5), 870-4, 1978.

39. Chekhovskoi, V.Ya. and Berezin, B.Ya., High Temp. (Engl. Transl.), 8(6), 1244-6, 1970.

40. West, E.D. and Churney, K.C., J. Appl. Phys., 39(9), 4206-15, 1968.

41. Oetting, F.L., J. Chem. Thermodyn., 2, 727-39, 1970.

42. Gunn, S.R., J. Chem. Thermodyn., 3, 19-34, 1971.

43. Olette, M., 'An Adiabatic Dropping Calorimeter for Enthalpy Measurements at High Temperatures. The Heat Content of Silicon from 1200 to 1550°C,' in The Physical Chemistry of Steelmaking (Elliot, J.F., Editor), [Proc. Conf. at Endicott House, Dedham, MA, May 28-June 3, 1956], John Wiley, New York, NY, 18-26, 1958.

44. Levinson, L.S., Rev. Sci. Instrum., 33, 639-42, 1962.

45. Tydlitat, V., Blazek, A., Halousek, J., Pietsch, K., and Prihoda, K., Czech. J. Phys., B21(8), 817-22, 1971.

46. Spedding, F.H. and Henderson, D.C., J. Chem. Phys., 54(6), 2476-83, 1971.

47. Grønvold, F., Acta Chem. Scand. Ser. A, 26, 2216-22, 1972.

48. Efremova, R.I. and Matizen, E.V., Metrologiya, 8, 56-63, 1972.

49. Arkhipov, V.A., Gutina, E.A., Dobretsov, V.N., and Ustinov, V.A., Radiokhimiya, 16(1), 123-6, 1974.

50. Ditmars, D.A., Cezairliyan, A., Ishihara, S., and Douglas, T.B., NBS Special Publication 260-55, 80 pp., 1977 (for sale by the Superintendent of Documents, U.S. Govt. Printing Office, Washington, D.C. 20402).

51. Maier, C.G., J. Phys. Chem., 34, 2860-8, 1930.

52. Vasil'ev, Ya.V. and Neermolov, A.F., Russ. J. Phys. Chem., 46(4), 589-90, 1972.

53. Ginnings, D.C. and Corruccini, R.J., J. Res. Natl. Bur. Stand., 38, RP1796, 593-600, 1947.

54. Jessup, R.S., J. Res. Natl. Bur. Stand., 55(6), RP2636, 317-22, 1955.

55. Lakhanpal, M.L. and Parashar, R.N., Indian J. Chem., Sec. A, 8(4), 368-9, 1970.

56. Davies, J.V. and Pritchard, H.O., J. Chem. Thermodyn., 4, 9-22, 1972.

57. Swietoslawski, W., in Microcalorimetry, Chapt. V, Reinhold, New York, NY, 1946.

58. Leake, L.E. and Turkdogan, E.G., J. Sci. Instrum., 31, 447-9, 1954.

59. Oriani, R.A. and Murphy, W.K., J. Am. Chem. Soc., 76, 343-5, 1954.

60. Deem, H.W. and Lucks, C.G., 'An Improved, All-Metal, Bunsen-Type Ice Calorimeter,' in Proc. 13th Annual Instrument-Automation Conference, Paper No. PPT-4-58, Vol. 13, Pt. II, Instrument Society of America, 1958.

61. Lucks, C.F., Deem, H.W., and Wood, W.D., Am. Ceram. Bull., 39(6), 313-9, 1960.

62. Spedding, F.H., McKeown, J.J., and Daane, A.H., J. Phys. Chem., 64(3), 289-94, 1960.

63. Smith, D.F., Kaylor, C.E., Walden, G.E., Taylor, A.R., Jr., and Gayle, J.B., U.S. Bur. Mines Rept. Invest. 5832, 1961.

64. Neel, D.S., Pears, C.D., and Oglesby, S., Jr., U.S. Air Force Technical Documentary Rept. WADD-60-924, 1962. [AD 275 536]

65. Opdycke, J., Gay, C., and Schmidt, H.H., Rev. Sci. Instrum., 37(8), 1010-3, 1966.

66. Welty, J.R. and Wicks, C.E., U.S. Bur. Mines Rept. Invest. 6028, 1962.

67. Maglic, K. and Parrot, J.E., Bundesministerium fur Bildung und Wissenschaft, Forschungs- bericht K 70-01, 263-82, 1970.

68. Ditmars, D.A., Int. J. Appl. Radiat. Isot., 27, 469-90, 1976.

69. Zakurenko, O.E. and Kuz'michev, V.M., Prib. Tekh. Eksp., 6, 167-8, 1978.

70. Giguere, P.A., Morissette, B.G., and Olmas, A.W., Can. J. Chem., 33, 657-64, 1955.

71. Hultgren, R., Newcombe, P., Orr, R.L., and Warner, L., 'Diphenyl Ether Calorimeter for Meas- uring High-Temperature Heat Contents of Metals and Alloys,' in The Physical Chemistry of Metallic Solutions and Intermetallic Compounds, NPL Symposium No. 9, Vol. I, Paper 1H, HMSO, London, England, 1958.

72. Peters, H. and Tappe, E., Monatsber. Dtsch. Akad. Wiss. Berlin, 10(2), 88-101, 1968.

73. Shpil'rain, E.E., Kagan, D.N., and Barkhatov, L.S., High Temp-High Pressures, 4, 605-9, 1972.

74. Stephens, H.P., High Temp. Sci., 6, 156-66, 1974.

75. Gilbreath, W.P. and Wilson, D.E., Rev. Sci. Instrum., 41(7), 969-73, 1970.

76. Stephens, H.P., Sandia Labs. Rept. SAND-76-0635, 21 pp., 1977.

77. Stephens, H.P., Sandia Labs. Rept. SAND-77-0105, 38 pp., 1978.

78. Stull, D.R. and Prophet, H. (Editors), JANAF Thermochemical Tables, Second Ed., NSRDS-NBS 37, 1971. (For sale by the Superintendent of Documents, U.S. Govt. Printing Office, Washington, D.C., 20402. Semiannual supplements to this compilation are issued by and obtainable from Dow Chemical U.S.A., Midland, MI 48640.)

79. Hultgren, R., Desai, P.D., Hawkins, D.T., Gleiser, M., Kelley, K.K., and Wagman, D.D., Se- lected Values for the Thermodynamic Properties of the Elements, American Society for Metals, Metals Park, OH, 1973.

80. Kelley, K.K., Contributions to the Data on Theoretical Metallurgy XIII. High-Temperature Heat-Content, Heat-Capacity, and Entropy Data for the Elements and Inorganic Compounds, U.S. Bureau of Mines Bull. 584, 1960.

81. Touloukian, Y.S. and Ho, C.Y. (Editors), Thermophysical Properties of Matter - The TPRC Data Series, 13 Vols., IFI/Plenum Press, New York, NY, 1970.

82. Hoge, H.J., J. Res. Natl. Bur. Stand., 36, 111-8, 1946.

83. Westrum, E.F., Jr., Furukawa, G.T., and McCullough, J.P., 'Vaporization Corrections,' in Ex- perimental Thermodynamics (McCullough, J.P. and Scott, D.W., Editors), Vol. I, Plenum Press, New York, NY, 191-4, 1968.

84. Osborne, N.S., Stimson, H.F., Sligh, T.S., and Cragoe, C.S., BS Sci. Pap. No. 501, 20, 65- 110, 1925.

85. Greville, T.N.E., Theory and Applications of Spline Functions, Academic Press, New York, NY, 1969.

86. Lathrop, J.F., 'DATFIT User's Manual - An Interactive Constrained Least-Squares Data-Fitting Program,' Sandia Labs. Technical Publication SAND 80-8204, 1980.

87. Natrella, M.G., Experimental Statistics, Natl. Bur. Stand. Handbook 91, 1963.

88. Dixon, W.J. and Massey, F.J., Jr., Introduction to Statistical Analysis, McGraw-Hill, New York, NY, 1957.

89. Ku, H.H. (Editor), Precision Measurement and Calibration, Natl. Bur. Stand. Spec. Pub. 300, Vol. I, 1969.

90. Ku, H.H. (Editor), Precision Measurement and Calibration, Natl. Bur. Stand. Spec. Pub. 300, Vol. I, 73-8, 1969.

91. Ku, H.H. (Editor), Precision Measurement and Calibration, Natl. Bur. Stand. Spec. Pub. 300, Vol. I, 296-330, 1969.

CHAPTER 7

Levitation Calorimetry

*D. W. BONNELL, R. L. MONTGOMERY, B. STEPHENSON,
P. C. SUNDARESWARAN, and J. L. MARGRAVE*

7.1. INTRODUCTION

The determination of high-temperature thermodynamic properties is of fundamental importance in chemistry, physics, and engineering both from a practical standpoint and from the viewpoint of one interested in a basic understanding of the behavior of matter. The temperature region above 2000 K has been largely a no-man's land, where even landmarks are rare. Basic properties of many chemical elements are often obtained by extrapolations. For compounds, they are almost totally unknown.

The nature of the liquid state is known primarily from studies at or near room temperature and from measurements above the melting points of only a few high-melting substances. Theoretical treatments of the liquid state are so unsatisfactory that available heats of fusion of metals are often only estimated using the Tammann correlation [1], $\Delta S_{fus} = 2.2\text{--}2.3$ cal$\cdot$K$^{-1}\cdot$mol^{-1}. Exact theoretical studies of the liquid state face many problems. The virial formulation breaks down in the liquid case owing to the singularity which must occur at $T = T_c$, the condensation temperature. A solution from first principles involves a direct evaluation of the partition function of the canonical or grand canonical ensemble in statistical thermodynamics. Such an evaluation requires detailed calculation of the energy levels of the system as they depend upon the positions and velocities of each of the atoms. Each atom may interact significantly with numerous others. High-speed computers have been applied [2–5,93] to systems of several hundred particles, but a direct solution for macroscopic systems seems unlikely even with present-day facilities. Many theories based on model descriptions of the liquid state exist [6–8] and show the qualitative features desired, but, so far, they have not been significantly more successful than classical estimation methods [9,10] for quantitative calculations of high-temperature thermodynamic functions.

The conspicuous dearth of reliable experimental high-temperature data for liquids can be attributed primarily to a single problem: 'At high temperatures, everything reacts with everything!' [11]. Thus, in addition to merely reaching higher temperatures, one must isolate the materials to be studied from potential reactants. For solids, this is not too difficult. Solid-solid reactions are generally slower, and often a container can be found which is inert to the substance in question, or the material itself can be used without a container. Liquids, on the other hand, must be contained, at least for static measurements. The container, then, creates a very real problem, resulting in systematic absolute errors which must be recognized and corrected.

These corrections are rarely negligible and can often contribute a sizable percentage of the measured quantity. For instance, in drop calorimetry, where measurements of total heat content are made, the container always contributes a substantial part of the heat evolved, even if reactions are negligible.

The need for a 'containerless' calorimetric technique has been apparent, and the various levitation techniques offer an ideal solution. It is this combination of levitation and calorimetry, i.e., levitation calorimetry, which has greatly extended our experimental knowledge of the thermodynamic properties of liquid metals.

7.2. LEVITATION METHODS

Levitation, i.e., supporting a sample in opposition to gravity so that it floats freely in a vacuum or in a gas, can be accomplished by a variety of techniques [12-15]. The electrical conductivity, density and vapor pressure of the sample, and the temperature desired are among the parameters which lead one to select a specific technique for a given experiment. In general, electromagnetic levitation is most convenient for high-temperature studies on earth, i.e., in a 1 g environment [12]. In special circumstances, it may be desirable to use acoustic levitation with standing sound waves [14], aerodynamic (Bernoulli) levitation (as practiced, for example, in carnival ping-pong ball displays) [13], electrostatic levitation (as in the Millikan oil-drop experiments), or a sounding rocket or drop-tower experiment (free-fall of liquid drops to yield spherical pellets), etc. In the space environment, far from earth, one has only microgravity and everything levitates, i.e., floats freely unless tied down or provided an external acceleration. Supports and containers are unnecessary, but 'drop calorimetry' will require special techniques to transfer the sample to the calorimeter.

7.2.1. Electromagnetic Levitation

Electromagnetic levitation is an eddy current phenomenon and has been known for many years. The presence of a conductor near a source of alternating current results in an opposing eddy current being generated in the conductor. The interaction of the magnetic field induced in the conductor with that of the source results in a repulsive force. The geometric form of the source determines the effect of this force on the sample. For a suitable choice of geometry, this force can be directed so as to oppose gravitational forces and, thus, to 'float' or maintain the conductor in a desired position. The basic geometry of Fig. 7.1 is the starting point for most levitation coil designs. The upper counter-wound coil turn serves to create a minimum field point to ensure stability and remove the constraint of exact balance between lifting force and gravity.

In 1952, Wroughton et al. [12] described a levitation technique which utilized an electromagnetic field for suspending and heating conducting materials. It offered a 'clean' alternative to the use of a containing vessel in the preparation of high-purity titanium metal.

Comenetz et al. [16-18] presented a formula for this type of configuration as a function of sample and coil characteristics (Fig. 7.1). If $d(g \cdot cm^{-3})$ is the sample density, $m(g)$ is its mass,

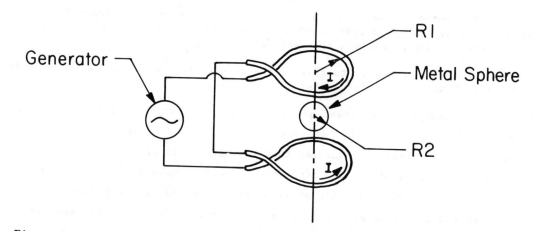

Figure 7.1. Simplified schematic showing principles of levitation apparatus [35].

ρ(ohm·cm) is the sample resistivity,   (Hz) is the impressed frequency, I (rms A) is the coil current, g (980 cm·s^{-2}) is the acceleration due to gravity, R_1 (cm) is a properly-selected characteristic dimension of the coil (typically, its radius), and R_2 (cm) is the sample radius, the levitation force to sample mass ratio can be expressed as

$$\frac{F_L}{m} = \delta \; \frac{I^2 G(X)}{dR_1^3}$$

(1)

where $G(X) = 1 - (3/4X)[(\sinh 2X - \sin 2X)/(\sinh^2 X + \sin^2 X)]$ and the dimensionless quantity, X, given by $X = 2\pi R_2 \sqrt{(\nu/10^9\rho)}$ represents the ratio of sample radius to 'skin depth' [19]. The quantity δ is a dimensionless parameter depending on coil shape and the point in the coil where the sample is suspended. For a given coil shape and point of suspension, δ can be determined by a single experiment in which the other parameters are measured, such as at low current where the lifting force is measured by a balance. More recently, Frost and Chang [92] presented a new theoretical analysis of levitation forces and suggested a different formulation. Hatch and Smith [20] give a generalized formula for any shape in any coil system in terms of a potential function which is determined by measurements of the inductance of the coil system.

These formulations give moderately good results for small samples in relatively large coils [21], but when the sample radius becomes comparable with that of the coils the errors become large. These calculations are suitable for determining feasibility but, in application, empirical rules are often more useful. Additional factors apply to levitation of liquids. For instance, in the levitation of molten Au, the decisive factor [22] seems to be a high ratio of density to surface tension and/or viscosity, since gold can be lifted but is very difficult to contain while molten. Sidorov and Mezdrogina [23] have reported a semiempirical equation connecting surface tension and density with the voltage (across a coil) required to levitate a given mass of material.

Empirical considerations for optimizing coil design can be summarized as follows:

(1) Sharp changes in field gradient near the point of sample flotation increase the lifting force and maximum temperature.

(2) The closer the sample is physically to the coil, the greater the maximum temperature. A design in which the sample floats completely out of the coil (in the gap between upper and lower turns) provides maximum lift with minimum heating.

(3) The nearer the coil diameter is to that of the sample, the better the power transfer and, thus, the higher the maximum temperature.

The levitation technique provides other advantages as well. High temperatures (in excess of 3000 K) are attained very quickly. Stable temperatures can often be reached in less than a minute. Electromagnetic stirring generally ensures that liquid samples are homogeneous and that temperature gradients are minimized. Since the reaction products of metals with gases are often more volatile than the bulk metal, electromagnetic stirring usually helps purify the metals. In actual operation, most metals are recovered with a surface appearance indicating that even the surface oxidation present before levitation has disappeared.

In order to take full advantage of the levitation technique, temperatures must be measured by non-contact means. In the case of liquids, it is not feasible to create blackbody conditions, so spectral emissivities must be independently measured to establish the temperature by optical pyrometry. The high field gradients in the coil area can cause significant temperature gradients, particularly in unsymmetrical samples; so temperature determination is more difficult for solids.

The first adaptation of the levitation technique for isothermal drop calorimetry was by Jenkins et al. [24], who reported construction of a Bunsen ice calorimeter to be used in conjunction with levitation heating. Satisfactory results were not obtained because of difficulties with temperature measurements. Apparently the project was not pursued further. At Rice University, our long experience in traditional drop calorimetry and high-temperature thermodynamics led us to undertake the development of a 'levitation calorimeter' which was first operated successfully in 1968 [25]. The first experimental data were published in 1970 [26,27]. At essentially the same time in the USSR, a system for levitation calorimetry was developed [28-30].

In the period 1970-1982 there was a growing interest in the levitation technique* and a recognition of its versatility so that in November, 1981, an entire session of the Materials Research Society Conference was devoted to reports of levitation research programs [31]. Table 7.1 shows the breadth of techniques currently used for levitation (including the utilization of microgravity environments in space) and some of the kinds of measurements/applications under study. The major emphasis of this chapter will be on calorimetric applications of levitation techniques.

7.2.2. Laboratory Levitation Systems

Electromagnetic levitation studies have been reported with alternating current (AC) frequencies ranging from 60 Hz to 25 MHz and power inputs up to at least 100 kW. Industrial systems for materials processing could utilize megawatts of power for levitation heating. Most levitation calorimetry has been done with electronic oscillator systems at frequencies of 100-500 kHz and power in the 5-25 kW range.

---

*For example, NASA has activated several advisory committees for studying electrostatic, acoustic, electrodynamic, and other levitation methods, as well as investigating possible applications on earth and in space.

Table 7.1. Current Levitation Research Programs [91]

Levitation Technique	Active Investigators	Location	Recent Research Publications/Interests
(A) Electromagnetic Levitation	J.L. Margrave and Assoc.	Rice University Houston, TX USA	1. Thermodynamic Properties of Liquid Metals
	D.W. Bonnell	U.S. Natl. Bur. Standards Gaithersburg, MD USA	2. Emissivities/Reflectivities
	B.Ya. Chekhovskoi A.E. Sheindlin and Assoc.	High Temperature Institute Moscow, USSR	3. Densities of Liquid Metals
			4. Calorimeter Design
	R.G. Bautista	Iowa State University Ames, IO USA	5. Thermodynamic Properties of Alloys
	H.P. Stephens	Sandia Laboratories Albuquerque, NM USA	6. Theory of Levitation Devices
			7. Crystal Growth
	M. Frohberg and Assoc.	Technical Univ. Berlin Berlin, West Germany	8. Industrial Applications of Levitation
	R.T. Frost and Assoc.	General Electric Co. Philadelphia, PA USA	
(B) Electrostatic Levitation	D.D. Elleman M. Saffren and Assoc.	Jet Propulsion Laboratory Pasadena, CA USA	1. Manipulation/Positioning in Vacuum
	C.D. Hendricks	Lawrence Livermore Lab. Livermore, CA USA	2. Coating of Small Particles
(C) Aerodynamic Levitation	P.C. Nordine and Assoc.	Midwest Research Inst. Kansas City, MO USA (formerly Yale Univ.)	1. Kinetics of High-Temperature Reactions
	E.C. Ethridge and S.L. Dunn	Marshall Space Flight Center, NASA Huntsville, AL USA	2. Thermal Gradients in Levitated Systems
			3. Temperature Measurement Techniques

Table 7.1. Current Levitation Research Programs [91] (Continued)

Levitation Technique	Active Investigators	Location	Recent Research Publications/Interests
(D) Acoustic Levitation	W. Oran and Assoc.	Marshall Space Flight Center, NASA Huntsville, AL USA	1. Design of Apparatus 2. Heating of Samples to ~900°C
	Intersonics, Inc.	Northbrook, IL USA	3. Glass-Formation Processes
	T.G. Wang, M. Saffren, D.D. Elleman, and Assoc.	Jet Propulsion Laboratory Pasadena, CA USA	4. Transport of Molten Samples 5. Coating of Microballoons
(E) Molecular Beam Levitation	C.D. Hendricks	Lawrence Livermore Lab. Livermore, CA USA	1. Coating of Small Particles
(F) Drop Tube Studies	NASA Scientists and Staff	Marshall Space Flight Center, NASA Huntsville, AL USA	1. Super-Cooling and Solidification Levitation 2. Ex-Solution of Gases and Bubble Formation
	Jet Propulsion Lab Staff	Pasadena, CA USA	3. Super-Cooling Phenomena
	A.J. Drehman and D. Turnbull	Harvard University Cambridge, MA USA	4. Kinetics of Crystallization
(G) Aircraft and Rocket Flights	NASA Scientists and Staff	KC-135 Cargo Planes, VC and SPAR Rockets in Flight	1. Equilibration of Liquid Drops 2. Containerless Processing of Materials
(H) Levitation Experiments in Space	NASA Scientists and Sponsored Researchers	NASA Shuttle Flights	FUTURE EXPERIMENTS PLANNED 1. Containerless Processing in Space 2. GULP Calorimetry 3. Alloy Formation 4. Super-Cooling Phenomena 5. Kinetics of Crystallization

At Rice University, the main power supplies employed were 450 kHz units with 15–25 kW power
ratings from General Electric and from the Lepel Company. These units were originally designed
for heating large susceptors, and efficient coupling of the generator to the very small coils used
for levitation calorimetry required special apparatus. Since the oscillator functions as an AC
power supply, the impedance of the load must be near that of the supply for maximum power transmission to occur. Thus, the problem of efficient coupling is simply that of providing an impedance match of the load coil to the generator. For simple coils of standard design, e.g., one or
two cylindrical turns, heating loads which are fixed, the necessary impedances can be calculated.
However, this is rarely practical, even for simple heating without levitation, and final adjustments must still be made under operating conditions. In practice, tank circuit capacities and
inductances must be adjusted to obtain maximum output, and for a new work coil this is usually a
simple iterative process. The most difficult problems arise when it is desired to use very small
work coils which are poorly coupled to the load piece.

For example, the G.E. generator at Rice was designed to use work coils of about 10 turns at
7.6–10.2 cm (3–4 in.) diameter. An impedance matching transformer was made by winding 12 turns of
6.4 mm (1/4 in.) copper tubing as the primary on a convenient form 16.5 cm (6 1/2 in.) in diameter. The secondary was three turns of 13 mm (1/2 in.) copper tubing, and 10 layers of 0.2 mm
polyethylene film served as insulation from the primary. The secondary and work coil were supplied cooling water from a separate booster pump independent of the generator water supply. This
simple transformer allowed about 1.8 A of the 3.5 A maximum plate current, i.e., about a 50% coupling effectiveness. This performance was equal to that obtained with a commercial impedance
matching transformer obtained from G.E. for the generator.

Alternative coil designs (Figs. 7.2 and 7.3) have been tested and the effects of various
physical and geometrical parameters have been thoroughly explored. Current industrial technology
can provide apparatus capable of levitating multikilogram samples in sizes up to a few feet at
temperatures up to 2000°C.

7.2.3. Design of Levitation Coils

Even when a satisfactory impedance match to the generator for a particular work coil can be
achieved, the actual coil used will depend on the size of sample used and the desired temperature
range. The design of such coils has been largely empirical, with experience being the best guide.
There are several theoretical treatments of ideal coil designs which are helpful in developing
prototype coils.

The design problem can be separated into two areas of concern: (1) producing stable levitation, and (2) achieving the desired temperature. An ideal way to gain independent control of
these two factors would be to use two independent generators at two different frequencies, since
different frequencies provide different ratios of heating power to lifting force, with lifting
force favored at lower frequencies and heating favored at higher frequencies. There is a minimum
frequency which is effective in levitating a particular material; this frequency is below 10 kHz
for metals and becomes higher for poorer conductors. The heating effect for a given material

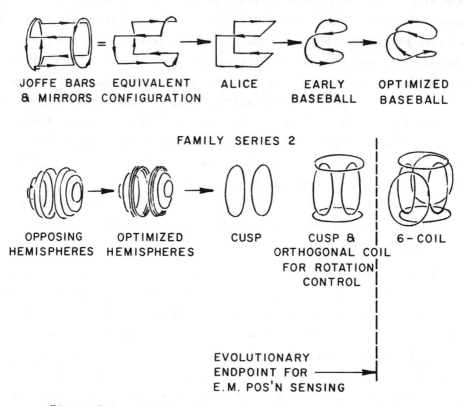

Figure 7.2. Relationship among various coil types [36].

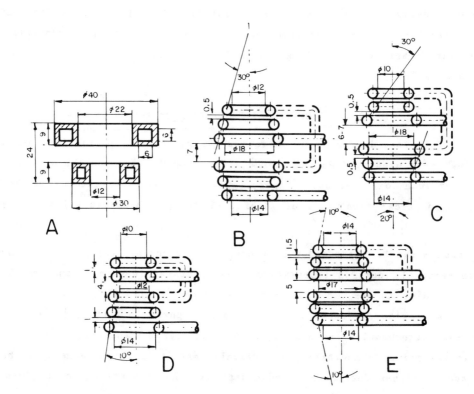

Figure 7.3. Various coil designs [28].

increases with increasing frequency. Equation (1) indicates approach to a constant levitation force as the frequency increases, but experience shows that very high frequencies are less effective in levitating metals*. Dual frequency levitation has been discussed in the literature [32].

Since the actual process of melting a material results in a change in the density and shape of the material and requires input of the heat of fusion, this is usually the most critical point in the procedure. A coil maintaining stable levitation at the time of melting will usually hold the sample at any temperature the system can attain.

The following factors determine levitation forces [24]:

(1) Field strength, which depends on number of turns per unit length, diameter of the coil, and the total current circulating in the coil;

(2) Field divergence, which depends on the shape and separation of the top and bottom coils;

(3) Sample properties, such as size, mass, shape, conductivity, liquid surface tension and viscosity, etc.;

(4) Position of the load with respect to the coil; and

(5) The frequency of the current in the coil.

The maximum temperature also depends on these factors--though not necessarily in the same way that the lifting force does--and, in addition, depends on the total power transmission and modes of heat loss. Since the lifting force per se is not the primary problem for small solid samples (1-100 g), the primary considerations in coil design are the temperature desired, the range of temperatures afforded by power control, sample stability, and the physical manipulation of the sample to be accomplished.

The levitation force is a direct function of field strength, i.e., the greater the field strength, the greater the lift. Thus, coils are usually wound of the smallest tubing that can still be adequately cooled. Round copper tubing, 3.2 mm (1/8 in.) OD is typically adequate at low power. For higher power (>15 kW), 3.2 x 6.4 mm (1/8 x 1/4 in.) oval tubing, made by flattening 4.8 mm (3/16 in.) round tubing, is necessary to provide passages large enough for cooling water flow. Current adjustment is provided by the phase shifting circuits of the generator. Since temperature is a function primarily of power transmitted, and transmission increases with proximity of sample to coil, reducing the power control setting may actually give a temperature increase as the sample settles into the coil. (This occurs with relatively heavy samples at relatively low currents.) The best compromise occurs when the weight plus repulsion from the upper coils just balance repulsion from the lower coils for a relatively wide range of total applied power.

Field divergence primarily affects the lift. Divergence is controlled by coil geometry. Figure 7.3 shows some typical coil designs. Since a good balance of lifting force to heating is provided by the solenoid design, and these coils are easiest to wind, this geometry is preferred, with modification toward the other forms being used where advantage dictates. Separation of the

---

*As the frequency increases, the impedance of an inductive load (such as a coil) increases. To maintain the same current, the voltage-current product, i.e., the generator capacity, would have to be increased. The possibility of arcing or other electrical discharges would be greater. This limits the effectiveness of very high frequencies in levitating metals.

upper and lower sets of turns is a convenient way of 'fine tuning' the divergence effect and is accomplished by experiment. Often the same coil can be used over a much wider range of conditions by slight adjustment of this coil gap. Magnetic field contours for a typical coil are shown in Fig. 7.4. From this diagram it is obvious that minor adjustments of both shape and spacing can have large effects on field gradients.

The sample itself is a major factor in coil design. The conductivity directly affects the final temperature since (at high frequencies) as the bulk resistance increases, the power dissipated as heat in the sample increases. At frequencies so low that there is almost no 'skin effect,' the relation between conductivity and heating is reversed. Since the magnitude of the circulating eddy currents goes down with increased resistivity, the lifting force decreases. The total mass of sample has a strong effect on the final temperature. Since mass increases with the cube of the radius, an increase in mass by a factor of 2 will only increase the radius by about 25%. Thus, the lifting force does not increase enough to restore the sample to an identical flotation point. This results in the sample occupying a lower point in the coil, causing it to intercept more of the field, and thus attaining a higher temperature. This is the same apparently paradoxical effect noted

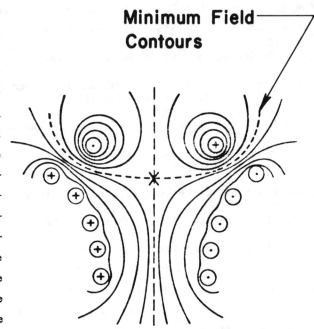

Figure 7.4.   Magnetic field contours in typical levitation apparatus [24,32].

previously, i.e., lowering the power control setting produces an increase in final temperature of the levitated sample. The option of changing the sample mass offers one of the more convenient ways to vary the temperature attainable in a particular coil.

Altering the coil size is a useful method of controlling the temperature characteristics of the system. However, a minimum coil size is reached when small sample oscillations may cause the sample to strike the coil. This can be alleviated with a minor loss in power delivered by slightly increasing the diameter of the top turn of the main coil. Design of the top reverse turns is also critical, and several techniques have been used with varying degrees of success. The minimum temperature (when other parameters are held constant) is achieved when the top turns are as large or larger than the main coil. Treverton and Margrave [33] report a novel adaptation of the above considerations by winding coils with inverted conical top turns. Increased power then forces the sample into this upper region where the field is shaped for more effective heating. An alternative method of accomplishing this effect is to increase the field divergence by winding top turns on a very small form, typically 6.4 mm (0.25 in.) ID. Finally, where the maximum power interception is important, winding the top coil down from the top (preserving the opposing field effect) produces coils where there is almost no gap between upper and lower turns. The top turns can be eliminated altogether by means of a 'dock'--a water-cooled ring above the main coil isolated from ground. Eddy currents are induced in this plate 180° out of phase with the main coil, providing the necessary minimum field point. For achieving high temperatures, however, the advantage of field coils carrying power is indispensable. In extreme cases, as in recent studies of liquid tungsten, it may be necessary to supplement the RF heating with electron beam [34,35], or with CO_2-laser heating for studies of Al_2O_3 [13].

Altering the current frequency offers an alternative and powerful technique for achieving the desired versatility in heating and levitating samples. For each sample, there is a range of frequencies which will provide adequate levitation forces. Resistive heating occurs even at low frequencies and generally increases as the frequency increases because of decreasing skin depth. Thus, by adjusting the RF frequency one has another option for temperature control.

Altering the environment of the sample changes the rate of conduction loss. Vacuum operation eliminates conduction losses and leads to higher maximum temperatures. However, in poor vacuum, corona discharges from the high-voltage coils may limit the use of this technique. In the Rice University system, argon gas at 0.1 MPa (one atmosphere) provides reasonably low thermal conductivity [32]. Helium has been used when lower temperatures were required with samples where the lifting force dictated a coil design which did not allow a sufficiently low temperature in argon. The increase in thermal conductivity obtained by using helium is about a factor of 10 at 0.1 MPa (one atmosphere) pressure. Gas cooling of levitated samples has been found to be very effective and liquid metals can be supercooled several hundred degrees, then nucleated and resolidified while levitated. One can inject the helium directly at the sample for maximum heat exchange when necessary.

7.3. CALORIMETER DESIGNS

Although the initial designs of apparatus for levitation calorimetry were implemented on a 'cut and try' basis, because the practicality of the technique had not been established, there have been extensive basic studies in recent years by Frost and associates at General Electric [36] and by others [37].

Several unique prototype designs/tests are being developed for experiments utilizing natural levitation in microgravity environments for containerless studies of high-temperature materials in a space laboratory. For example, a 'gulp calorimeter' to leap forward under pneumatic pressure and engulf the sample to be studied has been developed [38].

A 'Levitation Calorimeter System' such as that at Rice University [32] consists of the following components: (1) induction furnace (RF power supply), (2) levitation chamber, (3) radiation isolation gate, (4) calorimeter, block and isothermal jacket, (5) isothermal bath and controls, (6) block temperature readout, and (7) optical pyrometer. Figure 7.5 shows a schematic of this device. Figure 7.6 shows a similar system [39] employing an isothermal (phase change of argon) calorimeter instead of the isoperibol type shown in Fig. 7.5. Systems similar to the Rice system

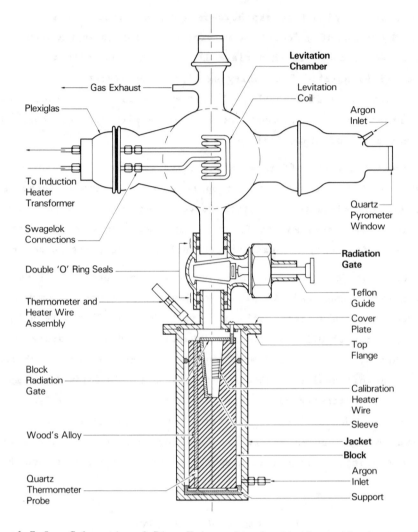

Figuré 7.5.   Schematic of Rice University levitation calorimeter [32].

have been constructed and employed by Stretz and Bautista [40], by Chekhovskoi and associates [28,29], and by Betz and Frohberg [41]. Figures 7.7 and 7.8 illustrate two additional levitation approaches which have potential application to calorimetry.

7.3.1. The Rice University Levitation Calorimeter

In the following sections some of the design details of the Rice University levitation calorimeter are presented. For further information, the reader should refer to the Ph.D. thesis of D. W. Bonnell [32].

7.3.1.1. The Levitation Chamber

The levitation chamber in the early Rice apparatus consisted of a 1 liter round glass flask with a port at each of the six points of emergence of the Cartesian axes (Fig. 7.5). The rear port was for generator-to-work coil connections. The front was a quartz pyrometer port, with a gas inlet angled to sweep the port clear of fumes, and the top port was for sample introduction

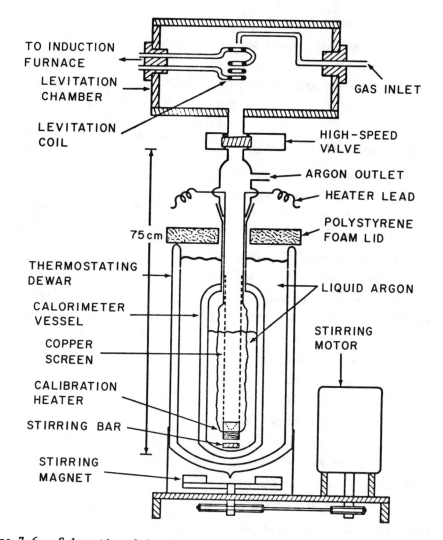

Figure 7.6. Schematic of levitation calorimeter at Sandia Laboratories [39].

and assembly alignment. Flow gas was exhausted to a Nujol bubbler from this connection. Two other openings were for direct gas injection or a viewing port and for the sample handling arm. The bottom opening was a straight tube to which the O-ring joint of the radiation gate sealed. The entire chamber was supported by the generator leads and the connection to the calorimeter.

7.3.1.2. The Radiation Gate

The radiation gate was a modified commercial 3.8 cm (1-1/2 in.) gate-type water valve. Double O-ring grooves were machined directly in each fitting, one to accept the flask and the other to fit a tube on the calorimeter cover flange. This allowed immediate assembly and disassembly (no bolts) and a slight amount of play, making vertical alignment of the entire system easy. This was a necessity, since the hand-wound coils were not made to high tolerances and run-to-run adjustment of the work coils was necessary. The valve stem screw threads were turned to a smooth shaft and bore, and a Teflon plug installed as the guide to allow rapid opening and closing of the gate (~1 s·cycle^{-1} compared with the 0.2 s drop time). Ideally, this gate should be solenoid operated as part of a completely automatic drop cycle. Such a system has been implemented [34,35].

7.3.1.3. Calorimeter Block and Jacket Assembly

The calorimeter assembly was of the standard isothermal block design. The block proper was a pure (99.9%) aluminum cylinder 20.3 cm (8 in.) long by 7.6 cm (3 in.) diameter, bright nickel plated. The central receiving well was tapered (4°35') and accepted sleeves made of 6061 aluminum alloy which has a specific heat of 0.8941 J·g^{-1}·K^{-1} (0.2137 cal·g^{-1}·K^{-1}) at 25°C. The need for replacement sleeves in the receiving well was anticipated in view of the likelihood of damage being caused to the receiving well by hot samples. The ability to remove the well also made it convenient to determine the weights of samples dropped. The sleeve was further lined with metal foils of the same material as the sample being dropped, to avoid alloying of the sample with the metal of the calorimeter.

Electrical calibrations were achieved by using an approximately 100 Ω heater made from #40 B&S gauge manganin wire wound in a double helix about a groove in the sleeve. The wire was painted with Glyptal paint to hold it in place. Since normal runs used sleeves without the heater wires, Glyptal, or groove, their specific heats were deducted in calculations.

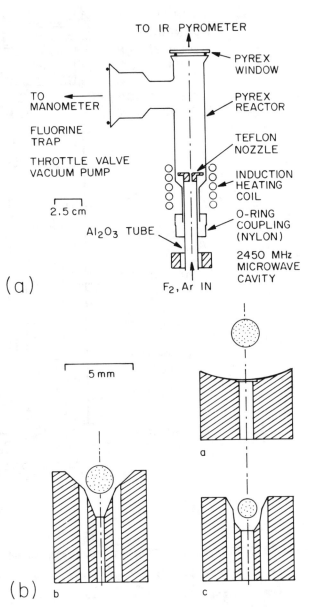

(a)

(b)

Figure 7.7. Schematic of aerodynamic levitation apparatus [13].

The quartz thermometer probe was placed in a cylindrical well drilled 2.5 cm (1 in.) from the side wall of the block and 9.6 mm (3/8 in.) diameter by 17.8 cm (7 in.) deep. This depth was chosen empirically (by successive trials) to eliminate transient thermal responses which would have produced overshoot of the final temperature in calorimetric experiments. The probe was held in good thermal contact with the block by casting Wood's alloy to fill the well. The thermometer cable was sheathed in mesh sleeve shielding attached to the jacket and grounded at the thermometer readout end. This was necessary as electrical pickup from the induction heater, with plate current applied, induced a slight (~0.001°C) power-dependent shift in the thermometer reading. The extra shielding completely eliminated the problem.

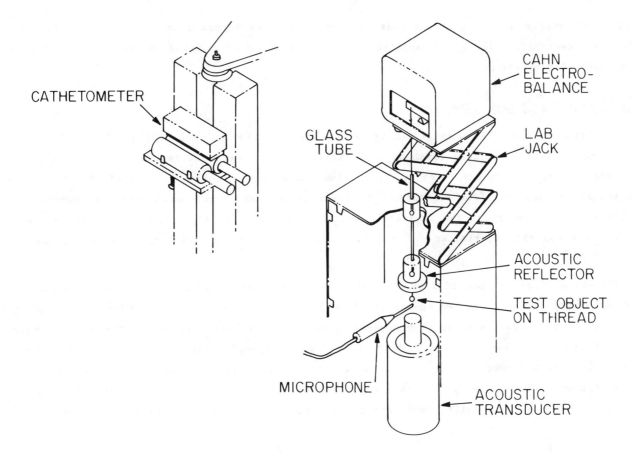

Figure 7.8. Schematic of acoustic levitation apparatus [14].

An indentation opposite the thermometer well accepted a nylon rod to which the block gate, a 6.4 mm (1/4 in.) thick 6061 aluminum plate, was attached. It was rotated to cover the receiving well to minimize conduction and radiation losses after the sample struck. Another source of measurement error is thought to be heat loss by gas escape past this gate. The extent of this loss is estimated to be negligible for the 1 g sample size used. As a confirmation, improved sealing produced no detectable changes in measured enthalpy functions for nickel.

The calorimeter jacket was a chrome-plated brass tube, 10.2 cm (4 in.) inside diameter by 4.8 mm (3/16 in.) wall, 23.50 cm (9.25 in.) from the top of the flange to inside bottom, with the top flange and bottom plate silver soldered to the jacket tube prior to plating. The cover flange, containing necessary feedthroughs, was mated to the top flange of the jacket body with an O-ring gasket. The thermometer and calibration leads were brought through the cover flange by a feed-through set at a 45° angle over the thermometer well position. The block gate actuator passed through the cover flange and was sealed by means of an O-ring inset in the flange. Attached to the jacket below the top flange by the same nylon bolts used to seal the cover flange to the top flanges was a circular brass ring with three arms which supported the calorimeter assembly in the bath. Leveling screws through the ends of the arms aided alignment of the calorimeter.

The entire unit--levitation chamber, radiation gate, and calorimeter--was designed for easy movement and disassembly, since it was necessary to disassemble the apparatus after every drop to measure the dropped mass. Also, in order to obtain various temperatures, changing levitation coils was a necessity. The entire apparatus could be assembled gas tight and aligned by one worker in about 20 minutes.

7.3.1.4. Isothermal Bath and Controls

The bath for the calorimeter assembly was constructed from a large 0.04 m^3 (10 gal.) ceramic crock, supported on a platform which could be raised from a lower carriage by a hydraulic jack. The lower carriage was on wheels which ran in aluminum channel tracks. This provided the means of removing the calorimeter from the chamber by lowering bath and calorimeter from the chamber and rolling them out from under the levitation coil. A wooden cover for the bath was used as the support base for the calorimeter and the attachment base for the stirrer, bath heater, and thermoswitch.

Thermal control was provided by a simple thyratron switch controlled by a thermoswitch in which a mercury column contacted an adjustable wire (Precision Scientific Co. C4747). The thyratron supply was controlled by an autotransformer and the bath heater was a 250 W blade-type heater. Cooling was obtained with a slow flow of 10°C water through a single coil of 6.4 mm (1/4 in.) copper tubing in the bottom of the bath. This control loop provided bath temperature regulation better than ±0.01 K long term, more than adequate for this work. More sophisticated controllers experienced control point shifts much larger than this due to circuit interferences from the RF supply.

7.3.1.5. Flow Gas Purification

Since the only source of contamination for the sample was the surrounding gas, care had to be taken to allow no reactive species in the gas flow. Molten metals in the temperature region of 2000°C are quite reactive and will oxidize or form nitrides in the presence of trace amounts of gas. In many cases, the reaction products are far more volatile than the bulk material and are removed by the flow of gas. On the other hand, zirconium forms both oxides and nitrides very easily, and these reaction products are quite refractory. Thus, a purification train for scrubbing the gas used was required.

First, the gas was passed through a glass tube containing magnesium perchlorate to remove water, and Ascarite, which removed CO_2 by formation of sodium carbonate. A quartz tube (wound with a Nichrome heater strip) which contained fine copper ribbon heated to 450°C was used to remove oxygen. Since heated copper catalyzes the production of water if trace amounts of hydrogen are also present, a further train element was another tube of perchlorate. When nitridation was a problem, an additional element in the train was a calcium trap constructed of stainless steel. Calcium at 600°C reacts readily with nitrogen. A final trap provided a means of mechanically impeding transfer of pyrophoric calcium dust to the calorimeter. A thick plug of glass wool followed by a drying agent was chosen as the packing material in this essentially mechanical trap, to eliminate any possible water, since at 600°C any calcium hydroxide present decomposes to H_2O + CaO. The effluent from the gas train was split, and a very low flow (~3 cm$^3 \cdot$min^{-1}) went to the calorimeter while a flow 10-100 times this amount went to the levitation chamber.

7.3.1.6. The Quartz Thermometer

The convenience of the quartz thermometer (DYMEC-2801A, Hewlett-Packard) was the primary factor in its choice as the thermometer for these experiments. In this thermometer, a quartz crystal has been cut so that, for moderate temperature spans, the resonant frequency shift is linear with temperature. This frequency is compared to a stable crystal-controlled reference source by digital counting techniques, and digitally scaled by a slope factor to give a direct digital display of temperature in Celsius units. Ten second counting periods were sufficient to give a resolution of 0.0001 K and this precision, over temperature spans of a few degrees near room temperature, was acceptable. Absolute accuracy was less important, and the relative precision (long term stability) was checked periodically against the ice point and indicated no detectable changes.

A primary advantage of this Hewlett-Packard or similar quartz thermometer is that data are collected automatically at equal time intervals with high precision. This allows virtually error-free data acquisition and convenient high accuracy interpretation.

7.4. OPTICAL PYROMETRY

Optical pyrometry is ideal for the high-temperature measurements in levitation experiments, and various wavelengths of radiation (ultraviolet/visible/infrared) can be selected [42]. At Rice University, both a manual L&N 8622C disappearing filament pyrometer and an automatic L&N 8641 pyrometer have been used at a wavelength of 650 nm.

In the automatic pyrometer, a mechanical chopper allows radiation from a standard lamp and the unknown source to fall alternately on a phototube. Phase-sensitive circuitry detects the difference in the signals from the two sources, and the current to the standard lamp is continuously adjusted to make the difference vanish. The current required by the standard lamp at the null point was measured as the potential drop across a standard resistor in the standard lamp circuit, and this signal was displayed on a recorder. The chart divisions were read to ± 0.01 mv. The temperature as a function of chart recorder reading is established by calibration of this system. The temperature error (precision) was on the order of 2–5 degrees Celsius. Absolute accuracy for brightness temperatures was 0.5–1.0% or better in the temperature range 1500–3300 K.

The manual pyrometer was calibrated against secondary standards on the IPTS-68 scale and corrections to the slide wire values were derived for each range. At no point did the errors amount to more than 7 K and they generally were less than 3 K. The automatic pyrometer was supplied with calibration charts which plotted mv output vs. temperature ($^\circ$F), and the tabular data were fitted by a computer program to within an RMS error of less than 2 K. The program then generated tables of millivolts vs. temperature (K) at 0.01 mv intervals. Direct comparison of the automatic pyrometer with the calibrated manual device was then undertaken at 1 mv intervals (~30–80 K) over the medium range of the automatic pyrometer for calibration and correction to IPTS-68.

Emissivity corrections to pyrometric temperature measurements for the metals under investigation should be determined in the laboratory with the test materials, if possible [43,44]. If it can be assumed that emissivity is constant over the range of measurement, a few determinations at a single temperature may be sufficient to establish this value. A unique and convenient point for this measurement is the melting point of the metal. The use of this point gives a parameter independent of temperature scales and of actually measured melting points. When emissivities are referred to such a thermodynamically-defined state and the observed brightness temperature is used to calculate emissivities from reported melting points, then any later redetermination of the measured melting point simply changes the emissivity, and the data can be recalculated for this change. Publications reporting temperatures referred to the melting point should contain enough information so that the reader can recalculate the temperatures.

During the levitation and melting of a metal while under observation by a continuous reading automatic pyrometer, it is possible to have chosen the coil and other operating parameters so that the steady-state temperature attained is slightly above the melting point. Under these conditions, a melting plateau in the heating curve of the sample will be observed which can be made long compared to the response time of the pyrometer. Typically, this period for the process of melting is 10 s or more. A typical plot of brightness temperature vs. time is shown for a molybdenum sample in Fig. 7.9.

Actual computation of emissivity may be based on Wien's approximation to Planck's law, i.e.,

$$J_\lambda = \varepsilon_{N,\lambda} C_1 \lambda^{-5} e^{-C_2/\lambda T_o} = C_1 \lambda^{-5} e^{-C_2/\lambda T_B^*} \tag{2}$$

where J_λ is the unit bandwidth flux of radiation from a surface of monochromatic emissivity $\varepsilon_{N,\lambda}$ at wavelength λ centimeters. C_1 and C_2 are the first and second radiation constants, respec-

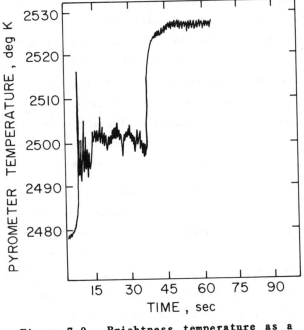

tively, $C_1 = 3.741832 \times 10^{-12}$ W·cm^2 and $C_2 = 1.438786$ cm·K, IPTS-68, T_o is the thermodynamic temperature, and T_B^* is the observed brightness temperature which one obtains by evaluating Eq. (2) for T_o assuming $\varepsilon_{N,\lambda} = 1$. When T_o is known for a measurement of T_B^*, $\varepsilon_{N,\lambda}$ is given by

$$\varepsilon_{N,\lambda} = \exp\left(\frac{C_2}{\lambda}\left(\frac{1}{T_o} - \frac{1}{T_B^*}\right)\right). \tag{3}$$

The error associated with such measurements is found by taking differentials of both sides of Eq. (3), giving

$$d\varepsilon_{N,\lambda} = \exp\left[\frac{C_2}{\lambda}\left(\frac{1}{T_o} - \frac{1}{T_B^*}\right)\right]\left(-\frac{C_2}{\lambda}\right)\left(\frac{-1}{T_B^{*2}}\right) dT_B^*. \tag{4}$$

Thus,

$$\frac{d\varepsilon_{N,\lambda}}{\varepsilon_{N,\lambda}} = \frac{C_2}{T_B^*} \cdot \frac{dT_B^*}{T_B^*} \tag{5}$$

where $C_2/\lambda = 22310$ K with $\lambda = 650$ nm.

Figure 7.9. Brightness temperature as a function of time for a levitated molybdenum sample at Rice University [33].

Use of Eq. (5) shows the sensitivity of emissivity values to small errors in brightness temperatures. For example, in the neighborhood of 2000 K, if the brightness temperature is in error by 10 K (a typical accuracy), and if the nominal emissivity is 0.30, $d\varepsilon_{N,\lambda} = 0.017$, an error of more than 5%. Conversely, relatively large emissivity errors from whatever source cause relatively minor errors in blackbody (i.e., true) temperatures derived from brightness temperatures. At 2000 K, an emissivity error of 5% implies a temperature error of 0.5% which is of the order of the best experimental uncertainty attained.

A method which can be used for determining emissivities directly depends on the relation between specular reflectivity (r_λ) from a polished (i.e., smooth) opaque surface and emissivity [45].

$$\varepsilon_{N,\lambda} = 1 - r_\lambda. \tag{6}$$

If the brightness of the sample due to its thermal radiation, $J_{1,\lambda}$, the directly measured brightness of some light source, $J_{2,\lambda}$ and the brightness, $J_{3,\lambda}$ observed when that light source is reflected from the sample's surface are measured, the reflectivity is then given by

$$r_\lambda = \frac{J_{3,\lambda} - J_{1,\lambda}}{J_{2,\lambda}} \tag{7}$$

from which emissivity can be obtained. Use of this technique has been recently reported [94].

Multicolor pyrometry offers an alternative method for temperature measurement which does not require independently obtained emissivity values [35]. For example, if it can be assumed that the emissivity is independent of wavelength, then simultaneous measurements of T_B^* at two wavelengths allow both $\varepsilon_{N,\lambda}$ and T_0 to be determined. Measurements at additional wavelengths allow fewer restrictions to be placed on the functional form of $\varepsilon_{N,\lambda}$. However, at higher temperatures, the ratios become less sensitive unless there are wide intervals between the wavelengths employed.

Another approach to the measurement of the temperature of levitated metals has been described by Nordine and Schiffman [46]. In aerodynamic levitation experiments, with added mercury vapor in the levitation jet (see Figs. 7.7 and 7.8), laser fluorescence can be used to determine gas densities. As in static gas thermometry, comparison of the gas density at an unknown temperature with that in a known temperature region allows the unknown temperature to be determined from the ideal gas law (constant density-temperature product constraint). They have demonstrated 3% temperature accuracy with molten alumina/sapphire specimens, and anticipate improved accuracy in low-convection microgravity experiments. Substantial system pressures are necessary, and the technique is still in the laboratory development stage.

7.5. DATA REDUCTION

The reduction and processing of calorimetric data are straightforward and offer an ideal opportunity for computerized techniques. Since the desired result is a state function of one parameter, the data are represented by a function [H(T)−H(T$_{REF}$)] versus T, where T$_{REF}$ is a reference temperature, usually 298.15 K. The enthalpy values are determined by operating the calorimeter, and the temperature is independently measured by pyrometry. Thus, initial reduction of the data handles the two variables independently.

7.5.1. Enthalpy Determination

The process under consideration is that a known mass of material at some temperature T_0 and pressure P is placed in contact with a much larger mass of material of precisely known total heat capacity, c_p (calorimeter), at temperature T_1, and the two are allowed to come to thermal equilibrium at temperature T_2. This results in a change of the temperature of the calorimeter from T_1 to T_2. A knowledge of this temperature change, (T_2-T_1), allows one to calculate the enthalpy of the material, which is simply

$$\Delta H^* = H^*(T_0) - H^*(T_2) = (T_2-T_1) \cdot c_p (\text{calorimeter}). \tag{8}$$

For accurate calorimetry, (T_2-T_1) should be kept small, and T_1 and T_2 must be measured with very high precision. The absolute accuracy of T_1 and T_2 does not need to be as good as this precision. To avoid implying an extremely high accuracy for T_1 and T_2 on the absolute thermodynamic temperature scale, one can call them θ_1 and θ_2, respectively, and use any convenient temperature units. This convention is followed in Fig. 7.10, which shows a typical calorimeter heating curve.

The calorimetric procedure followed in these measurements is called isoperibol calorimetry. The block must be surrounded by a jacket at some fixed temperature, so situated that the energy exchange of the block with the jacket obeys Newton's Law of cooling, i.e., the rate of temperature rise due to thermal leakage, $d\theta/dt$, is proportional to the difference between the temperatures of the block (θ) and jacket (θ_j). Here θ represents temperature in any convenient units. The total rate of temperature rise due to thermal leakage is then given by

$$\frac{d\theta}{dt} = k(\theta_j-\theta) \tag{9}$$

where k is the Newton's Law heat exchange constant.

Let g_f represent $d\theta/dt$ evaluated at $t = t_f$, the time of the mean temperature in the final temperature drift period of the calorimeter after sample and block have come to equilibrium. Let g_i represent $d\theta/dt$ at $t = t_i$, the time of the mean temperature in the initial drift period before the sample is dropped. The reason for evaluation at the mean can be seen by a simple argument. To the accuracy that drift is Newtonian, the determination of $\Delta\theta$ (the temperature change caused by thermal leakage, $\theta_2-\theta_1$) will

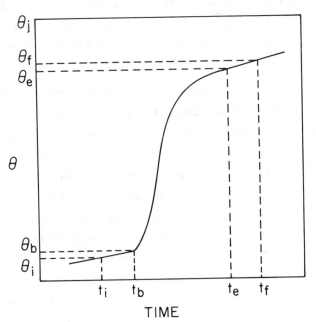

Figure 7.10. Typical calorimeter heating curve [32].

not change for any choice of t_i and t_f within the respective drift periods. However, the true function θ versus t, which will be represented by a linear fit, is nonlinear. To the second order, the tangent to the true function at its mean will be closest to the average slope determined from the linear fit. Thus, by using the mean value, one should obtain final results as precise as the linear fit allows. For the Rice levitation calorimeter, θ versus t is very nearly linear; the question of selecting t_i and t_f was trivial, and the choice was primarily one of standardization. Substituting these values in Eq. (9) and solving for k gives

$$k = \frac{g_i - g_f}{\theta_f - \theta_i} .$$ (10)

Since $g_f = k(\theta_j - \theta_f)$, Eq. (9) also leads to

$$\frac{d\theta}{dt} = g_f + k(\theta_f - \theta).$$ (11)

Integration of Eq. (11) during the mean period (i.e., from the time the sample is dropped until the final drift period begins) gives the total temperature correction. This thermal leakage correction to the observed temperature rise $(\theta_e - \theta_b)$ (subscripts e and b representing ending and beginning of the mean period, respectively), is

$$\Delta\theta = -g_f(t_e - t_b) - k \int_{t_b}^{t_e} (\theta_f - \theta)dt.$$ (12)

To this point the only assumption has been that the calorimeter thermal leakage obeys Newton's Law of cooling. For very small thermal leaks, it is convenient to treat the initial and final drift periods as though $\theta = \theta(t)$ were a linear function. For the Rice apparatus the error involved in this assumption is of the order of 0.0001 K average deviation of measured θ from a linear fit to the entire period. Thus, θ_i, θ_f, g_i, and g_f can be determined and k calculated. Random errors in the measured values θ_e and θ_b can be smoothed by the linear fit to increase the precision of these values.

Since no analytical expression for the behavior of θ during the main period is available, and, in fact, it is certainly different for different calorimeters, it is necessary to evaluate the integral of θdt numerically. The two-point trapezoidal rule provides a precision which is at least as good as the temperature data (i.e., six-figure precision for degrees C) and numerical simplicity. Therefore, it was chosen. The error can be estimated using the error term in the quadrature formula. The trapezoidal rule for two points of a function y (of x) is

$$\int_{x_o}^{x_1} ydx = \frac{h}{2} (y_o + y_1) - \frac{y''h^3}{12}$$ (13)

where h is the interval size and y'' is the second derivative of the function evaluated somewhere in the interval. For n equally-spaced points the form of integral I_1 is

$$I_1 = \int_{x_o}^{x_n} ydx = h \left[\sum_1^{n-1} y_i + \frac{y_o + y_n}{2} \right] - (n-1) \frac{y''h^3}{12}$$ (14)

where y'' has been assumed constant. If y'' is small, or constant, or a symmetric function about the midpoint of the domain of integration, then it can be treated as constant, and estimated in the following manner. Let n be odd, and evaluate the integral again at twice the step size. Then integral I_2 is

$$I_2 = h \left[2 \left(\sum_1^{(n-1)/2} y_{2i} + \frac{y_o + y_n}{2} \right) \right] - \frac{n-1}{2} \cdot \frac{y''(2h)^3}{12} . \tag{15}$$

Let the quantity in square brackets in Eq. (15) be J_2 and the corresponding quantity in Eq. (14) be J_1. Setting $I_1 = I_2$, one obtains

$$hJ_1 - \frac{(n-1)y''h^3}{12} = hJ_2 - \frac{n-1}{2} \frac{8y''h^3}{12} . \tag{16}$$

Solving for y'' gives

$$y'' = \frac{12(J_2 - J_1)}{3(n-1)h^2} . \tag{17}$$

For these experiments one obtains typical values of y'' of the order of 10^{-6} to 10^{-7}. This agrees well with the expectation of high precision, and carrying through the error estimate as

$$\text{ERROR} = \frac{-(n-1)y''h^3}{12} = - \frac{h(J_2 - J_1)}{3} \tag{18}$$

gives error estimates in the neighborhood of 10^{-4} to 10^{-5}. The total integral of θdt is of the order of 2×10^4 degree-seconds. Since the first two significant figures of each term (temperature x Δ time) in the integral are observed to be constant, this error implies the integral is accurate to six significant figures, which is the precision of the thermometer reading.

The final corrected temperature rise for each experiment, $\Delta\theta_{corr}$ is given by

$$\Delta\theta_{corr} = \theta_e - \theta_b + \Delta\theta \tag{19}$$

where $\Delta\theta$ is now obtained from Eq. (12).

There is no requirement that curves of θ versus t in electrical or other calibrations be similar to those in the experimental runs themselves. Implicit in the errors quoted, however, are the assumptions that initial and final periods have certain characteristics (low thermal leak and a length sufficient to reduce error to a minimum) and that the main period is in the vicinity of 1000 s for $\Delta\theta_{corr}$ values of 1–2°C. These constraints show that in practice the precision will be best if the electrical calibration runs and the experiments themselves produce curves of θ versus t which are similar. Systematic errors also tend to be smaller when calibrations and experiments with unknowns are similar to each other.

The FORTRAN program which implements the algorithm described has the name CALOR and was presented in detail by Bonnell [32].

7.5.2. Temperature Determination

To convert observed temperatures in millivolts from the pyrometer output into absolute temperature on the International Practical Temperature Scale [47-49], one must first correct for any chart recorder error and then convert into kelvin temperature with the calibration tables. These temperatures were corrected further by the error graphs to give corrected observed temperatures (T^*) on the International Practical Temperature Scale. A correction for incomplete transmission of light through the window was then applied to obtain the true brightness temperature (T_B^*) and finally the emissivity correction was made to obtain T_0, the true thermodynamic temperature.

Actually, the temperature, T_0, derived from measurement of radiation intensity at the moment of drop is not in general the true temperature of the sample on impact in the calorimeter. During the drop, the sample radiates energy away and undergoes conductive losses to the protective atmosphere. It is convenient to make the correction to the enthalpy increments as energy lost, while retaining the initial measured temperatures for calculational purposes.

7.5.3. Radiation Loss Correction

Consider a body of surface area A and total hemispheric emissivity ε. At a temperature T, the radiative heat loss dH_r^* during time dt is given by the Stefan-Boltzmann Law as

$$dH_r^* = \varepsilon A \sigma T^4 dt \tag{20}$$

where σ is the Stefan-Boltzmann radiation constant. If c_p^* is the constant pressure heat capacity for a sample of mass m, then

$$dH_r^* = -c_p^* dT. \tag{21}$$

Upon setting Eqs. (21) and (22) equal and rearranging,

$$dt = -\frac{c_p^*}{A\sigma\varepsilon} \cdot \frac{dT}{T^4} . \tag{22}$$

It can be assumed that c_p^* is independent of temperature (from the current evidence for linearity of liquid enthalpy functions near the melting point) and that there are no temperature gradients in the body (assured by electromagnetic stirring of the melt).

With the initial conditions that at t = 0, T = T_0, Eq. (22) may be integrated (limits T_0 to T) to yield

$$T = \left[\frac{c_p^* T_o^3}{3A\sigma t\varepsilon T_o^3 + c_p^*} \right]^{1/3} . \tag{23}$$

Substituting Eq. (23) into Eq. (20) and integrating gives the total heat loss in time t as

$$\Delta H_r^* = c_p^* T_o \left[1 - \left(\frac{1}{1+y} \right) \right]^{1/3} \tag{24}$$

where

$$y = \frac{3\varepsilon A \sigma t T_o^3}{c_p^*} . \tag{25}$$

Assuming the body of mass m and density d to be spherical in shape, the surface area is given by

$$A = 4\pi \left(\frac{3m}{4\pi d}\right)^{2/3} . \tag{26}$$

Also, if the body falls a distance x under the influence of an acceleration g, due to gravity, the time of fall is

$$t = \left(\frac{2x}{g}\right)^{1/2} . \tag{27}$$

If C_p is the constant pressure molar heat capacity for a material of molecular weight M,

$$c_p^* = \frac{m C_p}{M} . \tag{28}$$

On substitution of the above equations in Eq. (25), one finds

$$\Delta H_r = C_p T_o \left[1 - \left(\frac{1}{1+y}\right)^{1/3} \right] \tag{29}$$

where

$$y = \frac{12\pi\varepsilon\sigma M T_o^3}{C_p} \left(\frac{2x}{g}\right)^{1/2} \left(\frac{3}{4\pi d}\right)^{2/3} \left(\frac{1}{m}\right)^{1/3} \tag{30}$$

$$= \frac{3.716 \cdot 10^{-12} \varepsilon M T_o^3 x^{1/2}}{C_p (d^2 m)^{1/3}} \qquad (C_p \text{ in Joule units}). \tag{31}$$

The expression in square brackets in Eq. (29) may be expanded by the binomial formula as

$$\left[1 - \left(\frac{1}{1+y}\right)^{1/3} \right] = \frac{1}{3} y - \frac{2}{9} y^2 + \frac{13}{81} y^3 - \ldots \tag{32}$$

Taking only the leading term of this expansion, Eqs. (30) and (32) yield

$$\Delta H_r = \frac{1}{3} y C_p T_o = 4\pi\varepsilon M T_o^4 \sigma \left(\frac{2x}{g}\right)^{1/2} \left(\frac{3}{4\pi\rho}\right)^{2/3} \left(\frac{1}{m}\right)^{1/3} \tag{33}$$

which is identical to the expression obtained by direct integration of Eq. (23) assuming constant temperature.

For a body with m = 1 g, d = 10 g·cm^{-3}, ε = 0.35, C_p = 41.84 J·mol^{-1}·K^{-1}, M = 100 g·mol^{-1}, T = 3000 K, and x = 25 cm, all typical figures for a levitation experiment, y = 0.09. It is obvious that the leading term is the dominant one, and, therefore, the radiation heat loss is not highly sensitive to small changes in C_p as a function of T. The second term, however, does contribute about 8% of the correction, or more than 0.5% of the total enthalpy at the higher

temperatures.   Thus, the first order form is not completely satisfactory for data gathered at
3000 K or higher temperatures.

Calculation of the radiation loss is easily performed by an iterative process which converges
rapidly due to the lack of sensitivity of the correction to changes in C_p. The only data required
for this algorithm are experimental enthalpy increments, temperature at the start of each drop,
sample masses and densities, and drop distances for each experiment, plus the total hemispheric
emissivity of the material. Unfortunately, total hemispheric emissivity measurements are few and
of rather poor reliability, and high-temperature density data for liquid metals are almost unknown
[50-53]. Densities of liquids estimated to be 10% less than for room-temperature solids are consistent with available data.

7.5.4. Conduction Loss Correction

Energy loss by conduction has been a more difficult problem to resolve. Initially, the
observations of Kirillin et al. [54] that the major mode of heat loss from a sample during a drop
was radiative, and of Hultgren et al. [55] that the T^4 temperature dependence of radiative energy
losses outweighed the expected linear dependence of conduction losses from platinum to argon gas
at high temperatures (1500 K and above), were used as a basis for neglecting conduction loss. A
simple calculation for a static sphere assuming Newtonian (linear gradient) heat flow for the drop
conditions of Section 7.5.3 in argon gas also indicated conduction losses to be much less than
radiation losses.

This model is, however, not very realistic since the drop conditions are far from static. A
more realistic model, for which a solution exists, uses the engineering solution for the smooth
flow of gas past a sphere [56], which gives results good to at least a few percent. Since this
problem amounts to accelerated motion, an integral form of this equation was developed as

$$\Delta H_c^* = \int_0^t h A \Delta T dt \tag{34}$$

where

ΔH_c^* is the heat loss by conduction, and

$h = 1.1 \left(\dfrac{c_p \mu_f}{\lambda_f} \right)^{0.3} \left(\dfrac{\lambda_f}{3D} \right) \left(\dfrac{DG}{\mu_f} \right)^{0.6}$, the heat transfer coefficient,

$A = 4\pi \left(\dfrac{3m}{4\pi d} \right)^{2/3}$, cross-sectional area,

$\Delta T = T_o - T_g$, temperature difference between sample and gas,

M_g = molecular weight of the gas,

$G = \dfrac{g M_g P t}{R T_g}$, mass flow (at gas pressure P),

$D = 2 \left(\dfrac{3m}{4\pi d} \right)^{1/3}$, sample diameter,

$$c_p = C_p / M_g = \frac{5R}{2M_g}, \text{ heat capacity of gas,}$$

λ_f = gas thermal conductivity evaluated at $\dfrac{T_o + T_g}{2}$,

and μ_f = gas viscosity evaluated at $\dfrac{T_o + T_g}{2}$.

Therefore,

$$\frac{\Delta H_c^*}{\Delta T} = 1.1 \left(\frac{c_p \mu_f}{\lambda_f}\right)^{0.3} \left(\frac{\lambda_f}{3D}\right) \left(\frac{D}{\mu_f}\right)^{0.6} \left(\frac{gM_g P}{RT_g}\right)^{0.6} \int_0^t t^{0.6} dt. \tag{35}$$

Integration and substitution of constants gives (in $J \cdot K^{-1}$)

$$\frac{\Delta H_c^*}{\Delta T} = 7.09 \cdot 10^{-3} \ M_g^{0.3} \ x^{0.8} \left(\frac{m}{d}\right)^{0.533} \left(\frac{\mu_f}{\lambda_f}\right)^{0.3} \frac{\lambda_f}{\mu_f^{0.6}} . \tag{36}$$

In this treatment, the heat exchange coefficient h extrapolates to zero for zero velocity, so that static conduction at the beginning of the drop is neglected. This mathematical simplification introduces negligible error under typical conditions, as noted above. For samples much smaller than the approximately 6 mm diameter samples considered here, both static and dynamic conduction losses could become significant.

For the conditions of a typical experiment, the drop time was ~0.23 s and the percentage loss at 2000 K in argon, for the sample used in the radiation loss calculation, was about 0.4%. This correction is nearly the same at other temperatures in the region of interest since the change caused by a moderate change in temperature is small relative to [H(T) − H(298.15 K)]. The correction using an average velocity for the drop is about 25% lower than the integrated correction for the conditions stated. The small size of the correction is sufficient justification for ignoring the change in T with time during the drop. The correction can generally be added to the calorimetrically-measured enthalpy at T_o before consideration of the radiation losses.

7.5.5. The Enthalpy Function

For a series of measurements of H and T_o it is desired to represent H as a continuous function of T for interpolation purposes and for use in analytical expressions to derive other thermodynamic state functions. Since there is no special theoretical form for this function, a power series representation is usually employed. For most metals studied in the liquid state by levitation calorimetry, with the exception of zirconium, a linear fit to the data produced a standard deviation of the order expected from random errors (±0.5%) and, thus, represented the data to the precision of measurement.

7.6. ENTHALPIES AND HEAT CAPACITIES OF LIQUID METALS

A selection of metals from across the Periodic Table, but mainly transition metals, has been studied in various laboratories by levitation techniques. Although details of experimental apparatus differ slightly, the usual results of experiments include (H_T-H_{298}), $C_p(T)$, and ΔH_{fusion}. Minor inconsistencies are noted because of the use of different emissivities in temperature calculations, different reference data for solids in calculating heats of fusion, and different temperature measurement techniques (manual/automatic pyrometry with one or more colors being monitored). In all cases studied by the levitation technique, except for zirconium, the plots of (H_T-H_{298}) versus T were linear, implying that C_p^{liq} is constant up to 700 degrees above the melting points and, from supercooling studies, down to at least 200 degrees below the melting points of various high-melting transition metals.

The data for nickel in Table 7.2 are typical of those obtained at Rice University for well-behaved metals. The sizes of the conduction and radiation corrections are stated. The average absolute deviation of the measured enthalpies from a linear function of temperatures is 0.5%. Special symbols used in Table 7.2 are OBS, indicating uncorrected data; CORR, indicating data corrected for conduction and radiation heat losses; CALC, indicating values calculated from a linear relationship between enthalpy and temperature; ℓ, indicating the liquid state; and T_m, indicating the melting point.

Levitation studies of most of the transition metals, including several rare-earth metals, can be found in the literature [26,32,33,39–41,57–80]. Table 7.3 shows which elements have been studied in the liquid state by levitation calorimetry.

A few alloys have also been studied [58,81,82]. Table 7.4 presents an example of a detailed investigation of heat capacity changes as a function of composition in the alloy system cerium—

Table 7.2. Enthalpy Data for Liquid Nickel [57]

Run #	Temperature K (IPTS-68)	Mass g	H_T-H_{298}(OBS) J·mol^{-1}	$\Delta H_{cond} + \Delta H_{rad}$ J·mol^{-1} (LOSS)	H_T-H_{298}(CORR) J·mol^{-1}	H_T-H_{298}(CALC) J·mol^{-1}	Deviation %
R	1786	0.9494	64790	295 + 255	65430	64980	0.56
S	1823	0.8458	66440	300 + 230	66970	66610	0.53
O	1866	0.9311	67540	315 + 275	68130	68510	−0.56
T	1887	1.0333	68840	330 + 280	69450	69430	0.02
GG	1920	1.1707	70070	325 + 315	70710	70890	−0.26
EE	1923	1.2083	70510	330 + 310	71150	71020	0.17
P	1967	1.4905	71120	380 + 320	71820	72970	−1.57
FF	2003	1.2569	73750	360 + 360	74470	74550	−0.11
CC	2008	1.4486	74590	385 + 345	75320	74780	0.73
U	2091	2.2011	77610	510 + 360	78480	78440	0.05
V	2112	1.5770	78880	435 + 415	79730	79370	0.45

$C_p = 44.14 \pm 1.5$ J·mol^{-1}·K^{-1}
$= (10.55 \pm 0.36$ cal·mol^{-1}·K$^{-1})$

$T_m = 1728$ K (IPTS-68)

Average Deviation 0.46%

$H(\ell,T_m) = 62420 \pm 360$ J·mol^{-1}
$= (14915 \pm 85$ cal·mol$^{-1})$

$E_\lambda = 0.398$ ($\lambda = 645$ nm)

Standard Deviation 500 J·mol^{-1} (120 cal·mol^{-1})

$\Delta H_{fus} = 14900 \pm 360$ J·mol^{-1}
$= (3560 \pm 85$ cal·mol$^{-1})$

$E_T = 0.285$

Corresponds to a Temperature Error of ~10 K.

$\Delta S_{fus} = 8.6 \pm 0.2$ J·mol^{-1}·K^{-1}
$= (2.06 \pm 0.05$ cal·mol^{-1}·K$^{-1})$

$\rho(\ell) = 8.01$ g·cm^{-3} (est.)

Table 7.3. Elements Studied in the Liquid State by Levitation Calorimetry

Sc	Ti	V	Cr	Mn	Fe	Co	Ni	Cu
------	[60,61,62]	[60,63,64, 65]	------	------	[60]	[66]	[57]	[26,39,40, 58,67]
Y	Zr	Nb	Mo	Tc	Ru	Rh	Pd	Ag
[40]	[32]	[32,65,68, 69,70]	[33,41,71, 72]	------	[59,73]	[74]	[66]	[95]
La	Hf	Ta	W	Re	Os	Ir	Pt	Au
[75]	------	[78]	[79]	------	------	[80]	[26]	------
Ce	Pr	Nd	Pm	Sm	Eu	· · ·	· · ·	Lu
[67]	[76]	[77]	------	------	------			------
Th	Pa	U	Np	Pu	Am	· · ·	· · ·	Lr
------	------	[39]	------	------	------			------

Table 7.4. Heat Capacities of Liquid Cerium-Copper Alloys [58]

Atomic Percent Cerium	Atomic Percent Copper	Measured C_p $J \cdot mol^{-1} \cdot K^{-1}$	Estimated C_p[a] $J \cdot mol^{-1} \cdot K^{-1}$	Temperature Range, K
100.0	0.0	33.36 ± 0.46	------	1531-2407
89.64	10.36	32.66 ± 0.69	33.03	1703-2247
80.24	19.76	32.43 ± 0.83	32.74	1605-2271
71.27	28.73	32.41 ± 1.57	32.45	1644-2218
61.28	38.72	32.67 ± 1.28	32.14	1683-2200
50.09	49.91	33.25 ± 1.33	31.79	1690-2206
40.70	59.30	35.24 ± 0.72	31.49	1420-2088
30.27	69.73	35.77 ± 1.46	31.16	1655-2092
19.95	80.05	32.99 ± 0.61	30.39	1659-2201
0.0	100.0	30.21 ± 0.39	------	1415-2048

[a]Molar average C_p values.

copper [58]. A comparison of the data with molar additivity estimates illustrates the unreliability of such estimates even for materials of similar molar heat capacity. In the Technical University of Berlin, levitation studies of alloys have been conducted using a coil with two reversals of winding direction and, thus, two minimum field points where two samples can be levitated at the same time [81]. The upper sample can be dropped into the lower sample by electrically shorting out the upper part of the coil.

There have also been studies of the cooling rates of levitated spheres of liquid Be and of liquid W [83]. An analysis of the cooling curve approach was prepared by Margrave et al. [84]. This technique is attractive because it yields thermodynamic data directly from measurements of temperature as a function of time, potentially valuable for space microgravity experiments.

Finally, in the exploding wire method [85-88] one attains the liquid phase for only a few microseconds, but this is long enough for electrical and temperature measurements from which thermodynamic and transport parameters can be derived. At the Lawrence Livermore Laboratories [85-87] there have been exploding wire studies of liquid metals which overlap the temperature range of some static levitation experiments and allow the preparation of tables of estimated thermodynamic properties of liquid metals to 5000 K [58,89], as summarized in Table 7.5.

Table 7.5. A Comparison of Liquid Metal Heat Capacities ($J \cdot mol^{-1} \cdot K^{-1}$) at High Temperatures*

Element	MP, K	Near MP	3000 K	4000 K	5000 K
Sc	1812	(42)	(50)	(59)	(67)
Ti	1943	46.4 [L]	(59)	(71)	(92)
V	2175	46.9 [L]; 47.3 [E]	61.1 [E]	69.5 [E]	(92)
Cr	2130	(39)	(50)	(59)	(67)
Mn	1517	(46)	(54)	(63)	(71)
Fe	1809	46.4 [L]	(59)	(71)	(92)
Co	1768	50.2 [L]	(63)	(75)	(96)
Ni	1726	39.3 [L]	(54)	(63)	(71)
Cu	1357	34.7 [L]	(50)		
Zn	693	31.4 [D]			
Y	1799	43.1 [D]; 39.8 [L]	(50)	(59)	(67)
Zr	2125	40.6 [L]	64.0 [L]	(71)	(84)
Nb	2740	40.6 [L]	(42)	(54)	(67)
Mo	2890	37.7 [L]	(50)	(67)	(84)
Tc	(2473)	(38)	(42)	(50)	(63)
Ru	2523	(36)	(42)	(54)	(71)
Rh	2233	(42)	(50)	(63)	(75)
Pd	1825	41.4 [L]	(50)	(63)	(75)
Ag	1234	(30.5)	(42)		
Cd	594	29.7 [D]			
La	1193	32.8 [L]	(40)	(45)	(50)
Ce	1071	33.4 [L]	(41)	(46)	(51)
Pr	1208	41.6 [L]	(49)	(54)	(59)
Nd	1297	44.0 [L]	(52)	(57)	(62)
Hf	2500	(40)	(50)	(63)	(75)
Ta	3287	42.2 [L]		(54)	(75)
W	3695	53.0 [L]		(54)	(67)
Re	3453	(45)		(54)	(67)
Os	3300	(36)		(50)	(63)
Ir	2716	30.1 [E]; 59.4 [L]	(60)	42.3 [E]	46.4 [E]
Pt	2042	38.9 [L]	(46)	(54)	(63)
Au	1336	37.7 [D]	(38)	(42)	
Hg	234	28.5 [D]			
Th	2028	(50)	(67)	(75)	(92)
U	1405	47.3 [L]; 48.1 [E]	67.8 [E]	73.2 [E]	96.7 [E]

*Letters following numbers indicate:

    [L] = Levitation measurements
    [E] = Exploding wire measurements
    [D] = Drop calorimetry measurements
    ( ) = Estimated values.

7.7.　ACKNOWLEDGMENTS

Levitation calorimetry at Rice University has been supported by funds from the Robert A. Welch Foundation, the National Science Foundation (B. S. under Grant PDF-8166044), and the National Aeronautics and Space Administration.

7.8.　REFERENCES

1. Tammann, G., Z. Phys. Chem., 85, 273-96, 1913.

2. Alder, B.J. and Wainwright, T.E., J. Chem. Phys., 31, 459-66, 1959.

3. Ho, P.S., Phys. Rev. B, 3, 4035-43, 1971.

4. Takeuchi, S., Tanaka, M., Fukui, Y., Watabe, M., and Hasegawa, M., in The Properties of Liquid Metals (Takeuchi, S., Editor), John Wiley and Sons, New York, NY, 143-50, 1973.

5. Hiwatari, Y., Matsuda, H., Ogawa, T., Ueda, A., and Ogita, N., in The Properties of Liquid Metals (Takeuchi, S., Editor), John Wiley and Sons, New York, NY, 571-5, 1973.

6. Egelstaff, P., An Introduction to the Liquid State, Academic Press, New York, NY, 9-10, Chap. 4, 86-9, 1967.

7. Hill, T.L., Introduction to Statistical Thermodynamics, Addison-Wesley, Reading, MA, 1962.

8. Ubbelohde, A.R., The Molten State of Matter, John Wiley and Sons, New York, NY, 1978.

9. Kelley, K.K., 'Contributions to the Data on Theoretical Metallurgy XIII,' U.S. Bureau of Mines, Bulletin No. 584, 1960.

10. Hultgren, R., Desai, P.D., Hawkins, D.T., Gleiser, M., Kelley, K.K., and Wagman, D.D., Selected Values of the Thermodynamic Properties of the Elements, American Society for Metals, Metals Park, OH, 1973.

11. Searcy, A.W., Proceedings of an International Symposium on High Temperature Technology, McGraw-Hill, New York, NY, 1960.

12. Wroughton, D.M., Okress, E.C., Brace, P.H., Comenetz, G., and Kelley, J.C.R., J. Electrochem. Soc., 99, 205-11, 1952.

13. Winborne, D.A., Nordine, P.C., Rosner, D.E., and Marley, N.F., Metall. Trans., B7, 711-3, 1976.

14. Oran, W.A., Berge, L.H., and Parker, H.W., Rev. Sci. Instrum., 51, 626-31, 1980.

15. Margrave, J.L., in Materials Processing in the Reduced Gravity Environment of Space (Rindone, G., Editor), Elsevier, New York, NY, 39-42, 1982.

16. Comenetz, G. and Salatka, J.W., J. Electrochem. Soc., 105, 673-6, 1958.

17. Begley, R.J., Comenetz, G., Flinn, P.A., and Salatka, J.W., Rev. Sci. Instrum., 30, 38, 1959.

18. Okress, E.C., Wroughton, D.M., Comenetz, G., Brace, P.H., and Kelley, J.C.R., J. Appl. Phys., 23, 542-52, 1952.

19. Stoll, R.L., The Analysis of Eddy Currents, Clarendon Press, Oxford, England, xi-xii, 1-14, 1974.

20. Hatch, A.J. and Smith, W.E., J. Appl. Phys., 38, 742-5, 1967.

21. Peifer, W.A., J. Metals, 17, 487-93, 1965.

22. Weisberg, L.R., Rev. Sci. Instrum., 30, 135, 1959.

23. Sidorov, T.A. and Mezdrogina, M.M., in Fiz. Khim. Granits Razdela Kontaktiruyushchikh Faz (Eremenko, V.N., Editor), Naukova Dumka, Kiev, USSR, 45-7, 1976.

24. Jenkins, A.E., Harris, B., and Becker, L., Symposium on Metallurgy at High Pressures and High Temperatures, Met. Soc. AIME Conf., Vol. 22, Gordon and Breach, New York, NY, 23-43, 1954.

25. Margrave, J.L. and Bonnell, D.W., Chem. Eng. News, p. 36, Oct. 28, 1968.

26. Chaudhuri, A.K., Bonnell, D.W., Ford, A.L., and Margrave, J.L., High Temp. Sci., _2_, 203-12, 1970.

27. Margrave, J.L., High Temp.-High Pressures, _2_, 583-5, 1970.

28. Chekhovskoi, V.Ya., in Compendium on Thermophysical Properties Measurement Methods (Maglic, K.D., Cezairliyan, A., and Peletskii, V.E., Editors), Plenum Press, New York, NY, 1984.

29. Chekhovskoi, V.Ya., Sheindlin, A.E., and Berezin, B.Ya., High Temp.-High Pressures, _2_, 301-7, 1970.

30. Chekhovskoi, V.Ya. and Berezin, B.Ya., Authorship Certificate No. 218492, Invention Bulletin No. 17, 1968.

31. Rindone, G.E., Materials Processing in the Reduced Gravity Environment of Space, (North-Holland, New York, NY, 1-676, 1982.

32. Bonnell, D.W., 'Property Measurement at High Temperatures - Levitation Calorimetry Studies of Liquid Metals,' Rice Univ., Ph.D. Thesis, 1972.

33. Treverton, J. and Margrave, J.L., in Proceedings of the Fifth Symposium on Thermophysical Properties (Bonilla, C., Editor), ASME, New York, NY, 489-94, 1970.

34. Frost, R.T., Bonnell, D.W., and Margrave, J.L., unpublished work, 1980-1984.

35. Bonnell, D.W., in NBS: Materials Measurements (Manning, J.R., Editor), NBS Rept. NBSIR 81-2295, 1981.

36. Frost, R.T., Final Report for NASA Contract NAS8-29680 (General Electric Space Div., Valley Forge, PA), May, 1974.

37. Rony, P.R., 'The Electromagnetic Levitation of Metals,' USAEC Rept. UCRL-16073, 1-90, 1965.

38. Frost, R.T., Margrave, J.L., Bonnell, D.W., et al., NASA Project, Contract No. NAS8-33199, 1981-1983.

39. Stephens, H.P., High Temp. Sci., _6_, 156-66, 1974.

40. Stretz, L.A. and Bautista, R.G., Metall. Trans., _5_, 921-8, 1974.

41. Betz, G. and Frohberg, M.G., High Temp.-High Pressures, _12_, 169-78, 1980.

42. Margrave, J.L. and Hauge, R.H., in Chemical Experimentation Under Extreme Conditions (Weissberger, A. and Rossiter, B., Editors), Vol. 9, John Wiley and Sons, New York, NY, 277-360, 1980.

43. Bonnell, D.W., Treverton, J.A., Valerga, A.J., and Margrave, J.L., in Temperature, Its Measurement and Control in Science and Industry (Plumb, H.H., Editor), Vol. 4, Pt. 1, Instrument Society of America, Pittsburgh, PA, 483-7, 1972.

44. Moscowitz, C.M., Stretz, L.A., and Bautista, R.G., High Temp. Sci., _4_, 372-8, 1972.

45. Kingery, W.D., Property Measurements at High Temperatures, John Wiley and Sons, New York, NY, 121, 1959.

46. Nordine, P.C. and Schiffman, R.A., 'Containerless High Temperature Property Measurements by Atomic Fluorescence,' NASA Annual Technical Report on Contract NAS8-34383, 1982.

47. Thirteenth General Conference on Weights and Measures, Metrologia, _5_, 35-44, 1969.

48. Rossini, F.D., J. Chem. Thermodyn., _2_, 447-59, 1970.

49. Mangum, B.W. and Furukawa, G.T., Metrologia, _18_, 161-8, 1982.

50. Ivashchenko, Yu.N. and Martsenyuk, P.S., Teplofiz. Vys. Temp., _11_, 1285-7, 1973.

51. Saito, T., Shiraishi, Y., and Sakuma, Y., Trans. Iron Steel Inst. Jpn., _9_, 118-26, 1969.

52. Saito, T. and Sakuma, Y., Nippon Kinzoku Gakkai, _31_, 1140-4, 1967.

53. Shpil'rain, E.E., Fomin, V.A., and Kachalov, V.V., High Temp. (Engl. Transl.), _20_, 49-53, 1982.

54. Kirillin, V.A., Sheindlin, A., and Chekhovskoi, V.Ya., High Temp. Technol. Proc. Int. Symp. 1963, Pacific Grove, CA, 471-84, 1964.

55. Hultgren, R., Newcomb, P., Orr, R.L., and Warner, L., in <u>Proceedings of the Physical Chemistry of Metallic Solutions and Intermetallic Compounds Symposium No. 9</u>, Vol. 1, Paper No. 1H, National Physical Lab., London, England, 1–8, 1959.

56. McAdams, W.H., <u>Heat Transmission</u>, 2nd Ed., McGraw-Hill, New York, NY, 236, 1942.

57. Bonnell, D.W. .and Margrave, J.L., Proc. Third Internatl. Conf. on Chem. Therm., Vienna, Austria, Sept. 3–7, 1973, 105–11, 1973.

58. Dokko, W. and Bautista, R.G., Metall. Trans., <u>B11</u>, 511–8, 1980.

59. Sheindlin, A.E., Kats, S.A., Berezin, B.Ya., and Chekhovskoi, V.Ya., Rev. Int. Hautes Temp. Refract., <u>12</u>, 12–15, 1975.

60. Treverton, J.A. and Margrave, J.L., J. Chem. Thermodyn., <u>3</u>, 473–81, 1971.

61. Berezin, B.Ya., Chekhovskoi, V.Ya., Kats, S.A., and Kenisarin, M.M., in <u>Proceedings of the Sixth Symposium on Thermophysical Properties</u> (Liley, P.E., Editor), ASME, New York, NY, 263–72, 1973.

62. Berezin, B.Ya., Kats, S.A., Kenisarin, M.M., and Chekhovskoi, V.Ya., Teplofiz. Vys. Temp., <u>12</u>, 524–9, 1974.

63. Berezin, B.Ya., Chekhovskoi, V.Ya., and Sheindlin, A.E., High Temp. Sci., <u>4</u>, 478–86, 1972.

64. Berezin, B.Ya., Chekhovskoi, V.Ya., and Sheindlin, A.E., Proc. Acad. Sci. USSR, Appl. Phys., <u>201</u>, 583–5, 1971.

65. Berezin, B.Ya. and Chekhovskoi, V.Ya., Teplofiz. Vys. Temp., <u>14</u>, 772–8, 1977.

66. Treverton, J.A. and Margrave, J.L., J. Phys. Chem., <u>75</u>, 3737–40, 1971.

67. Kuntz, L.L. and Bautista, R.G., Metall. Trans., <u>B7</u>, 107–13, 1976.

68. Sheindlin, A.E., Berezin, B.Ya., and Chekhovskoi, V.Ya., High Temp.-High Pressures, <u>4</u>, 611–9, 1972.

69. Betz, G. and Frohberg, M.G., Scripta Met., <u>15</u>, 269–72, 1981.

70. Betz, G. and Frohberg, M.G., Z. Metallkd., <u>71</u>, 451–5, 1980.

71. Berezin, B.Ya., Chekhovskoi, V.Ya., and Sheindlin, A.E., High Temp.-High Pressures, <u>3</u>, 287–97, 1971.

72. Chekhovskoi, V.Ya., Sheindlin, A.E., and Berezin, B.Ya., High Temp.-High Pressures, <u>2</u>, 301–7, 1970.

73. Kats, S.A., Berezin, B.Ya., Gorina, N.B., Polyakova, V.P., Savitski, E.M., and Chekhovskoi, V.Ya., Izv. Akad. Nauk SSSR, Met., 6, 87–9, 1974.

74. Kats, S.A., Chekhovskoi, V.Ya., Gorina, N.B., Polyakova, V.P., and Savitski, E.M., Trans. 7th All-Union Sci. Conf. on Calorimetry, Moscow, Chernogolovka, Jan. 31–Feb. 3, 1977, 351–7, 1977.

75. Stretz, L.A. and Bautista, R.G., J. Chem. Thermodyn., <u>7</u>, 83–8, 1975.

76. Stretz, L.A. and Bautista, R.G., J. Chem. Eng. Data, <u>21</u>, 13–5, 1976.

77. Stretz, L.A. and Bautista, R.G., High Temp. Sci., <u>7</u>, 197–203, 1975.

78. Arpaci, E. and Frohberg, M., Z. Metallkd., <u>73</u>, 548–51, 1982.

79. Margrave, J.L., Bonnell, D.W., Frost, R.T., and Grow, R.T., unpublished work, 1981–1984.

80. Kats, S.A., Abstract of Dissertation for Cand. Sci., Institute for High Temp., Acad. Sci. USSR, 1978.

81. Frohberg, M. and Betz, G., Arch. Eisenhuettenwes., <u>51</u>, 235–40, 1980.

82. Betz, G. and Frohberg, M., Metall (Berlin), <u>35</u>, 299–303, 1981.

83. Wouch, G., Gray, E.L., Frost, R.T., and Lord, A.E., Jr., High Temp. Sci., <u>10</u>, 241–59, 1978.

84. Margrave, J.L., Bonnell, D.W., and Weingarten, J.S., 'Thermodynamic Property Determination in Low Gravity,' NASA Final Report on Contract NAS8-32030, 35 pp., 1977.

85. Shaner, J.W., Gathers, G.R., and Minichino, C., High Temp.-High Pressures, <u>8</u>, 425–9, 1976.

86. Shaner, J.W., Gathers, G.R., and Minichino, C., High Temp.-High Pressures, <u>9</u>, 331-63, 1977.

87. Gathers, G.R., Shaner, J.W., and Hodgson, W.M., High Temp.-High Pressures, <u>11</u>, 529-38, 1979.

88. Cezairliyan, A., Morse, M.S., Foley, G.M., and Erickson, N.E., in <u>Proceedings of the Eighth Symposium on Thermophysical Properties</u> (Sengers, J.V., Editor), Vol. II, ASME, New York, NY, 45-50, 1982.

89. Margrave, J.L., in <u>Proceedings of the Eighth Symposium on Thermophysical Properties</u> (Sengers, J.V., Editor), Vol. II, ASME, New York, NY, 31-44, 1982.

90. Montgomery, R.L., Sundareswaran, P.C., and Margrave, J.L., unpublished work, Rice Univ., 1983.

91. Barmatz, M., in <u>Materials Processing in the Reduced Gravity Environment of Space</u> (Rindone, G., Editor), Vol. 9, Elsevier, New York, NY, 25-37, 1982.

92. Frost, R.T. and Chang, C.W., in <u>Materials Processing in the Reduced Gravity Environment of Space</u> (Rindone, G., Editor), Vol. 9, Elsevier, New York, NY, 71-80, 1982.

93. Munro, R.G. and Mountain, R.D., Phys. Rev., <u>B28</u>, 2261-3, 1983.

94. Iuchi, T., Symposium on Applications of Radiation Thermometry, Sponsored by ASTM and NBS, May 8, 1984.

95. Montgomery, R.L., Sundareswaran, P.C., and Margrave, J.L., High Temp. Sci., in press.

7.9. APPENDIX: NOMENCLATURE

Symbol	Description	Unit
AC	alternating current	
D	diameter	cm
F_L	levitation force	dyne
G	mass flow rate	$g \cdot s^{-1} \cdot cm^{-2}$
G(X)	a function in Eq. (1)	
g	acceleration due to gravity	$cm \cdot s^{-2}$
g_i, g_f	temperature drift rates	$K \cdot s^{-1}$
H	enthalpy	$J \cdot mol^{-1}$
h	heat transfer coefficient; interval size in the trapezoidal rule	$J \cdot cm^{-2} \cdot s^{-1} \cdot K^{-1}$
ID	inside diameter	cm
I_1, I_2	integrals	
J_1, J_2	quantities in brackets in Eqs. (14) and (15)	
J_λ	radiation flux per unit band width	$W \cdot cm^{-2}$
k	Boltzmann constant	$erg \cdot K^{-1}$
k	constant in Newton's law for heat exchange	s^{-1}
OD	outside diameter	cm
RF	radio frequency	
R_1	characteristic length in Eq. (1)	cm
R_2	radius of a levitated sample, Eq. (1)	cm
r_λ	reflectivity	
T_B^*	observed brightness temperature	K
T_0	thermodynamic temperature	K

T_{REF}	reference temperature	K
v	velocity	$cm \cdot s^{-1}$
X	a dimensionless quantity in Eq. (1)	
x	drop distance	cm
y	a function in general; a function defined in Eq. (24)	
*	superscript indicating values for the actual sample size in an experiment	

Greek Symbols

ΔH_c^*	heat lost by conduction	J
ΔT	temperature difference	K
θ	temperature in arbitrary units	
λ	wavelength	cm
λ_f	gas film thermal conductivity	$W \cdot m^{-1} \cdot K^{-1}$
μ_f	gas film viscosity	$g \cdot cm^{-1} \cdot s^{-1}$
π	ratio of circumference to diameter	

CHAPTER 8

Modulation Calorimetry

YA. A. KRAFTMAKHER

8.1. INTRODUCTION

The method of modulation calorimetry is based on modulating the power which heats the sample and measuring the arising oscillations of the sample temperature around its mean value. The use of periodic temperature oscillations provides important advantages. When the modulation frequency is sufficiently high, the correction for heat exchange can be made small even at the highest temperatures. Another important advantage of modulation calorimetry is that the harmonic temperature oscillations can be measured by using lock-in amplifiers and selective amplifiers tuned to the modulation frequency. This feature becomes very important when a good temperature resolution is required. The modulation technique provides a unique method of performing measurements within temperature oscillations of the order of 0.001–0.01 K.

Various modifications of modulation calorimetry differ in the method by which the modulation of heating power is achieved (heating by alternating current or by direct current with a small a.c. component, radiation or electron-bombardment heating, induction heating) and in the methods of measuring the oscillations in the sample temperature (by the resistance of the sample, by thermal radiation of the sample, or by using a thermocouple or a resistance thermometer). Modulation calorimetry enables measurements to be performed with high sensitivity over a wide temperature range, from fractions of a kelvin up to the melting points of refractory metals. In many cases it is possible to develop compensation schemes whose balance does not depend on the amplitude of power oscillations, or to employ automatic recording of the quantities being measured.

In modulation calorimetry, it is assumed that the mean temperature and the amplitude of temperature oscillations are the same over the entire sample; in this respect, this method differs from the method of temperature waves. As a rule, the measurements are performed in a regime where the amplitude of temperature oscillations is inversely proportional to the specific heat of the sample.

The first measurements of specific heat by the modulation method were made by Corbino [1], who originally developed the theory of the method. In his measurements, the amplitude of temperature oscillations was determined from the oscillations of the sample resistance. Since then, considerable advances in the method have been made, particularly in recent years, as a result of the rapid progress in electronic instrumentation. During the first stage of development, the method was used exclusively for high-temperature measurements. The attractive feature of the method was the smallness of the correction for the heat exchange. The samples were metals, usually in the form of a wire or a rod, which were heated by passing an electric current through them. The temperature oscillations were determined from measurements of sample resistance or from measurements of thermal radiation emitted by the sample. This method was used to measure the specific heats of a number of refractory metals at high temperatures. In later studies, the method was also used for studying the specific heat near phase transitions in metals where good temperature resolution is essential.

The second stage of development of modulation calorimetry involved the application of the method to measurements at low and moderate temperatures and on nonconducting materials. A significant advance in the method has been the development of modulated-light heating techniques and the use of thermocouples and resistance thermometers for measuring the temperature oscillations. It turns out that at low and moderate temperatures, the traditional domain of adiabatic calorimetry, the modulation method can give better temperature resolution and higher sensitivity although the absolute accuracy of the method is lower than that of the adiabatic calorimetry. (Under the most favorable conditions, the uncertainty in measuring the specific heat by the modulation method is about 1-2%.) Consequently, the method is widely used for studying the nature of anomalies in the specific heat at phase transition points (ferromagnets, antiferromagnets, ferroelectrics, ordering alloys, superconductors, and liquid crystals). The modulation method is particularly useful in studies where a small size of sample is important and has also been successfully used for calorimetry under high pressures.

A compilation of studies [1-22] important in the development of modulation calorimetry is presented in Table 8.1. References [23-25] are recent review papers on the technique.

8.2. THEORY OF MODULATION CALORIMETRY

The basic theory of modulation calorimetry is quite simple. If the heating power is modulated by a sine wave $(p_0 + p \sin\omega t)$ then the sample temperature starts to oscillate around a mean value T_0. The heat balance equation takes the form

$$mcT' + P(T) = p_0 + p \sin\omega t, \tag{1}$$

where m and c are the mass and specific heat of the sample, $P(T)$ is the rate of heat loss from the sample, $T' = dT/dt$, and ω is the modulation frequency $(\omega = 2\pi f)$. Assuming $T = T_0 + \theta$, $\theta \ll T_0$, and taking $P(T) = P(T_0) + P'\theta$ (P' is the heat transfer coefficient), one obtains the steady-state solution of Eq. (1):

$$\left.\begin{aligned} T &= T_0 + \theta_0 \sin(\omega t - \phi), \\ P(T_0) &= p_0, \\ \theta_0 &= (p/mc\omega) \sin\phi, \\ \tan\phi &= mc\omega/P'. \end{aligned}\right\} \tag{2}$$

The condition $\tan\phi \gg 1$ ($\sin\phi \simeq 1$) is the criterion of the adiabatic regime of measurements. If the angle ϕ is close to 90°, the corrections for heat exchange can be neglected. The

Table 8.1. Development of Modulation Calorimetry

Item	Investigators	Year	Ref.
General theory, supplementary-current method, third-harmonic method	Corbino	1910	[1]
Use of thermionic-current oscillations	Smith and Bigler	1922	[2]
Development of the third-harmonic method	Filippov	1960	[3]
	Rosenthal	1961	[4]
Bridge circuit for wire samples	Kraftmakher	1962	[5]
Photodetectors	Lowenthal	1963	[6]
Electron-bombardment heating	Filippov and Yurchak	1965	[7]
Liquid metals	Akhmatova	1965	[8]
Account of temperature gradients	Holland and Smith	1966	[9]
Account of thermal resistances, measurements at low temperatures	Sullivan and Seidel	1966	[10]
Modulated-light heating	Handler et al.	1967	[11]
Induction heating	Filippov and Makarenko	1968	[12]
Nonconducting materials	Glass	1968	[13]
Measurement of the temperature coefficient of specific heat	Kraftmakher and Tonaevskii	1972	[14]
Nonadiabatic regime	Varchenko and Kraftmakher	1973	[15]
High pressures	Bonilla and Garland	1974	[16]
Microcalorimeters for organic materials	Schantz and Johnson	1978	[17]
	Smaardyk and Mochel	1978	[18]
	Tanasijczuk and Oja	1978	[19]
Improvement of the modulated-light method	Ikeda and Ishikawa	1979	[20]
Measurement of specific heat using temperature fluctuations	Kraftmakher and Krylov	1980	[21]
Frequencies up to 10^5-10^6 Hz	Kraftmakher	1981	[22]

adiabatic regime means that the oscillations of the heating power are much larger than the oscillations of the heat loss from the sample due to the temperature oscillations. Although the heat transfer coefficient grows rapidly with increasing temperature, the regime of measurements can be kept adiabatic by increasing the modulation frequency. Under adiabatic conditions, $mc = p/\omega\theta_0$ and, thus, it is sufficient to measure the quantities p, ω, and θ_0 in order to calculate the heat capacity of the sample.

With a separate heater and thermometer, one has to take into account thermal resistances in the system, as well as a finite thermal conductivity of the sample. Sullivan and Seidel [10] have considered a heater, sample, and thermometer with heat capacities C_h, C_s, and C_t, interconnected by thermal conductances K_h and K_t as shown in Fig. 8.1. If the thermal conductivity of the sample is assumed to be infinite, the heat balance equations for the system are given by

$$C_h T'_h = p_o + p \sin\omega t - K_h(T_h - T_s),$$

$$C_s T'_s = K_h(T_h - T_s) - K_b(T_s - T_b) - K_t(T_s - T_t), \qquad (3)$$

$$C_t T'_t = K_t(T_s - T_t).$$

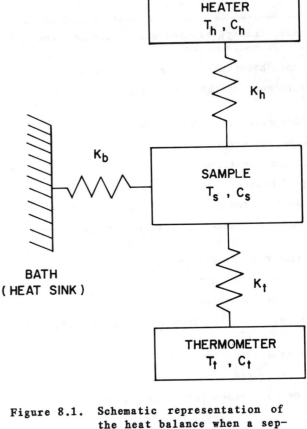

If the temperature variations are sufficiently small, the parameters C and K may be considered constant. The steady-state solution of these simultaneous equations consists of two terms, one a constant that depends upon K_b and the other an oscillatory term inversely proportional to the heat capacity:

$$T_t = T_b + p_o/K_b + (pB/\omega C)\sin(\omega t - \phi), \qquad (4)$$

where $C = C_s + C_t + C_h$, and B is a complicated expression involving quantities from Eq. (3); ϕ is a phase angle approximately equal to 90° under the conditions discussed below.

If (1) the heat capacities of the heater and of the thermometer are much less than that of the sample, (2) the sample, heater, and thermometer come to equilibrium in a time much shorter than the modulation period, and (3) the period of modulation is much shorter than the sample-to-bath relaxation time, then to first order in $1/\omega^2 \tau_s^2$ and $\omega^2(\tau_h^2 + \tau_t^2)$,

Figure 8.1. Schematic representation of the heat balance when a separate heater and thermometer are used in modulation calorimetry. The a.c. power applied to the heater provides an oscillating heat input to the sample. The heat flows through the sample and out to the heat sink. The internal relaxation times must be much shorter than the sample-to-bath relaxation time (Sullivan and Seidel [10]).

$$B = [1 + 1/\omega^2 \tau_s^2 + \omega^2(\tau_h^2 + \tau_t^2)]^{-1/2},$$

$$\cot\phi = 1/\omega\tau_s - \omega(\tau_h + \tau_t), \qquad (5)$$

where the relaxation times are defined as $\tau_s = C/K_b$, $\tau_h = C_h/K_h$, and $\tau_t = C_t/K_t$.

For determining the effect of the finite thermal conductivity of the sample, Sullivan and Seidel have considered a sample in the form of a slab which is heated uniformly on one side by a sinusoidal heat flux. The other side of the slab is coupled uniformly to the bath through the thermal conductance K_b. In this case, the amplitude of the temperature oscillations of the sample can be expressed in the form

$$\theta_o = (p/\omega C)[1 + 1/\omega^2\tau_s^2 + \omega^2\tau^2 + 2K_b/3K_s]^{-1/2}, \qquad (6)$$

where various time constants have been lumped into τ, such that $\tau^2 = \tau_h^2 + \tau_t^2 + \tau_{int}^2$; the quantity

τ_{int} is the internal relaxation time of the sample which depends on its thermal diffusivity and thickness, and K_s is the thermal conductance of the sample in the direction of the heat flow.

Under actual operation conditions, τ_s is about 2-3 orders of magnitude larger than τ so that condition $\omega^2\tau_s^2 \gg 1 \gg \omega^2\tau^2$ is satisfied by a proper choice of the modulation frequency. The term $2K_b/3K_s$ is very small due to the small thickness of the sample and can be neglected. Most modulation measurements are performed under conditions when the above assumptions are well satisfied so that the expression for calculating the heat capacity of the sample takes the simple form:

$$C_s = p/\omega\theta_o. \qquad\qquad (7)$$

8.3.   MODULATION OF HEATING POWER

8.3.1.   Direct Heating

Samples with sufficiently high electrical conductivity, in the form of a wire, a foil, or a rod, can be conveniently heated by passing an electric current through them. Three methods of modulation of the heating power have been used:

(1) Heating by alternating current. The modulation frequency is twice the frequency of the current. The amplitude of power oscillations can be easily determined from the effective power.

(2) Heating by direct current with a small a.c. component. The modulation frequency is equal to the frequency of the a.c. component. In this case the mean temperature and the amplitude of temperature oscillations can be changed independently.

(3) Heating by high-frequency current modulated with a suitable frequency. Such a method is useful when the sample temperature and its oscillations are measured with a thermocouple connected electrically with the sample. In this case, the low-frequency signal corresponding to the temperature oscillations can be easily separated from the high-frequency voltage.

8.3.2.   Electron-Bombardment Heating

In this case the heating power can be modulated in several ways:

(1) In the saturation regime, the accelerating voltage is modulated but the electron beam current remains constant.

(2) The temperature of the cathode is varied or a control electrode is used to produce the modulation of the electron beam current with the accelerating voltage remaining constant.

(3) The accelerating voltage is periodically switched on and off so that the form of the heating-power pulses is rectangular.

8.3.3.   Use of Electrical Heaters

Separate heaters for a.c. heating have been used mainly at low temperatures and in microcalorimeters used for measurements on organic materials. The main requirements for the heater are that it must have a small heat capacity and a good thermal contact with the sample. Often for this purpose microresistors or specially deposited thin films are used as heaters. With this method, high accuracy in the determination of the heating-power oscillations can be achieved.

8.3.4. Modulated-Light Heating

This elegant method was first used in the measurements of specific heat of nickel near the Curie point [11] and is widely employed now. A sample in the form of a foil or a thin slab is placed in a furnace which controls the mean temperature (Fig. 8.2). The sample is illuminated by the light from an incandescent lamp passed through a chopper. The temperature oscillations are measured with a thermocouple and a lock-in amplifier. The output signal from the lock-in amplifier is a d.c. voltage proportional to the input a.c. voltage, i.e., inversely proportional to the specific heat of the sample. The output signal is fed into the Y-input of a recorder and the X-input is connected to a thermocouple which measures the mean temperature in the furnace. The measurements are made with the furnace temperature gradually changing.

In the majority of studies using this method, the absolute magnitude of the heating-power oscillations was not determined since steps were taken to make it independent of the sample temperature. This was usually accomplished by coating the samples with a thin layer of graphite in order to make the absorptivity of the sample surface essentially independent of temperature.

8.3.5. Induction Heating

Induction heating of the samples is achieved by placing them in an induction furnace whose high-frequency current is modulated with a low frequency. The applied power is determined by measuring the voltage induced in a separate coil. The electrical conductivity of the sample, which is required for calculating the heating-power oscillations, is measured by means of current and potential probes [12].

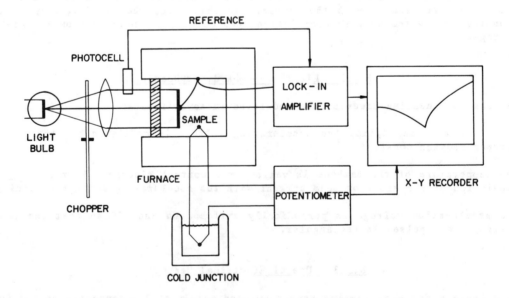

Figure 8.2. Functional diagram of a measurement system employing the modulated-light heating method. Modulated light falls onto the surface of a thin sample causing small temperature oscillations. Their amplitude, of the order of 0.01 K, is measured with a thermocouple and a lock-in amplifier and is recorded as a function of the mean temperature (Handler et al. [11]).

8.4. MEASUREMENT OF TEMPERATURE OSCILLATIONS

8.4.1. Methods Based on the Temperature Dependence of Sample Resistance

Determination of the temperature oscillations of the sample from the oscillations of its electrical resistance has been used by many workers. The advantage of this method is that measurements can be performed at relatively low temperatures where the radiation from the samples is too small to be measured pyrometrically; also, the modulation frequencies can be high since the problem of thermal inertia of the temperature sensor is eliminated here. The only, but very serious, drawback of the method is the necessity of knowing the temperature dependence of both the sample resistance and its temperature derivative. Therefore, the method should not be used when these quantities exhibit anomalies. Sample resistance oscillations can be used to detect temperature oscillations in three different ways: by the supplementary-current method, by the third-harmonic method, and by the equivalent-impedance method.

8.4.1.1. Supplementary-Current Method

This method was first used by Corbino [1]. He observed the appearance of a d.c. voltage across a sample when a supplementary a.c. current was passed through it with a frequency twice that of the heating current, i.e., with the frequency of sample temperature oscillations. A modified version of the method was used for measuring the temperature coefficient of specific heat [14]. In another version of the method [26], a current with a frequency of 1 Hz was passed through the sample, setting up temperature oscillations at 2 Hz. The temperature oscillations were detected by means of a supplementary current with the frequency 320 Hz.

8.4.1.2. Third-Harmonic Method

Corbino [1] has shown that when an a.c. voltage is applied to a sample the current through it contains a third-harmonic component related to the sample temperature oscillations. If the a.c. current is passed through the sample in series with a sufficiently high resistance, then the corresponding amplitude of the third-harmonic voltage appearing across the sample under adiabatic conditions is

$$V_3 = I^3 R_o R' / 8mc\omega = V^3 R' / 8R_o^2 mc\omega, \tag{8}$$

where I and V are the amplitudes of the fundamental current and voltage across the sample, R_o and R' are the sample resistance and its temperature derivative at the mean temperature.

Filippov [3] and Rosenthal [4] used a bridge circuit with the sample as one of the arms to study third-harmonic generation (Fig. 8.3). The bridge was balanced at the fundamental frequency and the third-harmonic signal at the bridge output was then measured.

8.4.1.3. Equivalent-Impedance Method

Radio engineers have known for a long time that a temperature sensitive resistor through which a direct current is flowing can be represented by an equivalent impedance which is, in

particular, a function of the specific heat of
the resistor [27,28]. The method of equivalent
impedance has proved to be very convenient for
measuring the specific heat of wire samples at
high temperatures [5].

Consider a wire sample with a current I_0 +
$i \sin \omega t$ ($i \ll I_0$) passing through it. Sample
temperature oscillations about a mean value are
described by expressions:

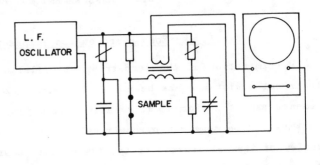

$$\left. \begin{array}{l} \theta = \theta_0 \sin(\omega t - \phi), \\ \theta_0 = (2I_0 iR_0/mc\omega) \sin\phi, \\ \tan \phi = mc\omega/(P'-I_0^2 R'). \end{array} \right\} \quad (9)$$

Here ϕ is the phase shift between the a.c. component of the current and the temperature oscillations.

Figure 8.3. Bridge circuit employed to
study third-harmonic generation. The fundamental voltage is compensated, and the
third-harmonic signal is seen
on the oscilloscope screen
(Filippov [3], Rosenthal
[4]).

The resistance of the sample oscillates and is equal to

$$R(T) = R_0 + R'\theta = R_0 + R'\theta_0 \cos\phi \sin\omega t - R'\theta_0 \sin\phi \cos\omega t. \qquad (10)$$

Thus, an a.c. voltage appears across the sample which, if the smallest terms are neglected, is
given by

$$V = iR_0 \sin\omega t + I_0 R'\theta_0 \cos\phi \sin\omega t - I_0 R'\theta_0 \sin\phi \cos\omega t. \qquad (11)$$

The impedance of the sample, which describes the amplitude and phase relation between the a.c.
component of the current and the a.c. voltage across the sample, can therefore be written as $Z = R_0+A-jB$. The quantities A and B can be readily determined by dividing the a.c. voltage across the
sample given in Eq. (11) by the a.c. component of the current and then using the expression for θ_0
in Eq. (9):

$$Z = R_0 + (2I_0^2 R_0 R'/mc\omega)\sin\phi \cos\phi - j(2I_0^2 R_0 R'/mc\omega)\sin^2\phi. \qquad (12)$$

An analogous impedance is displayed by a circuit of resistance R and capacitance C connected
in parallel:

$$Z_1 = R/(1 + \omega^2 R^2 C^2) - j\omega R^2 C/(1 + \omega^2 R^2 C^2). \qquad (13)$$

At sufficiently high modulation frequencies such that $\phi \simeq 90°$ and that $\omega^2 R^2 C^2 \ll 1$ is still satisfied, $A \ll R_0$. Then $R = R_0$, $B = \omega R^2 C$, and the sample heat capacity may be expressed as:

$$mc = 2I_0^2 R'/\omega^2 RC. \qquad (14)$$

One can either estimate the errors due to the assumptions made or replace the approximate expressions by the exact ones in calculating mc.

The specific heat of the sample is thus directly related to the parameters of the equivalent impedance, R and C. Consequently, the specific heat can be measured by the use of a bridge circuit in which one arm is shunted by a variable capacitor (Fig. 8.4). A selective amplifier tuned to the modulation frequency permits high sensitivity to be obtained. For very small temperature oscillations, it is necessary to use a lock-in detector.

The bridge circuit does not allow measurements to be made at relatively low temperatures (below 1000 K) since the cold-end effects become significant. In this case it is necessary to use very long samples or to heat the current leads clamped to the ends of the samples. More convenient is a potentiometric scheme which makes it possible to measure the specific heat of the central part of a sample defined by fine potential probes [29].

The temperature of wire samples heated by an electric current can also be determined by their thermal noise [30].

8.4.2. Photoelectric Detectors

Determination of the sample temperature oscillations using photoelectric detectors has been widely employed. Lowenthal [6] used this method to measure the specific heat of refractory metals in the 1200-2400 K range. The samples were heated by an a.c. current and a photomultiplier was used as the temperature sensor (Fig. 8.5). The dependence of the photomultiplier current on the sample temperature was assumed to have the form $I = AT^n$, with n weakly dependent on temperature. The amplitude of the temperature oscillations can then be determined from the expression

$$\theta_o = TV/nV_o, \tag{15}$$

where V_o and V are the amplitudes of the d.c. and a.c. components of the photomultiplier output signal.

Filippov and Yurchak [7] assumed the current-temperature dependence in the form $I = B \exp(-A/T)$. In this case

$$\theta_o = T^2V/AV_o. \tag{16}$$

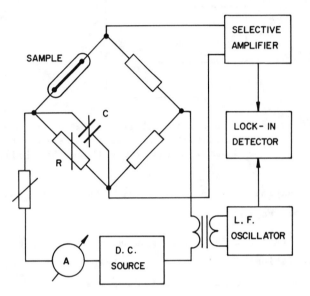

Figure 8.4.   Bridge circuit for measuring the specific heat of wire samples. The balance of the bridge does not depend on the amplitude of the a.c. component of the heating current. The frequency of temperature oscillations is of the order of 10^2 Hz. For temperature oscillations of the order of 1 K, the sensitivity is 0.1%. The accuracy of measurements is determined by the data on the temperature dependence of the sample resistance [5].

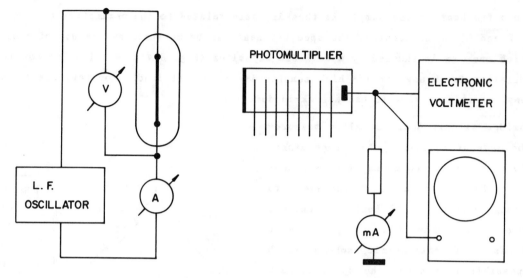

Figure 8.5. Functional diagram of a system for measuring the temperature oscillations with a photodetector. It is necessary to know the dependence of the photocurrent on the sample temperature (Lowenthal [6]).

This expression can only be used when the temperature dependence of quantity A can be established in some way.

Akhmatova [8] determined the specific heat of some molten metals by comparing the temperature oscillations (as measured by a photoelectric detector) of two niobium capillaries, one filled with the metal under study and the other empty.

8.4.3. Thermocouples and Resistance Thermometers

Thermocouples and resistance thermometers provide the most reliable method of measuring the sample temperature and the amplitude of its oscillations and should be used whenever possible. Thermocouples are widely used when the modulated—light heating method or when a separate heater is employed. They are made from thin wires or strips, 1–10 μm thick or, in some cases, are formed by depositing thin films, 10–100 nm thick. Thus, a good thermal contact with the sample and low thermal inertia can be achieved.

8.4.4. Use of Phase—Locked Detectors

Phase—locked detectors are widely used now for measuring weak periodic signals in the presence of noise. The method is based on the fact that the frequency of the expected signal is known. A periodic signal is detected with a parametric detector controlled by a reference voltage. The reference voltage is taken from the same oscillator which controls the process under study. Therefore, the frequency of the signal always coincides with the reference frequency while the phase shift between them remains constant. In modulation calorimetry the reference voltage is supplied either by the source of the modulated power or by a special sensor (for example, a photocell when modulated—light heating is used). The output signal from the phase—locked detector is averaged over a time sufficiently long for suppression of the noise. The effective bandwidth of the phase-locked detector is inversely proportional to the time of averaging. Thus, the phase-locked detector is always tuned to the signal frequency and has a bandwidth which can be easily adjusted by suitable selection of averaging time.

Amplifiers employing phase-locked detection are called lock-in amplifiers. Due to the narrow bandwidth, they have a high sensitivity and are immune to noise. When used with the thermocouples, lock-in amplifiers make it possible to measure temperature oscillations as small as 10^{-4}K [16]. Some lock-in amplifiers employ two phase-locked detectors which are controlled by reference voltages whose phases are shifted by 90°. This allows the measurements of both amplitude and phase of the measured signal. A double phase-locked detector is necessary, for example, in bridge measurements where the signal may contain both in-phase and quadrature components.

8.5. MEASUREMENTS AT HIGH MODULATION FREQUENCIES

In modulation calorimetry the frequency of the temperature oscillations can be varied over a wide range. Therefore, the method is suitable for studying relaxation effects in specific heat. Such effects may be expected, for example, in the ordering of alloys, in the dissolving of gases in metals, and in the formation of equilibrium vacancies in the crystal lattice.

The relaxation method for studying vacancy formation in solids was proposed a long time ago [31,32]. It involves creating such rapid changes of the sample temperature that the vacancy concentration cannot follow the temperature. Hence, when specific heat is measured at a sufficiently high modulation frequency the result must correspond to the specific heat of a vacancy-free crystal (the changes in specific heat due to the presence of static defects are much smaller than the effect due to the changes in their concentration). In metals, however, the short relaxation time of the vacancies makes implementation of the method very difficult.

The relaxation effect in specific heat manifests itself in a variation of the amplitude and phase of the temperature oscillations caused by the modulation of the heating power [33]. When the observation of relaxation effects requires very high modulation frequencies, of the order of 10^5–10^6 Hz, serious problems associated with diminishing amplitude of the temperature oscillations and with uncontrollable phase shifts in the measuring scheme begin to appear. These problems can be overcome, however, by using one of two methods of heating the sample such that the specific heat can be determined simultaneously from low and high frequency temperature oscillations (Fig. 8.6). In the first method, the high-frequency current heating the sample is modulated slightly by a low frequency (modulation coefficient of about 1%) thereby giving rise to simultaneous temperature oscillations at both low and high frequencies. The relation between the oscillations of the power heating the sample at the two frequencies remains constant when the heating current is changed. The second method of heating consists of passing through the sample a direct current with two a.c. components, of high and low frequencies, superimposed on it. In this case the relation between the amplitudes of modulation at the two frequencies must remain constant when the direct current is varied.

The oscillations of the sample temperature are measured with a photomultiplier. The low-frequency component of its output signal is amplified by a selective amplifier. The high-frequency component is selected with a resonant circuit, amplified and fed to a frequency converter. A quartz oscillator is used as a heterodyne. The signal with the difference frequency is then amplified with a selective amplifier and fed to the lock-in detector. The frequency of the current heating the sample is set by the frequency synthesizer, the source of the reference frequency being the heterodyne. This arrangement enables any value of the difference frequency to be selected and insures its stability. The reference frequency for the lock-in detector is taken from an auxiliary frequency converter.

The observation of relaxation effects in specific heat can be conveniently carried out on thin foil samples. The sample has the form of a strip narrowing towards one of its ends. When a current is passed through the sample its different areas have different mean temperatures. The temperature of each area is determined by comparing its brightness with the brightness of an auxiliary sample from the same foil whose temperature is measured with a thermocouple. The measurements are simplified by successive projection of different areas of the sample on the cathode of the photomultiplier. The low-fre-

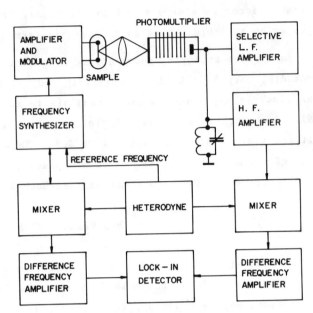

Figure 8.6.   Functional diagram of a system for measurements at high modulation frequencies. High-frequency current heating the sample is slightly modulated by a low frequency. For observing relaxation effects in specific heat it is sufficient to compare the temperature oscillations at high and low frequencies when the mean temperature of the sample is gradually changed [22].

quency component of the photocurrent is maintained constant by means of a diaphragm. The mean current of the photomultiplier is also maintained constant by using additional illumination of the cathode from an incandescent lamp. The signal from the lock-in detector is recorded for each area of the sample. Thus, the measurements in the whole temperature range are made without changing the heating current, the photocurrent, or the signal amplitude. This has proved to be very useful for excluding undesirable phase shifts.

This method considerably increases the possibility of observation of relaxation effects in specific heat. In measurements on samples 0.01 mm thick at temperatures above 2300 K and a modulation frequency of 3×10^5 Hz, a resolution of 1% in amplitude and $0.2°$ in phase of the temperature oscillations was achieved [22]. Earlier, similar measurements were made at modulation frequencies below 10^3 Hz [34,35].

8.6.   ACCURACY OF MODULATION MEASUREMENTS

The measurement errors can be grouped into two categories: errors arising from differences between the theoretical model for the modulation method and experimental conditions, and instrumentation errors due to the inaccuracy of the measuring instruments. The theoretical model assumes the following conditions:

(1) The mean temperature is the same throughout the sample.

(2) The amplitude and the phase of temperature oscillations do not vary throughout the sample.

(3) The modulation frequency is sufficiently high to meet the criterion of adiabatic conditions: the changes in the heat loss from the sample during oscillations of its temperature are much smaller than the oscillations of the heating power.

(4) The heat capacity of the sample is much larger than that of the heater and the thermometer.

Condition (2) supposes that, when a separate heater and thermometer are used, the time for establishing thermal equilibrium between them and the sample is much smaller than the period of modulation. Hence, conditions (2) and (3) pose contradictory requirements for the modulation frequency.

The theoretical conditions are best satisfied when the sample is heated by an electric current and the temperature oscillations are determined from measurements of the oscillations in sample resistance or surface brightness. In this case, condition (4) does not apply and condition (3) is easily met by increasing the modulation frequency. In order to exclude cold-end effects, the measurements can be performed only on the central part of the sample where the axial temperature gradients are small. Radial temperature differences are insignificant due to high thermal conductivity of conducting samples and their small thickness. Therefore, the conditions (1) and (2) are also easily satisfied with direct electrical heating.

In other methods of heating the conditions (1) and (2) are met by decreasing the sample thickness and the modulation power. When the modulated power is applied to the sample, the mean temperature of the sample increases. This temperature increase is much larger than the amplitude of the temperature oscillations and can result in a nonuniformity of the temperature in the sample. To make this effect negligible, only a small modulated power is supplied to the sample while the mean sample temperature is controlled with a furnace. This method of heating is necessary in phase transition studies when a good temperature resolution is one of the important requirements. Thickness of the samples with low thermal conductivity usually does not exceed 0.1-0.2 mm. Decreased sample thickness is favorable for satisfying conditions (1) and (2) but contradicts requirement (4); this contradiction is removed by making the heater and the thermometer in the form of thin films. Clearly, condition (4) cannot be met when specific heat of a thin film is measured by methods other than direct electrical heating.

In some applications, the increase in modulation frequency needed to achieve adiabatic conditions is undesirable (for example, when heat is not uniformly generated in the entire volume of the sample or when a temperature sensor with a long time constant is used). In the case of radiative heat transfer, it is possible to decrease the modulation frequency and to determine the specific heat in the nonadiabatic regime [15]. Another possible method consists of reducing the effects of the losses by surrounding the sample with a heated shield, the temperature of which oscillates in phase with the temperature oscillations of the sample [36]. These methods, however, cannot be used when other mechanisms of heat exchange contribute significantly to the heat loss from the sample. In such cases the modulation frequency should be chosen carefully. Generally, one should experimentally verify that there is a frequency range where the product $\omega\theta_0$ is constant; this means that conditions (2) and (3) are satisfied. It turns out that these conditions can usually be met simultaneously. The only exception is when samples with low thermal diffusivity are studied under high pressures. In this case, in order to obtain the specific heat value, one has to introduce large corrections [37-39].

The main sources of error in modulation calorimetry are the determinations of the amplitudes of the heating power oscillations and the sample temperature oscillations. Modern measuring instruments can accurately determine the oscillations of the heating power when direct heating or resistive heaters are used. However, when using other methods of heating, the situation is not as favorable.

Temperature oscillations can be measured with highest accuracy by means of thermocouples or resistance thermometers. In other methods (temperature dependence of resistivity, photoelectric detectors) the errors are determined by the reliability of the data used or by the accuracy of calibration.

Estimates of errors in modulation calorimetry are presented in Table 8.2. The term 'imprecision' refers to the differences of single data points from the smoothed values, and the term 'inaccuracy' refers to the total error including random and systematic errors.

Table 8.2. Estimates of Errors in Modulation Calorimetry

Source of Error	Imprecision (%)	Inaccuracy (%)
Mean temperature of the sample		0.01-1[a]
Mass of the sample		0.1-1[b]
Modulation frequency		$\leq$0.1[c]
Oscillations of heating power:		
Direct resistive heating		0.2-1[c]
Electron bombardment		1-2[c]
Indirect resistive heating		0.2-1[c]
Modulated-light heating	0.1	2[d]
Induction heating		4-6[e]
Temperature oscillations:		
Supplementary-current method	1	$\geq$2[f]
Third-harmonic method	1	$\geq$2[f]
Equivalent-impedance method	0.1	$\geq$1[f]
Photoelectric detectors	0.5	$\geq$1[g]
Thermocouples	0.1	0.2-1[a]
Resistance thermometers	0.01-0.1	0.2-1[a]

[a]Depends on the temperature.
[b]May be greater in microcalorimeters.
[c]Depends on the equipment used.
[d]According to [20].
[e]According to [12].
[f]Depends on the accuracy of R' values.
[g]Depends on the accuracy of calibration.

8.7. APPLICATIONS OF MODULATION CALORIMETRY

There have been numerous studies utilizing modulation calorimetry and their number is rapidly increasing. The main advantages of the method are: wide temperature range, good temperature resolution, and high sensitivity to the changes in specific heat. Some areas in which modulation calorimetry has been used are summarized below.

8.7.1. Specific Heat of Metals at High Temperatures

The interest in this temperature range was enhanced after the observation of a substantial nonlinear rise in the specific heat of molybdenum, tantalum, and graphite [40]. This effect can be explained by formation of equilibrium vacancies in the crystal lattice. A number of specific heat measurements have been performed for the purpose of determining the formation energies and equilibrium concentrations of the vacancies in metals [41–49]. The formation of point defects in solids was predicted long ago by Frenkel [50] when he showed that the equilibrium vacancy concentration is equal to $A \exp(-E_f/k_B T)$, where E_f is the formation energy, k_B is the Boltzmann constant, T is the absolute temperature, and A is a pre-exponential factor. The excess specific heat caused by the vacancy formation is then given by $(E_f^2 A/k_B T^2)\exp(-E_f/k_B T)$. It should be noted that this effect is related not to static defects, but to changes of their concentration with temperature.

For determining the formation energy and the equilibrium vacancy concentration it is necessary to correctly separate the vacancy contribution from the observed specific heat. For most metals the melting point is appreciably higher than the Debye temperature and, therefore, the temperature dependence of the specific heat of a vacancy-free crystal is determined by the temperature dependence of the difference between specific heats C_p and C_v, the electronic specific heat γT, and a possible modification of C_v due to anharmonicity. Theoretical calculations of anharmonicity predict a weak rise of the specific heat (some calculations predict a decrease) which is essentially linear with the temperature. The linear temperature dependence of the specific heat at moderate temperatures is of the order of 0.01 $J \cdot mol^{-1} \cdot K^{-2}$ while the nonlinear rise in specific heat at high temperatures amounts to 10–20 $J \cdot mol^{-1} \cdot K^{-1}$ (Fig. 8.7). The rise in the specific heat due to vacancy formation can be easily distinguished from premelting phenomena. The latter effects have a stronger temperature dependence and exist only in the vicinity of the melting point.

The equilibrium vacancy concentrations obtained from the specific heat data are of the order of 1% at melting points. This result is contrary to the widely-held opinion that the vacancy concentrations in metals are small and cannot cause any observable contribution to the specific heat [56,57]. Many investigators believe that the nonlinear rise in specific heat of metals has another origin, e.g., anharmonicity [58,59]. However, there are numerous arguments favoring high vacancy concentrations in metals [60,61]. One of them is the correctness of the vacancy formation energies derived from specific heat data (Fig. 8.8). The values obtained coincide practically in all cases either with the results obtained by other methods or with the theoretical or empirical estimates.

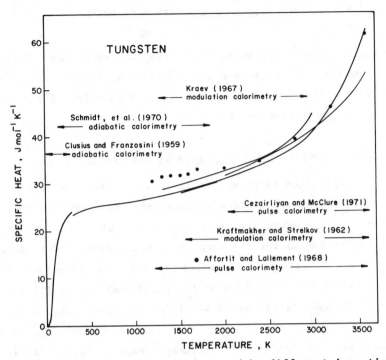

Figure 8.7.   The specific heat of tungsten as measured by different investigators. Only a part of
               the existing data is presented [41,51-55]. The substantial nonlinear rise of specif-
               ic heat still does not get an unambiguous interpretation.

Among the other modulation measurements at high temperatures are the measurements of the specific heat of refractory metals [62-65], of germanium [66] and gold [67], of molten cesium under high pressures [68], and the study of the effect of deformation on the specific heat [69].

8.7.2. Specific Heat Anomalies at Phase Transitions

The excellent temperature resolution and high sensitivity of modulation calorimetry have been successfully used for studying the character of the specific heat anomalies at phase transitions. In such measurements, the exact absolute values of specific heat are often not of interest. The modulated-light heating method is most often used in these studies. Since the electrical resistivity of metals exhibits anomalies at the phase-transition points, the temperature oscillations are usually measured by means of photodetectors, thermocouples, or resistance thermometers.

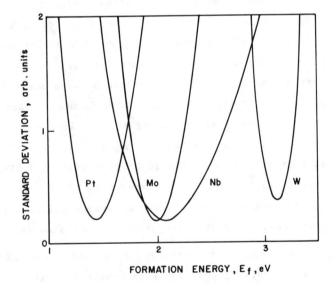

Figure 8.8.   Energies of vacancy formation in metals [61]. To determine the vacancy formation energies, the specific heat data are approximated by the expression $A + BT + (C/T^2)\exp(-E_f/k_BT)$. Among the possible values of E_f, the most probable value is that which provides the best approximation.

Modulation measurements have also contributed to the development of the theory of phase transitions [70–73]. According to theory, the specific heat at a second-order phase transition has a power-law singularity: $C_p = A + Bt^{-\alpha}$, where $t = |T - T_c|/T_c$, T_c is the transition temperature, and α is the critical exponent describing the specific heat anomaly. Thus, the equation for fitting the data contains four parameters: A, B, T_c, and α. In general, the parameters may not be equal for both branches of the specific heat curve (for $T < T_c$ and $T > T_c$). In such cases the critical exponent can be determined if both branches are assumed to have the same value of T_c (Fig. 8.9).

The specific heats of iron [74–77], nickel [11,78–81], cobalt [82], and gadolinium [83–85] have been studied near their Curie points. Modulation calorimetry also has been employed for studying antiferromagnets near the Néel points [86–96], β–brass near the ordering temperature [97,98], phase transitions in metals [99–101] and inorganic substances [102–112], amorphous ferromagnets [113–115], solid electrolytes [116,117], and ferroelectrics [118–120].

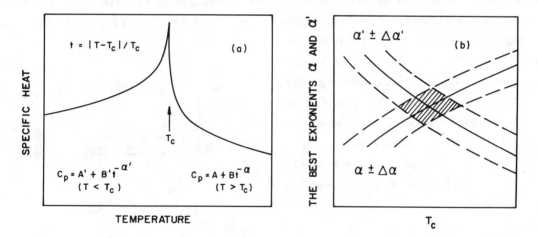

Figure 8.9.   Specific heat at a second-order phase transition. For each branch of the specific heat curve the data are fitted by a power-law singularity. The best values of the critical exponents are determined for every possible T_c. To unambiguously estimate α, one is required to use the same values of T_c and α for both branches. The dashed region shows the uncertainty in the derived values for T_c and α (Connelly et al. [79]).

8.7.3.   Specific Heat at Low Temperatures

For a long time adiabatic calorimetry was the best method of measuring specific heat at low and moderate temperatures, and the modulation method seemed to have no application in this temperature range. However, the importance of low-temperature modulation calorimetry was established by the well-known measurements of Sullivan and Seidel [10] on the specific heat of indium and beryllium at liquid helium temperatures as a function of a magnetic field. They were able to measure temperature oscillations of the order of 0.001 K and changes in the specific heat with a sensitivity of about 0.01%.

Low-temperature modulation calorimetry has been employed for studying thin films [121–131], the effects of high pressures [132–135], and of high magnetic fields [136–138]. The lowest temperature attained in these studies was 0.3 K [122,123], the highest pressure was 35×10^8 Pa

[135], and the strongest magnetic field was 7.5 T [138]. Phase transitions of various kinds [139-143] and properties of elements, alloys, and chemical compounds [144-150] have also been investigated.

8.7.4. Specific Heat of Organic Substances

In recent years, modulation calorimetry has been used for studying organic substances [17-19, 151-160], particularly liquid crystals. In different temperature intervals there exist different phases of the liquid crystal. Accordingly, transitions from one phase to another are accompanied by anomalies in the specific heat. Using modulation calorimetry, it is possible to study these anomalies more precisely than with any other method.

8.7.5. Microcalorimetry

The modulation method has also been applied to microcalorimetric studies of thin films and of organic substances. A microcalorimeter used for the investigation of thin films at liquid helium temperatures [121] is illustrated in Fig. 8.10(a). The total heat capacity is of the order of $10^{-7} J \cdot K^{-1}$ and the resolution which can be attained is about $10^{-10} J \cdot K^{-1}$. Figure 8.10(b) presents a schematic diagram of a microcalorimeter used in the study of organic substances [18]. The calorimeter, which has a sample volume of approximately 0.1 mm^3, can detect heat capacity changes of 1 part in 10^4 with a temperature resolution of better than 0.001 K in the temperature range 20-200°C.

8.8. CONCLUSION

During the last two decades modulation calorimetry has been successfully used over a wide temperature range for studying metals and alloys, semiconductors, dielectrics, and organic substances. Due to the modern measuring techniques, modulation measurements can be performed with good temperature resolution and sensitivity unattainable with other methods. Among the already-developed variants of modulation calorimetry, it is possible to find the most suitable one for each specific case. At the present time, the modulation method has become firmly established among the different methods of performing specific heat measurements.

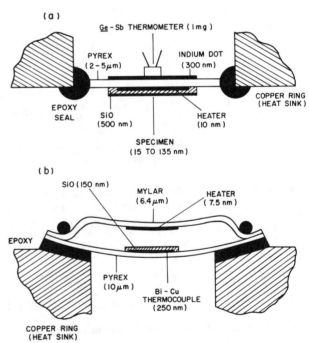

Figure 8.10. Schematic diagrams of microcalorimeters used for (a) the study of thin films (Zally and Mochel [121]) and for (b) the study of organic substances (Smaardyk and Mochel [18]).

8.9. REFERENCES

1. Corbino, O.M., Phys. Z., 11, 413-7, 1910; 12, 292-5, 1911.

2. Smith, K.K. and Bigler, P.W., Phys. Rev., 19, 268-70, 1922.

3. Filippov, L.P., Inzh.-Fiz. Zh., 3(7), 121-3, 1960.

4. Rosenthal, L.A., Rev. Sci. Instrum., 32, 1033-6, 1961; 36, 1179-82, 1965.

5. Kraftmakher, Ya.A., Zh. Prikl. Mekh. Tekhn. Fiz., 5, 176-80, 1962.

6. Lowenthal, G.C., Austral. J. Phys., 16, 47-67, 1963.

7. Filippov, L.P. and Yurchak, R.P., Teplofiz. Vys. Temp., 3, 901-9, 1965.

8. Akhmatova, I.A., Dokl. Akad. Nauk SSSR, 162, 127-9, 1965; Izmer. Tekh., 8, 14-7, 1967.

9. Holland, L.R. and Smith, R.C., J. Appl. Phys., 37, 4528-36, 1966.

10. Sullivan, P.F. and Seidel, G., Annales Academiae Scientiarum Fennicae, Ser. A, VI. Physica, No. 210, 58-62, 1966; Phys. Lett., A25, 229-30, 1967; Phys. Rev., 173, 679-85, 1968.

11. Handler, P., Mapother, D.E., and Rayl, M., Phys. Rev. Lett., 19, 356-8, 1967.

12. Filippov, L.P. and Makarenko, I.N., Teplofiz. Vys. Temp., 6, 149-56, 1968.

13. Glass, A.M., Phys. Rev., 172, 564-71, 1968.

14. Kraftmakher, Ya.A. and Tonaevskii, V.L., Phys. Status Solidi A, 9, 573-9, 1972.

15. Varchenko, A.A. and Kraftmakher, Ya.A., Phys. Status Solidi A, 20, 387-93, 1973.

16. Bonilla, A. and Garland, C.W., J. Phys. Chem. Solids, 35, 871-7, 1974.

17. Schantz, C.A. and Johnson, D.L., Phys. Rev., A17, 1504-12, 1978.

18. Smaardyk, J.E. and Mochel, J.M., Rev. Sci. Instrum., 49, 988-93, 1978.

19. Tanasijczuk, O.S. and Oja, T., Rev. Sci. Instrum., 49, 1545-8, 1978.

20. Ikeda, S. and Ishikawa, Y., Jpn. J. Appl. Phys., 18, 1367-72, 1979.

21. Kraftmakher, Ya.A. and Krylov, S.D., Teplofiz. Vys. Temp., 18, 317-21, 1980; Fiz. Tverd. Tela, 22, 3157-9, 1980.

22. Kraftmakher, Ya.A., Teplofiz. Vys. Temp., 19, 656-8, 1981.

23. Kraftmakher, Ya.A., High Temp.-High Pressures, 5, 433-54, 645-56, 1973.

24. Filippov, L.P., Izmer. Tekh., 5, 45-7, 1980.

25. Hatta, I. and Ikushima, A.J., Jpn. J. Appl. Phys., 20, 1995-2011, 1981.

26. Gerlich, D., Abeles, B., and Miller, R.E., J. Appl. Phys., 36, 76-9, 1965.

27. Griesheimer, R.N., in Technique of Microwave Measurements (Montgomery, C.G., Editor), McGraw-Hill Book Co., New York, NY, 79-220 (see p. 101), 1947.

28. Jones, R.C., in Advances in Electronics (Marton, L., Editor), Vol. 5, Academic Press, New York, NY, 1-96 (see p. 48), 1953.

29. Kraftmakher, Ya.A., Zh. Prikl. Mekh. Tekhn. Fiz., 2, 144, 1966.

30. Kraftmakher, Ya.A. and Cherevko, A.G., Prib. Tekh. Eksp., 4, 150-1, 1972; Phys. Status Solidi A, 14, K35-8, 1972; Phys. Status Solidi A, 25, 691-5, 1974.

31. Jackson, J.J. and Koehler, J.S., Bull. Am. Phys. Soc., 5, 154, 1960.

32. Korostoff, E., J. Appl. Phys., 33, 2078-9, 1962.

33. Van den Sype, J., Phys. Status Solidi, 39, 659-64, 1970.

34. Skelskey, D.A. and Van den Sype, J., Solid State Commun., 15, 1257-62, 1974.

35. Seville, A.H., Phys. Status Solidi A, 21, 649-58, 1974.

36. Kraftmakher, Ya.A. and Cherepanov, V.Ya., Teplofiz. Vys. Temp., 16, 647-9, 1978.

37. Baloga, J.D. and Garland, C.W., Rev. Sci. Instrum., $\underline{48}$, 105-10, 1977; Phys. Rev., $\underline{B16}$, 331-9, 1977.

38. Itskevich, E.S., Kraidenov, V.F., and Syzranov, V.S., Prib. Tekh. Eksp., 3, 221-5, 1977; Cryogenics, $\underline{18}$, 281-4, 1978.

39. Kasting, G.B., Lushington, K.J., and Garland, C.W., Phys. Rev., $\underline{B22}$, 321-31, 1980.

40. Rasor, N.S. and McClelland, J.D., J. Phys. Chem. Solids, $\underline{15}$, 17-26, 1960.

41. Kraftmakher, Ya.A. and Strelkov, P.G., Fiz. Tverd. Tela, $\underline{4}$, 2271-4, 1962.

42. Kraftmakher, Ya.A., Zh. Prikl. Mekh. Tekhn. Fiz., 2, 158-60, 1963; Fiz. Tverd. Tela, $\underline{5}$, 950-1, 1963; $\underline{6}$, 503-5, 1964; $\underline{9}$, 1850-1, 1967.

43. Kraftmakher, Ya.A. and Lanina, E.B., Fiz. Tverd. Tela, $\underline{7}$, 123-6, 1965.

44. Shestopal, V.O., Fiz. Tverd. Tela, $\underline{7}$, 3461-3, 1965.

45. Kanel', O.M. and Kraftmakher, Ya.A., Fiz. Tverd. Tela, $\underline{8}$, 283-4, 1966.

46. Kraftmakher, Ya.A. and Strelkov, P.G., Fiz. Tverd. Tela, $\underline{8}$, 580-2, 1966.

47. Sukhovei, K.S., Fiz. Tverd. Tela, $\underline{9}$, 3660-3, 1967.

48. Akimov, A.I. and Kraftmakher, Ya.A., Phys. Status Solidi, $\underline{42}$, K41-2, 1970.

49. Kraftmakher, Ya.A. and Cherevko, A.G., High Temp.-High Pressures, $\underline{7}$, 283-6, 1975.

50. Frenkel, J.I., Z. Phys., $\underline{35}$, 652-69, 1926.

51. Kraev, O.A., Teplofiz. Vys. Temp., $\underline{5}$, 817-20, 1967.

52. Affortit, C. and Lallement, R., Rev. Int. Hautes Tempér. Refract., $\underline{5}$, 19-26, 1968.

53. Schmidt, U., Vollmer, O., and Kohlhaas, R., Z. Naturforsch., $\underline{A25}$, 1258-64, 1970.

54. Cezairliyan, A. and McClure, J.L., J. Res. Natl. Bur. Stand., $\underline{75A}$, 283-90, 1971.

55. Clusius, K. and Franzosini, P., Z. Naturforsch., $\underline{A14}$, 99-105, 1959.

56. Seeger, A., Crys. Lattice Defects, $\underline{4}$, 221-53, 1973.

57. Siegel, R.W., J. Nucl. Mater., $\underline{69-70}$, 117-46, 1978.

58. Ida, Y., Phys. Rev., $\underline{B1}$, 2488-96, 1970.

59. MacDonald, R.A., Mountain, R.D., and Shukla, R.C., Phys. Rev., $\underline{B20}$, 4012-7, 1979.

60. Kraftmakher, Ya.A. and Strelkov, P.G., Fiz. Tverd. Tela, $\underline{8}$, 1049-52, 1966; in Vacancies and Interstitials in Metals (Seeger, A., Schumacher, D., Schilling, W., and Diehl, J., Editors), North-Holland, Amsterdam, 59-78, 1970.

61. Kraftmakher, Ya.A., Phys. Status Solidi B, $\underline{48}$, K39-43, 1971; J. Sci. Ind. Res., $\underline{32}$, 626-32, 1973; in Defect Interactions in Solids (Vasu, K.I., Raman, K.S., Sastry, D.H., and Prasad, Y.V.R.K., Editors), Indian Institute of Science, Bangalore, 64-70, 1974; in Proceedings of the 7th Symposium on Thermophysical Properties, ASME, New York, NY, 160-8, 1977; Scripta Metall., $\underline{11}$, 1033-8, 1977.

62. Pigal'skaya, L.A., Yurchak, R.P., Makarenko, I.N., and Filippov, L.P., Teplofiz. Vys. Temp., $\underline{4}$, 144-7, 1966.

63. Makarenko, I.N., Trukhanova, L.N., and Filippov, L.P., Teplofiz. Vys. Temp., $\underline{8}$, 445-7, 667-70, 1970.

64. Trukhanova, L.N. and Filippov, L.P., Teplofiz. Vys. Temp., $\underline{8}$, 919-20, 1970.

65. Trukhanova, L.N. and Banchila, S.N., Vestnik Mosk. Univ., Ser. Fiz. Astron., $\underline{15}$, 599-601, 1974.

66. Smith, R.C., J. Appl. Phys., $\underline{37}$, 4860-5, 1966.

67. Skelskey, D. and Van den Sype, J., J. Appl. Phys., $\underline{41}$, 4750-1, 1970.

68. Filippov, L.P., Blagonravov, L.A., and Alekseev, V.A., High Temp.-High Pressures, $\underline{8}$, 658-9, 1976.

69. Novikov, I.I., High Temp.-High Pressures, $\underline{8}$, 483-92, 1976.

70. Fisher, M.E., The Nature of Critical Points, Univ. of Colorado Press, Boulder, CO, 1965.

71. Stanley, H.E., <u>Introduction to Phase Transitions and Critical Phenomena</u>, Clarendon Press, Oxford, 1971.

72. Ma, S., <u>Modern Theory of Critical Phenomena</u>, W. A. Benjamin, Inc., London-Amsterdam-Don Mills-Sydney-Tokyo, 1976.

73. Patashinskii, A.Z. and Pokrovskii, V.L., <u>Fluctuation Theory of Phase Transitions</u>, Pergamon Press, New York, NY, 1979.

74. Kraftmakher, Ya.A. and Romashina, T.Yu., Fiz. Tverd. Tela, <u>7</u>, 2532-3, 1965.

75. Lederman, F.L., Salamon, M.B., and Shacklette, L.W., Phys. Rev., <u>B9</u>, 2981-8, 1974.

76. Shacklette, L.W., Phys. Rev., <u>B9</u>, 3789-92, 1974.

77. Varchenko, A.A., Kraftmakher, Ya.A., and Pinegina, T.Yu., Teplofiz. Vys. Temp., <u>16</u>, 844-7, 1978.

78. Kraftmakher, Ya.A., Fiz. Tverd. Tela, <u>8</u>, 1306-8, 1966.

79. Connelly, D.L., Loomis, J.S., and Mapother, D.E., Phys. Rev., <u>B3</u>, 924-34, 1971.

80. Maszkiewicz, M., Phys. Status Solidi A, <u>47</u>, K77-80, 1978.

81. Maszkiewicz, M., Mrygoń, B., and Wentowska, K., Phys. Status Solidi A, <u>54</u>, 111-5, 1979.

82. Kraftmakher, Ya.A. and Romashina, T.Yu., Fiz. Tverd. Tela, <u>8</u>, 1966-7, 1966.

83. Lewis, E.A.S., Phys. Rev., <u>B1</u>, 4368-77, 1970.

84. Simons, D.S. and Salamon, M.B., Phys. Rev., <u>B10</u>, 4680-6, 1974.

85. Wantenaar, G.H.J., Campbell, S.J., Chaplin, D.H., and Wilson, G.V.H., J. Phys. E, <u>10</u>, 825-8, 1977.

86. Salamon, M.B., Simons, D.S., and Garnier, P.R., Solid State Commun., <u>7</u>, 1035-8, 1969.

87. Salamon, M.B., Phys. Rev., <u>B2</u>, 214-20, 1970.

88. Garnier, P.R. and Salamon, M.B., Phys. Rev. Lett., <u>27</u>, 1523-6, 1971.

89. Salamon, M.B. and Hatta, I., Phys. Lett., <u>A36</u>, 85-6, 1971.

90. Salamon, M.B. and Ikeda, H., Phys. Rev., <u>B7</u>, 2017-24, 1973.

91. Salamon, M.B., Garnier, P.R., Golding, B., and Buehler, E., J. Phys. Chem. Solids, <u>35</u>, 851-9, 1974.

92. Lederman, F.L. and Salamon, M.B., Solid State Commun., <u>15</u>, 1373-6, 1974.

93. Ikeda, H., Hatta, I., and Tanaka, M., J. Phys. Soc. Jpn., <u>40</u>, 334-9, 1976.

94. Ikeda, H., Okamura, N., Kato, K., and Ikushima, A., J. Phys. C, <u>11</u>, L231-5, 1978.

95. Suzuki, M. and Ikeda, H., J. Phys. C, <u>11</u>, 3679-85, 1978.

96. Hatta, I. and Ikeda, H., J. Phys. Soc. Jpn., <u>48</u>, 77-85, 1980.

97. Ashman, J. and Handler, P., Phys. Rev. Lett., <u>23</u>, 642-4, 1969.

98. Simons, D.S. and Salamon, M.B., Phys. Rev. Lett., <u>26</u>, 750-2, 1971.

99. Holland, L.R., J. Appl. Phys., <u>34</u>, 2350-7, 1963.

100. Zaitseva, G.G. and Kraftmakher, Ya.A., Zh. Prikl. Mekh. Tekhn. Fiz., 3, 117, 1965.

101. Boyarskii, S.V. and Novikov, I.I., Teplofiz. Vys. Temp., <u>16</u>, 534-6, 1978; <u>19</u>, 201-3, 1981.

102. Schwartz, P., Phys. Rev., <u>B4</u>, 920-8, 1971.

103. Garnier, P.R., Phys. Lett., <u>A35</u>, 413-4, 1971.

104. Hatta, I. and Ikushima, A., Phys. Lett., <u>A40</u>, 235-6, 1972; J. Phys. Chem. Solids, <u>34</u>, 57-66, 1973; J. Phys. Soc. Jpn., <u>41</u>, 558-64, 1976.

105. Hatta, I., Shiroishi, Y., Müller, K.A., and Berlinger, W., Phys. Rev., <u>B16</u>, 1138-45, 1977.

106. Hatta, I. and Kobayashi, K.L.I., Solid State Commun., <u>22</u>, 775-7, 1977.

107. Hatta, I. and Rehwald, W., J. Phys. C, <u>10</u>, 2075-81, 1977.

108. Sugimoto, N., Matsuda, T., and Hatta, I., J. Phys. Soc. Jpn., 50, 1555-9, 1981.

109. Stokka, S., Fossheim, K., and Ziolkiewicz, S., Phys. Rev., B24, 2807-11, 1981.

110. Stokka, S. and Fossheim, K., J. Phys. E, 15, 123-7, 1982; J. Phys. C, 15, 1161-76, 1982.

111. Stokka, S., Fossheim, K., Johansen, T., and Feder, J., J. Phys. C, 15, 3053-8, 1982.

112. Robinson, D.S. and Salamon, M.B., Phys. Rev. Lett., 48, 156-9, 1982.

113. Schowalter, L.J., Salamon, M.B., Tsuei, C.C., and Craven, R.A., Solid State Commun., 24, 525-9, 1977.

114. Craven, R.A., Tsuei, C.C., and Stephens, R., Phys. Rev., B17, 2206-11, 1978.

115. Ikeda, S. and Ishikawa, Y., J. Phys. Soc. Jpn., 49, 950-6, 1980.

116. Lederman, F.L., Salamon, M.B., and Peisl, H., Solid State Commun., 19, 147-50, 1976.

117. Vargas, R., Salamon, M.B., and Flynn, C.P., Phys. Rev. Lett., 37, 1550-3, 1976; Phys. Rev., B17, 269-81, 1978.

118. Cheung, K.M. and Ullman, F.G., Phys. Rev., B10, 4760-4, 1974.

119. Ema, K., Hamano, K., Kurihara, K., and Hatta, I., J. Phys. Soc. Jpn., 43, 1954-61, 1977.

120. Ema, K., Hamano, K., and Ikeda, Y., J. Phys. Soc. Jpn., 46, 345-6, 1979.

121. Zally, G.D. and Mochel, J.M., Phys. Rev. Lett., 27, 1710-2, 1971; Phys. Rev., B6, 4142-50, 1972.

122. Manuel, P., Niedoba, H., and Veyssié, J.J., Rev. Phys. Appl., 7, 107-16, 1972.

123. Manuel, P. and Veyssié, J.J., Phys. Lett., A41, 235-6, 1972; Solid State Commun., 13, 1819-23, 1973; Phys. Rev., B14, 78-88, 1976.

124. Greene, R.L., King, C.N., Zubeck, R.B., and Hauser, J.J., Phys. Rev., B6, 3297-305, 1972.

125. Krauss, G. and Buckel, W., Z. Phys., B20, 147-53, 1975.

126. Kämpf, G. and Buckel, W., Z. Phys., B27, 315-9, 1977.

127. Gibson, B.C., Ginsberg, D.M., and Tai, P.C.L., Phys. Rev., B19, 1409-19, 1979.

128. Rao, N.A.H.K. and Goldman, A.M., J. Low Temp. Phys., 42, 253-76, 1981.

129. Tsuboi, T. and Suzuki, T., J. Phys. Soc. Jpn., 42, 437-44, 1977.

130. Suzuki, T. and Tsuboi, T., J. Phys. Soc. Jpn., 43, 444-50, 1977.

131. Suzuki, T., Tsuboi, T., and Takaki, H., Jpn. J. Appl. Phys., 21, 368-72, 1982.

132. Chu, C.W. and Testardi, L.R., Phys. Rev. Lett., 32, 766-9, 1974.

133. Chu, C.W., Phys. Rev. Lett., 33, 1283-6, 1974.

134. Eichler, A. and Gey, W., Rev. Sci. Instrum., 50, 1445-52, 1979.

135. Eichler, A., Bohn, H., and Gey, W., Z. Phys., B38, 21-5, 1980.

136. Farrant, S.P. and Gough, C.E., Phys. Rev. Lett., 34, 943-6, 1975.

137. Huang, C.C., Goldman, A.M., and Toth, L.E., Solid State Commun., 33, 581-4, 1980.

138. Garoche, P. and Johnson, W.L., Solid State Commun., 39, 403-6, 1981.

139. Hempstead, R.D. and Mochel, J.M., Phys. Rev., B7, 287-99, 1973.

140. Ogawa, S. and Yamadaya, T., Phys. Lett., A47, 213-4, 1974.

141. Craven, R.A. and Meyer, S.F., Phys. Rev., B16, 4583-93, 1977.

142. Ikeda, H., J. Phys. C, 10, L469-72, 1977.

143. Wei, T., Heeger, A.J., Salamon, M.B., and Delker, G.E., Solid State Commun., 21, 595-8, 1977.

144. Viswanathan, R., Wu, C.T., Luo, H.L., and Webb, G.W., Solid State Commun., 14, 1051-4, 1974.

145. Viswanathan, R., Lawson, A.C., and Pande, C.S., J. Phys. Chem. Solids, 37, 341-3, 1976.

146. Eno, H.F., Tyler, E.H., and Luo, H.L., J. Low Temp. Phys., 28, 443–8, 1977.

147. Lannin, J.S., Eno, H.F., and Luo, H.L., Solid State Commun., 25, 81–4, 1978.

148. Hatta, I., Matsuda, T., Doi, H., Nagasawa, H., Ishiguro, T., and Kagoshima, S., Solid State Commun., 27, 479–81, 1978.

149. Dawes, D.G. and Coles, B.R., J. Phys. F, 9, L215–20, 1979.

150. Suematsu, H., Suzuki, M., and Ikeda, H., J. Phys. Soc. Jpn., 49, 835–6, 1980.

151. Craven, R.A., Salamon, M.B., DePasquali, G., Herman, R.M., Stucky, G., and Schultz, A., Phys. Rev. Lett., 32, 769–72, 1974.

152. Viswanathan, R. and Johnston, D.C., J. Phys. Chem. Solids, 36, 1093–6, 1975.

153. Johnson, D.L., Hayes, C.F., deHoff, R.J., and Schantz, C.A., Phys. Rev., B18, 4902–12, 1978.

154. Garland, C.W., Kasting, G.B., and Lushington, K.J., Phys. Rev. Lett., 43, 1420–3, 1979.

155. LeGrange, J.D. and Mochel, J.M., Phys. Rev. Lett., 45, 35–8, 1980.

156. Lushington, K.J., Kasting, G.B., and Garland, C.W., Phys. Rev., B22, 2569–72, 1980; J. Physique Lett., 41, L419–22, 1980.

157. Huang, C.C., Viner, J.M., Pindak, R., and Goodby, J.W., Phys. Rev. Lett., 46, 1289–92, 1981.

158. Viner, J.M. and Huang, C.C., Solid State Commun., 39, 789–91, 1981.

159. Bloemen, E. and Garland, C.W., J. Phys. (Paris), 42, 1299–302, 1981.

160. Garoche, P., Brusetti, R., Jérome, D., and Bechgaard, K., J. Physique Lett., 43, L147–52, 1982.

CHAPTER 9

Pulse Calorimetry

A. Cezairliyan

9.1. INTRODUCTION

Conventional steady-state and quasi-steady-state techniques for the accurate measurement of specific heat are generally limited to temperatures below about 2000 K. This limitation is the result of severe problems (chemical reactions, heat transfer, evaporation, specimen containment, loss of mechanical strength and electrical insulation, etc.) which are created by the exposure of the specimen and its immediate environment to high temperatures for extended periods of time (minutes to hours). An approach to minimize the effect of these problems and, thus, to permit the extension of the measurements to higher temperatures is to perform the entire experiment in a very short period of time (less than a second). It is in this context that most of the pulse techniques for the measurement of specific heat at high temperatures were developed.

Although the advantages of the pulse measurement methods were realized for over 60 years, the development had been slow for a number of reasons. The most important of these was the lack of adequate electrical pulse generators and control, measurement, and recording equipment with short response times. Rapid advances in the electronics field during the last two decades, coupled with the increased demand for accurate data on specific heat of substances at high temperatures, revived the interest in the pulse measurement techniques.

The objective of this chapter is to present the description and a general survey of the pulse techniques used for the measurement of specific heat of electrically conducting substances. The presentation is limited to methods in which energy is imparted uniformly to the entire volume of the specimen through resistive self-heating of the specimen. Methods that utilize surface heating (laser pulse or other energy source), cyclic heating (modulation methods), and free cooling are excluded. Techniques which yield specific heat as a by-product, such as some pulse thermal diffusivity measurements, also are not included.

In this writing, 'pulse' refers to experiments which are of subsecond duration. However, for the sake of completeness and continuity between pulse and conventional techniques, quasi-dynamic experiments with durations greater than one second are also included. Although present primary interest in pulse experiments is for measurements at high temperatures (above 1000 K), some earlier investigations, near room or at moderate temperatures, are also discussed.

9.2. DESCRIPTION OF METHODS

9.2.1. General Description

Pulse methods for specific heat measurements are based on rapid resistive self-heating of the specimen by the passage of an electrical current pulse through it and on measuring the pertinent experimental quantities with appropriate time resolution. The required quantities are: current through the specimen, voltage across the specimen, and specimen temperature.

Basically the measurement system consists of an electrical power-pulsing circuit and associated high-speed measuring circuits. The power-pulsing circuit includes the specimen in series with an electric pulse power source, an adjustable resistance, a standard resistance, and a fast-acting switch. The specimen is contained in a controlled-environment chamber. The high-speed measuring circuits include detectors and recording systems. A simplified block diagram of a generalized system for specific heat measurements by the pulse method is shown in Fig. 9.1.

9.2.2. Classification of Methods

Although the general principle of all the pulse methods is the same, there are considerable variations that are used by different investigators. In order to be able to compare the techniques and discuss their advantages as well as disadvantages, it may be appropriate to classify them into the following categories:

I. The specimen is initially at a temperature above room temperature under steady-state conditions. Then, a current pulse is sent through it to raise its temperature in a short time. The techniques in this category may be divided into two groups:

    a. The specimen is initially under steady-state conditions at a high temperature as the result of resistive self-heating.

    b. The specimen is initially under steady-state conditions at a high temperature as the result of being in a high-temperature environment (furnace).

II. The specimen is initially at or near room temperature. Then, a current pulse is sent through it to raise its temperature to the desired level in a short time.

A graphical representation of the specimen temperature excursions during experiments in the above categories is shown in Fig. 9.2. For purposes of clarity and simplicity, the shapes of the curves are exaggerated.

In the following paragraphs, the techniques that belong to the above categories are discussed in detail.

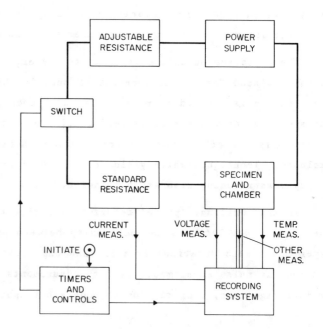

Figure 9.1. Block diagram of a typical system for the measurement of specific heat by the pulse method.

The methods in category I are a combination of steady-state and pulse experiments. The desired temperature level is achieved by steady-state power dissipation which allows the measurement of initial temperature and other pertinent quantities with conventional equipment. Then a current pulse is sent through the specimen to raise its temperature. Quantities required for the determination of power and temperature are measured during this time. Usually the temperature increase is of the order of a few degrees. Because of the small temperature rise, the magnitude of the pulse current is usually small, less than 10 A. Of course, this depends on the specimen geometry. In the case where the specimen is brought to a steady-state temperature by resistive self-heating (category Ia), radiative heat losses during the pulse heating period become appreciable. This implies that a correction to imparted power has to be made, based on either an estimated emittance value for the specimen or on supplementary data which can be obtained in the course of experiments. For example, from the measurements of voltage and current during the steady-state period, power loss from the specimen can be evaluated. An alternative technique is to determine the power loss from data obtained during the initial cooling period following the heating period. Because of the additional uncertainties resulting from large power loss corrections, experiments in category Ia have to be rapid. However, in the case of those in category Ib, where the specimen is initially in thermal equilibrium with its surroundings, the net temperature dif-

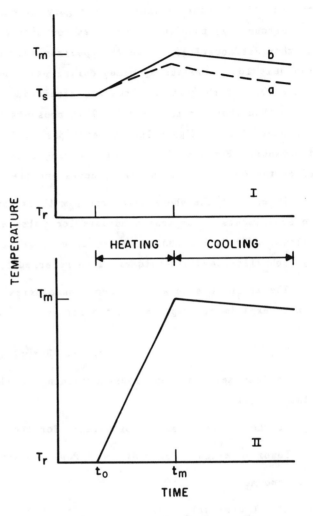

Figure 9.2. Variation of the specimen temperature during different pulse experiments. Explanation of categories Ia, Ib, and II is given in the text (Section 9.2.2). T_r is room temperature; T_s specimen initial high steady-state temperature; T_m specimen maximum temperature after pulse; t_o time indicating start of pulse; and t_m time indicating end of pulse.

ference between the specimen and its surroundings during the pulse heating period is small; thus, radiative heat loss is very small. This allows the performance of slower pulse experiments (slower heating rates than those in category Ia), which have certain advantages from experimental as well as from theoretical viewpoints.

The methods in category II represent truly fast experiments, in the sense that the specimen and its immediate environment are initially at or near room temperature and only the specimen is heated to high temperatures during the very short pulse period. Thus, these techniques do not have the limitations of those in category I, which result from the exposure of the specimen and its surroundings to high temperatures for extended periods. However, these techniques are subject to other difficulties. Since the specimen surrounding is maintained near room temperature, radiation heat loss from the specimen during pulse heating becomes appreciable. In order to reduce this effect, high heating rates are required. In general, experiments in category II are much faster than those in category I. This presents difficulties in measuring and recording the pertinent quantities. High pulse currents pose additional problems from both design and operational viewpoints. For example, rapidly varying currents create varying electromagnetic fields which affect the operation of the transducers and the entire measuring equipment.

In each of the above categories, methods may be further classified according to the speed of the experiments. For subsecond-duration pulse experiments a convenient classification may be as follows: (1) 'subsecond' techniques, also referred to as millisecond-resolution techniques, and (2) 'submillisecond' techniques, also referred to as microsecond-resolution techniques.

The techniques in all the above categories will be briefly discussed in Section 9.3 in relation to experiments reported in the literature.

9.2.3. Formulation of Equations

In this section, equations pertaining to the determination of specific heat by the pulse method are given.

In the general case, power balance for the specimen may be expressed as

Power Imparted = Power Stored + Power Losses

which becomes

$$EI = C_p m(dT/dt)_h + Q. \tag{1}$$

Solving Eq. (1) for C_p, one obtains

$$C_p = \frac{EI - Q}{m(dT/dt)_h} \tag{2}$$

where

E: voltage across the effective specimen, in V
I: current through the specimen, in A
C_p: specific heat of specimen in $J \cdot mol^{-1} \cdot K^{-1}$, at constant pressure
m: effective quantity of specimen, in mol
Q: total power loss from effective specimen, in W
$(dT/dt)_h$: heating rate of specimen, in $K \cdot s^{-1}$

In pulse experiments in which the temperature rise is only a few degrees, the quantity dT/dt in Eq. (2) may be replaced by $\Delta T / \Delta t$.

The quantity Q may be obtained from data during the initial cooling period. Power balance for this period gives

$$-C_p m (dT/dt)_c = Q \tag{3}$$

where $(dT/dt)_c$ is the cooling rate of the specimen in $K \cdot s^{-1}$. Substituting Eq. (3) in Eq. (2) for Q one obtains

$$C_p = \frac{EI}{m(dT/dt)_h(1 + 1/M)} \tag{4}$$

where

$$M = -\frac{(dT/dt)_h}{(dT/dt)_c} . \tag{5}$$

Specific heat can be obtained from Eq. (4) provided that experimental data on voltage, current, and temperature during the heating period and temperature during the initial cooling period are available, all as a function of time.

At temperatures above 1500 K, in high-speed experiments of 0.001–1 s duration, thermal radiation is the major source of power loss. When data during the initial cooling period are not available, power loss may be estimated using the relation for thermal radiation

$$Q_r = \varepsilon \, \sigma \, A_s (T^4 - T_o^4) \tag{6}$$

where

- ε: hemispherical total emittance
- σ: Stephan–Boltzmann constant ($5.67032 \times 10^{-8} \ W \cdot m^{-2} \cdot K^{-4}$)
- A_s: effective surface area in m^2
- T: specimen temperature in K
- T_o: ambient temperature in K

Equation (6) requires a knowledge of hemispherical total emittance for the specimen. However, if power loss by thermal radiation is only a few percent of imparted power, the effect of even an appreciable error (10%) in estimated emittance on specific heat is much less than 1%.

9.2.4. Measurement of Experimental Quantities

In this section, a summary is given of the various techniques that were reported in the literature for the measurement, including recording, of the experimental quantities. Because of the recent advances in electronics and optics, most of the measurement and recording techniques used by investigators before 1970 have become obsolete. However, they are included here for completeness, for their historical interest, and to demonstrate the magnitude of the advances in pulse measurement methods.

9.2.4.1. Measurement of Power

In general, power imparted to the specimen during the heating period was determined from the measurement of current through the specimen and voltage across it as a function of time. During

the early developmental years, in some slow experiments, power was measured with specially designed wattmeters; in others, the specimen was placed in one of the arms of a bridge (Wheatstone
or Kelvin) and the sudden deflection of the galvanometer resulting from the passage of pulse current was used to determine power input to the specimen. The selection of a particular power
source was governed by the desired duration of the pulse experiment. Batteries were used for most
of the experiments in the 1–1000 ms range. Capacitors were used in experiments shorter than 1 ms.
In some of the slow experiments, regulated d.c. power supplies were used.

9.2.4.2. Measurement of Temperature

Most investigators determined specimen temperature either directly with thermocouples or
indirectly by measuring its resistance as a function of time. In general, thermocouples were
formed by spot welding the wires individually to the specimen at a plane perpendicular to the current flow. In the resistance method, a separate steady-state experiment was required to obtain
the resistance-temperature relation for the specimen, which was then used to convert experimental-
ly-obtained resistance-time relation to the required temperature-time relation. During the last
decade, most investigators have used optical methods to measure specimen temperature.

9.2.4.3. Recording of Quantities

During the early years of pulse experiments, recording of electrical quantities under transient conditions was achieved by relatively crude methods. These methods were based on the observation or recording of the deflection of a ballistic instrument (galvanometer) placed either in a
bridge (Wheatstone, Kelvin) or a potentiometric circuit. Several variations of this method were
used by different investigators. In most cases, the magnitude of galvanometer deflection gave the
magnitude of the quantity to be measured. In other cases the bridge or the potentiometer circuit
was preadjusted, after several trial experiments, to give zero galvanometer deflection under pulse
conditions. The preadjusted value gave the magnitude of the quantity to be measured. In two
cases, voltage and current were recorded using a fast two-channel recorder. A few investigators
have reported the use of oscillographs as a means of recording the electrical quantities. All of
the above recording techniques are relatively crude and are limited in speed; thus, they are
applicable only to slow experiments in which specimen temperature increases slowly.

The first truly high-speed recording started around 1950 with the use of oscilloscopic techniques. Because of their very fast response characteristics (as low as nanoseconds), oscilloscopes were used for a wide range of experimental conditions, from millisecond to nanosecond
resolution. However, they too have limitations which stem primarily from their relatively poor
recording resolution. Using oscilloscopes, one may not expect to have better than 1–2% full-scale
resolution per recorded quantity under the most favorable conditions.

During the last decade, major advances in high-speed digital recording techniques and their
application to pulse measurements in the millisecond and microsecond time regimes have improved
the recording resolution by approximately one to two orders of magnitude (yielding 0.01% in millisecond-resolution experiments and 0.1% in microsecond-resolution experiments).

9.3. CHRONOLOGY OF DEVELOPMENTS

9.3.1. Developments Prior to 1970

In the literature, pulse methods of measuring specific heat are referred to as 'dynamic,' 'high-speed,' 'pulse,' 'transient,' etc. Although the proper usage may sometimes depend on the details of the techniques, the wording has been somewhat arbitrary. In this section, all such methods are referred to as 'pulse.'

Historically, the earliest attempts in using pulse techniques were confined to those in category I (classification is given in Section 9.2.2). The choice was most likely dictated by the instruments available at that time, which meant that smaller (in amplitude) and longer pulses had to be used for the measurement of the pertinent quantities. During the last decade, development of instruments with faster response times and increased accuracy, and also growing interest in the measurement of properties at temperatures above the reliable operating range of the transducers under steady-state conditions, have encouraged the concentration of efforts on methods belonging to category II.

In the following paragraphs a summary of the pulse experiments performed by various investigators for the measurement of specific heat of electrical conductors is presented. In order to avoid repetitions, some of the items relevant to all the techniques and experiments reported in the literature are given in Table 9.1. The presentation of the experiments is classified under categories discussed in Section 9.2.2.

9.3.1.1. Category Ia

The first successful attempt to use a pulse technique for the measurement of specific heat of electrical conductors may be attributed to Worthing [1]. The technique, reported in 1918, was based upon the measurement of absorbed power by the specimen (tungsten and carbon filaments) while going from one steady-state condition to another as the result of the passage of electric current through it in a short time (1 s). Batteries were used as the pulse power source. Electrical measurements were made with a modified potentiometer-galvanometer technique, in which the galvanometer was preadjusted to give zero deflection under pulse conditions. This required a number of trials and was severely limited by the time constant of the galvanometer. The temperature of the specimen was determined from the measurement of specimen resistance during the pulse experiment and from the knowledge of the resistance-temperature relation, which was obtained separately under steady-state conditions.

This general concept was used, after considerable modifications and refinements, by Pasternak and co-workers [17,18], Parker [20], and Kollie [21].

Pasternak et al. [18] made an exploratory study of a method in which total hemispherical emittance could be obtained in addition to specific heat. This required the performance of two pulse experiments with different heating rates. One experiment was made at such a high rate that radiative heat loss was small relative to absorbed power by the specimen. A second experiment was made at a slower rate for which radiative heat loss was comparable in magnitude to absorbed power. Results of these two experiments gave specific heat and total hemispherical emittance.

Parker [20] used a high-speed capacitive discharge system, primarily for the determination of heats of transformation and the rates involved in solid-solid transformations. The same system was used for specific heat measurements.

Kollie [21] described a pulse calorimeter capable of operating up to 1200 K. The novel feature of this was the utilization of a digital voltmeter and a memory which permitted the recording of 400 readings per second with 0.1% full-scale resolution. Relatively low heating rates (10–20 $K \cdot s^{-1}$) combined with the high time resolution of the recordings made this system a very attractive one for the measurement of specific heats around transition points at moderately high temperatures.

9.3.1.2. Category Ib

The difference between categories Ia and Ib is that in Ib the specimen is initially in a furnace at a constant temperature. Apparently, Lapp [2] was the first investigator to use this approach. Measurements were made on nickel at temperatures up to 730 K. A small current pulse was used, which increased the specimen temperature by approximately 1–2 K. Duration of the pulse was of the order of 30 s. Power imparted to the specimen was measured with a specially designed wattmeter. Temperature of the specimen was measured with a thermocouple. Since the experiments were slow, a correction for heat losses was required. This correction was based on data obtained during the cooling period following the heating period.

This technique was used, after considerable modification and improvement, by several other investigators: Grew [3], Néel and Persoz [5], Kurrelmeyer et al. [6], Pallister [8], Pochapsky [11,12], Rasor and McClelland [14,59], Wallace et al. [15], Wallace [16], and Kollie et al. [26].

Grew's [3] approach was almost identical with that of Lapp [2], except that he measured pulse current and specimen resistance separately instead of determining the power using a wattmeter.

Néel and Persoz [5] used a method similar to that of Lapp [2]; however, the specimen temperature was determined from its resistance-temperature relation.

Kurrelmeyer et al. [6] placed the specimen in the arm of a Wheatstone bridge and balanced the bridge with the specimen in a furnace at constant temperature. Then they discharged a small capacitor through the specimen. Specific heat was determined from the resulting ballistic deflection of the galvanometer. An alternate method was also tried. In this case, the bridge was initially unbalanced in such a manner that the discharge of the capacitor gave zero deflection. The measurements were of a preliminary nature and were conducted near room temperature.

Pallister [8] used a larger specimen than any other used before. An interesting addition, in this case, was a concentric shield placed around the specimen. The shield was connected electrically in series with the specimen and was designed in such a way that the passage of pulse current heated both the specimen and the shield coincidentally in magnitude and time. This scheme greatly reduced the heat loss from the specimen.

Pochapsky [11,12] conducted experiments with a pulse calorimeter in which the current was supplied by a thyratron-controlled capacitive pulse generator.

Rasor and McClelland [14,59] described a system in which temperature rise during the pulse was measured optically. In this system a photomultiplier was used to detect radiation emitted from a slot in the specimen. A typical temperature rise was about 5 K. The steady-state output of the photomultiplier (when the specimen was at a constant high temperature in the furnace) was suppressed potentiometrically and only the relatively small signals arising from temperature changes during the pulse were recorded.

Wallace et al. [15] and Wallace [16] reported the use of a Kelvin bridge as a means of measuring specimen resistance during the pulse. An oscilloscope was connected to the galvanometer terminals of the bridge to record the unbalanced voltage.

Kollie et al. [26] described a pulse calorimeter which essentially is an extension of an earlier version [21]. A furnace was added to the system and the data acquisition system was modified to yield a full-scale resolution of 0.01%, with a time resolution of 2000 readings per second. Measurements up to 1500 K were made. Thermocouples were used for temperature measurements. The heating rate of the specimen was 20 $K \cdot s^{-1}$. Additional results (iron, nickel, and Ni_3Fe) with this measurement system were reported.

9.3.1.3. Category II

As mentioned earlier, the technique in this category is a true pulse technique and is free, to a very large extent, from problems that result from subjecting the specimen to high temperatures for prolonged periods, as in techniques belonging to Categories Ia and Ib. Apparently, Avramescu [4] was the first investigator to use this technique. The specimen (rod) was placed in series with a battery bank and a switch. The entire system was initially at room temperature. When the switch was closed, high currents (up to 2000 A) passed through the specimen, heating it to its melting point in a few seconds. Voltage and current were recorded with an oscillograph. Specimen temperature was determined from its resistance-temperature relation. The principle of this technique was used by several other investigators: Baxter [7], Khotkovich and Bagrov [10], Nathan [9], Strittmater et al. [13], Taylor and Finch [19], Finch and Taylor [24], Affortit and Lallement [22], Affortit [23], and Jura and Stark [25].

Baxter [7], in a short note, described a technique for heating metallic wires to their melting point in about 50 ms.

Khotkovich and Bagrov [10] reported experiments on metallic wires in which pulse lengths as short as 10 ms were used. The general approach in the measurements was the same as that of Avramescu [4].

Nathan [9] departed from the above by using thermocouples (spot welded to the specimen) for temperature measurements. He studied the behavior of specific heat around the transition points of steels. Electrical quantities were recorded with a dual-beam oscilloscope; one of the channels was used for the thermocouple signal and the other channel was used to display voltage and current signals chopped with an electronic switch.

Strittmater et al. [13] conducted pulse experiments on platinum and nickel in the range 300–720 K. Specimen temperature was determined from resistance measurements. A dual-beam oscilloscope was used for recording voltage and current signals.

Taylor and Finch [19] pulse-heated metallic wires. Heating rates ranged from 1000–60,000 $K \cdot s^{-1}$. The resistance of the specimen, obtained from the instantaneous values of voltage and current, served as the means of determining the specimen temperature.

A similar technique with a few modifications was used by Finch and Taylor [24]. A thermocouple, spring-loaded against the specimen, was used to measure final specimen temperature. In addition to this, specimen temperatures during certain times in the heating period were obtained from the resistance-temperature relation. For temperatures above 1400 K, the use of a pyrometer was mentioned, but no details were given regarding its construction and operational characteristics.

Affortit and Lallement [22] reported measurements up to 3600 K. Pulse lengths were in the range 0.1–1 s. In the range 300–1500 K, temperature measurements were made with a thermocouple spot welded to the specimen. At higher temperatures, a photomultiplier was used. In the determination of specific heat, a correction for radiative heat loss was made on the basis of an assumed value for the hemispherical total emittance of the specimen.

Affortit [23] used the above technique to measure specific heats of some uranium compounds. Since at room temperature these compounds were relatively poor electrical conductors, they were heated to 1300 K, either using a furnace or passing alternating current through the specimen, before applying the pulse current. Because of the initial high steady-state temperature, this might have been classified under category I. However, since the only purpose of preheating was to decrease the specimen resistance, and since the system was designed primarily for experiments belonging to category II, it was included in this section.

An interesting extension of the pulse methods was given by Jura and Stark [25]. They described a technique for measuring specific heat of metals under high pressure. The technique is basically the same as that presented above with the exception that the specimen was placed in a high-pressure environment (anvil device). The specimen was in the form of a thin wire. Recordings of voltage and current were made with an oscilloscope. Results on iron up to 10 GPa (about 100 kilobars) and in the temperature range 77–300 K were given.

9.3.2. Developments Since 1970

During the decade starting with 1970, major advances in the pulse methods of measuring specific heat took place. For convenience of presentation in this section, they are divided into two groups: millisecond and microsecond resolution techniques.

9.3.2.1. Millisecond-Resolution Techniques

Cezairliyan et al. [27], at the National Bureau of Standards, developed a technique for measuring specific heat and other properties of electrical conductors above about 1500 K and up to the melting point of the specimen. The technique belongs to category II, in that the specimen was initially at room temperature and then a current pulse was sent through the specimen to heat it to any desired high temperature in less than one second. The technique, which incorporated a high-speed photoelectric pyrometer and a high-speed digital data acquisition system, is considered to be the first accurate millisecond-resolution pulse method for the measurement of specific heat and other related properties at high temperatures. The technique is described in Section 9.4. The system was used to measure the specific heat of several refractory substances. The measurements are summarized in Table 9.1.

A millisecond-resolution system for the measurement of specific heat and other properties developed recently at the Istituto di Metrologia 'G. Colonnetti' (IMGC) in Italy is very similar to that at the National Bureau of Standards (NBS). The IMGC system was described in the literature by Righini et al. [65]. Some modifications and additions to the system may be found in a recent publication [49]. The main difference between the system at IMGC and that at NBS is the pyrometer. The NBS pyrometer [70] uses a photomultiplier tube as a detector and operates in the chopped mode near 0.65 μm. The IMGC pyrometer uses a silicon photodiode as a detector and operates in the d.c. mode near 1 μm. The design of the IMGC instrument was based on the original work by Ruffino et al. [66] on pyrometry with silicon detectors. The high-speed pyrometer at IMGC was described by Coslovi et al. [67]; a more recent modified version with a shorter time resolution (5 μs) was presented by Coslovi et al. [74]. The measurements performed with the IMGC system are summarized in Table 9.1.

The above-described systems belong to category II. Two systems that belong to category Ib have been developed by Petrova and Chekhovskoi [50] and by Naito et al. [52].

In the system described by Petrova and Chekhovskoi [50], the specimen was initially heated in a furnace to a desired high temperature and was then rapidly pulse-heated to a higher temperature. The temperature change in the specimen as the result of pulse heating was of the order of 50 K. The steady-state temperature and the temperature of the specimen during the pulse heating period were measured optically. The signal corresponding to the temperature pulse was recorded with a digital voltmeter. The signals corresponding to pulse current and voltage were recorded with an oscilloscope.

The system described by Naito et al. [52] also incorporated a steady-state furnace to establish the desired high temperature. An added feature was an adiabatic shield that enclosed the specimen in the furnace. During the pulse heating period, current was passed through the shield as well as through the specimen (separate circuits) in such a way that the temperature rise in the shield was approximately the same as that in the specimen. This arrangement greatly reduced the heat loss from the specimen during the pulse period. The temperature rise in the specimen during the pulse was in the range 2-4 K, and was measured with a thermocouple.

Measurements of specific heat with a system developed primarily for thermal conductivity measurements was described by Taylor [51]. The technique belongs to category Ia. The specimen was initially under a steady-state condition at a high temperature as the result of resistive self-heating of the specimen. The specific heat measurements were made by subjecting the specimen

Table 9.1. Summary of Pulse Experiments for the Measurement of Specific Heat of Electrical Conductors*

No.	Investigator	Ref.	Year	Category	Power Source	Pulse Length s	Heating Rate K·s⁻¹	Substance	Temp. Range K	Spec. Geom.	Power Meas.	Temp. Meas.	Recording	Uncertainty
1	Worthing	1	1918	Ia	B	1		W,C	1200–2400	W	EI	R	V	2
2	Lapp	2	1929	Ib	B			Ni	100–730	R	W	TC	V	2
3	Grew	3	1934	Ib	B			Ni	90–720	R	EI	TC	V	2
4	Avramescu	4	1939	II	B	1–2		Al,Cu	400–1300	R	EI	R	G	2
5	Néel and Persoz	5	1939	Ib	B	0.1		Cu,Ni,Pt	300–1300	W	W	R	V	2
6	Kurrelmeyer et al.	6	1943	Ib	C	0.002–0.05		Pt	300	W	B	R	V	0.5
7	Baxter	7	1944	II		0.05				W	EI	R	G	
8	Pallister	8	1949	Ib	B		1	Fe	273–1500	R	EI	TC	V	2
9	Nathan	9	1951	II	B		15–1000	Steel	770	R	EI	TC	S	
10	Khotkovich and Bagrov	10	1951	II	B	0.01	$10^4 - 5 \times 10^4$	Cu,W,Mo,Cd		W	EI	R	G	3
11	Pochapsky	11	1953	Ib	C	0.001		Al,Pb	273–920	W	B	R	V	5
12	Pochapsky	12	1954	Ib	C	0.001		Pt	273–900	W	B	R	V	5
13	Strittmater et al.	13	1957	II	B	0.05–0.15	3000–9000	Pt,Ni	300–720	W	EI	R	S	5–10
14	Rasor and McClelland	14	1960	Ib	B		50	Mo,Ta,C	1300–3920	R	O	O	G	5
15	Wallace et al.	15	1960	Ib	B	0.04		Fe	300–1300	W	B	R	S	2
16	Wallace	16	1960	Ib	B	0.03		Th	300–1300	W	B	R	S	2
17	Pasternak et al.	17	1962	Ia	B	1–10	100–1000	Pt	300–1100	W	EI	R	R	
18	Pasternak et al.	18	1963	Ia	B		35–2000	Pt	400–1300	W	EI	R	R	4–7
19	Taylor and Finch	19	1964	II	B		$10^3 - 6 \times 10^4$	Mo,Ta	100–3200	W	EI	R	S	
20	Parker	20	1965	Ia	C	10^{-5}	$10^4 - 10^9$	Ti	300–1100	S	C	TC	S	
21	Kollie	21	1967	Ia	R	4–60	10–20	Fe	300–1200	R	EI	TC	D	1
22	Affortit and Lallement	22	1968	II	B	0.1–1		Nb,W	300–3600	W	EI	TC,O	S	3–5
23	Affortit	23	1969	II	B	<1		UN,UC,UO₂	700–3100	R	EI	TC,O	S	5
24	Finch and Taylor	24	1969	II	B	$<5 \times 10^{-5}$	$7 \times 10^3 - 1.6 \times 10^5$	ZrU$_{0.04}$H$_{0.5}$	300–800	R	EI	R,O	S	5
25	Jura and Stark	25	1969	II	B	0.1		Fe	80–300	W	EI	TC	S	5
26	Kollie et al.	26	1969	Ib	R	10–35	20	Fe	300–1500	R	EI	TC	D	1–2
27	Cezairliyan et al.	27	1970	II	B	0.3–0.7	4000–8000	Mo	1900–2800	T	EI	O	D	2–3
28	Dikhter and Lebedev	28	1970	II	C	$<5 \times 10^{-5}$		W	2600–4500	W	EI	O	S	
29	Dikhter and Lebedev	29	1971	II	C			Mo	2200–3700	W	EI	O	S	
30	Cezairliyan et al.	30	1971	II	B	0.3–0.5	4000–6000	Ta	1900–3200	T	EI	O	D	2–3
31	Cezairliyan and McClure	31	1971	II	B	0.4–0.6	4000–7000	W	2000–3600	T	EI	O	D	2–3
32	Cezairliyan	32	1971	II	B	0.4–0.5	5000	Nb	1500–2700	T	EI	O	D	2
33	Cezairliyan	33	1972	II	B	0.4–0.5	4000–7000	Ta-10W	1500–3200	T	EI	O	D	3
34	Cezairliyan	34	1973	II	B	0.4	5000–7000	Nb-1Zr	1500–2700	T	EI	O	D	3
35	Cezairliyan and McClure	35	1974	II	B	0.5–0.9	2000–4000	Fe	1500–1800	T	EI	O	D	3
36	Cezairliyan et al.	36	1974	II	B	0.4	4000–5000	V	1500–2100	T	EI	O	D	3

Table 9.1. Summary of Pulse Experiments for the Measurement of Specific Heat of Electrical Conductors* (Continued)

No.	Investigator	Ref.	Year	Category	Power Source	Pulse Length s	Heating Rate K·s⁻¹	Substance	Temp. Range K	Spec. Geom.	Power Meas.	Temp. Meas.	Recording	Uncertainty
37	Cezairliyan and Righini	37	1974	II	B	0.4-0.5	3000-5000	Zr	1500-2100	T	EI	O	D	3
38	Cezairliyan	38	1974	II	B	0.4-0.5	4000-5000	Nb-10Ta-10W	1500-2800	T	EI	O	D	3
39	Cezairliyan and Righini	39	1975	II	B	0.2-0.5	3000-8000	C	1500-3000	T	EI	O	D	3
40	Cezairliyan and McClure	40	1975	II	B	0.3-0.5	4000-8000	Hf-3Zr	1500-2400	T	EI	O	D	3
41	Lebedev et al.	41	1976	II	C		10^8	Mo,W	1500-3700	W	EI	O	S	
42	Gathers et al.	42	1976	II	C	$<10^{-4}$		Ta	2500-8000	W	EI	O	S	
43	Shaner et al.	43	1976	II	C	$<10^{-4}$		Mo,Nb,Ta,W	2500-5000	W	EI	O	S	
44	Shaner et al.	44	1977	II	C	$<10^{-4}$		Mo,Ta	2000-7500	W	EI	O	S	
45	Shaner et al.	45	1977	II	C	$<10^{-4}$		Nb,Pb	2000-6000	W	EI	O	S	
46	Lebedev and Mozharov	46	1977	II	C		10^8-10^9	Ta	2300-3900	T	EI	O	S	8
47	Cezairliyan and Miiller	47	1977	II	B	0.5-0.6	3000-4000	Ti	1500-1900	T	EI	O	D	3
48	Cezairliyan et al.	48	1977	II	B	0.4-0.5	3000-4000	Ti-6Al-4V	1500-1900	T	EI	O	D	3
49	Righini et al.	49	1977	II	B	0.6-1.5	1000-3000	Zircaloy-2	1300-2000	T	EI	O	D	4
50	Petrova and Chekhovskoi	50	1978	Ib	B	300-500		NbC,TaC,ZrC	1600-2300	R	EI	O	S,D	3-4
51	Taylor	51	1978	Ia	R	6		W	1500-2400	R	EI	R	D	
52	Naito et al.	52	1979	Ib				C,SiC	300-1200	R	EI	TC		
53	Gathers et al.	53	1979	II	C	$<5\times10^{-5}$		Pt	2000-7500	W	EI	O	S	3-5
54	Seydel et al.	54	1979	II	C		10^9	Mo	2900-6000	W	EI	O	S	
55	Cezairliyan and Miiller	55	1980	II	B	0.6	2000	Steel	1400-1700	T	EI	O	D	3
56	Miiller and Cezairliyan	56	1980	II	B	0.5-0.6	3000-4000	Pd	1400-1800	T	EI	O	D	3
57	Cezairliyan and Miiller	57	1980	II	B	0.4-0.8	4000	C-composite	1500-3000	T	EI	O	D	3
58	Righini and Rosso	58	1980	II	B	0.7-1.1	1700-3000	Pt	1000-2000	T	EI	O	D	3

*Abbreviations and Notes

Category: Designation of categories are described in Section 9.2.2.
Power Source: B = Battery, C = Capacitor, R = Regulated DC power supply.
Temp. Range: Range between two extreme temperatures covered by the investigator regardless of particular substance.
Spec. Geom.:   R = Rod (diam. > 1 mm), S = Strip, T = Tube, W = Wire (diam. ≤ 1 mm).
Power Meas.:   B = Bridge, C = Capacitor energy, EI = Voltage-current, W = Wattmeter.
Temp. Meas.:   O = Optical, R = Resistive, TC = Thermocouple.
Recording: D = Digital, G = Oscillographic, R = Chart recording, S = Oscilloscopic, V = Visual-manual.
Uncertainty: Total error in specific heat as reported in the literature by the investigators. (In some instances where the measured quantity was
                enthalpy and the authors gave only the error in enthalpy the entry in this column is left blank.)

335

to a small step-function change in the electrical power.   Change in the specimen temperature was determined from the change in its resistivity.

9.3.2.2. Microsecond-Resolution Techniques

During the last decade, increased interest in properties of high-melting-point substances, especially in their liquid phase, resulted in the exploration of microsecond-resolution techniques.   Primarily, the following four laboratories were involved in research related to specific heat measurements: Institute for High Temperatures (USSR), Lawrence Livermore Laboratory (USA), University of Kiel (FRG), and National Bureau of Standards (USA).

In the Institute for High Temperatures, Dikhter and Lebedev [28,64] developed a system for the measurement of specific heat of metals in their solid and liquid phases. The system included a capacitor bank (5.7 μF), a series resistance (48.8 Ω), a standard resistance (0.445 Ω), and a thyratron for switching the current. The specimen was in the form of a wire 0.08 mm in diameter and 20 mm in length. Energy imparted to the specimen was obtained from the integral of the product of the current through the specimen and the voltage across it. The specimen temperature was obtained from measurements of radiation at two wavelengths with the use of photomultipliers. The experimental quantities were recorded with oscilloscopes. Later on, Lebedev and Mozharov [46] introduced refinements in their temperature measurements by utilizing tubular specimens which approximated blackbody conditions. A summary of the measurements of specific heat performed with the microsecond-resolution system in the Institute for High Temperatures is given in Table 9.1.

The system developed in the Lawrence Livermore Laboratory started with the work of Henry et al. [72], who developed a capacitor discharge system for thermodynamic measurements on liquid metals. The system included a capacitor bank (20 kV, 17 kJ) and a high pressure (200 MPa, about 2000 atm.) cell. Modifications and refinements of the system were described by Gathers et al. [63]. The changes included a larger capacitor bank (45 kJ), an improved voltage probe assembly, and a streaking camera. Other improvements to the above system were made by Gathers et al. [42]. The new features were a rapid temperature-measurement capability and a new pressure cell (400 MPa) to accommodate an optical window. The temperature measurements were based on three-color pyrometry. To ensure that each channel views the same spot on the specimen, a randomized trifurcated glass fiber-optic bundle was used to divide the signal. The output of each of the three fiber bundles was passed through interference filters (50 nm bandwidth) centered at 450, 600, and 700 nm, respectively. Silicon photodiodes were used as detectors, signal amplification was achieved with logarithmic amplifiers, and the outputs were recorded with oscilloscopes. In subsequent publications, Shaner et al. [43-45] and Gathers et al. [53] have discussed the operational characteristics of the system and have reported the results of the measurements of the enthalpy of selected metals in their solid and liquid phases. A summary of the work performed in the Lawrence Livermore Laboratory is given in Table 9.1.

The system developed in the University of Kiel was described by Seydel et al. [60] and by Seydel and Fucke [61]. The system had a 5.1 μF capacitor bank rated at 35 kV. The electrical characteristics of the system were: short-circuit ringing period, 6 μs; circuit resistance,

20 mΩ; circuit inductance, 200 nH. The heating rate of the specimen (thin wire) was typically 10^9 K·s^{-1}. The electrical measuring circuits had a rise-time of about 8 ns. In another paper, Seydel et al. [62] have described modifications and refinements to their system which included the addition of a fast pyrometer operating at two wavelengths (in the range 450–950 nm), and an experiment chamber for pressures up to about 400 MPa (about 4000 atm.). Measurements of the enthalpy of liquid molybdenum with this system were discussed by Seydel et al. [54].

At the National Bureau of Standards, a system was developed for the measurement of specific heat of electrically conducting substances in their solid and liquid phases up to about 10,000 K [73]. The technique is described in Section 9.5.

An extension of the pulse technique for measuring specific heat under very high pressures, originally described by Jura and Stark [25], was given by Loriers-Susse et al. [71]. They made significant improvements in the technique and performed measurements on copper up to 10 GPa (about 100 kbars) in experiments of 100–200 μs duration. Investigations in this line are described in detail in the chapter 'Calorimetry at Very High Pressures' of this volume.

9.4. A MILLISECOND-RESOLUTION TECHNIQUE

In this section, a description is given of a millisecond-resolution technique developed at the National Bureau of Standards [27] for the measurement of specific heat and other related properties of solid electrical conductors.

The technique is based on resistive self-heating of the specimen from room temperature to high temperatures (in the range 1500 K to its melting temperature) in less than one second by the passage of an electrical current pulse through it; and on measuring, with millisecond resolution, power imparted to the specimen and the specimen temperature.

The measurement system consists of an electric power pulsing circuit and associated measuring circuits. The power pulsing circuit includes the specimen in series with a battery bank, a standard resistance, an adjustable resistance, and a switching system. The measuring circuits include potential probes, amplifiers for the measurement of current and voltage (to yield power), a high-speed pyrometer, and a digital data acquisition system. A functional diagram of the complete system is presented in Fig. 9.3. The details of the measurement system and its operational characteristics are given in earlier publications [27,68]. In the following subsections, a summary of some of the important features is given.

9.4.1. Pulse Circuit

The battery bank consists of 14 series-connected 2V batteries each having approximately 1100 A-h capacity. The adjustable resistance (total resistance, 30 mΩ) controls the heating rate of the specimen and the shape of the current pulse. The switching system consists of two series-connected, fast-acting switches. The second switch is used as a back-up in the event the first one fails to open at the end of the heating period.

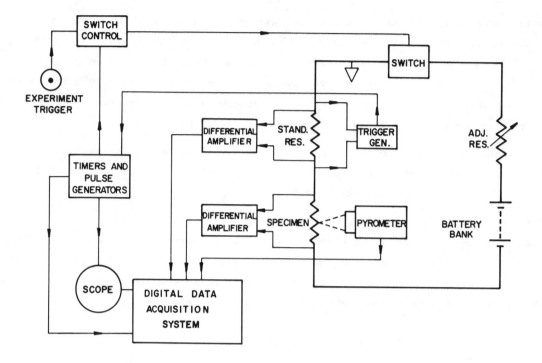

Figure 9.3. Functional diagram of the millisecond-resolution pulse calorimeter [27].

The specimen is a tube of the following nominal dimensions: length, 75-100 mm; outside diameter, 6 mm; and wall thickness, 0.5 mm. A small rectangular hole (1 x 0.5 mm) is fabricated in the wall at the middle of the specimen to approximate blackbody conditions for pyrometric temperature measurements. For the above geometry, the blackbody quality of the sighting hole is estimated to be approximately 0.99. Depending on the investigation, specimens of other geometrical forms (rods, strips) may be used. The specimen is mounted vertically 6 mm off-center with respect to the axis of the experiment chamber to reduce the effect of internal reflections. The chamber wall, as well as the specimen clamps, are water cooled. Thermocouples are connected (electrically insulated) to the two end clamps to measure the specimen temperature before each pulse experiment. An expansion joint allows the expansion of the specimen in the downward direction. The chamber is designed for conducting experiments with the specimen in either vacuum or a controlled atmosphere. A schematic diagram of the experiment chamber is shown in Fig. 9.4.

9.4.2. Measurement of Power

Power imparted to the specimen is determined from the product of measured current through the specimen and voltage across the specimen. Current is determined from the measurement of the potential difference across the standard resistance (1 mΩ) placed in series with the specimen. Knife-edge probes are placed on the specimen approximately 13 mm from the end clamps for voltage measurements. The knife-edges define a portion of the specimen which should be free from axial temperature gradients for the duration of the pulse experiment. The probes are made of the specimen material to eliminate thermoelectric effects on the measurements. Differential amplifiers for signals corresponding to both current and voltage are used in order to avoid inaccuracies arising

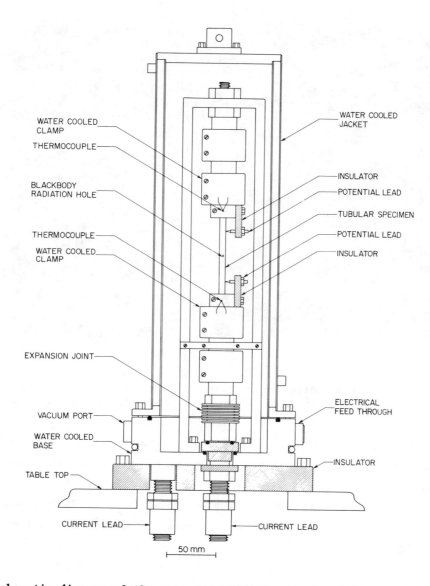

Figure 9.4. Schematic diagram of the experiment chamber of the millisecond-resolution pulse calorimeter [68].

from electrical ground problems. The total error in either current or voltage measurements is about 0.1%. The precision is at least three times better. Oscilloscope trace photographs of typical voltage and current pulses are shown in Fig. 9.5.

9.4.3. Measurement of Temperature

The temperature of the specimen during the pulse experiment is measured with a high-speed photoelectric pyrometer, which permits 1200 evaluations of the specimen temperature per second. The pyrometer alternately passes precisely-timed samples of radiance from the specimen and from a reference source (gas-filled tungsten filament lamp) through an interference filter (wavelength 653 nm, bandwidth 10 nm) to a photomultiplier. During each exposure the photomultiplier output is integrated and is recorded. For measurements at temperatures above 2500 K, calibrated optical

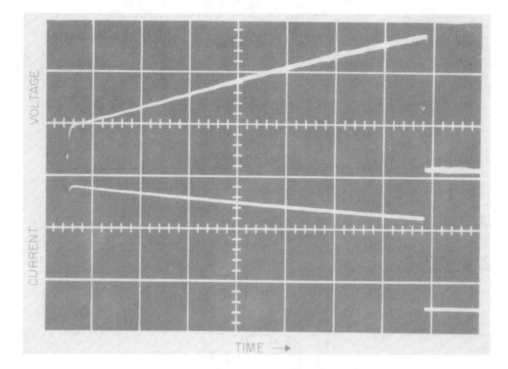

Figure 9.5. Oscilloscope trace photograph of typical voltage and current pulses. Equivalence of
each major division is: time, 50 ms; voltage, 2 V; current, 1000 A.

attenuators are placed in the path of the radiation from the specimen. The pyrometer target is a
circular area 0.2 mm in diameter. Radiance from the blackbody hole in the specimen and from the
reference lamp, as seen by the pyrometer during a typical pulse experiment, are shown in Fig. 9.6.
The total error and precision in temperature measurements (at about 2000 K) are about 5 K and
0.5 K, respectively. Details regarding the construction and operation of the pyrometer are given
in the literature [70].

9.4.4. Data Acquisition System

Data corresponding to temperature, current, and voltage are recorded with a high-speed digital data acquisition system, which consists of a multiplexer, analog-to-digital converter, and a
core memory together with control and interfacing equipment. The multiplexed signals go to the
analog-to-digital converter, which has a full-scale reading of ± 10 V and a full-scale resolution
of one part in 8192 ($8192 = 2^{13}$). Digital output from the converter consists of 13 binary bits
plus a sign bit. This output is stored in a core memory having a capacity of 2048 words of 16
bits each. The data acquisition system is capable of recording a set of signals corresponding to
temperature, voltage, and current approximately every 0.4 ms. At the end of the pulse experiment,
data stored in the memory are transferred to a minicomputer for processing. Oscilloscopes are
used only to monitor the general pattern of the experimental results and to detect any anomalies.

9.4.5. General Comments

The operational characteristics of the system depend primarily on the electrical circuit
parameters, properties of the specimen, and specimen maximum temperature. The system

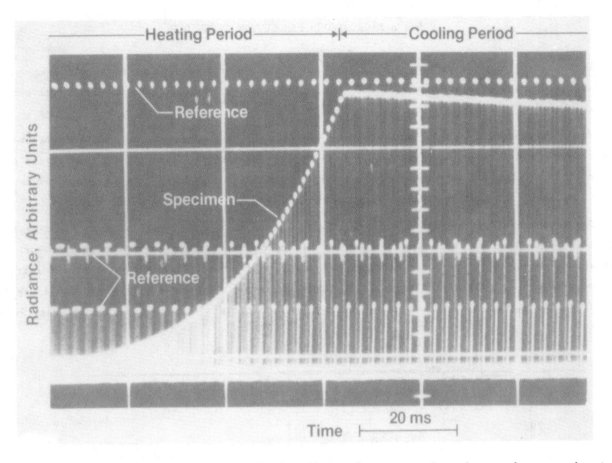

Figure 9.6. Oscilloscope trace photograph of radiance from a specimen in a pulse experiment. Dots forming the long horizontal lines correspond to radiances from the reference source. Equivalence of each major division is: time, 20 ms; radiance, arbitrary unit.

characteristics for typical experiments are as follows: current through specimen, 1000–3000 A; power imparted to specimen, 5–15 kW; specimen heating rate, 3000–9000 $K \cdot s^{-1}$; specimen heating duration, 0.3–0.9 s; ratio of heating rate to cooling rate, 10–100.

Formulation of the relation for specific heat in terms of the experimental quantities is given by Eq. (4) in Section 9.2.3. In Eq. (4) there is a built-in correction for the heat loss from the specimen due to thermal radiation. Uncertainty (total error, random and systematic) in specific heat determined by the millisecond-resolution technique is estimated to be not more than 2% at about 2000 K and not more than 3% at about 3000 K. A complete discussion of sources and magnitudes of errors is given in the literature [27]. The measurements performed on several refractory substances are summarized in Table 9.1. In addition to specific heat, the system can measure electrical resistivity, hemispherical total emittance, normal spectral emittance, thermal expansion, melting point, heat of fusion, and temperature and energy of solid–solid phase transformations. A summary of these measurements is given in the literature [69].

9.5. A MICROSECOND–RESOLUTION TECHNIQUE

In this section, the description is given of a microsecond–resolution technique developed at the National Bureau of Standards [73] for the measurement of specific heat of electrical conductors in their solid and liquid phases.

The technique is based on resistive self–heating of the specimen from room temperature to high temperatures (in the range 2000–10,000 K) in less than one millisecond by the passage of an electrical current pulse through it; and on measuring, with microsecond resolution, power imparted to the specimen, and the specimen temperature.

The measurement system consists of an electric power pulsing circuit and associated measuring circuits. The power pulsing circuit includes the specimen in series with a capacitor bank, a fast–acting switch, and a crowbar circuit. The measuring circuits include pulse transformers for the measurements of current and voltage (to yield power), a high–speed pyrometer, and a digital data acquisition system. A functional diagram of the complete system is presented in Fig. 9.7.

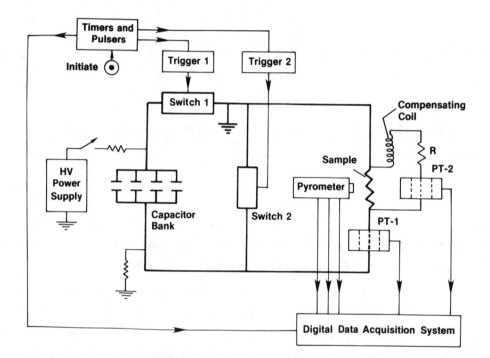

Figure 9.7. Functional diagram of the microsecond–resolution pulse calorimeter [73].

9.5.1. Pulse Circuit

The capacitor bank consists of eight parallel–connected capacitors (20 kV, 15 μF, each) and is capable of storing 24 kJ of energy. The main switch (Switch 1 in Fig. 9.7) is an ignitron, and when it is triggered (closed), high currents of the order of 10^4 A flow through the specimen. Because of the low resistance of the circuit with a metallic specimen, the discharge is generally oscillatory (upper trace in Fig. 9.8). Large oscillations cause difficulties in the measurements of the experimental quantities. In order to minimize the oscillatory behavior of low resistance circuits, a crowbar circuit is incorporated in the main circuit. The principle behind the crowbar operation is that as the current begins to fall from its peak value and the capacitor voltage is almost zero, all the energy initially stored in the capacitor which was not dissipated during the initial current surge now resides in the magnetic field of the inductive portion of the circuit. When the crowbar ignitron (Switch 2 in Fig. 9.7) is triggered, this magnetically-stored energy is returned as unidirectional current in the circuit consisting of the crowbar ignitron and load. The residual undulatory portion of the current waveform is quite modest (lower trace in Fig. 9.8) compared to the non-crowbarred discharge and simplifies inductive corrections to the electrical measurements.

The specimen is typically in the form of a solid cylinder (about 25 mm in length, and 1.5 mm in diameter), and is placed between two heavy brass electrodes. The entire assembly is placed in a chamber for operation in a controlled environment, usually argon gas. A photograph of the experiment chamber is shown in Fig. 9.9.

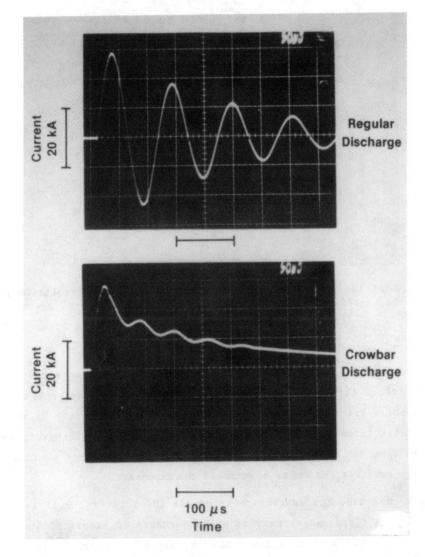

Figure 9.8. Oscilloscope trace photographs of current waveform for regular (upper trace) and crowbar (lower trace) discharge conditions.

Figure 9.9. Photograph of the experiment chamber of the microsecond-resolution pulse calorimeter showing the specimen and the electrodes. The front cover is removed.

9.5.2. Measurement of Power

Power imparted to the specimen is determined from the product of measured current through the specimen and voltage across the specimen. Pulse transformers are used for both current and voltage measurements. A pulse transformer is a toroidal device which generates a voltage proportional to the current that flows through a conductor placed along its axis. The fact that this device is electrically isolated from the discharge circuit, coupled with its other performance characteristics, makes it ideally suitable for pulse electrical measurements.

The pulse transformer used for current measurements (PT-1 in Fig. 9.7) is designed to have a voltage output of $0.01 \text{ V} \cdot \text{A}^{-1}$ and can operate at pulse currents in excess of 5×10^4 A. The output of the transformer is attenuated (with a noninductive voltage divider) so that the input voltage to the digital recording system stays below 10 V under all operating conditions.

Voltage across the specimen is determined indirectly with a pulse transformer (PT-2 in Fig. 9.7) that measures the current through the resistor R, connected in parallel with the specimen. From the knowledge of the value of the parallel resistance and the current through the resistor, voltage across the specimen is determined. The present arrangement uses a 1 V·A^{-1} pulse transformer and a 50 Ω resistor for the parallel path. The inductive effects between the main discharge circuit and the voltage measuring circuit create complications. The key to making accurate power measurements is to cancel out the inductive component in the voltage measurement. This is accomplished by introducing a compensating coil in the parallel resistance path (Fig. 9.7) and adjusting the coil in such a way that the induced voltage in the coil compensates all other undesirable induced voltages. The near perfect compensation is established by tuning the coil so that there is no phase shift between the current and voltage signals. The details of the voltage measurement technique are given in another publication [76]. The estimated total error in either current or voltage measurements is about 1%.

9.5.3. Measurement of Temperature

A special microsecond-resolution pyrometer is used for the accurate measurement of the temperature of a rapidly-heating specimen continuously in the range from 2000–6000 K [75]. For higher temperatures, calibrated neutral density filters are used. The salient features of the pyrometer are:

(1) It measures temperatures in two narrow spectral regions; one near 0.65 μm, and the other near 0.9 μm. The bandwidth of each channel is about 0.03 μm.

(2) The electronics are entirely outside of the shielding of the room in which capacitors and all of the discharge circuits are placed, in order to minimize the effect of electromagnetic interference. Radiation selected by the pyrometer optics in each of the two spectral regions is transmitted by fiber-optic cables through the shielding walls to the silicon diode photodetectors and their associated electronic circuits.

(3) Each channel of the pyrometer has three series-connected linear amplifiers, and has high-speed gain switching, thus providing high stability, high resolution, and wide temperature range.

(4) The signals are recorded with a resolution of about 0.1% by a digital data acquisition system as rapidly as every 1.5 μs.

(5) The target of the pyrometer is a circular area 0.5 mm in diameter.

Radiance temperature (at two wavelengths) of a rapidly-heating niobium specimen measured with the pyrometer is shown in Fig. 9.10. Duration of the entire experiment, time taken to heat the specimen from room temperature to near 3500 K, was about 100 μs. The total error in radiance temperature measurements is estimated to be about 5 K at 2000 K, increasing to 30 K at 6000 K.

9.5.4. Data Acquisition System

The data acquisition system enables the simultaneous recording of five experimental quantities at intervals as short as 1.5 μs, with a full-scale resolution of about 0.1%. Each of the five channels utilizes a high-speed sample-and-hold amplifier, and a 10 bit analog-to-digital converter. The aperture time of the sample-and-hold amplifiers is a nominal 50 ns. The output from the analog-to-digital converters is in the form of parallel binary data which is stored in a high-speed semiconductor memory. The memory has a total capacity of 5000 data points. After an experiment, stored data are converted to serial ASCII form and are transmitted to a minicomputer for processing. Oscilloscopes are used only to monitor the general pattern of the experimental results and to detect any anomalies.

9.5.5. General Comments

The operational characteristics of the system depend primarily on the electrical circuit parameters, properties of the specimen, and specimen maximum temperature. The system characteristics for a typical experiment are as follows: current through specimen, 3×10^4 A; specimen heating rate, 5×10^7 K·s^{-1}; specimen heating duration, 100 μs. Formulation of the relation for specific heat in terms of the experimental quantities is given by Eq. (2) in Section 9.2.3. In submillisecond-duration experiments, power loss from the specimen is almost negligible below 5000 K. At higher temperatures, a satisfactory correction can be made by assuming a value for the hemispherical total

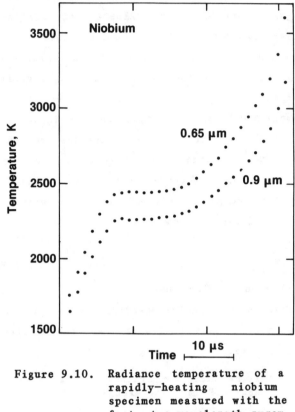

Figure 9.10.   Radiance temperature of a rapidly-heating niobium specimen measured with the fast, two-wavelength pyrometer. Plateaus in temperature indicate melting of niobium. Data on the right of the plateaus are for liquid niobium.

emittance of the specimen and by using Eq. (6) for the heat loss term that appears in Eq. (2). The uncertainty in specific heat determined by the microsecond-resolution technique is estimated to be about 5% at temperatures below 5000 K. The uncertainty in specific heat at higher temperatures is not well established yet.

9.6. EXPERIMENTAL DIFFICULTIES AND SOURCES OF ERROR

Although the basic principle of the pulse method of measuring specific heat is simple, the development of systems and performance of accurate experiments present difficulties and require the solution of serious problems. In this section, some of the experimental difficulties which give rise to errors in the measurements are discussed briefly.

9.6.1. Errors in Directly-Measured Quantities

9.6.1.1. Errors in Electrical Measurements

Identifiable sources of errors that affect the measurement of electrical quantities such as current and voltage are: skin effect, inductive effects, thermoelectric effects, and errors due to calibrations. The contributions of the above depend on the specific systems and on the speed of the experiments, in particular.

The contribution of the skin effect in millisecond-resolution experiments is generally negligible. However, it should be taken into consideration in the design of microsecond-resolution systems and especially in the selection of the specimen geometry.

The inductive effects should be considered in all pulse experiments. In millisecond-resolution experiments their contribution is generally small and can be eliminated by proper shielding of pertinent cables and components. In microsecond-resolution experiments, both self and mutual inductive effects can cause serious measurement problems by creating large induced voltages in the electrical circuits. In these cases, conventional shielding is usually inadequate and other techniques, such as compensation, have to be employed.

The thermoelectric effects can be neglected in almost all pulse experiments, since either they take place outside the 'effective' specimen or their contribution is small compared to the resolution of the measurements.

Normally, calibrations of the electrical measuring circuits, which include amplifiers, resistors, etc., do not present any serious problems. If they are performed under steady-state conditions, it is essential to establish the adequacy of the time-response of the circuits.

At present, techniques have been developed to permit the millisecond-resolution measurements of electrical quantities (current, voltage) with an uncertainty of about 0.1%. Measurements of the same quantities with microsecond-resolution are considerably less accurate, to about 1% under the most favorable conditions.

9.6.1.2. Errors in Temperature Measurements

Temperature is, by far, the most critical experimental quantity, as it appears in the form dT/dt in the relation for specific heat [Eq. (4)]. Each method of temperature measurement (resistive, thermoelectric, optical) has its own unique sources of error. Since most of the modern pulse experiments utilize optical techniques for temperature measurements, some of the problems associated with such techniques are discussed in the following paragraphs.

Optical pyrometry with appropriate time resolution (millisecond or microsecond) requires sophistication in both optics and electronics. Because of the short times and the small target sizes involved, the pyrometer should have a large aperture and minimum internal attenuation as well as high-quality optical elements to minimize scattering of light.

A source of temperature error is the temperature nonuniformities along the specimen during its rapid heating. This may be due mainly to nonuniformities in the specimen's cross-sectional area. Temperature gradients, both radial and axial, may also contribute to the temperature error if the dimensions of the specimen and/or the speed of the experiments are not carefully selected.

Ability to measure the true temperature of the specimen is probably the most important issue in optical pyrometry. When the pyrometer is sighted at the surface of the specimen, the resultant

temperature is the surface radiance temperature. Conversion to true temperature requires a knowledge of the normal spectral emittance of the specimen at the wavelength of the measurements. Usually, normal spectral emittance data are not readily or easily available, and one cannot rely on
literature values (even if available) since this property depends very strongly on the surface
conditions of the specimen. The most direct approach to measuring the true temperature is to have
a specimen that approximates blackbody conditions. This can be achieved if the specimen is in the
form of a tube, with a small hole fabricated in the wall at the middle of the length. The blackbody quality for a realistic size specimen (length, 75 mm; outside diameter, 6 mm; wall thickness,
0.5 mm) may be as high as 99% when the sighting hole is 0.5 x 1 mm. Such a specimen can be used
in millisecond-resolution experiments. In cases where specimens in the form of a solid rod are to
be used (in most of the microsecond-resolution experiments), an alternate approach for obtaining
true temperature, based on some assumptions regarding the dependence of normal spectral emittance
on wavelength, is to measure radiance temperature at more than one wavelength. Simultaneous
measurements at two and three wavelengths have been tried by a few investigators. However, the
reliability of true temperature determinations with the multi-wavelength measurements approach is
not as good as that obtained from one wavelength measurements on a blackbody.

In addition to the above items, accurate pyrometry requires careful calibration and data
reduction procedures. Calibration procedures are generally based on the use of calibrated tungsten filament lamps. However, such lamp calibrations are usually performed at 0.65 μm, the wavelength at which conventional slow pyrometers operate. Calibration of lamps at wavelengths other
than 0.65 μm presents difficulties, and is not readily available except at some standards laboratories. In some cases, it may be desirable to perform pyrometer calibrations directly with blackbody radiation sources.

At present, pyrometry techniques have been developed to permit the millisecond-resolution
measurements of temperature with an uncertainty of about 5 K at 2000 K, and 10 K at about 3500 K.
The existing microsecond-resolution pyrometry techniques are considerably less accurate and are
likely to yield results with an uncertainty of about 30 to over 100 K at 6000 K. Extensive research is under way to improve the accuracy of microsecond-resolution pyrometry.

9.6.1.3. Errors in Data Acquisition

In most of the modern pulse experiments, recordings of the experimental quantities are done
either with oscilloscopes or digital data acquisition systems. In millisecond-resolution experiments, one may use digital data acquisition systems which have a full-scale signal resolution of
about 0.01%. In microsecond-resolution experiments researchers generally use oscilloscopes, which
under the best conditions may have a resolution of about 1%. Recent advances in the electronics
field have brought about the capability of developing microsecond-resolution digital data acquisition systems with a full-scale signal resolution of about 0.1%. While the above figures refer to
measurement resolution, accuracy of the data recording systems may be lower, depending on factors
such as calibration procedures and the effect of external electromagnetic fields, etc.

9.6.2. Errors Due to Departure from Assumed Conditions

The accuracy with which specific heat can be determined depends also on the degree of experimental realization of the conditions imposed in the formulation of the equation [Eq. (4)] for specific heat. This formulation assumes accurate accounting of heat loss, no specimen evaporation, no thermionic emission from or near the specimen, and that the specimen is in thermodynamic equilibrium at all times during its rapid heating period.

Because of the speed of the pulse experiments and the geometry of the specimen, the only significant heat loss from the specimen is due to thermal radiation. In millisecond-resolution experiments, this can be a significant portion (about 10%) of the imparted power at temperatures near 3000 K. However, a correction for radiation heat loss can be made, either based on data obtained from the free radiative cooling period, or from a literature value for the hemispherical total emittance. Even a significant uncertainty (10%) in the correction does not introduce more than 1% error in the specific heat results.

Other errors may result from specimen evaporation. However, because of the short times involved, this can be neglected for metals up to their melting points. For experiments on liquid metals considerably above their melting points, evaporation can introduce significant errors. In order to avoid this, experiments should be made with the specimen in a pressurized environment such as argon at pressures 200 MPa (about 2000 atm.) and higher, and in some cases in water or some other transparent liquid.

Thermionic emission from the specimen usually does not introduce any serious problems, since either the temperature is low when the specimen is in a vacuum environment or the specimen is surrounded with a pressurized fluid when the temperature is high.

Whether a specimen is in thermodynamic equilibrium while measurements are taken during a rapid heating period depends upon several factors, such as crystalline structure, magnitude of impurities and imperfections, temperature, etc. In pure electrical conductors at high temperatures, relaxation time of electron-phonon interactions is of the order of 10^{-10} s. However, other processes, such as formation and migration of vacancies, may have relaxation times of the order of 10^{-6} to 10^{-3} s.

9.6.3. Errors in Specific Heat

Errors in measured specific heat reported in the literature by most investigators (as listed in Table 9.1) are based on crude estimates and probably are somewhat optimistic. An analysis of sources and magnitudes of errors in the best pulse experiments indicate that in the range 1000-2000 K specific heat can be measured as accurately as with conventional methods; above 2000 K the pulse methods may yield an accuracy that surpasses the accuracy of any conventional method.

With the best millisecond-resolution pulse methods available at present, it is possible to measure specific heat with an uncertainty of about 2% at 2000 K and 3% at 3000 K. Reproducibility of measurements may be approximately 0.5% over this temperature range.

Uncertainties in the specific heat obtained from microsecond-resolution experiments are considerably greater than those given above. The estimates indicate that, in the best experiments reported in the literature, the uncertainty in specific heat is not better than 10%, and is likely that it is considerably worse than 10%. However, new developments indicate that it may be possible to measure specific heat with an uncertainty of about 5% up to about 5000 K.

9.7. SUMMARY AND CONCLUSIONS

In the previous sections, the description, classification, and a summary of the experiments for the measurement of specific heat of electrically-conducting substances with the use of pulse heating techniques were presented. The experimental difficulties and the sources of errors associated with such experiments were also discussed. It was pointed out that the main advantage of the pulse techniques is in measurements at high temperatures (above about 2000 K), above the limits of reliable operation of accurate steady-state systems.

During the last decade, major advances were made in developing pulse techniques for the measurements of specific heat. These advances may be attributed primarily to two factors: (1) the need for high-temperature properties of materials in various applications, such as in aerospace, nuclear energy, etc., and (2) the emergence of highly-sophisticated electronics, particularly in the area of fast digital data acquisition systems.

The millisecond-resolution techniques have been perfected to a level of competitiveness with the most accurate conventional steady-state techniques in the overlapping temperature ranges (1000-2000 K). In this temperature range, the pulse techniques have the added advantage of covering a wide temperature range in a single experiment and of yielding high-resolution temperature measurements when compared with most of the conventional techniques, such as drop calorimetry. This is an important feature, especially in measurements near and at phase transitions. While the low temperature limit of millisecond-resolution pulse techniques is governed by the operation of the millisecond-resolution pyrometers, the upper temperature limit is governed by the melting temperature of the specimen.

For measurements above the melting temperatures of the specimens, microsecond-resolution techniques show great promise. Because of the severity of the measurement conditions, these techniques have not yet been perfected to the level of millisecond-resolution techniques. However, the results of the preliminary research indicate the potential of the microsecond-resolution techniques as a means of obtaining accurate data at temperatures approaching 10,000 K. Their lower limit is governed by the operation of the microsecond-resolution pyrometers, which may be about 2000 K. This indicates that for a great number of refractory materials there is a considerable overlapping temperature range between the millisecond and the microsecond resolution techniques, which is important in establishing the reliability of the measurements.

Methods belonging to category I, where the specimen is initially at a high steady-state temperature, are primarily used for measurements below 2500 K. In general, small electrical pulses are used and the resultant temperature rise is only a few degrees. These are suited for investigations that require high resolution in temperature measurements.

Methods belonging to category II, where the specimen is initially at room temperature, are useful for measurements above 2000 K. Rapid measurements over a wide temperature range makes this very attractive for research at very high temperatures.

In conclusion, the pulse techniques, both millisecond and microsecond resolution, have a great potential in extending accurate measurements of specific heat to temperatures beyond the limit of steady-state techniques. In addition to measuring specific heat, the pulse techniques may be used to measure other thermal, electrical, and related properties in the solid and liquid phases. Also, in addition to simple metallic specimens, the pulse techniques may be used for measurements on other more complex electrically-conducting substances, such as certain carbides, oxides, borides, and nitrides.

9.8. REFERENCES

1. Worthing, A.G., Phys. Rev., 12, 199-225, 1918.

2. Lapp, E., Ann. Phys., 12, 442-521, 1929.

3. Grew, K.E., Proc. R. Soc. London, 145, 509-22, 1934.

4. Avramescu, A., Z. Tech. Phys., 20, 213-17, 1939.

5. Néel, L. and Persoz, B., C.R. Hebd. Seances Acad. Sci., 208, 642-3, 1939.

6. Kurrelmeyer, B., Mais, W.H., and Green, E.H., Rev. Sci. Instrum., 14, 349-55, 1943.

7. Baxter, H.W., Nature, 153, 316, 1944.

8. Pallister, P.R., J. Iron Steel Inst., 161, 87-90, 1949.

9. Nathan, A.M., J. Appl. Phys., 22, 234-5, 1951.

10. Khotkovich, V.I. and Bagrov, N.N., Dokl. Akad. Nauk, 81, 1055-7, 1951; Engl. Transl.: USAEC Rept. 1817.

11. Pochapsky, T.E., Acta Met., 1, 747-51, 1953.

12. Pochapsky, T.E., Rev. Sci. Instrum., 25, 238-42, 1954.

13. Strittmater, R.C., Pearson, G.J., and Danielson, G.C., Proc. Iowa Acad. Sci., 64, 466-70, 1957.

14. Rasor, N.S. and McClelland, J.D., J. Phys. Chem. Solids, 15, 17-26, 1960.

15. Wallace, D.C., Sidles, P.H., and Danielson, G.C., J. Appl. Phys., 31, 168-76, 1960.

16. Wallace, D.C., Phys. Rev., 120, 84-8, 1960.

17. Pasternak, R.A., Fraser, E.C., Hansen, B.B., and Wiesendanger, H.U.D., Rev. Sci. Instrum., 33, 1320-3, 1962.

18. Pasternak, R.A., Wiesendanger, H.U.D., and Hansen, B.B., J. Appl. Phys., 34, 3416-7, 1963.

19. Taylor, R.E. and Finch, R.A., J. Less-Common Met., 6, 283-94, 1964.

20. Parker, R., Trans. Met. Soc. AIME, 233, 1545-9, 1965.

21. Kollie, T.G., Rev. Sci. Instrum., 38, 1452-63, 1967.

22. Affortit, C. and Lallement, R., Rev. Int. Hautes Temp. Refract., 5, 19-26, 1968.

23. Affortit, C., High Temp.-High Pressures, 1, 27-33, 1969.

24. Finch, R.A. and Taylor, R.E., Rev. Sci. Instrum., 40, 1195-9, 1969.

25. Jura, G. and Stark, W.A., Rev. Sci. Instrum., 40, 656-60, 1969.

26. Kollie, T.G., Barisoni, M., McElroy, D.L., and Brooks, C.R., High Temp.-High Pressures, 1, 167-84, 1969.

27. Cezairliyan, A., Morse, M.S., Berman, H.A., and Beckett, C.W., J. Res. Natl. Bur. Stand., 74A, 65-92, 1970.

28. Dikhter, I.Ya. and Lebedev, S.V., High Temp., 8, 51-4, 1970.

29. Dikhter, I.Ya. and Lebedev, S.V., High Temp., 9, 845-9, 1971.

30. Cezairliyan, A., McClure, J.L., and Beckett, C.W., J. Res. Natl. Bur. Stand., 75A, 1-13, 1971.

31. Cezairliyan, A. and McClure, J.L., J. Res. Natl. Bur. Stand., 75A, 283-90, 1971.

32. Cezairliyan, A., J. Res. Natl. Bur. Stand., 75A, 565-71, 1971.

33. Cezairliyan, A., High Temp.-High Pressures, 4, 541-50, 1972.

34. Cezairliyan, A., J. Res. Natl. Bur. Stand., 77A, 45-8, 1973.

35. Cezairliyan, A. and McClure, J.L., J. Res. Natl. Bur. Stand., 78A, 1-4, 1974.

36. Cezairliyan, A., Righini, F., and McClure, J.L., J. Res. Natl. Bur. Stand., 78A, 143-7, 1974.

37. Cezairliyan, A. and Righini, F., J. Res. Natl. Bur. Stand., 78A, 509-14, 1974.

38. Cezairliyan, A., J. Chem. Thermodyn., 6, 735-42, 1974.

39. Cezairliyan, A. and Righini, F., Rev. Int. Hautes Temp. Refract., 12, 124-31, 1975.

40. Cezairliyan, A. and McClure, J.L., J. Res. Natl. Bur. Stand., 79A, 431-6, 1975.

41. Lebedev, S.V., Savvatimskii, A.I., and Sheindlin, M.A., High Temp., 14, 259-63, 1976.

42. Gathers, G.R., Shaner, J.W., and Brier, R.L., Rev. Sci. Instrum., 47, 471-9, 1976.

43. Shaner, J.W., Gathers, G.R., and Minichino, C., High Temp.-High Pressures, 8, 425-9, 1976.

44. Shaner, J.W., Gathers, G.R., and Minichino, C., High Temp.-High Pressures, 9, 331-43, 1977.

45. Shaner, J.W., Gathers, G.R., and Hogson, W.M., in Proceedings of the Seventh Symposium on Thermophysical Properties (Cezairliyan, A., Editor), ASME, New York, NY, 896-903, 1977.

46. Lebedev, S.V. and Mozharov, G.I., High Temp., 15, 45-8, 1977.

47. Cezairliyan, A. and Miiller, A.P., High Temp.-High Pressures, 9, 319-24, 1977.

48. Cezairliyan, A., McClure, J.L., and Taylor, R., J. Res. Natl. Bur. Stand., 81A, 251-6, 1977.

49. Righini, F., Rosso, A., and Coslovi, L., in Proceedings of the Seventh Symposium on Thermophysical Properties (Cezairliyan, A., Editor), ASME, New York, NY, 358-68, 1977.

50. Petrova, I.I. and Chekhovskoi, V.Ya., High Temp., 16, 1045-50, 1978.

51. Taylor, R.E., J. Heat Transfer, 100, 330-3, 1978.

52. Naito, K., Inaba, H., Ishida, M., and Seta, K., J. Phys. E: Sci. Instrum., 12, 712-8, 1979.

53. Gathers, G.R., Shaner, J.W., and Hodgson, W.M., High Temp.-High Pressures, 11, 529-38, 1979.

54. Seydel, U., Bauhof, H., Fucke, W., and Wadle, H., High Temp.-High Pressures, 11, 635-42, 1979.

55. Cezairliyan, A. and Miiller, A.P., Int. J. Thermophys., 1, 83-95, 1980.

56. Miiller, A.P. and Cezairliyan, A., Int. J. Thermophys., 1, 217-24, 1980.

57. Cezairliyan, A. and Miiller, A.P., Int. J. Thermophys., 1, 317-31, 1980.

58. Righini, F. and Rosso, A., High Temp.-High Pressures, 12. 335-49, 1980.

59. Rasor, N.S. and McClelland, J.D., Rev. Sci. Instrum., 31, 595-604, 1960.

60. Seydel, U., Fucke, W., and Moller, B., Z. Naturforsch., 32a, 147-51, 1977.

61. Seydel, U. and Fucke, W., Z. Naturforsch., 32a, 994-1002, 1977.

62. Seydel, U., Bauhof, H., Fucke, W., and Wadle, H., High Temp.-High Pressures, 11, 35-42, 1979.

63. Gathers, G.R., Shaner, J.W., and Young, D.A., Phys. Rev. Lett., 33, 70-2, 1974.

64. Dikhter, I.Ya. and Lebedev, S.V., High Temp.-High Pressures, 2, 55-8, 1970.

65. Righini, F., Rosso, A., and Ruffino, G., High Temp.-High Pressures, 4, 597-603, 1972.

66. Ruffino, G., Righini, F., and Rosso, A., in Temperature: Its Measurement and Control in Science and Industry (Plumb, H.H., Editor), Vol. 4, Part 1, ISA, Pittsburgh, PA, 531-7, 1972.

67. Coslovi, L., Righini, F., and Rosso, A., Alta Frequenza, 44, 592-6, 1975.

68. Cezairliyan, A., J. Res. Natl. Bur. Stand., 75C, 7-18, 1971.

69. Cezairliyan, A., High Temp. Sci., 13, 117-33, 1980; also most recent version in Int. J. Thermophys., 5, 177-93, 1984.

70. Foley, G.M., Rev. Sci. Instrum., 41, 827-34, 1970.

71. Loriers-Susse, C., Bastide, J.P., and Bäckström, G., Rev. Sci. Instrum., 44, 1344-9, 1973.

72. Henry, K.W., Stephens, D.R., Steinberg, D.J., and Royce, E.B., Rev. Sci. Instrum., 43, 1777-84, 1972.

73. Cezairliyan, A., Morse, M.S., Foley, G.M., and Erickson, N.E., in Proceedings of the Eighth Symposium on Thermophysical Properties (Sengers, J.V., Editor), Vol. II, ASME, New York, NY, 45-50, 1982.

74. Coslovi, L., Righini, F., and Rosso, A., J. Phys. E: Sci. Instrum., 12, 216-24, 1979.

75. Foley, G.M., Morse, M.S., and Cezairliyan, A., in Temperature: Its Measurement and Control in Science and Industry (Schooley, J.F., Editor), Vol. 5, American Institute of Physics, New York, NY, 447-52, 1982.

76. Cezairliyan, A. and McClure, J.L., Time-Resolved Measurement of High Voltages, in preparation.

CHAPTER 10

Calorimetry at Very High Pressures

C. LORIERS-SUSSE

10.1. INTRODUCTION

The application of conventional calorimetric techniques to the measurement of the specific heat of solids at high pressures encounters several major difficulties. Among these, the most hindering is the impossibility of insulating the sample from the surrounding pressure medium. Other difficulties lie in the generally small available volume, in the pressure gradients created in the sample, and in the measurements of pressure and temperature. Although high pressure may also have some positive effects, such as ensuring good thermal contact, eliminating convection phenomena (when a solid medium is used), and reducing the time to achieve equilibrium in the steady-state methods, the difficulties account for the scarcity of the measurements reported, so far, in the literature.

In this chapter the various solutions proposed by the investigators to adapt the conventional techniques to high pressures are described and their advantages and basic limitations as regards to pressure, temperature, and accuracy are roughly evaluated, in view of the fact that a large proportion of the experiments reported were of a preliminary nature. A classification according to the method is adopted, although a classification separating the measurements made at very low temperatures from the others would also have been suitable. In each case, the principle of the method is given together with the practical applications achieved by each investigator. Chronological summaries of the experiments in each category are given in separate tables.

The emphasis is on measurements at very high pressures (>1 GPa) but experiments at lower pressures are also considered for the sake of completeness and for their importance in the historical development of the methods. Differential techniques, such as differential thermal analysis and differential scanning calorimetry, which are used to measure the transition enthalpies at phase transitions, are not considered.

10.2. ADIABATIC CALORIMETRY

10.2.1. Principle

Conventional adiabatic calorimetry is impractical at high pressure since the essential requirement of thermal insulation of the sample from its surroundings cannot be fulfilled due to the large heat losses through the pressure transmitting medium. However, the method can be used with some success in specific cases where the following conditions are satisfied: (a) the pressure vessel is autonomous and thus can be put into a calorimeter without creating heat losses, (b) the heat capacity of this vessel plus the medium is not too large compared to that of the sample, and (c) the measurements of heat capacity and temperature are of sufficient accuracy. The heat capacity of the sample C_s is then obtained from the relation:

$$C_s = C_{tot} - C_{cell} \qquad (1)$$

the total heat capacity C_{tot} and the heat capacity of the empty cell C_{cell} being measured separately at high pressures.

Condition (b) is best satisfied when the temperature is so low that the lattice contribution in C_{cell} is small and that C_{cell} can be made a reasonably small fraction of C_{tot}. The majority of measurements using adiabatic calorimetry at high pressures relates to this case which will be considered first. Measurements performed at temperatures close to ambient relate to lower pressures and will be considered separately since the techniques involved in each category are different in many respects. The main parameters of the experiments in both categories are given in Table 10.1.

10.2.2. Measurements at Low Temperatures

Experiments at low temperatures were prompted by interest in the electronic structure of high-pressure phases and in its change with pressure for particular substances. The method was used, for the first time, by Ho et al. [1] to measure the specific heat of uranium between 0.3 and 6 K at ~1 GPa in zero magnetic field and at 2000 Oe and later by the same authors (Phillips et al. [2]) in an investigation on cerium under the same conditions.

The high-pressure cell of the so-called 'clamp' type is shown schematically in Fig. 10.1; it consists of a small piston and cylinder device weighing ~150 g and constructed entirely from a high-purity Cu-Be alloy (2.1 wt.% Be) which was heat treated for maximum hardness. The alloy has a very small heat capacity at low temperature, a condition imposed by eq. (1) if a reasonable accuracy in C_s is looked for. Commercial grade Cu-Be alloys were found unsuitable by Smith and Phillips [2] as they exhibited large magnetic field dependent heat capacities below 1.5 K. The weighed sample is entirely surrounded by Teflon (polytetrafluorethylene-PTFE) which acts as a pressure transmitting medium. The cell is supported by an external screw-on collar A and pressure is applied to the sample at room temperature through the internal piston B and the removable piston C in a

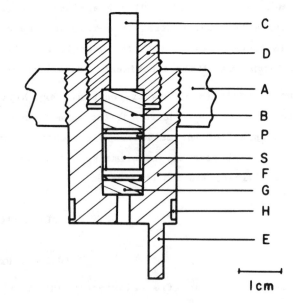

A, C - removable parts; B, D, E, F, G - Cu-Be parts; S - sample; P - PTFE; H - heater

Figure 10.1. High-pressure calorimeter (Smith et al. [2b]).

standard press. At the required pressure, roughly calculated from the ratio of the load to the area of the piston, the locking screw D is tightened down onto the internal piston B and the cell, separated from the removable parts A and C, is placed in a He^3 cryostat. The small post E provides a thermal link to the thermometer and to the cooling bath, through a mechanical heat switch. The performance of the cell is best represented by the ratio C_s/C_{tot} for a material with a given low temperature heat capacity. For uranium for which $C_s = 0.7$ T in $mJ \cdot cm^{-3} \cdot K^{-1}$, T being the temperature, it was 3–5% in the temperature range 0.3 to 6 K. The temperature was measured with a germanium resistance thermometer which retained its calibration from run to run. Such a reproducibility is an essential feature of the method since C_s is obtained from C_{tot} and C_{cell} which are measured separately. The heat capacities were measured by the heat pulse method with a relative inaccuracy of a few tenths of 1%. Taking the value of C_s/C_{tot} given above, one can estimate that the uncertainty in C_s was ~10%. The scatter in the values of C_s/T at 1 GPa, and T > 0.5 K was 1–2% for U and 5–10% for Ce. The value of pressure calculated from the ratio load/(area of the piston) was overestimated since it accounted neither for the frictional forces in the cell nor for the pressure loss during cooling arising from the different expansion coefficients of the various parts of the cell. In the case of uranium, however, pressure was determined from the magnetically measured superconducting transition temperature T_c of the sample by use of the known pressure dependence of T_c. For cerium, monitoring of the piston displacement during the application of pressure allowed one to observe the first order $\gamma \rightarrow \alpha$ transition and to estimate the pressure during the measurements, (1.1 ± 0.1) GPa.

Subsequent experiments were performed by different authors with some modifications to the technique described above. The modifications affected mainly the design and materials of the cell, the method of measuring pressure, and the determination of the heat capacity C_{cell}. McWhan et al. [3], whose aim was to measure the specific heat of the metallic phase of V_2O_3 (stable above ~2.5 GPa), redesigned the apparatus to achieve higher pressures. Their cell is shown in Fig. 10.2. The body of the cell F was made of a hardened commercial Be–Cu alloy, and the mushroom-shaped piston B was made of tungsten carbide. In addition, an insulated electrical lead G was passed through the end-plug H to allow monitoring of the resistance of the sample. The cell was prestretched to 2.8 GPa. The masses of the sample and of the filled cell were ~3 and 260 g, respectively. The pressure locked-in after loading to 2.8 GPa at room temperature was estimated to be 2.5 GPa by comparison with strain-gauge measurements on similar cells. From the thermal expansions of the materials, it was estimated that no pressure loss occurred on cooling.

The heat capacity of the sample was determined at 0.1 and 2.5 GPa between 0.3 and 10 K in the same way as was done previously [1]. The heat capacity of the empty cell was determined at these two pressures by measuring the heat capacity of the cell filled with pure diamond powder, the contribution of which was negligible compared to C_{cell}. The experimental values of C_{cell} were expressed as:

$$C_{cell} = C_c \left(1 + \frac{\Delta C}{C_c} \right) \quad \text{with } C_c \equiv A_o + \sum_{i=1}^{4} A_{2i+1} T^{2i+1} \tag{2}$$

The coefficients in Eq. (2) were obtained by a
least-squares fit to the 0.1 GPa data and the
deviations $\Delta C/C_c$ were tabulated. A nearly systematic shift in $\Delta C/C_c$ of ~0.2% was observed
between 1 and 2.5 GPa at T > 0.5 K. The introduction of carbide and steel parts in the cell
resulted in an increase of C_{cell}/unit volume of
sample compared to the cell of Ho et al. [1],
but the accuracy of the results remained acceptable because of the high specific heat of the
samples. The error in the specific heat due to
an estimated ±0.5% error in the heat capacities
was ~±3% at 0.4 K for both phases of V_2O_3 (i.e.,
at both pressures) and increased to ±15% for the
insulating phase (at 0.1 GPa) and to ±8% for the
metallic phase (at 2.5 GPa) at 0.9 K. This difference in the errors, which arises from the
fact that the last phase has a specific heat ~60
times that of the first one, illustrates the
effect of the specific heat of the sample on the
accuracy. Indeed, Eq. (1) shows that the best
results are obtained with samples having a large
specific heat at low temperature.

The same technique [3] was applied to the
measurement of the heat capacity of the metallic
phase of SmS at 1.5 GPa between 0.3 and 20 K by
Bader et al. [4]. Although no error limits were
given by the authors, probably because of an
anomaly in C_s near 3 K coming from impurities,
metallic SmS was also a favorable substance
since its electronic coefficient of specific
heat $\gamma = (C/T)_{LT}$ was found to be high (145
$mJ \cdot mol^{-1} \cdot K^{-2}$).

Another version of the original cell of
Fig. 10.1 was later described by Berton et al.
[5] in which the piston and end plugs were constructed in Maraging steel and the body of the
cell in a Cu-Be alloy containing less than 0.22%
(Co+Ni+Fe). The pressure was determined from
the superconducting transition temperature T_c of
a small piece of tin enclosed with the sample.

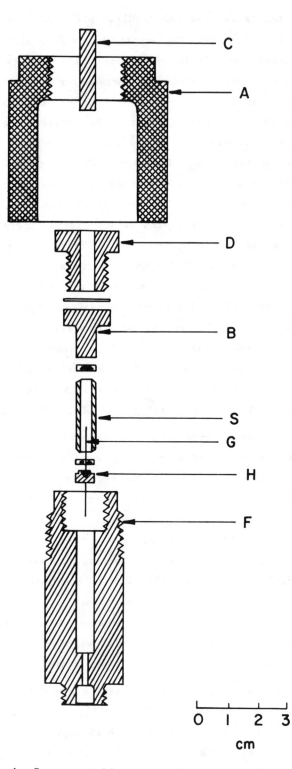

A, C - removable parts; B - piston; D -
locking nut; F - cylinder; H - end-plug;
G - electrical lead; S - sample in PTFE
sleeve

Figure 10.2.   High-pressure   calorimeter
(McWhan et al. [3]).

The temperature T_c was characterized magnetically and measured between 1 and 4 K with a Ge resistor using a double-wall calorimeter with temperature stabilized within 10^{-4} K. The inaccuracy in pressure measurements was estimated to be 2% and the maximum pressure achieved at 4 K was 1.3 GPa. The heat capacity of the empty cell under pressure was obtained from measurements at ~0.8 and 1.2 GPa with the cell loaded with a pure copper sample after subtracting the calculated copper contribution. C_{cell} was represented by the well-known relation

$$C_{cell} = \gamma T + jT^3 \tag{3}$$

which was found to fit the results between 1 and 4 K. The technique was used by Berton et al. [6] to measure the specific heat of $CeAl_2$ and $CeAl_3$ at 0 and 0.7 GPa between 1 and 15 K. The claimed inaccuracy of the specific heats which varied from 400 to 7000 $mJ \cdot mol^{-1} \cdot K^{-1}$ in the temperature range was ~0.5%, and the pressure was probably only estimated in these early measurements since an error bar of ± 0.1 GPa was given.

From the preceding experiments, the following remarks can be made concerning the use of the adiabatic calorimeter method to measure the specific heat at high pressures and low temperatures. The method has three main advantages: it yields absolute values of the specific heat, it can be used for measurements on conductors as well as insulators, and finally the apparatus is a relatively simple and inexpensive accessory of a conventional adiabatic calorimeter. Its major drawback is that, unless the low-temperature specific heat of the sample is exceptionally high, its accuracy is low. This follows from Table 10.1 where the approximate values of C_s (1 K) under pressure can be correlated with the inaccuracy given by the authors for the different substances. The minimum observed C_s (1 K) is seen to be 10 $mJ \cdot mol^{-1} \cdot K^{-1}$ which is already high and the inaccuracy which is then ~10% only allows measurement of large pressure effects on C_s. The limiting capabilities of the method as regards to very high pressures also stem from the necessity of keeping the ratio C_{cell}/C_s sufficiently low. For a given substance, the accuracy will decrease when high strength materials (having large low-temperature specific heat) are used for the construction of the cell in order to achieve these higher pressures. The pressure limit will depend on the substance studied.

Finally the method has the limitation common to all experiments performed at very low temperatures and high pressures arising from the use of a solid pressure medium*. The finite shear strength of PTFE only ensures a quasi-hydrostatic pressure around the sample and may create stresses in it, a particularly unfavorable circumstance for measurements on anisotropic samples. It is not clear whether a better medium could be found for the extremely severe conditions of the experiment. Finally, it is difficult to estimate the pressure under these conditions and it can be precisely determined only in the vicinity of a superconducting transition temperature.

10.2.3. Measurements Near Room Temperature

In this case, if the pressure range is moderate and the temperature not too low, a pressure medium can be found which remains fluid during the measurements. This circumstance offers two

---

*It must be remembered that helium at 80 atm solidifies at 3 K.

advantages; first, the sample is submitted to a truly hydrostatic pressure with no shearing stress introduced and second, this pressure can be continuously measured with precision. However, the ratio R (mass of the pressure container/mass of the sample) remains high and limits the accuracy of the measurements as in the methods considered in Section 10.2.2. On the other hand, the difference C_p-C_v is no longer negligible, as it is at very low temperatures and the experimental data which generally refer to neither isobaric nor isochoric conditions must be reduced to C_p or C_v.

In the method described by Amitin et al. [7] the sample was immersed in polysiloxane inside a steel bomb and the ratio R was ~1.5%. The bomb calorimeter is shown in Fig. 10.3. The heater A, a constantan wire, is wound uniformly on the surface of the bomb B and the adiabatic shield C, which surrounds it in the vacuum calorimeter. The temperature is measured with a calibrated platinum resistor D located in a copper sheath E soldered to the bomb.

The heat capacity was measured to within 0.1-0.2%. The pressure was measured with an estimated error of ~2% using a Manganin gauge calibrated at room temperature, assuming a constant temperature coefficient of resistance under pressure. The total measured heat capacity was:

$$C = C_{cell} + C_L + C_{se} \qquad (4)$$

The heat capacity of the bomb C_{cell} was measured at normal pressure and its change with pressure was calculated and found negligible in the pressure range explored ($\Delta C_{cell} < 0.05\% \, C_s$). The heat capacity contribution from the liquid C_L was determined in experiments without the sample at different pressures. But, since the thermal expansion of steel is much smaller than that of polysiloxane, the experimental value obtained for C_L corresponded to constant volume and had to be reduced to the experimental conditions of the measurements with sample, before introduction in Eq. (4). This was done with use of the thermodynamic relation:

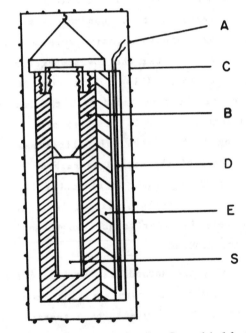

A – constantan; B – bomb; C – shield; D – Pt resistor; E – copper; S – sample

Figure 10.3. Bomb calorimeter (Amitin et al. [7]).

$$C_L = C_L' + T \left(\frac{\partial V}{\partial T}\right)_P \left[\left(\frac{dP}{dT}\right)_e' - \left(\frac{dP}{dT}\right)_e\right] \qquad (5)$$

where $(dP/dT)_e'$ and $(dP/dT)_e$ were determined in experiments without and with the sample while $(\partial V/\partial T)_P$ was taken from published data. The experimental value of the heat capacity of the sample C_{se} was then reduced to isobaric conditions using the formula:

$$C_{sp} = C_{se} + T \left(\frac{\partial V}{\partial T}\right)_P \left(\frac{dP}{dT}\right)_e \tag{6}$$

Using this method, Amitin et al. [7] were able to measure the specific heat of NH_4Cl in the vicinity of its order-disorder transition between 220 and 310 K in the pressure range 0-0.4 GPa. The overall error in C_p was estimated to be 2-2.5%, more than 50% of which arose from reduction of the data through errors in $(\partial V/\partial T)_P$ and dP/dT in Eqs. (5) and (6). The scatter of the data was 0.1%.

The limitations of the method using a bomb calorimeter towards higher pressures are obvious; a first limit is set by the freezing of the possible liquids at high pressures. The pressure limit will increase with the temperature of the measurements. Moreover, the ratio R increases with increasing pressure and, as a result, reduces the measurement accuracy. Another limitation arises from the necessity of reducing the experimental data to C_p (or C_v) which requires additional measurements and a knowledge of the properties of the liquid and the sample as functions of pressure and temperature which in most cases is not available. As was pointed out above, this is an important source of error in the results.

The advantages of the method are also evident and have been quoted at the beginning of this section: hydrostaticity of the pressure, possibility of a continuous and precise determination of pressure. These will make the method particularly adaptable to the study of the specific heat in the vicinity of phase transitions at low pressure.

10.3. STEADY-STATE, a.c.-TEMPERATURE CALORIMETRY

The methods in this section are based on a steady-state calorimetry technique employing a.c. heating which was first developed for measuring heat capacity in vacuo or at atmospheric pressure (Sullivan and Seidel [8], Handler et al. [9], Salamon et al. [10]). The method permits precise measurements of changes in heat capacity as a function of an external parameter and has an absolute accuracy comparable to that obtainable with adiabatic calorimetry techniques. Furthermore, it allows thermal coupling of the sample to a heat reservoir and is thus particularly suited for measurements at high pressures where such coupling is unavoidable. Another advantage for its use at high pressures is that the sample can be quite small.

10.3.1. Principle of the Method

The principle of the method can be understood by means of Fig. 10.4. The sample of heat capacity C_s is loosely coupled to a reservoir at constant temperature T_0 by a thermal resistance Z. An electric heater H and a thermometer θ, of heat capacities C_H and C_θ small compared to C_s, are attached to the sample with thermal resistances Z_H and Z_θ much smaller than Z. An a.c. power $Q(t) = Q_0(1-\sin\omega t)$ is fed to the heater. The resulting temperature change of the sample is the sum of a time-independent temperature offset ΔT_{dc} and of an oscillatory variation $\Delta T_{ac}(t)$:

$$\Delta T = \Delta T_{dc} + \Delta T_{ac}(t) = \Delta T_{dc} + \tilde{T} \sin(\omega t + \phi). \tag{7}$$

For the steady-state conditions ($t \gg \tau_1$) where $\tau_1 = CZ$, ΔT_{dc} and $\tilde{T}$ can be expressed as:

$$\Delta T_{dc} = Q_o Z$$
$$\tilde{T} = (Q_o/\omega C) \, [1 + (\omega\tau_1)^{-2} + (\omega\tau_2)^2 + K]^{-1/2} \equiv (Q_o/\omega C) \, F(\omega) \qquad (8)$$

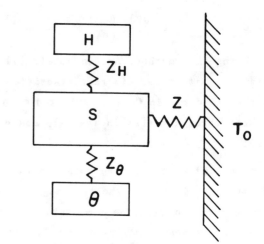

S – sample; H – heater; θ – thermometer

Figure 10.4.   Sample arrangement for a.c. calorimetry (schematic).

where C is the sum of the heat capacities of the sample, heater, thermometer, and a certain portion (~1/3) of the thermal resistance Z. The quantity τ_1 is the relaxation time which characterizes the thermal coupling between the sample and the reservoir and τ_2 is the relaxation time describing the coupling between the sample and the measuring elements heater and thermometer. The quantity K takes account of the finite thermal conductivity of the sample and is generally negligible in the case of metals and small sample dimensions at low temperatures [8]. Equation (8) is valid under the condition that the sample dimensions (L) are small with respect to the characteristic thermal length in the sample, $L \ll \ell_s = (2a/\omega)^{1/2}$ where a is the thermal diffusivity of the sample. The principle of the original method is to use a frequency ω such that:

$$(\omega\tau_1)^2 \gg 1 \quad \text{and} \quad (\omega\tau_2)^2 \ll 1 \qquad (9)$$

with the necessary condition: $\tau_1/\tau_2 \gg 1$. $F(\omega)$ can then be approximated by unity in Eq. (8) and C can be calculated from $\tilde{T}$ by:

$$C = Q_o/\omega\tilde{T} \qquad (10)$$

Different versions of the method have been proposed for the measurement of specific heat at high pressures. In the following subsections we will discuss the measurements on metals and on insulators which require a different treatment. A summary of the measurements performed by the a.c. temperature method is given in Table 10.2.

10.3.2.   a.c.–Temperature Method Applied to Metals

As mentioned in Section 10.3.1, one important assumption of the a.c. method is that the sample comes to thermal equilibrium very rapidly compared with the period of the heating signal ($\omega\tau_2 \ll 1$). This condition is best satisfied with metallic samples; for this reason most of the investigations with a.c.–calorimetry have been conducted on metals.

Among the early studies using a.c.–calorimetry at high pressure one finds the experiments of Chu et al. [11] on U, Nb_3Sn, and V_3Si in which the authors employed the technique to obtain an approximate value of the relative specific heat as a source of information about the changes, with pressure, of the superconducting transition or of some other phase transition occurring in the sample. These experiments demonstrated the great utility of the method, but since no discussion of the errors in the specific heat was given, they will not be considered here.

Table 10.1. Summary of Experiments for the Measurement of Specific Heat of Solids at High Pressure by the Adiabatic-Calorimeter Method

Investigator	Year	Substance	Maximum Pressure (GPa)	Temperature Range (K)	Temperature Measurement (Resistor)	Medium	Approximate Specific Heat at 1 K - P_{max} $mJ \cdot mol^{-1} \cdot K^{-1}$	Inaccuracy* %
Ho et al.	1966	U	1	0.3-6	Ge	PTFE	12	10
Phillips et al.	1968	Ce	1	0.3-6	Ge	PTFE	10	10
McWhan et al.	1973	V_2O_3	2.5	0.3-10	Ge	PTFE	23	3-8
Bader et al.	1973	SmS	1.5	0.3-20	Ge	PTFE	145	-
Berton et al.	1977	$CeAl_2$,$CeAl_3$	0.7	1.3-12	Ge	PTFE	400-800	0.5
Amitin et al.	1976	NH_4Cl	0.4	220-310	Pt	polysiloxane	-	2-2.5

*As reported by the investigators.

Table 10.2. Summary of Experiments for the Measurement of Specific Heat of Solids at High Pressure by the a.c.-Temperature Method

Investigator	Year	Substance	Maximum Pressure GPa	Temperature Range K	Temperature Measurement	Medium	Sample Geometry mm	Heating*	Frequency s^{-1}	$\tilde{T}$ mK	Inaccuracy %
Bonilla and Garland	1974	Cr	0.4	~300	thermocouple	argon	0.05 dia.	S	44	0.1	-
Baloga and Garland	1977	NH_4Cl, ND_4Cl	0.4	240-300	thermistor	argon	0.5x10x10	I	0.42	1-2	4
Itskevich et al.	1978	Sn	1	3.1-4	thermocouple	kerosene + oil	9 dia.x20	I	240,470	-	3
Eichler and Gey	1979	In	0.8	1.3-4.2	C resistor	diamond powder	3 dia.x0.5	I	120	2-6	3
Eichler et al.	1980	Ga	3.5	1.3-4.2	C resistor	diamond powder + PTFE	2.5 dia.x3	I	-	~5	3
Eichler et al.	1981	In	0.8	1.3-4.2	C resistor	diamond powder	3 dia.x0.5	I	120	2-6	3
Bleckwedel et al.	1981	YbCuAl	~1	1.6-7	C resistor	diamond powder+Ga			<15		3

*S self-heating, I independent.

At nearly the same time, the work of Bonilla and Garland [12] on Cr represents an attempt to use the technique to measure precisely the heat capacity of a sample at high pressures. The authors investigated the changes of C_P in the vicinity of the Néel transition ($T_N \sim 300$ K) up to ~ 0.3 GPa. In order to reduce τ_2 and the thermal gradients in the sample, they used a thin wire (0.12 mm in diameter, 5 mm in length) which was heated by passing through it an electric current of frequency $\omega/4\pi$. The temperature of the wire was measured with a differential chromel-constantan thermocouple (0.05 mm in diameter). The hot junction of the thermocouple was attached to the center of the wire by a thin coating of varnish, the cold junction being in good thermal contact with the sample holder. The role of reservoir was played by the high-pressure bomb and that of the thermal resistance Z by a pressure-transmitting gas of low thermal conductivity (pure argon). The assembly was immersed in a thermostated bath. The temperature of the bomb T_0 was measured with an inaccuracy of ± 0.02 K. The pressure was measured with an absolute inaccuracy of ± 0.4 MPa and a precision and stability of ± 0.3 MPa. The oscillating heating signal had a frequency $\nu/2 = 3.5$ Hz and a long term amplitude stability of better than 0.1%. For measurements of $\tilde{T}$, the differential thermocouple was connected to a lock-in amplifier. Since ohmic heating causes a temperature oscillation at twice the frequency of the applied voltage, a frequency doubler was used. ΔT_{dc} could also be measured by connecting the differential thermocouple to a precision potentiometer.

The thermal conductivity of argon increased by a factor of ~ 10 in a typical pressure run, increasing the coupling of the sample with the reservoir. To ensure that the assumption $(\omega\tau_1)^2 \gg 1$ was verified in the entire pressure range, tests were performed to calculate $\omega\tau_1$ from $\tau_1 = C\,\Delta T_{dc}/Q_0$, $\omega\tau_1$ was 66 at ~ 0.3 GPa. It was also verified that $\tilde{T}$ was inversely proportional to ω as predicted by Eq. (8) when $F(\omega) = 1$, and that the calculated and observed ΔT_{dc} values were in good agreement.

Introducing the different parameters of the experiments, Eq. (8) could be written ($F(\omega) = 1$):

$$\tilde{v} = \frac{k(T)\ B\ V^2 R_S}{\sqrt{2}\ 2\pi\ \nu\ R_L^2\ C_P} \tag{11}$$

where $\tilde{v}$ is the r.m.s. voltage from the differential thermocouple, $k(T)$ the thermoelectric power coefficient of the thermocouple, V the heating voltage, R_S the resistance of the sample, R_L the load resistor in series with the sample ($R_L \gg R_S$) and B an attenuation constant due to imperfect thermal coupling between sample and thermocouple. It was assumed that B was independent of temperature and pressure, and values of $C_P(P,T)$ relative to a reference temperature at atmospheric pressure were obtained through the relation:

$$\frac{C_P(P,T)}{C_P(1\ \text{atm},T_{ref})} = \frac{C_P(1\ \text{atm},T)}{C_P(1\ \text{atm},T_{ref})} \times \frac{k(P,T)R_S(P,T)\tilde{v}(1\ \text{atm},T)}{k(1\ \text{atm},T)R_S(1\ \text{atm},T)\tilde{v}(P,T)} \tag{12}$$

The standard deviation of the $\tilde{v}$ signal was ± 0.05 nV, contributing a random error of $\sim 1\%$ to the heat capacity. The resistance of the sample R_S was measured with a high-precision a.c. bridge as a function of pressure at ~ 310 K and $R_S(P,T)$ evaluated from these measurements and other R(P,T) data in the literature. The errors in Cp due to this evaluation were estimated to be less than 1%. The function k(P,T) was taken from previous data. The accuracy of the relative specific heat, which was not given by the authors, can be roughly estimated to be a few percent if B is assumed constant.

The weak link of the method seems to be the varnish bond between sample and thermocouple; this caused unexpected drifts of $\tilde{v}$ between high-pressure runs perhaps originating in dissolved argon and cast some doubt on the assumption of B being independent of pressure and temperature.

Another adaptation of the principle of a.c.-calorimetry was made later by Itskevich et al. [13] for the measurement of specific heat of metals at high pressures and at low temperatures. It required a modification of one of the basic assumptions of Sullivan and Seidel [8]. In the new version, the sample was no longer loosely coupled to a reservoir by a gas of low thermal conductivity as in the experiments just described, but was immersed into a mixture of kerosene and oil used as a pressure-transmitting medium. In these conditions, it was not possible to find a frequency for which the inequalities in Eq. (9) were satisfied and $\tilde{T}$ could no longer be represented by Eq. (8) with $F(\omega) \cong 1$. The method developed by Itskevich et al. consisted in operating under conditions such that the amplitude and phase of ΔT_{ac}, $\tilde{T}$, and ϕ could be related simply to the heat capacity of the sample. The condition found was $\ell_m \ll L, D$, where $\ell_m = 2\pi/(\omega/2a)^{1/2}$ is the thermal characteristic length in the medium, a the thermal diffusivity of the medium, L the dimensions of the sample, and D the distance of the sample to the walls of the high-pressure container (bath). However, the other assumptions remained the same:

$$C_H, C_\theta \ll C_S \qquad Z_H, Z_\theta \ll Z \qquad (\omega\tau_2)^2 \ll 1.$$

The amplitude $\tilde{T}$ could then be expressed as:

$$\tilde{T} = (Q_o/\omega C) \ G(\phi) \tag{13}$$

with

$$G(\phi) = (\cot\zeta - 1) \ (\cot^2\zeta + 1)^{-1/2} \qquad \zeta = \phi + \pi/2 \tag{14}$$

ϕ being the phase of ΔT_{ac} with respect to the heating signal. $\tilde{T}$ and ϕ were measured and C obtained from Eq. (13).

The experimental arrangement consisted of a sample made of two cylinders (length 9 mm, diameter 8.5 mm), separated from the walls of the high-pressure chamber by $D \simeq 5$ mm, with a heater made of Bi film deposited on mica (~10 µm thick) glued between the cylinders. At 4.2 K, ℓ_m was ~0.5 mm and the assumption regarding the medium was thus correctly satisfied. The alternating component of the temperature $\tilde{T}$ was measured by two differential thermocouples (Au + 0.07% Fe)-Cu in series and soldered to both halves of the specimen at ~5 mm from the heater. The mean temperature T_{dc} was determined by a thermocouple (Cu + 0.15% Fe)-Cu led into the chamber without discontinuity.

A block diagram of the measuring system is shown in Fig. 10.5. The voltage from the oscillator O of frequency $\omega/2$ was fed to the heater H and to the frequency doubler FD and the reference channel of the phase meter PH. The oscillating emf from the thermocouple was measured by a compensation method, the amplifier A acting as a null instrument. The balanced-off voltage passed

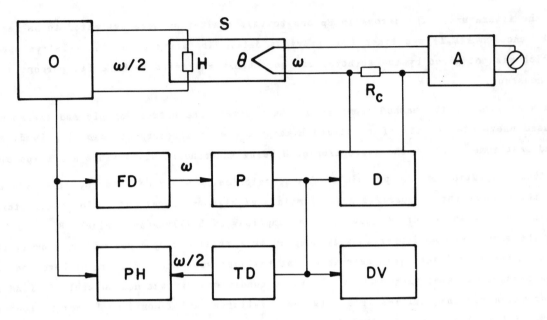

S – sample; H – heater; θ – thermocouple; O – oscillator; FD – frequency doubler; P – phase shifter; D – divider; A – amplifier; PH – phase-meter; TD – frequency divider; DV – digital voltmeter

Figure 10.5.   Block diagram of measurement system in the a.c.-temperature method of Itskevich et al. [13].

from a voltage doubler through the phase shifter P and divider D and was measured with a digital voltmeter DV.   The voltage from the divider was fed to the second channel of the phase-meter through the triggered frequency divider TD in order to measure the phase of $\tilde{T}$.   The measurements were performed at two frequencies, 39 and 75 Hz, in order to correct for the extra phase shift $\Delta\zeta_h$ coming from the heater which was included with ζ in the measured phase shift.   This correction was weakly pressure and temperature dependent and amounted to 30% of ζ at 39 Hz in the experiments described.   An iteration method was given to calculate $\Delta\zeta_h$ from the measured phase shifts at the two frequencies.

The error in the determination of ζ, however, remained to be the largest contributor to the error in the heat capacity, 1.5%.   The error in Q_0 was estimated as ~1% and that of the thermo-couple emf ~0.3%.   The heat capacity of the addenda amounted to less than 1% of C_s.   The resulting inaccuracy of C_s was estimated to be ~3%.   No information was given concerning the error in the mean temperature or in the thermocouple sensitivity.

The method was tested by measuring the specific heat of Sn between 3.2 and 4 K at various pressures up to 1 GPa.   Although the method was said to yield absolute specific heats, no compari-son was made with previous data at zero pressure.   However, the magnitude of the jump ΔC at the superconducting temperature was in agreement with previous data within less than 1%.

In conclusion, the method of Itskevich et al. [13] has the merit to show that a.c.-calorime-try can be used to measure accurately the specific heat of metals at low temperatures and high pressures even when appreciable thermal losses in the medium are present.   Since the pressure medium used is a fluid, it guarantees the homogeneity of pressure in the sample.   Compared to the adiabatic calorimetry techniques described in Section 10.2, it offers the advantage of greater sensitivity and accuracy and thus, although it is restricted to metals, allows investigations on materials having small specific heat at low temperatures. The pressures attainable are, however, limited by the freezing of the medium and the large dimensions of the vessel.

A different a.c.-method for measurements on metals at low temperature and high pressure was described by Eichler and Gey [14]. In contrast to the method discussed above, all the basic assumptions of the model of Sullivan and Seidel [8] were retained; the characteristic times τ_1 and τ_2 were determined from the measured frequency dependence of $\tilde{T}$ and the optimum frequency derived from τ_1 and τ_2. It follows from Eq. (8) that $F(\omega)$, which is proportional to $\omega\tilde{T}$, has a maximum value

$$F(\omega_o) = [1 + 2(\tau_2/\tau_1)]^{-1/2} \qquad \text{at} \qquad \omega_o = (\tau_1\tau_2)^{-1/2} \tag{15}$$

The normalized experimental values $\omega\tilde{T}(\omega)/(\omega\tilde{T})_{max}$ were fitted by the function $F(\omega)/F(\omega_o)$, τ_1 and τ_2 being the fit parameters. If the ratio τ_2/τ_1 so obtained was smaller than ~10^{-3}, ω_o satisfied the inequalities given by Eq. (9) to a good approximation and $F(\omega_o) \approx 1$. For a measuring frequency $\omega_m \neq \omega_o$ and/or higher values of τ_2/τ_1, a correction to the heat capacity, $F(\omega_m)$, could be derived from Eqs. (8) and (15). Typical values of τ_1, τ_2 and ω_o obtained in the experiments were $\tau_1 = 0.4$ s, $\tau_2 = 0.6 \times 10^{-3}$ s and $\omega_o = 65$ s^{-1}. It was found that τ_1 and τ_2 were pressure independent but that they decreased with decreasing temperature, their ratio remaining approximately constant. To compensate for the resulting increase of ω_o, the measuring frequency ω_m was then chosen a little higher than the optimum frequency measured at the maximum temperature of the experiments. This way, the correction to the specific heat $\Delta C/C = F(\omega_m) - 1$ was kept below 1% (it was 0.3 and 0.4% at 4.2 and 1.5 K, respectively, in a typical experiment) and, therefore, was neglected.

The question as to whether the model of Sullivan and Seidel had to be adapted to three dimensions in order to be used in their own experiments was discussed by Eichler et al. [15], who pointed out that since the linear dimensions of all the components were small compared to the thermal wavelength in the corresponding material, there should be no necessity for this extension.

The experimental arrangement is shown in Fig. 10.6. The high-pressure cell is a Cu-Be piston-in-cylinder AB cell which is part of a low-temperature elbow-lever pressure device (not shown). The sample (indium in the first experiments) S consists of three disks (diameter 3 mm and thickness 0.5 mm) weighing ~50 mg. It is heated by a heater of manganin wire H (diameter 0.03 mm, length 10 mm) bent into a loop and mounted between two of the sample disks, and insulated from these by epoxy resin. The temperature of the sample is measured with a circular slide D (diameter 1 mm, thickness 0.15 mm) cut from a standard carbon resistor insulated by epoxy on one face and mounted between two of the sample disks. The sample-heater-thermometer assembly is inserted into a hollow cylinder of compressed diamond powder E followed by two disks of the same material F. The diamond powder acts as the pressure transmitting medium and constitutes part of the thermal resistance Z between the sample and the cylinder which plays the role of the reservoir. Diamond powder was chosen because of its extremely small heat capacity (0.1% C_s if the total mass of diamond would contribute) and high thermal resistance which allows a loose coupling to the reservoir.

In fact, the measured thermal resistance Z was of the order of the calculated thermal resistance of the copper leads to the measuring elements alone; thermal contact between sample and reservoir was probably mainly due to these leads.

The calculated heat capacities C_H and C_θ were smaller than 1% of C_s, and the contact resistances between the sample and the heater or the thermometer Z_H, Z_θ obtained from the experimental τ_2 were found to be of the order $Z/10$. Thus, the requirements of the model were fulfilled to a good approximation.

The elbow lever force generating device allowed for changes of pressure up to 1 GPa at any temperature between 300 and 4.2 K. The apparatus contained in a vacuum chamber filled with helium gas was cooled in a He^4 cryostat and the measurements were performed between 1.3 and 4.2 K with the possibility of magnetically suppressing the superconducting transition of the samples.

The carbon thermometer was calibrated after completion of each experimental run by comparison with a Ge thermometer mounted on the outer ring of the pressure cell. Since the thermometer sensitivity was strongly dependent on pressure, the calibration had to be performed at each pressure. The pressure was determined by measuring the superconducting temperatures of the samples.

The a.c. voltages were measured with calibrated lock-in amplifiers; their output voltages were fed, together with the d.c. voltages, to a data acquisition system controlled by a desk-top calculator.

The method was tested by comparing the specific heat of indium in the normal state at zero pressure with previous data. The difference amounted to about $\pm 3\%$. The possible sources of error in the method were discussed by Eichler and Gey. The specific heat was calculated from Eq. (8):

$$C = (Q_o / \omega_m \tilde{T})\, F(\omega_m)$$

$$(16)$$

$$\text{where} \quad C = C_s + C_{addenda}$$

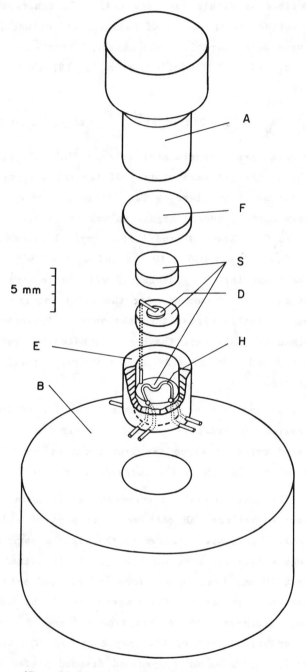

5 mm

AB – Cu-Be; H – manganin heater; D – carbon thermometer; EF – diamond powder; S – sample

Figure 10.6.   High-pressure   calorimeter (Eichler et al. [14]).

Q_o/ω_m was determined with an inaccuracy of ~0.5%. To evaluate the error in $\tilde{T}$ the thermometer sensitivity dr/dT was introduced:

$$\tilde{T} = (\tilde{v}/I_{dc}) \ (dr/dT) \tag{17}$$

where $\tilde{v}$ was the a.c. voltage amplitude at the thermometer and I_{dc} the constant current fed through the thermometer. The uncertainty in $\tilde{v}$ and I_{dc} was 0.3%. The total error in the mean sample temperature including the possible temperature difference between the sample and the reservoir, the calibration of the Ge thermometer, and the deviation from the polynomial fit used for $r(T)$ was about 0.5%. The relative error in dr/dT was estimated to be of the order of 1%. Finally, as was mentioned above, the measuring frequency ω_m was chosen such that the correction $\Delta C/C = F(\omega_m) - 1$ could be well approximated by unity throughout the entire temperature range. The heat capacity of the addenda was calculated and found nearly independent of temperature and definitely smaller than 1% of C_s. Thus, the total inaccuracy in C_s was found to be ~± 3%, which was in agreement with the observed differences between the data at zero pressure and previous data for In. To eliminate the introduction of systematic errors, such a comparison was made for each new pressure cell.

The authors [14] presented preliminary results on In at 0.3 and 0.8 GPa in the range 1.3–4.2 K. More extensive data on this element were obtained later by the same technique in the same pressure and temperature range by Eichler et al. [16].

The method was also used with slight modifications by Eichler et al. [15] for measuring the specific heat of a high-pressure, superconducting phase of gallium at 3.5 GPa. To attain this pressure a clamp device* replaced the cylinder of Fig. 10.6 and the diamond powder was replaced by a Teflon capsule except for a layer of ~0.4 mm thickness around the sample. This was sufficient to decouple the sample thermally from the surroundings and the Teflon capsule minimized the pressure gradients. A small indium disk placed outside the diamond layer was included in the cell and its superconducting transition, determined inductively, served as a low-temperature manometer. The pressure difference between the sample and the manometer was found to be smaller than 0.5 GPa. In view of the small heat capacity of Ga, the heat capacity of the addenda was no longer negligible; it was evaluated as a function of temperature from a comparison of the results at zero pressure with data from the literature and was assumed to be independent of pressure.

The method was later applied to the study of YbCuAl in the temperature range 1.6–7 K up to about 1 GPa by Bleckwedel et al. [17]. However, for this compound, which is hard and brittle compared to In and Ga, an intermediate medium was used between the diamond powder of the cell and the sample and measuring elements to provide a quasi-hydrostatic environment. Gallium metal was chosen because of its low shear strength, low thermal capacitance, and high thermal diffusivity. The low thermal diffusivity of YbCuAl resulted in a frequency dependence of $\tilde{T}$ different from that described by Eq. (16) and imposed the frequency condition $\omega < 15$ s^{-1}.

In conclusion, the method of Eichler and Gey [14] is an absolute method for measuring the specific heat of metals at high pressures and at low temperatures. Its accuracy allows one to measure small specific heats at low temperature. Compared to the method of Itskevich et al. [13],

---

*Clamp devices have been briefly described in Section 10.2.

the use of the thermometers prepared from carbon resistors allows a greater sensitivity of the temperature measurements even at very low temperatures. On the other hand the pressure media used, diamond powder and Teflon, and the smaller dimensions of the samples give access to higher pressures than the methods in [13]. Finally, the overall smaller dimensions of the apparatus lead to a lower consumption of refrigerating gas. The limitation of the technique, which is unavoidable in the extreme conditions of the experiments, as was previously noted in Section 10.2.2, is the use of a solid pressure transmitting medium. The pressure gradients created in the samples were observed to be small for soft metals like In and Ga but could lead to errors in the case of harder metals, especially where they are anisotropic.

10.3.3. a.c.-Temperature Method Applied to Thermal Insulators

An a.c. method for high-pressure measurements on samples with low thermal conductivity was developed by Baloga and Garland [18]. They showed that the a.c. technique could yield the specific heat even for samples of low thermal conductivity if one used sufficiently thin samples and low frequencies. The method nevertheless has limitations which will be discussed below.

The one-dimensional model on which the method is based is shown in Fig. 10.7. A thin slab S of cross-sectional area A is immersed in a gas at high pressure with a heater H attached to one face delivering a heat flux $Q/A = (Q_0/A)$ $(1-\sin\omega t)$ and a temperature sensor T attached to the opposite face. Based on the assumption that the distance L_1 from the sample to the wall of the reservoir was much larger than the thermal characteristic length in the gas ($L_1 \gg \ell_1$), a transcendental equation relating the heat capacity of the sample to measurable quantities was obtained. From this an analytical expression was derived for the heat capacity per unit area:

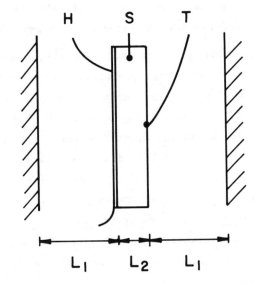

H – heater; T – thermometer; S – sample

Figure 10.7. One-dimensional heat flow model (Baloga et al. [18]).

$$\frac{C}{A} = \frac{-B\delta + [B^2\delta^2 + D(C*^2 - 2\ E\delta^2)]^{1/2}}{D} \qquad (18)$$

where $C* \equiv Q_0/A\omega\tilde{T}$ is the heat capacity per unit area in the ideal case of negligible thermal gradients in the sample and negligible heat capacity per unit volume of the gas; the quantities B, D, and E are polynomials in s where

$$s \equiv \omega(2d_1C_1\lambda_1/\omega)^{1/2} (2\ L_2/3\lambda_2)$$

and

$$\delta \equiv (2d_1C_1\lambda_1/\omega)^{1/2}$$

λ_i, d_i, C_i being the thermal conductivity, density, and specific heat of the gas (i=1) and sample (i=2). To take into account the differences between the model and the real three-dimensional case, two adjustable parameters were introduced, the effective area A_{eff} and a dimensionless calibration factor y defined by:

$$C^* = y(Q_o/A\omega\tilde{T}).$$
(19)

A_{eff}/A and y were considered independent of pressure and temperature. The calibration factor y was determined by comparison with some absolute heat capacity value at 1 atm. The factor A_{eff}/A was obtained by studying the pressure dependence of C^* at a temperature where the sample heat capacity was independent of the pressure or by studying a reference sample (such as Invar for which $\partial C_p/\partial P \simeq 0$) in the same geometrical configuration.

The model imposed the following restrictions on the frequency and the dimensions of the sample:

$$L_2\ \omega^{1/2}\quad small\quad (<1),\qquad L_2' \gg \ell_2$$

L_2' being the lateral dimension of the sample and ℓ_2 the thermal characteristic length in the sample.

The technique was used to measure the specific heat of NH_4Cl and ND_4Cl in the vicinity of their order-disorder transition in the range 240–300 K up to 0.32 GPa. Typical values of the parameters in the experiments were $\omega = 0.42$ s^{-1}, $L_2 = 0.5$ mm, $L_2' = 10$ mm, $L_1 = 20$ mm, and $\tilde{T} \sim 1$–2 mK. The inaccuracy of the relative C_p was about 4% of which 1% came from the error in $Q_o/\omega\tilde{T}$.

The method of Baloga and Garland offers the advantage of a high resolution in temperature. However, it requires knowledge of the properties of the sample and surrounding gas (entering as s and δ) as functions of pressure and temperature; generally the information concerning the sample is insufficient although the use of low enough frequencies decreases the importance of the corresponding correction terms. Since the pressure range of the method is limited to P < 1 GPa, and the inaccuracy is ~4%, the method is essentially adapted to measuring important variations of the specific heat, as occur in the vicinity of phase transitions.

10.4. METHOD USING A STEADY-STATE ALTERNATING HEAT FLOW

10.4.1. Principle

The specific heat of a solid is related to its thermal conductivity λ and thermal diffusivity a by the equation:

$$C_p = \lambda/ad$$
(20)

where d is the mass density. Since for many substances the change of d with pressure is known, at least near room temperature, a means of determining C_p is to measure λ and a at high pressure. If the two quantities are measured simultaneously, on the same sample, the accuracy on the specific heat is increased in comparison to the case of separate measurements. A review on the measurements of thermophysical properties of solids under high pressure prepared by Bäckström [19] gives a general view of the methods used to measure λ and a up to 1977. We shall consider here only the experiments where λ and a were measured simultaneously and in which the specific heat was calculated by the authors of the experiments. A summary of these experiments is given in Table 10.3.

10.4.2. The Method Developed by Andersson and Bäckström

The method developed by Andersson and Bäckström [20] consists of measuring simultaneously λ and a in a steady and oscillating temperature field. The procedure was developed for a solid pressure medium and is best adapted to insulators.

A cylindrical geometry was chosen for the sample and heater as shown in Fig. 10.8. The axial heater H delivered a power $Q = Q_o [1 + \exp(i\omega t)]$ and the temperature of the sample was measured by

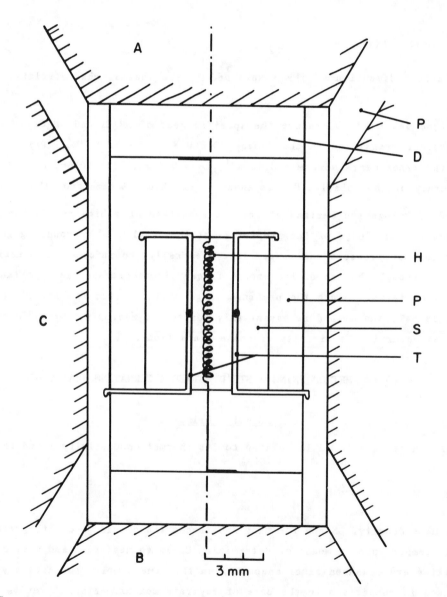

A, B – pistons; C – die; D – steel; H – heater; P – pyrophyllite; S – sample; T – thermocouples

Figure 10.8. The high-pressure cell used in the steady-state method of Andersson et al. [20].

thermocouples at two points at radii r_1 and r_2. In the ideal case of an infinite cylinder the steady-state solution for the temperature at radius r is well known:

$$T(r,t) = B - Q_0 \ln r/2\pi L\lambda + F[K_0(\xi\sqrt{i}) \exp(i\omega t) - GI_0(\xi\sqrt{i}) \exp(i\omega t)] \qquad (21)$$

where B and F are constants; L is the length of the heater, $\xi = r(\omega/a)^{1/2}$, and K_0 and I_0 are the modified Bessel functions. The second term in the bracket of Eq. (21) represents the reflected wave from the cylindrical boundary of the sample. It was found that the coefficient G in this term could be made negligible by choosing sufficiently large values of ω and of r_s-r_2, r_s being the radius of the sample. The attenuation α of the thermal wave between r_1 and r_2 was:

$$\alpha = \tilde{T}(r_2)/\tilde{T}(r_1) = K_0(\xi_2\sqrt{i})/K_0(\xi_1\sqrt{i}) = K_0(\xi_2\sqrt{i})/K_0(r_1\xi_2\sqrt{i}/r_2). \qquad (22)$$

For a homogeneous sample, the ratio r_1/r_2 could be considered constant during the compression and α could be written: $\alpha = \Psi(\xi_2) \exp[i\phi(\xi_2)]$. The amplitude ratio Ψ and the phase shift ϕ were calculated for various values of the parameter r_1/r_2 and compared to the experimental values of amplitude ratio and phase shift which gave two values of ξ_2, coincident in principle. The thermal diffusivity was calculated from $a = \omega r_2^2/\xi_2^2$. The agreement between the values of ξ_2 was a test of the validity of the model under the experimental conditions.

The thermal conductivity λ of the sample could be obtained from measurements of $\langle T(r_1,t) - T(r_2,t)\rangle$ and $\langle Q\rangle$ but it was found more convenient to use a constant power Q_0 and to measure the corresponding temperature drop $T(r_2) - T(r_1) = \Delta T$. Equation (21) then gives:

$$\lambda = Q_0 \ln(r_2/r_1)/2\pi L\Delta T \qquad (23)$$

and the specific heat is given by:

$$C_p = Q_0 \ln(r_2/r_1) \xi_2^2/2\omega\Delta Tm \qquad (24)$$

m being the mass of the sample inside radius r_2. The specific heat was measured relative to its atmospheric pressure value, thus eliminating the necessity of determining m.

The high-pressure cell is shown in Fig. 10.8. The measurements were performed in an apparatus of the girdle type, the medium being pyrophyllite. The pressure was measured with a manganin gauge calibrated in situ against the Bi I→Bi II transition (2.54 GPa). A current $I=I_0 \cos(\omega t/2)$ was driven through the heater yielding a heating power of frequency ω. The signals from the chromel-alumel thermocouples at r_1 and r_2 were amplified and displayed on an X-Y recorder. The amplitude ratio Ψ and phase shift ϕ were obtained from the measurements on the recorder.

Since the pressure was only quasi-hydrostatic within the cell, a distortion of the sample occurred during compression. The change in the diameter of the sample was measured as a function of pressure in a separate experiment, and the results were used with the known compressibility data to calculate the change in length L.

The method was first tested by measuring the relative changes in λ, a, and C_p of Teflon (PTFE) at 300 K up to 3 GPa. The samples were 6 mm in diameter and 9 mm in length. Satisfactory agreement was found between the values of ξ_2 obtained from the amplitude ratio and phase shift.

The time necessary to reach the steady-state was about one minute. The scatter of the data for one sample (increasing and decreasing pressure) was $\pm$2% and the scatter of all the data from the mean C_p was $\pm$7% at 3 GPa. A large part of this scatter came from the uncertainty in the zero pressure value. This in turn arose from the imperfect thermal contact between the different parts of the cell for virgin samples and from the difficulty to reach zero pressure after compression due to rupture of the measuring leads.

In a later investigation on polyethylene up to 2.5 GPa by the same authors (Andersson and Bäckström [21]), the method was unchanged except the length of the sample was increased in order to reduce the end effect. The error in C_p introduced by this end effect for the new length (15 mm) used was calculated and was found to be negligible. As the new samples occupied an increased proportion of the available volume, the pressure distribution in the cell was determined in separate experiments and the pressure scale corrected according to these measurements. The pressure inhomogeneity in the sample itself was found to be small ($<$0.1 GPa). The scatter of the data on relative specific heat between different samples was $\pm$2%.

The method was later improved, particularly concerning the acquisition of data, and used by Andersson and Sundquist [22] for measurements on four polymers (see Table 10.3) up to 3.7 GPa. The pressure calibration was made using a coaxial Ce-Bi-Tl wire having phase transitions at 0.75, 2.54, and 3.7 GPa and the change of resistance of the heater itself for interpolations. The signals from the thermocouples were amplified, integrated, sampled, converted to digital form, and fed into a desk computer programmed to calculate the amplitude ratio and the phase shift and finally the thermal diffusivity. The maximum scatter in the data of relative specific heat between samples was $\pm$5% at 2.5 GPa. As in the former measurements, most of the scatter came from the values at zero pressure.

In all these measurements (Refs. [20–22]) and particularly in the last two, the systematic errors were considered to be small relative to the random scatter of the data, and the estimated inaccuracy in C_p ranged from 5 to 10% depending on the samples.

The method has the advantage of yielding simultaneously λ and a (or dC_p). Owing to the increased heat transfer, the times necessary to reach steady-state conditions (~1 minute in the reported experiments) are greatly reduced compared to the situation at 1 atm. Regarding the specific heat, the method is capable of yielding relatively accurate values of dC_p relative to some reference value. A large part of its accuracy comes from the simultaneous measurement of λ and a which eliminates possible discrepancies between separate experiments as regards to pressure and chemical and geometrical factors. The method requires a solid medium since a film of liquid or gas between the heater and sample would influence the results. The limitations of the technique are that the samples must be prepared with great precision and that the zero pressure values of dC_p (which are used as a reference) are subject to an increased uncertainty coming from the thermal contact problem. Moreover, the assumptions made in the model require sufficiently large samples and this condition will probably limit the use of the method to the pressures attainable in apparatus of the girdle/belt or multi-anvil types having working volumes comparable to that shown in Fig. 10.8, that is to pressures below ~6 GPa. Finally, compressibility data for the samples are necessary for the calculation of C_p and this may lead to additional measurements of this property when previous data are lacking, especially at temperatures different from normal.

10.5. METHODS USING A HEAT PULSE

10.5.1. General Principle

In the pulse methods (also called transient methods), a heat pulse is sent to the sample either directly through ohmic heating, if the sample is a conductor, or indirectly by an independent heater (hot wire method) if the sample is an insulator and the temperature versus time variation in the sample is analyzed to yield its thermal properties. The advantages of the heat pulse method compared to the steady-state methods are essentially the rapidity, the easier preparation of samples and, in principle, the better maintenance of the boundary conditions since only the early part of the temperature history is used. Applied to high pressures, they have the further advantage that the temperature may be measured at a single point, which minimizes the number of electrical leads entering the apparatus.

10.5.2. Measurements on Conductors

Jura and Stark [23] were the first investigators to adapt the pulse method to the measurement of specific heat of metals (or semiconductors) at high pressures. Their sample was a wire of resistance R and heat capacity C_p through which a constant current I was passed. The variations in time of the current and of the voltage drop E = RI are shown schematically in Fig. 10.9. From the power balance equation the following relation was obtained:

$$dE/dt = (I^3 RR'/C_p) - (IR'/C_p)\, f(T-T_a) \qquad (25)$$

where $R' = dR/dT$ and $f(T-T_a)$ is a cooling term which was assumed to be dependent only on the difference between the sample temperature T and the temperature of the surroundings T_a. In the limit $t \to 0$, $T-T_a \to 0$, whence:

$$C_p = I^3 RR'/(dE/dt)_{t=0} \qquad (26)$$

For practical reasons it was not possible to reduce the measuring time below 10–20 µs and the value of $(dE/dt)_{t=0}$ had to be determined by extrapolation. As the experimental values of

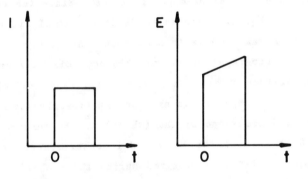

Figure 10.9. Schematized variations of the current and potential drop through the specimen in the pulse method.

dE/dt were found to be linear in t from 10 to 1000 µs and the theory for an ideal infinitely long wire predicted a nearly linear change of dE/dt at short times, $(dE/dt)_{t=0}$ was obtained by linear extrapolation to t = 0.

In the experiments, the samples were relatively short and it was difficult to estimate the exact length involved in the measurements, so that relative, rather than absolute, values for C_p were obtained.

The pressure cell used in the experiments is shown in Fig. 10.10. The sample was a hoop of wire, with four thin (0.13 mm diameter) platinum leads for the measurement of I and E, embedded in silver chloride used as the pressure-transmitting medium. The cell was pressed between opposed Bridgman anvils and calibrated in separate experiments at 300 K with the Bi I-II and VI-VIII transitions (2.54 and 7.7 GPa). Temperatures in the range 150-300 K were obtained by putting the entire system into liquid baths, and excursions to 77 K were made using massive heat sinks. The temperature was measured with thermocouples fastened to the anvil jackets.

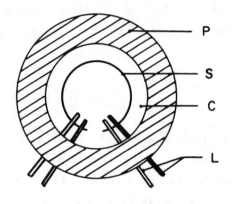

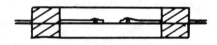

S − sample; P − pyrophyllite; C − AgCl; L − current and potential leads

Figure 10.10.   Schematic diagram of sample set-up used in the pulse method (Jura et al. [23]).

The resistance of the sample was determined with an uncertainty of ±5% for nominal R values of 0.1 Ω, by the four probe technique using a precision potentiometer and a static constant current of ~12 mA. In the dynamic measurements, a current I ~ 2A was used, having a stability of a few parts in a thousand during a 100 ms pulse. The switching circuit providing the square current pulses was allowed to dissipate the transient currents before the current was passed through the sample. The current I and voltage E were displayed on a dual beam oscilloscope. Since the knowledge of dE/dt alone was necessary to calculate C_p from Eq. (26) and since dE/dt was small (it was ~0.2% of E at t = 100 µs), a differential comparator was used (with a maximum sensitivity of 1 mV cm^{-1}) for measuring E. The data were taken from photographs of the oscilloscope displays and analyzed with a digitized protractor. At each temperature on a given isobar, a series of ~20 pictures were taken with several values of the current. A computer was used to fit linearly the measured $(dE/dt)_{t=0}$ data to the respective I^3 values and the slope of the $(dE/dt)_t$ vs I^3 function was computed graphically. The uncertainty was ~1% in each of the dE/dt and I measurements. The value of R' = dR/dT was computed graphically from the R-T data obtained during the isobaric runs. Two isothermal runs were made at ~300 and 195 K. The reported uncertainty in the relative specific heat was about 5%.

The method was first tested by examining the Curie transition of gadolinium up to 2 GPa, in the range 260-300 K, and the large anomaly of C_p at T_c was easily detected. An attempt was then made to measure the change of specific heat of iron between 2-10 GPa in the range 150-350 K and at 5 GPa from 77 K to 300 K. A minimum pressure of 2 GPa was necessary to insure good electrical contact between the sample and the leads. However, the experimental uncertainty was larger than the effect of pressure on C_p and thus the pressure effect could not be determined with certainty.

If one considers, following the authors themselves, that iron was a favorable substance to study because of its relatively large resistivity and temperature derivative* and of its high

---

*The factor controlling the magnitude of dE/dt is $g\rho^2/dC_p$ where ρ is the electrical resistivity and g = d(ℓn R)/dT, and this factor is about 60 times larger for Fe than for Cu.

Debye temperature, the method of Jura and Stark [23] appears better suited to the investigation of high-pressure phase transitions than to measurements of the specific heat of metals at high pressures.

A modified version of the method was later described by Loriers-Susse et al. [24]. They showed that, with suitable modifications, the heat pulse method of Jura and Stark [23] could permit measurements of small changes of specific heat at high pressures, even in the case of unfavorable substances such as copper (see Table 10.4).

One of the essential improvements provided by the new version consisted of increasing the sensitivity of the measurement of the slope of E(t). A differential technique was used in both methods but instead of subtracting a constant voltage from E and of measuring the slope of $E' = RI - R_0I_0$, the subscript zero referring to the state before the pulse, the combination $e = RI - R_0I$ was now used. The voltage e had the same slope as E and was much less sensitive to fluctuations of I than E. More precisely, the relative error introduced in e by a fluctuating current $I = I_0$ $[1 + \varepsilon(t)]$ was $\varepsilon(t)$ whereas the relative error in E was $\varepsilon(t)/g\Delta T(t)$, g being the temperature coefficient of the resistance and ΔT the temperature increase of the sample during the pulse, and this last value was much larger than $\varepsilon(t)$. The modification also eliminated the precision potentiometer used by Jura and Stark.

The circuit used for generating the voltage e was essentially equivalent to a Kelvin-Thomson bridge that would be used with a pulsed current source and operated off-balance. By properly adjusting a variable resistance in the bridge, the voltage e could be written simply:

$$e = I\eta + I (1-w) R_0 g\Delta T \tag{27}$$

where

$$\Delta T = T(t) - T_0$$

η being a remaining term after this adjustment, w a constant of the bridge ($w \simeq 0.1$), and T_0 the temperature of the sample before the pulse, at time t = 0. From Eq. (27) the time rate of change of e is:

$$\dot{e} = de/dt = dE/dt = I (1-w) R_0 g\dot{T} \tag{28}$$

and the specific heat is obtained as:

$$C_p = EI/m\dot{T} = (1-w)g R_0 I^2 E/m\dot{e} \tag{29}$$

m being the mass of the sample wire between the voltage probes. The quantity m was not measured since, because of possible end effects, only relative measurements of C_p were made. The value of the temperature coefficient g at high pressures was either determined experimentally or was taken from previous data.

A specially constructed current source delivered the current-stabilized pulses of 50 A of 100 μs duration. The quantities I, E, and e were measured using a high sensitivity preamplifier and were sampled. They were displayed on an oscilloscope for the necessary adjustments and finally recorded with a digital voltmeter. The stability of the amplitude of I was better than 2×10^{-4}

Table 10.3. Summary of Experiments for the Measurement of Specific Heat at High Pressure Using the Steady-State Alternating Heat Flow Method

Investigator	Year	Substance[c]	Maximum Pressure GPa	Temperature K	Temperature Measurement[a]	Medium[b]	Sample Geometry mm	Frequency s^{-1}	ΔT K	Inaccuracy %
Andersson and Backstrom	1972	PTFE	3	300	TC	Pyro.	φ 6 L 9	0.6	5	10
Andersson and Backstrom	1973	PE	2.5	300	TC	Pyro.	φ 6 L 14	1.2	2	5
Andersson and Sundquist	1975	PMMA,PS iPP,aPP	3.7	300	TC	Pyro.	φ 6 L 14	–	–	5-10

[a] TC thermocouple.
[b] Pyro.: pyrophyllite.
[c] PMMA poly (methyl methacrylate), PS polystyrene, i-a PP isostatic-atactic polypropylene.

Table 10.4. Summary of Experiments for the Measurement of Specific Heat of Metals at High Pressure Using the Pulse Method

Investigator	Year	Substance	Maximum Pressure GPa	Temperature Range K	Temperature Measurement[*]	Medium	Sample Geometry mm	Current A	Inaccuracy %
Jura and Stark	1969	Fe, Gd	10	77-300	R	AgCl	- L 15	1-3	5
Loriers-Susse et al.	1973	Cu	10	300	R	araldite	φ 0.1 L 15	50	1
Dzhavadov	1973	Bi	8.5	300	TC	AgCl	7x3.5 L 11	–	5
Bastide and Loriers-Susse	1975	Cu-Al-Ni	10	300	R	araldite	φ 0.1 L 15	50	1
Bastide et al.	1978	Ce	2	300	R	petrol+ pentane	φ 0.1 L 15	50	1.5

[*] R resistive, TC thermocouple.

over a few hours and the standard deviation in the digital measurements was 3×10^{-4} for the amplitudes and 10^{-3} for the slopes. This analog-to-digital conversion greatly improved the accuracy of the measurements compared to those of Ref. [23].

Another important refinement in the new method was the measurement in situ of the heat losses. Indeed, it was soon recognized that the assumption made by Jura and Stark [23] that the initial part of E(t) is linear was incorrect. With the increased sensitivity of the technique, the curvature of e(t), undetectable in air at 1 atm, became obvious as soon as the wire was immersed in a liquid or solid medium, even at t = 30 µs which was about the earliest possible time for the measurements since at lower times the curves e(t) were distorted by the switching transients. To survey the factors involved an approximate calculation of the heat loss correction was performed based on the assumption of one-dimensional propagation of heat (which was found justified a posteriori). The slope correction was found to be:

$$(\dot{\Delta T})/\dot{T}_o = 2.19 \ (d_m \lambda_m C_{p_m})^{1/2} \ \sqrt{t}/r_o dC_p \qquad (30)$$

where the subscript m refers to the medium. The slope correction amounted to 0.12 for a Cu wire with $2r_o = 0.15$ mm in water after 100 µs. The effect of the heat loss at the ends of the wire, treated in the same manner, was found to be negligible. Equation (30) was verified experimentally at 1 atm by putting a sample into several fluids having different values of $(\lambda dC_p)_m$. In order to correct the slopes at high pressures, a double pulse technique was developed according to the scheme of Fig. 10.11. First, a double pulse was sent through the sample (Fig. 10.11a). The voltage $e_o(t)$ was the voltage which would have been measured in the absence of heat losses; e(t) was the actual measured voltage and the difference $e_o(t_o) - e(t_o)$ was approximately proportional to the correction to be applied to the slope of e. The voltage $e_o(t)$ was obtained by the pulsing scheme shown in Fig. 10.11b. Here the interval τ was reported at the beginning of the pulse where there is no heat loss so that $e'(t_o) = e_o(t_o)$. The relation between $[e'(t_o)-e(t_o)]/e'(t_o)$ and the relative slope correction $[\dot{e}_o(t_o)-\dot{e}(t_o)]/\dot{e}(t_o)$ was established at 1 atm by putting the sample into various liquids, the reference medium (subscript zero) being air for which the calculated correction was negligible. This relation was found to be linear. The heat loss correction at high pressures could thus be determined without a knowledge of the thermal properties of the medium.

The first measurements of specific heat were performed on Cu samples ($2r_o = 0.15$ mm, L ≃ 12 mm) arranged in a manner similar to the one shown in Fig. 10.10 with the difference that the leads were soldered to the sample in order to eliminate eventual changes of m during compression and to allow measurements at low pressure, including P = 0. In addition, the thermocouple giving the average temperature of the sample was put inside the cell with its hot junction at the center of the sample loop. The medium used was araldite poured into the central part of the cell before polymerization. The high-pressure apparatus was of the belt type (a device somewhat similar to that shown on Fig. 10.8) and was pressure-calibrated at 300 K with the transitions of Bi (2.54 and 7.7 GPa) and Ba (5.5 GPa). No resistance measurements were made since g had been accurately measured previously in a liquid medium up to 3 GPa. As the change of g was an order of magnitude smaller than that of C_p, it was assumed that $d(\ell ng)/dP$ was constant in the range 0–10 GPa. Figure 10.12 shows the results obtained for the relative specific heat of copper up to 10 GPa, at 300 K,

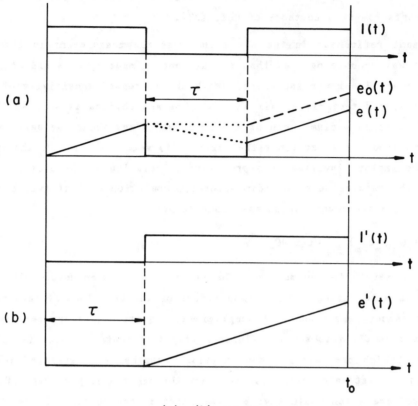

(a), (b) see text

Figure 10.11. Pulse scheme for measuring heat loss under pressure in the pulse method
              (Loriers-Susse et al. [24]).

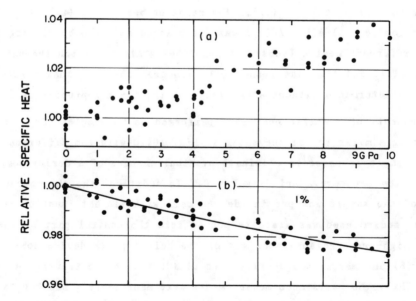

(a) uncorrected values; (b) values corrected for heat loss

Figure 10.12. Relative specific heat of Cu vs pressure at 300 K (Loriers-Susse et al. [24]).

before (upper curve) and after (lower curve) correction for the heat losses. The change of C_p with pressure for this metal is very small so that the effect of the change in the heat loss correction shows up particularly clearly; it results in an inversion of the sign of the pressure effect on C_p.

In Fig. 10.12 the data obtained at increasing and decreasing pressure or between different samples were indistinguishable; this shows that the precompaction of the cell up to ~2.5 GPa, which was made before the measurements in order to fill the voids in the cell, was efficient in reducing the distortion of the wire. There occurred a distortion of the sample during the pressure cycles which was revealed by the irreversible changes of resistance; however, due to the symmetry of the sample-cell-apparatus assembly, this distortion was essentially homogeneous along the length of the wire. An analysis of the error introduced in C_p by the distortion of the sample showed that in this case the error vanished, in agreement with the data of Fig. 10.12.

The method was later used by Bastide and Loriers-Susse [25] to measure the specific heat of Al and Ni up to 10 GPa at 300 K. For the three metals Cu, Ni, and Al the standard deviations of the relative values of C_p/g were smaller than 0.5% and the inaccuracy of C_p was estimated to be 1%.

In their investigation of the specific heat of cerium up to 2 GPa at 300 K, Bastide et al. [26] used the technique unchanged but preferred to put the sample into a fluid medium (petrol + 10% pentane) to make sure that the large volume change of the sample at the $\gamma \rightarrow \alpha$ transition would not cause an important distortion of the wire. The use of a fluid medium also increased the resolution and accuracy of the pressure measurements, a condition made necessary by the larger changes of C_p observed in the whole pressure range. The measurements were performed in an apparatus of the piston-in-cylinder type and pressure was measured with a manganin gauge with an inaccuracy of ± 2 MPa. The average temperature was measured to $\pm 0.03°$C with a copper-constantan thermocouple located in the immediate vicinity of the sample. The samples were flattened, opened rings of 0.2 mm thickness and diameters 3.8-5.8 mm. This shape was chosen to obtain a better temperature homogeneity in the wire. Since the previous measurements of resistance at high pressure were not accurate enough, the authors redetermined g between 0 and 2 GPa with a relative inaccuracy of 5 x 10^{-3} and the uncertainty in C_p was ~1.5%. The increased uncertainty compared to the previous metals was due to the influence of the history of the sample on g and C_p.

To summarize, the improvements of the new technique compared to the original one were an increase in the sensitivity and accuracy of the measurements and the introduction of a process for heat loss correction.

In a third version of the pulse method proposed by Dzhavadov [27], some basic modifications were made. Firstly, the temperature of the sample was no longer inferred from its change of resistance, but measured with a thermocouple, and secondly, an a.c. current was passed through the sample instead of a d.c. current. The sample was made of two bars 11 x 3.5 x 3.5 mm between which the thermocouple junction and two potential leads were pressed. Mica sheets (0.02 mm thick) were used to insulate the thermocouple leads from the sample. The current leads were soldered to the ends of the bars. The specific heat of the sample was calculated from the expression:

$$C_p = \frac{I^2 R}{m} \frac{k}{\omega} \Big/ \left(\frac{dv}{dn}\right)_{t \to 0} \tag{31}$$

k being the thermocouple constant, ω the frequency, n the number of periods in the pulse, and v the thermocouple signal. The function v(n) was linear in the initial region (near t=0). The resistance was measured with a d.c. potentiometric circuit. The signal from the thermocouple was registered by a galvanometer, previously calibrated for pulse conditions. The value of $(dv/dn)_{t \to 0}$ was derived from an analysis of the responses to a series of pulses of duration from 0.05 to 10 s with several values of the current I with an inaccuracy of ~1.5%. The specific heat of bismuth was measured up to ~8.5 GPa by this method with an uncertainty in the absolute value of C_p of 5%. This accuracy was not sufficient to allow measurement of the pressure effect on C_p or on the change of C_p at the phase transition.

In principle, the advantage of the method of Dzhavadov [27] is that it is not limited by the magnitude of the temperature coefficient of resistance of the sample. On the other hand, as noted by the author, Eq. (31) is valid only if the conductivity and current density remain constant in the sample between the potential leads, and this may lead to an additional error when a solid medium is used if the resistance depends on pressure in an appreciable manner. The problem of the distortion of the sample is probably the most crucial since m in Eq. (31) should not vary under pressure.

In conclusion, it has been demonstrated that the pulse method using resistive measurement of temperature is capable of yielding the relative specific heat of metals at high pressures with an inaccuracy of ~1%, even in the case of unfavorable substances such as copper. Such an accuracy is necessary if one wants to measure the effect of pressure on C_p which is very small for common metals. The method employs a thin wire as the sample which undergoes a homogeneous distortion, and is well suited for measurements at very high pressures, being easily adaptable, for instance, in devices of the Bridgman-anvil type which reach pressures of several tenths of GPa. It should also be applicable at moderately high or low temperatures, but generally not at very low temperatures, since for most metals $g \to 0$ when $T \to 0$. The major deficiency of the method is that it requires a precise knowledge of the temperature coefficient of resistance of the sample g; however, this can be measured at high pressures using the same equipment as for C_p/g.

10.5.3. Measurements on Insulators

In this section we consider methods in which a metal wire embedded in the sample is heated by a pulsed electrical current with constant power, and the temperature variation in time at one point of the sample is analyzed to obtain its thermal conductivity λ and diffusivity a, allowing one to calculate its specific heat from the relation $C_p = \lambda/da$. As in Section 10.4, only the experiments in which the authors have calculated C_p or dC_p from simultaneous measurements of λ and a will be considered. A summary of these experiments is given in Table 10.5. In the following subsections we shall distinguish between the methods where the temperature of the sample is measured by a thermocouple at a certain distance from the wire and the probe method where the temperature at the wire-sample interface is measured by the wire itself, used as a resistance thermometer.

Table 10.5. Summary of Experiments for the Measurements of the Specific Heat of Insulators at High Pressure Using Heat Pulse

Investigator	Year	Substance	Maximum Pressure GPa	Temperature Range K	Temperature Measurement[a]	ΔT K	Medium	Sample Geometry mm	Time[b] s	Inaccuracy %
Dzhavadov	1975	PTFE	4	300	TC	1	–	φ 20 L 3.5	5	6
Kieffer et al.	1976	PTFE	3-5	300	TC	13	Pb+Ag+PTFE	φ 13 L 50	30-60	–
Sandberg et al.	1977	glycerol	0.8	130-300	R	3-5	PTFE	φ 70 L 8	1	10
Andersson	1978	cyclohexane	1.5	120-340	R	3-5	PTFE	φ 45 L 10	1	10
Andersson	1978	cyclopentane	2.1	100-300	R	3-5	PTFE	φ 45 L 10	1	10
Ross and Sandberg	1979	NH_4Cl	2.0	140-520	R	3-5	PTFE	φ 45 L 10	1	10
Ross et al.	1979	benzene	2.4	105-375	R	3-5	PTFE	φ 45 L 10	1	10
Ross et al.	1979	benzene napthalene anthracene	2.3	110-480	R	3-5	PTFE	φ 45 L 10	1	10
Andersson and Ross	1980	CBr_4	2.0	170-425	R	3-5	PTFE	φ 45 L 10	1	10
Wigren and Andersson	1980	admantane HMT[c]	2.5	115-445	R	3-5	PTFE	φ 45 L 10	1	10
Andersson	1980	furan	2.8	120-300	R	3-5	PTFE	φ 45 L 10	1	10
Seipold and Gutzeit	1980	granulites	1.2	300	TC		pyrophyllite	φ 10 L 23	1	10
Ross and Andersson	1981	trichloro-ethanes	2.5	115-300	R	3-5	PTFE	φ 45 L 10	1	10

[a]TC thermocouple, R resistive.
[b]Time of the measurements.
[c]Hexamethylenetetramine.

383

10.5.3.1. Methods Where the Transient Temperature is Measured with a Thermocouple

Method Using a One-Dimensional Heating Pulse

A method using one-dimensional propagation of a heat pulse has been developed by Dzhavadov [28]. Its principle may be understood from Fig. 10.13. The sample was a plate of thickness D, the surface of which was maintained at a constant temperature T_0 (the temperature of the high-pressure apparatus). An internal flat heat source of cross-section S in plane B delivered a constant power Q during the time interval τ. The temperature was measured in plane C. Planes A, B, C, and D were equidistant. If T_0 was the initial temperature of the plate, its temperature increment during the pulse in plane C could be expressed as:

$$\Delta T = T - T_0 = \frac{2QD}{\pi^2 \lambda S} Y_2 \left(u, \frac{\pi^2 a}{D^2}, \tau, t\right); \qquad u = \frac{BC}{AD} = \frac{1}{3} \qquad (32)$$

where Y_2 is the Bessel function. The quantities λ and a were calculated by fitting the experimental values of the maximum temperature increment measured with the thermocouple, ΔT_{max}, and of the corresponding time, $t(\Delta T_{max})$ to Eq. (32). The quantities ΔT_{max} and $t(\Delta T_{max})$ were obtained from oscillograms of the thermocouple signal. The change of D with pressure was obtained from the distance between the platens of the press with an uncertainty of ~3%. The errors in the determination of λ and a were estimated to be

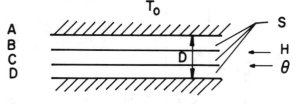

S – sample; θ – thermocouple; H – heater

Figure 10.13.   Specimen geometry for one-dimensional heat flow (Dzhavadov [28]).

approximately 7 and 8%, respectively. Since the sample apparently occupied the total available volume in the high-pressure apparatus, an increase in pressure during the pulse was measured with the manganin gauge located on the cold face of the sample. This gauge was calibrated in situ by observing the transitions of bismuth and thallium wires located in the plane of the thermocouple. The measured increase of pressure was found to be in agreement with that calculated from the normal pressure values of the compressibility and thermal expansion of the sample, assuming that the volume of the sample was constant during the pulse. The specific heat calculated from the measured λ and a was accordingly considered at constant volume and their inaccuracy estimated to be ~6%. It seems that the only substance investigated by this method is PTFE; ΔT_{max} and $t(\Delta T_{max})$ were ~1 K and ~2 s, respectively. The measurements were made up to 4 GPa at 300 K, and the C_v data obtained by using previous compression measurements for the function d(P) were in agreement with previous data at high pressure within the limits of the estimated errors.

It is difficult to evaluate the merits of the method from the one very short paper published so far on the subject. In particular, the resolution of ΔT_{max} and $t(\Delta T_{max})$ recorded on the oscillogram does not seem to be compatible with the claimed accuracy. On the other hand, one would like to have an estimation of the systematic error resulting from the assumption of one-dimensional heat flow; in that respect, comparison of the values of λ and a or dC_p obtained at normal pressure for standard substances with literature data would be useful.

Method Using a Radial Propagation of the Heat Pulse

This method, which utilizes a thin linear heater as the heat source, was first developed for measurements at atmospheric pressure by Jaeger and Sass [29] and later was modified for use at high pressures by Kieffer et al. [30]. In the original method the sample was assumed to be perfectly insulated or to have small heat losses, an assumption which is not justified in a high-pressure vessel. Kieffer et al. replaced this boundary condition by the condition that the cylindrical surface of the sample be as nearly isothermal as possible. The new method was adapted to the cylindrical geometry of a piston-in-cylinder, high-pressure apparatus.

The sample was a cylinder of radius r_o and length L having a linear heater parallel to its axis at radius r' and a thermocouple having its junction at radius r (see Fig. 10.14). The initial temperature of the sample T_o was assumed equal to that of the surroundings. For a heater generating heat at a constant rate QL, the temperature increase ΔT at the point with coordinates (r,θ) is:

$$\frac{\pi\lambda\Delta T(r)}{Q} = \frac{1}{2}\left[-\ell n\left(\frac{r}{r_o}\right) + \ell n\frac{[1+(r/r_o)^2]}{2}\right] - \sum_{n=0}^{\infty}\varepsilon_n \cos n\theta \sum_{m=0}^{\infty}\frac{\exp[-a\,\alpha_{n,m}^2 t/r_o]J_n^2[\alpha_{n,m}(r/r_o)]}{(\alpha_{n,m}^2/r_o^2)J_n'^2(\alpha_{n,m})} \tag{33}$$

where $\varepsilon_n = 1$ for $n = 0$, $\varepsilon_n = 2$ for all $n \geq 1$, and $\alpha_{n,m}$ are the positive roots of $J_n'(r_o) = 0$. The values of λ and a were obtained from Eq. (33) by the method of Jaeger [31] using the first part of the T(t) curves (over 30–60 s). From Eq. (33), ΔT may be written in the form:

$$\Delta T(t) = g(\lambda,a)\ f(at/r_o^2) \tag{34}$$

from which one obtains, assuming λ and a constant,

$$R = \frac{\Delta T(nt)}{\Delta T(t)} = \frac{f(nat/r_o^2)}{f(at/r_o^2)} \tag{35}$$

n being a positive number. The signal from the thermocouple was recorded with a strip chart recorder and the experimental values of R were compared with a theoretical curve or tables of the function $f(nat/r_o^2)/f(at/r_o^2)$; a value of a was obtained for each value of R. A systematic drift of the a-data was obtained with varying n except in the earliest part of the $\Delta T(t)$ curves and these were extrapolated back to the nominal value at t=0. The quantity λ was then calculated from Eq. (33), the resistance of the heater being obtained from measurements at normal and high pressures. The authors gave no indication of the method of determining the length L of the sample under pressure, a quantity necessary to the calculation of Q, which means that it was probably deduced from previous compression data and the assumption of hydrostatic conditions in the experiments. The a-values were found to depend on the values of Q and this dependence assigned to a change in the average temperature of the sample during the experiment. The highest temperature attained by the thermocouple at the end of the experiment (313 K) was taken as a reasonable approximation of this average temperature integrated over the duration of the experiment. A typical temperature increment of ~13 K during the experiments can be deduced from the reported value of T_o (300 K).

The high-pressure cell used in the measurements is shown in Fig. 10.14. The outside of the pressure vessel was cooled by circulating water. The sample was ~50 mm in length and ~12 mm in diameter with two diametrically opposed slots for the heater and the thermocouple ($\theta=\pi$). The silver cylinder A surrounding the sample insured good thermal coupling with the pressure vessel, the outer cylindrical shell of lead B assuring reasonable stress distribution. The ends of the cylinder were insulated by Teflon plates B giving a good approximation to the boundary condition $dT/dz = 0$.

The stress distribution in the cell was studied in separate experiments using bismuth foils placed at different levels and observing the Bi I → II and II → I transitions by monitoring electrical resistance. An axial pressure gradient of 0.24 GPa over the length of the sample was observed but as the thermocouple was found to be influenced by the thermal properties of only the central part of the sample (~25 mm), it was considered that the measured thermal properties were averaged over ±50 MPa. The pressure at the center of the sample was taken as the pressure at the midpoint of the hysteresis loops obtained in the measurements of increasing and decreasing pressure (i.e., the frictional effects were assumed to be symmetrical).

The method was used for measurements of λ and a for several substances up to 3.5 GPa, at ~310 K. The random scatter of the data was ±2% and the numerical 1-atm values were in agreement with previous data within a few percent. For one of the substances investigated, PTFE, the specific heat was calculated using previous data for d(P) and plotted relative to its 1-atm value. A minimum uncertainty of ~5% can be expected for dC_p taking into account the systematic errors arising at high pressures, from distortion of the sample, for instance.

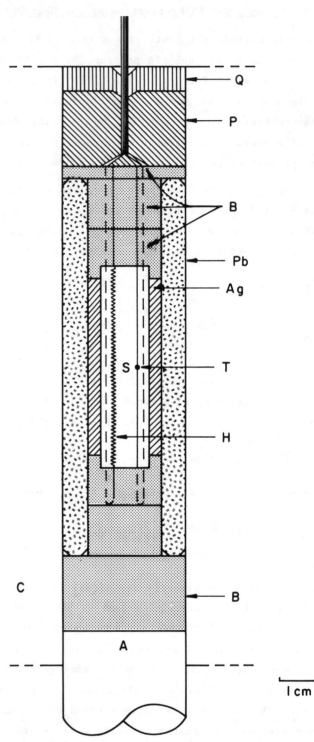

1 cm

A – piston; B – Teflon; C – cylinder; P – pyrophyllite; Q – steel; H – heater; S – sample; T – thermocouple

Figure 10.14.   High-pressure cell for measurement of λ, a and dC_p by the line source method (Kieffer et al. [30]).

The method uses relatively large samples for which the end-effects as well as the influence of the heater and thermocouple on the results are probably small. The heating times and the maximum temperatures attained are about ten times larger than those of Dzhavadov [28], but the rates of temperature rise of the sample are of the same order of magnitude. The problem of the true P,T trajectory followed by the sample during the heating was not discussed by Kieffer et al. [30]. However, from the reduced part of the total volume (~5%) of the cell occupied by the sample, this trajectory was certainly closer to an isobar than to an isochor. The weaker points of the technique might be the possible distortion of the sample leading to erroneous Q and the low resolution in temperature resulting from the large temperature rise during the measurements. The method is linked to the cylindrical geometry of the apparatus whence a maximum pressure of ~4 GPa can be attained in the experiments.

In a different version of the method, proposed by Seipold et al. [32], the heater wire was arranged along the axis of the cylindrical sample and the temperature recorded by a thermocouple at its outer surface. The pulse was assumed to be of infinitely short duration and the sample of infinite extension. The following expressions for the thermal diffusivity and conductivity were then obtained:

$$a = \frac{r^2}{4t_m} \tag{36}$$

$$\lambda = \frac{Q}{8\pi e \ell t_m T_m} \tag{37}$$

giving

$$C = \frac{Q}{2\pi e \ell r^2 dT_m} \tag{38}$$

where e is the base of natural logarithms, r and ℓ the radius and length of the sample, Q the total thermal energy of the pulse, and t_m and T_m the time and temperature at the maximum of the recorded T(t) curves. Although the effect of a pulse of finite duration was shown to be negligible in the experiments, the finite dimensions of the sample had a significant effect on the T(t) curves. The resulting systematic error on a was considered acceptable for relative measurements although no quantitative evaluation of this error was given.

The specific heat of several granulites calculated from the measured values of a and λ was given up to 1.2 GPa by Seipold and Gutzeit [33] without indication of the error limits.

The weakness of the method certainly lies in the uncorrected effect of the surrounding medium on the measurements.

10.5.3.2. The Probe Method

This method is currently used at normal pressures to determine the thermal conductivity of fluid substances for which there is a good thermal contact between sample and probe. In the case of solids, imperfect thermal contact may lead to large systematic errors at normal pressure but at sufficiently high pressures the ideal situation should be attained. This was expected by Andersson and Bäckström [34] who seem to have been the first to apply the probe technique to the measurement of λ under these conditions. The technique was later refined by Sandberg et al. [35] to obtain at the same time the thermal diffusivity and, therefore, the specific heat per unit volume dC_p.

The probe wire, of radius r and length L, was heated by a constant electrical power QL. Its temperature increase over the initial temperature T was given approximately by:

$$T - T_o = (Q/4\pi\lambda) \ [-0.5772 + \ln \ (4 \ at/r^2)] \tag{39}$$

Equation (39) is valid for $at/r^2 \gg 1$. If the temperature rise is sufficiently small, λ and a can be considered as constant and λ is obtained from the slope of the linear relation between $T-T_o$ and $\ln t$. The thermal diffusivity a can then be derived from Eq. (39):

$$a = (0.445 \ r^2/t) \ \exp[4\pi\lambda(T-T_o)/Q] \tag{40}$$

The problem in calculating λ or a was the determination of T_o since the resistance thermometer used in the measurements, being the wire itself, could not be calibrated from the knowledge of the common temperature of the medium and the wire before the heating pulse. Although the initial resistance could have been obtained using a current, small enough to produce negligible heating, a procedure allowing all resistance measurements to be made by the same technique was preferred. It consisted of determining the resistance of the probe at a small finite time t_1 and extrapolating to $t = 0$. An approximation of $T-T_o$ valid for small times was used:

$$T-T_o = (Qu/2\pi\lambda) \ f(\tau,u) \qquad \tau \equiv at/r^2 \tag{41}$$

where the properties of the probe come in through the parameter $u = 2\lambda/ad_wC_w$, the subscript w referring to the wire. An iteration procedure was used to calculate λ and a in which the temperature $T(t_1)$ was first set equal to T_o, the temperature of the medium before the heating pulse as measured with a thermocouple.

The possible sources of systematic errors in the measurements of λ and a were examined. It was first noted that the pressure transmitting medium should be a solid in order that no fluid film be introduced between the sample and the wire which would introduce additional unknown parameters. It was also found that for a solid medium the presence of a thermal resistance between the wire and the sample, due to an imperfect contact, led to an error in a and, therefore, in dC_p. To reduce this error, which was maximum at low pressures, a precompaction to ~0.5 GPa was made at the beginning of each run, except when the sample was liquid under ambient conditions.

The finite value of the sample thickness r_o led to an error in λ of less than 0.1% for $2r_o/(at)^{1/2} > 4$. This inequality was used to calculate the maximum value of r_o for a given sample for heat pulse durations of the order of 1 s used in the measurements. Upper limits of the error in λ arising from cooling of the ends through the potential leads were given. A minimum length of 40 mm for the wire was obtained for pulse durations of a few seconds.

The heat source used in the measurements was a nickel wire 0.1 mm in diameter and about 50 mm in length between the spot-welded potential leads made of the same metal. A simplified diagram of the electronics is shown in Fig. 10.15. The voltages $R_m I(t)$ and $R(t)I(t)$ at the potential leads of the manganin resistor R_m and of the probe wire R were amplified and fed to a multiplier M, the output of which was used to regulate the power dissipated in the selected part of the wire during the current pulse delivered by amplifier A_3. The voltages $R_m I$ and RI and their difference ΔE were sampled at equal time intervals, buffered in an analog memory, and measured with a digital voltmeter (DVM). The temperature T of the wire at time t and pressure P was obtained from the known resistivity of nickel $\rho(T,P) \equiv \rho(0,0) \; f(T,P)$ and from the measured relative change of resistance R between times t and 0

$$R \equiv \frac{(\Delta E/I)_t - (\Delta E/I)_0}{R(T_o,P,r,L)} = \alpha \left[\frac{\rho(T,P)}{\rho(T_o,P)} - 1 \right] \qquad (42)$$

where α is the amplification factor of A_1. For the resistivity of nickel, a simplified formula was used $f(T,P) = f(T)(1-\beta P)$, β being the compressibility. In terms of the inverse function of $f(T)$, $\theta(f)$, the temperature T was expressed as:

$$T = \theta[Q/(\alpha+1) \; f(T_o)] \qquad (43)$$

It was easily shown that Eq. (43) was valid even for a wire distorted by a nonuniform longitudinal strain, a particularly important point for measurements made at high pressures.

Since Q was the power per unit length of the wire, the knowledge of L was necessary. In practice, L was measured at P = 0 and its relative change at pressure derived from its resistance change, assuming a hydrostatic compression of the wire.

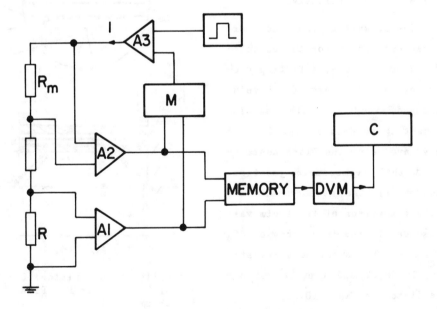

R — resistance of the sample; R_m — manganin resistance; M — multiplier;
DVM — digital voltmeter; C — desk calculator

Figure 10.15. Schematic of electronics for constant power generation (Andersson et al. [34]).

Tests of the method at normal pressure on several liquids showed a good linearity of T–T$_o$ versus ℓn t and a satisfactory agreement with previous values of λ (within 2%) and dC$_p$ (0–4%). Typical values of the maximum temperature increment and of the duration of the pulses were ΔT = 3–5 K, Δt ≃ 1 s.

The measurements at high pressures were made in a piston-in-cylinder apparatus which could be externally heated or cooled using the cell shown in Fig. 10.16. The pressure transmitting medium was PTFE and the sample dimensions were approximately 70 mm in diameter and 8 mm in thickness. The pressure was determined in situ by a Bi-Ce coaxial wire having transitions at 0.75 and 2.54 GPa and by strain gauge measurements on the outer surface of the cylinder. Glycerol was the first substance on which simultaneous measurements of λ and a (or dC$_p$) were made at high pressure. It was investigated between 130 and 300 K up to 0.8 GPa. Eight temperature values were recorded during the power pulse which had a typical duration of 1 s. A desk computer was used to store and process the data. No indication of the error limits on dC$_p$ was given for these preliminary measurements.

The method was later used by Andersson to study the changes of λ and dC$_p$ of cyclohexane [36] up to 1.5 GPa and cyclopentane [37] up to 2.1 GPa in the range ~100–300 K. Since these substances, like glycerol, were liquid at normal temperature and pressure, they were simply poured into the cell. Although the inaccuracy of the λ-data was 3%, the dC$_p$-data at normal pressure and temperature were found ~15% higher than those calculated from previous data. This deviation was attributed to systematic errors in the electronic measuring procedure and the dC$_p$-results were considered to be qualitative.

In subsequent investigations [38] the iterative procedure used previously for the calculation of λ and dC$_p$ was replaced by a fitting procedure by Gauss's method. The rest of the method remained unchanged except that the samples were made by compacting powders in the form of plates placed above and below the plane containing the probe and the thermocouple junction. The minimum pressure of the measurements was ~0.1 GPa. From the comparison of the 1 atm values with previous data, the inaccuracy of dC$_p$ was estimated to ~10%. The substances investigated, which include NH$_4$Cl and a number of organic solids, are listed in Table 10.5.

Compared to the other methods using a thermocouple described in Section 10.5.3.1, the

(a)

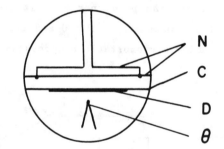

(b)

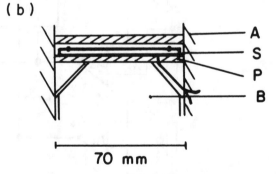

70 mm

A – cylinder; B – piston; C – Cu-lead; D – Bi-Ce; S – sample; P – PTFE; N – Ni wires; θ – thermocouple

Figure 10.16   High-pressure cell assembly used in the pulse method of Andersson et al. [34]).

probe method of Sandberg et al. [35] necessitates regulation of the power generated by the heater since its resistivity changes during the pulse, contrary to that of the manganin heater used by the other authors. On the other hand, it assumes perfect thermal contact between the probe and the sample, a condition which is difficult to realize at low pressures. However, when a thermocouple is used, one should consider the possible perturbation of the temperature field in the sample. The probe method of Sandberg et al. [35] has a better resolution in temperature than the method of Kieffer et al. [30] since $T-T_0$ is only 3–5 K. In view of the relatively large size of the samples, it is usable in about the same pressure range (i.e., up to ~4 GPa).

Compared to the steady–state technique described in Section 10.4 the methods in this section have a common advantage in that the preparation of the sample is simpler but have a common disadvantage which is the complexity in the reduction of data. Although this difficulty was largely alleviated in the probe method by complete automation of the acquisition and reduction of data, the complexity of the theory and the large number of measurements involved are probably responsible for the modest accuracy of the pulse methods in their present state.

10.6. DISCUSSION AND CONCLUSIONS

The difficulties of measuring the specific heat of solids at high pressures are reflected in the small number of experiments performed so far which have been summarized in Tables 10.1 to 10.5. The reasons for these difficulties, which were briefly discussed in the introduction, appear more clearly in the description of the methods given in the previous sections. At the present time, each of the major methods used at normal pressure has been applied to measurements at high pressures, with appropriate modifications. In many instances, the development of techniques was prompted by theoretical interest in the study of particular substances in particular pressure–temperature ranges, such as the study of high–pressure phases at low temperature or investigations in the vicinity of phase transitions, and a reasonable accuracy was only achieved under specific conditions. However, methods having a wider range of applicability and an accuracy comparable to that achieved at normal pressure have recently been proposed; their development should lead to an increase in investigations at high pressures.

The uncertainty of the measurements is most easily established for methods yielding absolute values of the specific heat since a direct comparison may then be made with previous results for standard substances at zero pressure. For relative methods, theoretical estimates and experimental tests are necessary to establish limits to the systematic errors; greater confidence can be assigned to the measurements when these estimates and tests have been made by the investigators themselves.

In the following paragraphs, methods having the potential of measurements above 1 GPa are summarized.

(1) At Low Temperatures

(a) For those substances, insulators, or conductors which have a high coefficient of electronic specific heat, $\gamma = C/T$, the adiabatic–calorimeter method can be used to obtain absolute values of C_p. The inaccuracy of the measurements is strongly dependent on γ and varies from 0.5 to 10% for C (1 K) values of ~600 and 10 $mJ \cdot mol^{-1} \cdot K^{-1}$, respectively. The apparatus used can be considered as a relatively simple and inexpensive accessory to a conventional calorimeter. The actual pressure limit of the measurements is 2.5 GPa, and it is improbable that this limit could be raised significantly in the future.

(b) For any metals, absolute values of C_p can be measured up to 3.5 GPa by a steady-state a.c. temperature method with an inaccuracy of ~3%, comparable to that obtained at normal pressures. The resolution in temperature is high since the amplitude of the temperature oscillations is only a few mK. Since the sample can be small, the pressure limit of 3.5 GPa in the actual measurements could probably be increased.

(2) At Room Temperature or Temperatures Not Very Far from Ambient (300 $\pm$ 200 K)

(a) For conductors, the pulse method using resistive self-heating of the sample can yield relative specific heat with an inaccuracy of ~1% up to 10 GPa. It requires a knowledge of the temperature coefficient of the resistance of the sample. This coefficient, which should preferably be measured on the same sample as the other measurements, can be obtained using the same equipment. The method uses very small samples and, consequently, could probably be extended to pressures above 10 GPa.

(b) For insulators, specific heat can be obtained from simultaneous measurements of thermal conductivity and thermal diffusivity through the relation $C_p = \lambda/da$, if the density d is known as a function of pressure. The proposed methods have the advantage that they yield at the same time λ, a and/or dC_p. However, the function d(P) is generally known with sufficient accuracy only for simple substances and essentially at room temperature so that the reduction to C_p is not always possible. Among the existing methods, the best results at the present time seem to be achieved by a steady-state, alternative, radial heat-flow method which has yielded relative specific heat up to 3.7 GPa with an inaccuracy of ~5% in favorable cases. The technique has the limitation of requiring very accurate machining of the high pressure cell. This probably explains the trend observed in the recent measurements on insulators towards the use of the pulse techniques in which the preparation of samples is simpler although the reduction of data is more complex. The pulse methods which allow absolute measurements of C_p seem to be still in a state of development, and their present performances (~5-10% inaccuracy) could probably be improved. Both methods necessitate a rather large sample size so that the maximum pressure of the measurements made so far (~4 GPa) could at most be multiplied by a factor of ~2.

Finally, in any of the methods using a solid pressure transmitting medium, that is, in most measurements performed at P > 1 GPa, it must be remembered that, whatever the accuracy of the measurements, a limit will be set to the interpretation of data by the pressure gradients created in the sample and in its surroundings. At best, this will result in an uncertainty in the pressure and in an averaging of the properties of the sample. Still more troublesome effects can be expected in the case of anisotropic substances.

10.7.   REFERENCES

1. Ho, J.C., Phillips, N.E., and Smith, T.F., Phys. Rev. Lett., 17, 694-6, 1966.

2. Phillips, N.E., Ho, J.C., and Smith, T.F., Phys. Lett., A27, 49-50, 1968; Smith, T.F. and Phillips, N.E., Colloque C.N.R.S., Paris, No. 188, 191-5, 1970.

3. McWhan, D.B., Remeika, J.P., Bader, S.D., Triplett, B.B., and Phillips, N.E., Phys. Rev., B7, 3079-83, 1973.

4. Bader, S.D., Phillips, N.E., and McWhan, D.B., Phys. Rev., B7, 4686-8, 1973.

5. Berton, A., Chaussy, J., Cornut, B., Odin, J., Paureau, J., and Peyrard, J., Cryogenics, 19, 543-6, 1979.

6. Berton, A., Chaussy, J., Chouteau, G., Cornut, B., Peyrard, J., and Tournier, R., in Proceedings of the International Conference on Valence Instabilities and Related Narrow Band Phenomena (Park, R.D., Editor), Plenum Press, New York, NY, 471-4, 1977.

7. Amitin, E.B., Kovalevskaya, Yu.A., and Paukov, I.E., Sov. Phys. Solid State, 14(11), 2902-3, 1973; Amitin, E.B., Kovalevskaya, Yu.A., Lebedeva, E.G., and Paukov, I.E., High Temp.-High Pressures, 7, 269-77, 1975; Amitin, E.B., Kovalevskaya, Yu.A., and Paukov, I.E., Sov. Phys. JETP, 44(2), 368-71, 1976.

8. Sullivan, P.F. and Seidel, G., Phys. Rev., B173, 679-85, 1968.

9. Handler, P., Mapother, D.E., and Rayl, M., Phys. Rev. Lett., 19, 356, 1967.

10. Salamon, M.B., Simons, D.S., and Garnier, P.R., Solid State Commun., 7, 1035, 1969.

11. Chu, C.W. and Knapp, G.S., Phys. Lett., 46A(1), 33-5, 1973; Chu, C.W. and Testardi, L.R., Phys. Rev. Lett., 32(14), 766-9, 1974; Chu, C.W., Phys. Rev. Lett., 33(21), 1283-6, 1974.

12. Bonilla, A. and Garland, C.W., J. Phys. Chem. Solids, 35, 871-7, 1974.

13. Itskevich, E.S., Kraidenov, V.F., and Syzranov, V.S., Cryogenics, 18, 281-4, 1978.

14. Eichler, A. and Gey, W., Rev. Sci. Instrum., 50(11), 1445-52, 1979.

15. Eichler, A., Bohn, H., and Gey, W., Z. Phys. B38, 21-5, 1980.

16. Eichler, A., Cieslik, J., and Gey, W., Physica, 108B, 1005-6, 1981.

17. Bleckwedel, A., Eichler, A., and Pott, R., Physica, 107B/C, 93, 1981.

18. Baloga, J.D. and Garland, C.W., Rev. Sci. Instrum., 48(2), 105-10, 1977; Garland, C.W. and Baloga, J.D., Phys. Rev., B16(1), 331-9, 1977.

19. Bäckström, G., in Proceedings of the Seventh Symposium on Thermophysical Properties (Cezairliyan, A., Editor), ASME, New York, NY, 169-80, 1977.

20. Andersson, P. and Bäckström, G., High Temp.-High Pressures, 4, 101-9, 1972.

21. Andersson, P. and Bäckström, G., J. Appl. Phys., 44, 2601-5, 1973.

22. Andersson, P. and Sundquist, B., J. Polym. Sci., 13, 243-51, 1975.

23. Jura, G. and Stark, W.A., Jr., Rev. Sci. Instrum., 40, 656-60, 1969.

24. Loriers-Susse, C., Bastide, J.P., and Bäckström, G., Rev. Sci. Instrum., 44, 1344-9, 1973.

25. Bastide, J.P. and Loriers-Susse, C., High Temp.-High Pressures, 7, 153-63, 1975.

26. Bastide, J.P., Loriers-Susse, C., Massat, H., and Coqblin, B., High Temp.-High Pressures, 10, 427-36, 1978.

27. Dzhavadov, L.N., High Temp.-High Pressures, 5, 455-8, 1973.

28. Dzhavadov, L.N., High Temp.-High Pressures, 7, 49-54, 1975.

29. Jaeger, J.C. and Sass, J.H., Br. J. Appl. Phys., 15, 1187-94, 1964.

30. Kieffer, S.W., Getting, I.C., and Kennedy, G.C., J. Geophys. Res., 81(17), 3018-24, 1976.

31. Jaeger, J.C., Aust. J. Phys., 12, 203, 1959.

32. Seipold, U., Gutzeit, W., and Stromeyer, D., in Physical Properties of Rocks and Minerals Under Extreme P-T-Conditions (Stiller, H. and Volarovich, M.P., Editors), Akademie Verlag, Berlin, 155-62, 1979.

33. Seipold, U. and Gutzeit, W., Phys. Earth Plan. Int., 22, 272-6, 1980.

34. Andersson, P. and Bäckström, G., Rev. Sci. Instrum., 47, 205-9, 1976.

35. Sandberg, O., Andersson, P., and Bäckström, G., J. Phys. E, 10, 474-7, 1977.

36. Andersson, P., Mol. Phys., 35, 587-91, 1978.

37. Andersson, P., J. Phys. Chem. Solids, 39, 65-8, 1978.

38. Ross, R.G. and Sandberg, O., J. Phys. Chem. Solids, 12, 3649-59, 1979; Ross, R.G., Andersson, P., and Bäckström, G., Molec. Phys., 38, 377-85, 1979; Ross, R.G., Andersson, P., and Bäckström, G., Molec. Phys., 38, 527-33, 1979; Andersson, P. and Ross, R.G., Molec. Phys., 39, 1359-68, 1980; Wigren, J. and Andersson, P., Molec. Cryst. Liq. Cryst., 59, 137-48, 1980; Andersson, P., High Temp.-High Pressures, 12, 655-62, 1980; Ross, R.G. and Andersson, P., Mol. Cryst., Liq. Cryst., 69, 145-59, 1981.

CHAPTER 11

Differential Scanning Calorimetry

S. C. MRAW

11.1. INTRODUCTION

This chapter will discuss in detail the apparatus and experimental methods by which the technique of differential scanning calorimetry (d.s.c.) can be used to measure the heat capacity of chemically stable, non-volatile, condensed phases*. Although this volume deals primarily with measurements on solids, the d.s.c. techniques to be described are equally applicable to measurements on non-volatile liquids, so there is generally no reason to distinguish between the two. Since the enthalpy and temperature of first-order solid-solid transitions and the solid-liquid transition are important properties of solids, their determination by d.s.c. will also be discussed. Broad transitions and other anomalies in heat capacity, such as the glass transition, will not be discussed separately, since these properties can generally be studied with modifications of the techniques for either heat capacity or first-order transition. Detailed considerations regarding experimental procedures for the glass transition have been analyzed by Garn and Menis [1].

Although d.s.c. and its predecessor, differential thermal analysis (d.t.a.), have long been used for the qualitative or semi-quantitative study of the behavior of materials, this chapter will concentrate on the use of d.s.c. as a truly quantitative calorimetric technique. In particular, concern for the accuracy, and not simply the precision or reproducibility, of the measuring techniques will be emphasized. The discussions will avoid highly detailed mathematical analyses of the d.s.c. technique and will concentrate instead on the practical set of procedures to be followed in order to acquire accurate heat capacity data.

Since most calorimetrists who use d.s.c. do so on commercially available apparatus, the question of the nature, capabilities, and performance of these instruments cannot be ignored. However, every effort will be made to be as objective as possible, and, in particular, the author makes no endorsement of one particular instrument manufacturer over another. Once the details of d.s.c. and of the various instruments have been presented, the ensuing discussions on measurement of heat capacity and enthalpy of transition will be quite general and could equally apply to many of the commercial instruments.

---

*In this chapter, we shall not use the abbreviation DSC as it is commonly used, but rather adopt the practice first established by the Journal of Chemical Thermodynamics. The abbreviation d.s.c. will be used for the general technique of differential scanning calorimetry, in order to distinguish between the technique itself and various commercial instruments, some of which bear names such as DSC-2, 910 DSC, etc.

References to the literature will be made frequently, particularly to demonstrate the success that many workers have reported in the measurement of heat capacity by d.s.c. with an accuracy of approximately $\pm 1\%$. The ability of d.s.c. to obtain this level of accuracy on sample sizes of the order of 10–100 mg should establish it firmly as a viable calorimetric technique for obtaining thermodynamic data, especially in those cases where other methods of calorimetry may prove too inconvenient, time-consuming, or expensive.

11.2. D.S.C. VERSUS D.T.A. – HISTORY, NOMENCLATURE, AND APPLICATIONS

11.2.1. Developments Leading to Current D.S.C. Techniques

The term 'differential scanning calorimetry' should be almost self-explanatory, i.e., a technique whereby a differential signal (between sample and reference) is monitored to obtain calorimetric data while scanning (in temperature). However, the history of d.t.a. and d.s.c. has led to much confusion over the proper use of the terms, so that a brief review is necessary.

A brief history of thermal analysis in general has been given by Pope and Judd [2]. The technique of differential thermal analysis (d.t.a.) has been practiced throughout this century and is depicted schematically in Fig. 11.1. A sample (s) and an inert reference material (r) are placed in a furnace and subjected to a single heat source (h). Since the sample and reference will not in general have identical heat capacity, let us assume that the heat capacity of s is greater than that of r. Then, as s and r are heated together at a constant rate, the temperature of s will lag behind that of r, and the temperature difference is monitored by the differential thermocouple (tc). In particular, when the sample undergoes a significant change, such as melting (endothermic) or decomposition (endo- or exothermic), the temperature difference deviates markedly from the quasi-steady-

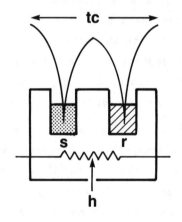

s – sample;   r – reference;   h – heat source; tc – differential thermocouple

Figure 11.1.   Schematic representation of a 'classical d.t.a.' apparatus. Sketch taken from Perkin-Elmer 'Thermal Analysis Newsletter No. 9' [5] and used by permission.

state value, producing a peak in the signal. A schematic trace from a d.t.a. experiment is shown in Fig. 11.2.

In principle, the temperature difference monitored in d.t.a. is proportional to a difference in heat flow between sample and reference, and thus the technique should be capable of yielding quantitative enthalpy data. However, the proportionality factor for this temperature difference is also dependent on the thermal conductivity of the sample and reference, including the effects of how well the thermocouple junctions make contact with the materials. This led to the situation that this type of d.t.a., now called 'classical d.t.a.,' while extremely useful for qualitative study of thermal events in materials, was capable of obtaining, at best, only semi-quantitative enthalpy data. Furthermore, it was almost never used for the rather exacting task of determining heat capacity in regions where no transition or decomposition occurs.

In 1955, Boersma [3] realized that a significant improvement could be made in d.t.a. by removing the thermocouple junctions from immersion in the sample and reference materials. Boersma's original design is shown schematically in Fig. 11.3, where the thermocouple junctions are shown mounted directly in nickel cups containing the sample and reference materials. The symbols s, r, h, and tc have the same meanings as in Fig. 11.1. Although Fig. 11.3 shows a differential thermocouple wire between sample and reference, in fact the plate itself has in later designs often been fashioned from a metal (e.g., constantan) that serves as the connecting thermocouple wire. After several previous workers had tried to improve d.t.a. by inventing ways of making better contact between thermocouple junctions and materials, Boersma's innovation at first seems contrary to intuition. However, as will become clear in Section 11.3, Boersma realized that, by this arrangement, the measured temperature difference is proportional to the difference in heat flow to the sample vs. reference, and the proportionality factor is now independent of the heat transfer or other properties of the sample. Since this factor can be determined as a function of temperature, Boersma-type d.t.a. is capable of truly quantitative enthalpy measurement and, in fact, is often referred to as 'quantitative differential thermal analysis' (q.d.t.a.).

In 1964, Watson et al. [4] introduced a technique which they called 'differential scanning calorimetry,' and their concept has been incorporated into the d.s.c. instruments marketed by the Perkin-Elmer Corporation [5]. A

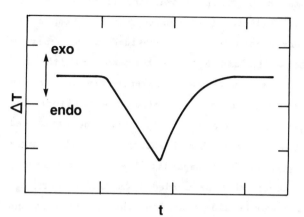

Figure 11.2.   Schematic diagram of the signal from a d.t.a. apparatus during a thermal event in the sample. The peak shown is the type which would result, for example, from the melting of a pure solid.

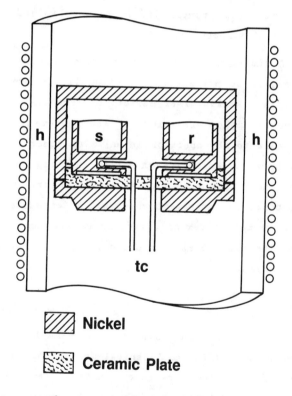

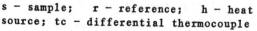

s – sample;   r – reference;   h – heat source; tc – differential thermocouple

Figure 11.3.   Schematic diagram of the original sample-holder design presented by Boersma [3]. Sketch used by permission of the American Ceramic Society.

schematic of their calorimeter is shown in Fig.
11.4. Individual platforms for the sample (s)
and reference (r) are provided with separate
heaters (h_s and h_r) and thermometers (th_s and
th_r). As sample and reference are scanned in
temperature, electronic circuitry monitors the
temperatures indicated by th_s and th_r and
strives to keep them identical. If the sample
has greater heat capacity than the reference,
more power will be needed from the heater (h_s)
on the sample side than from the one (h_r) on the
reference side. This differential power, required to keep th_s and th_r equal, is the primary
signal output and is directly proportional to
the difference in heat flow to sample vs. reference. The calorimetric sensitivity of the instrument (relating the millivolt signal to actual differential power) is, in principle, independent of temperature.

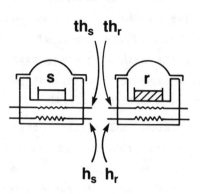

s – sample; r – reference; h_s – sample
heater; h_r – reference heater; th_s –sample
thermometer; th_r – reference thermometer

Figure 11.4.   Schematic diagram of the
               d.s.c. apparatus incorporat-
               ing the technique introduced
               by Watson et al. [4].
               Sketch taken from Perkin-
               Elmer 'Thermal Analysis
               Newsletter No. 9' [5] and
               used by permission.

A simple mathematical model which relates the three techniques of thermal analysis described
above will be presented in Section 11.3. However, it is worth noting that many manufacturers
other than Perkin-Elmer market 'differential scanning calorimeter' systems. Although these instruments employ variations of the Boersma technique, they produce an output which is directly
available to the operator in units of power rather than temperature difference. They accomplish
this by electronically converting the original differential emf signal first into a temperature
difference, then into a differential power. Furthermore, the conversion is done in such a way as
to produce an instrument which has a temperature-independent calorimetric sensitivity. Although
there is not universal agreement on the nomenclature [6], an instrument of this type is often
given the name 'heat-flux d.s.c.,' to distinguish it from the 'power-compensated d.s.c.' used by
Perkin-Elmer.

11.2.2. Applications of D.T.A. and D.S.C.

Both d.t.a. and d.s.c. have been applied in the study of a wide variety of problems in many
diverse scientific and technical areas. Since it is our purpose here to discuss these instruments
primarily for the measurement of heat capacity, the reader is referred to the articles by
McNaughton and Mortimer [7] and Gill [8], as well as to the book by Pope and Judd [2], for general
reviews of the applications of d.t.a. and d.s.c. It is worth noting that the various instrument
manufacturers (see Section 11.4.2) generally maintain bibliographies in this area and are often
excellent sources for review material concerning previous applications, particularly where their
own instrument was used.

11.3. SIMPLE THEORY OF HEAT FLOW IN D.T.A. AND D.S.C.

11.3.1. Introduction

The discussion in this section is adapted from the general treatment given by Gray [9], but, as has been done by Richardson [10], the present discussion will use only those principles necessary for understanding the measurement of heat capacity and of transition enthalpy. Neither of these works included specifically a discussion of the adaptations made by Boersma [3], which led to the type of instruments now available as heat-flux d.s.c.'s, although Baxter [11] has treated this type of instrument with an analysis similar in some respects to what follows. This section, therefore, will present a unified analysis which is applicable to all three techniques: 'classical d.t.a.,' 'power-compensated d.s.c.,' and 'heat-flux d.s.c.' The differences and similarities among these techniques will then be apparent.

11.3.2. General Equations

The construction of a thermal analysis instrument is depicted schematically by Fig. 11.5. The diagram in Fig. 11.5 is similar to that given by Gray [9] but has a very important difference. The thermal resistance, e.g., on the sample side, has been divided into two portions (R_s and R'_s) separated by a stage at the point where temperature is being monitored. The temperature of this point (T_{sm}) may or may not be equal to the actual temperature of the sample (T_s) or to the temperature of the heat source (T_h), depending on the design of a particular instrument. It is the inclusion of this added stage into the analysis which allows heat-flux d.s.c. to be considered along with classical d.t.a. and power-compensated d.s.c.

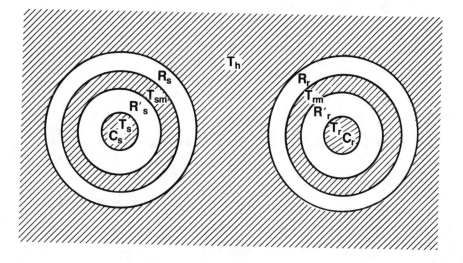

Figure 11.5.   Sketch illustrating general considerations common to almost any type of thermal analysis instrument. C_s, total heat capacity of sample plus sample pan; T_h, temperature of the heat source; T_{sm}, temperature at the point where thermometer is located on the sample side; T_s, actual temperature of the sample and its pan; R_s, thermal resistance to heat flow between temperatures T_h and T_{sm}; R'_s, thermal resistance to heat flow between temperatures T_{sm} and T_s. C_r, T_{rm}, T_r, R_r, and R'_r have analogous meanings for the reference side.

For the analysis, all the thermal resistances (R or R'), as well as the heat capacities (C_s and C_r), will be considered constant over a narrow temperature range (but certainly not over the complete operating range of an instrument). For simplicity, R_s will be taken equal to R_r, and R'_s equal to R'_r, reflecting the similarity of the sample and reference sides. However, the two separate thermal resistances on a given side (R and R') are definitely not equal, nor are the heat capacities of sample and reference (C_s will be assumed to be greater than C_r).

All heat flow in the system is approximately proportional to temperature difference, via an equation frequently called Newton's Law of Cooling

$$\frac{dq}{dt} = a\Delta T \tag{1}$$

where dq/dt is the heat flow between two stages separated in temperature (ΔT) by a barrier with a heat-transfer coefficient (a). The thermal resistance (R or R') in Fig. 11.5 is the inverse of the heat-transfer coefficient (a or a').

On the sample side, the heat flow from T_h to T_{sm} is assumed equal to that from T_{sm} to T_s, since it is assumed there are no other losses in the system. In principle, the sample monitoring stations themselves have finite heat capacities (i.e., C_{sm} and C_{rm}), and a portion of the heat flow to each side serves to heat the sample monitoring station [e.g., $C_{sm} (dT_{sm}/dt)$]. However, for all heat capacity measurements, the terms due to this effect on the sample and reference sides will always cancel, due to the assumption of identical construction. Thus, this effect can be ignored in the treatment which follows, although it will become important when the characteristics of a transition peak are discussed in later sections. With this simplification in mind, the equation for heat flow on the sample side is

$$\frac{dq_s}{dt} = \frac{1}{R_s} (T_h - T_{sm}) = \frac{1}{R'_s} (T_{sm} - T_s). \tag{2}$$

When the sample is being scanned at a temperature rate dT_s/dt, the heat flow is also proportional to the heat capacity

$$\frac{dq_s}{dt} = C_s \frac{dT_s}{dt}. \tag{3}$$

A similar set of considerations leads to the equations for the reference side

$$\frac{dq_r}{dt} = \frac{1}{R_r} (T_h - T_{rm}) = \frac{1}{R'_r} (T_{rm} - T_r) \tag{4}$$

$$\frac{dq_r}{dt} = C_r \frac{dT_r}{dt}. \tag{5}$$

In order to relate Fig. 11.5 and Eqs. (2)–(5) to a given thermal analysis technique, it is only necessary to identify the schematic components in Fig. 11.5 with the particular physical components of a given instrument. This is done in the following subsections. The important points are summarized in Table 11.1, the details of which will be made clear in the following discussions. Of course, no real measuring system can ever exactly correspond to the simplistic sketch

Table 11.1. Summary of the General Principles Involved in Thermal Analysis Measurements of Heat Capacity (for details, see Fig. 11.5 and Section 11.3 in text, including the summary Section 11.3.6)

General Considerations

$R_s = R_r \equiv R$, $R_s' = R_r'$, $R_s \neq R_s'$, $R_r \neq R_r'$, $C_s > C_r$. All R, R', and C are constant over a narrow temperature range, but not over entire instrument range. The simple approximation for heat flow [Eq. (1)] applies.

Specific Considerations

	Classical d.t.a.	Power-compensated d.s.c.	Heat-flux d.s.c.
Assumptions	$T_{sm} = T_s$, $T_{rm} = T_r$, $R_s = R_r \equiv R \neq 0$, $R_s' = R_r' = 0$	$T_{hs} = T_{sm} = T_{hr} = T_{rm}$, $R_s = R_r \equiv R = 0$. $R_s' = R_r' \neq 0$	$R_s = R_r \equiv R \neq 0$ $R_s' = R_r' \neq 0$
Equations			
primary signal	$(T_r - T_s) = R\frac{dT}{dt}(C_s - C_r)$ [Eq. (10)]	$\left(\frac{dq_s}{dt} - \frac{dq_r}{dt}\right) = \frac{dT}{dt}(C_s - C_r)$ [Eq. (12)]	$(T_{rm} - T_{sm}) = R\frac{dT}{dt}(C_s - C_r)$ [Eq. (15)]
sample temperature lag	none	$(T_{sm} - T_s) = R_s'C_s\frac{dT_s}{dt}$ [Eq. (14)]	$(T_{sm} - T_s) = R_s'C_s\frac{dT_s}{dt}$ [Eq. (14)]
Comments			
R	function of instrument, and of the characteristics of sample and reference materials; difficult to quantify	no R applies	function of instrument only; can be calibrated and compensated for
R_s'	no R' applies	produces sample temperature lag	produces sample temperature lag

of Fig. 11.5, which actually serves to illustrate only the overall principle of measurement. In particular, the above assumption that the sample material and its pan are at exactly the same temperature is often not valid, due to a further thermal resistance between the pan and the material itself. This point will be discussed in later sections, but the present analysis will be kept simple in order to illustrate the main principles of measurement in different systems.

11.3.3. Classical D.T.A.

In the 'classical' type of d.t.a., described in Section 11.2.1 and Fig. 11.1, there is only one heat source, at temperature T_h. Since the monitoring thermocouples are embedded in the sample and reference materials, $T_{sm} = T_s$, and $T_{rm} = T_r$. There is, thus, no thermal resistance R_s' (or R_r') in the instrument, and the controlling thermal resistances are R_s and R_r. In the simple analysis, these have been taken to be equal ($R_s = R_r \equiv R$).

In classical d.t.a., the monitored signal is the difference in temperature between the reference and sample thermocouples

$$\text{signal} \equiv T_{rm} - T_{sm} = T_r - T_s \tag{6}$$

and the expression for this signal can be obtained by subtracting Eq. (4) from Eq. (2)

$$R_s \frac{dq_s}{dt} - R_r \frac{dq_r}{dt} = T_{rm} - T_{sm}. \tag{7}$$

Combining this with Eq. (6) and the above assumptions about R_s and R_r

$$\text{signal} \equiv T_r - T_s = R \left(\frac{dq_s}{dt} - \frac{dq_r}{dt} \right) \tag{8}$$

and, using Eqs. (3) and (5)

$$\text{signal} \equiv T_r - T_s = R \left(C_s \frac{dT_s}{dt} - C_r \frac{dT_r}{dt} \right). \tag{9}$$

During a scan in temperature, when a dynamic equilibrium has been established (i.e., the quasi-steady-state), both sample and reference are changing temperature at the same rate, which is the operator-selected scan rate (dT/dt). Thus,

$$\text{signal} \equiv T_r - T_s = R \frac{dT}{dt} (C_s - C_r). \tag{10}$$

Equation (10) (as also obtained by Gray [9] and Richardson [10]) shows that the signal being monitored is directly proportional to the difference in heat capacity between sample and reference sides. However, although the scan rate (dT/dt) can usually be controlled by the operator, the thermal resistance R is a function both of the instrument itself and of the characteristics of the sample and reference materials, as well as being dependent on temperature. For this reason, the classical type of d.t.a. was historically a semi-quantitative technique at best. In order to extract the desired quantity ($C_s - C_r$) from the observed signal ($T_r - T_s$), the R value not only had to be calibrated for every temperature range but, in principle, had to be known for different configurations of sample mounting (i.e., powder vs. pellets vs. liquid, etc.).

The significance of these points will become apparent as the other instruments, power-compensated d.s.c. and heat-flux d.s.c., are described in the following subsections.

11.3.4. Power-Compensated D.S.C.

The Perkin-Elmer Corporation [5] introduced the concept of power-compensated d.s.c. and have used this technique in their DSC-1 and DSC-2 instruments. As described in Section 11.2.1 and Fig. 11.4, their DSC-2 instrument contains an independent heater under each of the sample and reference sides. Each heater is mounted with a corresponding thermometer under a platinum sample holder, and each assembly, in the simplest analysis, is considered to have no thermal gradients. Thus, on the sample side, $T_{hs} = T_{sm}$ and $R_s = 0$, and, on the reference side, $T_{hr} = T_{rm}$ and $R_r = 0$. Furthermore, the electronic monitoring circuit adjusts the power to the sample and reference sides in order to keep the sample and reference thermometers at the same temperature. Thus, $T_{sm} = T_{rm}$ as well. In the DSC-2, the directly measured quantity is the differential power generated in the sample vs. reference sides and is given by Eqs. (3) and (5)

$$\text{signal} \equiv \frac{dq_s}{dt} - \frac{dq_r}{dt} = C_s \frac{dT_s}{dt} - C_r \frac{dT_r}{dt} \; . \tag{11}$$

As in the previous case of d.t.a., a dynamic equilibrium is established during a scan, such that $dT_s/dt = dT_r/dt \equiv dT/dt$. Thus,

$$\text{signal} \equiv \frac{dq_s}{dt} - \frac{dq_r}{dt} = \frac{dT}{dt} (C_s - C_r). \tag{12}$$

Equation (12) shows that the signal being monitored is proportional to the difference in heat capacity between sample and reference sides. The proportionality constant is the operator-selected scan rate (dT/dt). However, although T_{sm} is maintained equal to T_{rm} by the circuitry, the actual sample temperature (T_s) will differ from the temperature at the thermometer (T_{sm}) by an amount which is determined by three factors: (1) the thermal resistance between the 'platform' and the sample pan (R'_s); (2) the scan rate (dT/dt); and (3) the sample heat capacity (C_s). Equations (2) and (3) can be combined to give this relationship as

$$C_s \frac{dT_s}{dt} = \frac{1}{R'_s} (T_{sm} - T_s) \tag{13}$$

$$T_{sm} - T_s = R'_s C_s \frac{dT_s}{dt} \; . \tag{14}$$

Equation (14) is the basic equation which determines the amount of sample temperature lag with respect to the monitored temperature (T_{sm}). As Gray [9] concluded, the calorimetric signal in power-compensated d.s.c. [Eq. (12)] is not dependent on thermal resistances in the system, but the sample temperature lag definitely is [Eq. (14)].

In the Perkin-Elmer DSC-2, typical values for the parameters in Eq. (14) are: $R'_s \sim 0.06$ $K \cdot s \cdot mJ^{-1}$ and $C_s \sim 40\text{--}60$ $mJ \cdot K^{-1}$. Thus, $R'_s C_s$, which is actually a 'time constant' for sample temperature lag, is of the order of two to three seconds. At a scan rate of 0.167 $K \cdot s^{-1}$, the sample temperature lag compared to conditions at zero scan rate should be $0.3\text{--}0.5$ K.

Since the calorimetric signal will still be correct, the consequence of the temperature lag, for moderate scan rates, is simply that the value of this signal may be assigned to the sample at a temperature which should generally be incorrect by no more than a few-tenths of a kelvin. Since heat capacity generally changes slowly with temperature (except at low temperatures or in the vicinity of a phase transition), this is not normally a problem. In special cases, for instance when the temperature of the sample itself lags considerably behind that of the pan [12], problems in temperature determination can become more serious. Further concerns regarding sample temperature will be discussed in later sections.

11.3.5. Boersma-Type D.T.A. or Heat-Flux D.S.C.

To this point, the analysis using Fig. 11.5 has followed Gray [9] and Richardson [10] quite closely, since the R' was not used in classical d.t.a., and the R was not used in power-compensated d.s.c. However, the point of including both an R and an R' in Fig. 11.5 in the present analysis is to preserve the distinction between them and to allow the 'heat-flux d.s.c.' instrument to be analyzed. In the past analyses, the identification of R with R' has led to the confusion that still exists over the meanings of terms such as 'instrument calibration constant' and 'sample temperature lag.'

Returning to Fig. 11.5, it is necessary to retain both the R and R' in order to see the great improvement that was made by Boersma [3] over the classical d.t.a. techniques. As was described in Section 11.2.1 and Fig. 11.3, a Boersma-type d.t.a. instrument contains the monitoring thermocouples mounted in cups or in a plate under the sample and reference materials, rather than immersed directly in these materials. Thus, there are __two__ temperature differences to consider, (1) that of the sample thermocouple (at T_{sm}) with respect to the reference thermocouple (at T_{rm}), and (2) that of the sample itself (at T_s) with respect to the sample thermocouple (at T_{sm}).

Repeating the sequence of Eqs. (7) through (10), it is only necessary __not__ to identify T_{sm} with T_s, and T_{rm} with T_r, as was done in Eq. (6), in order to arrive at an equation analogous to Eq. (10).

$$\text{signal} \equiv T_{rm} - T_{sm} = R \frac{dT}{dt} (C_s - C_r). \tag{15}$$

The important difference between Eqs. (10) and (15) is that the R which appears in Eq. (15) is now a __constant of the apparatus__ (being dependent on the furnace construction, plate thickness, etc.) and is not dependent on the sample configuration or thermal conductivity. R is still a function of temperature but can be calibrated for this dependence and incorporated into all subsequent measurements. In fact, the manufacturers of heat-flux d.s.c. instruments generally calibrate for this R and incorporate it electronically into the measuring circuit, such that the operator receives a signal already adjusted to be the differential power output. Since the dependence of R on temperature is also manipulated electronically, the instrument also has a 'temperature-independent calorimetric sensitivity.'

The question of the actual sample temperature (T_s) vs. the monitored thermocouple temperature (T_{sm}) is addressed in exactly the same fashion as for power-compensated d.s.c. Equation (14) describes temperature lag in heat-flux d.s.c. exactly as it did earlier for the power-compensated case. Because of this separation of the original R of classical d.t.a. into an R and R', it is in principle no more important to carefully reposition sample pans in their holders for a well-designed heat-flux d.s.c. instrument than for a well-designed power-compensated instrument. Confusion over this point in the past is more likely due to deficiencies in a particular instrument than to a fundamental deficiency in the actual principle.

11.3.6. Summary Comparison of Similarities and Differences Among the Instruments

Table 11.1 summarizes the main points of the preceding subsections.

In 'classical d.t.a.,' the fundamental equation for heat capacity is Eq. (10). Since the thermocouples are immersed in the materials, there is no temperature lag between the monitoring point (at T_{sm}) and the sample (at T_s). The R is dependent on both the instrument itself and the characteristics of the sample, is dependent on temperature, and is difficult to quantify.

In 'power-compensated d.s.c.,' the fundamental equation for heat capacity is Eq. (12), while sample temperature lag is given by Eq. (14). No thermal resistances appear in Eq. (12), so the calorimetric sensitivity of the instrument can be independent of temperature. The R_s' which appears in Eq. (14) does not affect calorimetric accuracy but determines the (usually small) error made in the assignment of temperature, and even this error can, in principle, be corrected for.

In Boersma d.t.a., which is the original principle incorporated in the 'heat-flux d.s.c.' instruments, the fundamental equation for heat capacity is Eq. (15), while sample temperature lag is again given by Eq. (14). Although the R which appears in Eq. (15) is dependent on temperature, it is a constant of the instrument design only and not of the sample placement or characteristics. Thus, its value can be determined and, in commercial instruments, is generally compensated electronically. The instrument then appears to the operator to have a temperature-independent calorimetric sensitivity and can be used in the same way as a power-compensated d.s.c. Sample temperature lag is dependent on the R_s' in Eq. (14) and is treated exactly as in the power-compensated instrument. Of course, the specific value of R_s' in either case will be dependent on the individual design of a particular instrument, and sample temperature lag may be more or less important in a given system.

11.3.7. Further Literature on Mathematical Treatments

Readers who wish further information concerning mathematical analyses of the d.t.a. or d.s.c. principles may find the following works of interest, in addition to those cited above [9,10].

O'Neill [13], in what is considered a classic paper, presented an analysis of the principles of differential scanning calorimetry and discussed the effects of these considerations on calorimeter design.

Flynn considered the nature of the time constants of d.s.c. instruments in general [14] and of the Perkin-Elmer DSC-1B in particular [15], as well as further questions of the theory and techniques involved in d.s.c. measurements [16]. This last paper [16] includes a mathematical analysis of the measurement of glass transition properties by d.s.c. More recently, Churney and Garvin [17] have considered many questions concerning time constants in d.s.c. in considerable detail.

Vallebona [18] has considered the effects of thermal gradients and temperature lag on heat-capacity measurements by d.s.c. Randzio and Sunner [19] have analyzed the question of proper 'sample position with respect to the temperature sensor, heater, and surroundings' in designing an effective d.s.c. instrument.

Heuvel and Lind [20], Smith [21], and Dumas [22] have all analyzed aspects of the shape of transition peaks in d.s.c.

Again, as was mentioned earlier, instrument manufacturers are often a good source for references to earlier work, particularly that concerning their own instrument.

11.4.   TYPES OF DIFFERENTIAL SCANNING CALORIMETERS – CUSTOM VERSUS COMMERCIAL

11.4.1.   Custom Built Scanning Calorimeters

It is expected that most experimentalists who use d.s.c. to study the properties of materials will continue to use commercially available instruments, as has been the case at least since the mid-1960's. However, several authors have described custom-designed scanning calorimeters, many of which operate on the principles of power-compensation in one way or another.

Buckingham et al. [23] have described what they call a 'scanning ratio calorimeter,' which is not truly a differential scanning calorimeter in the sense described in previous sections. However, they do point out some aspects of their system, with respect to classical adiabatic calorimetry, which are generally applicable to all types of d.s.c.'s:

'Instead of violently disturbing the equilibrium and then awaiting its return, one tolerates in the scanning method a much smaller but steady departure from equilibrium. This departure can be taken as unimportant so long as the observed heat capacity is independent of scan rate. Our technique permits negative (cooling) scans as well as positive and any dependence on rate is easily disclosed. Thus, lack of adequate equilibrium in the sample is readily apparent when it exists and may be overcome by a reduction in the scan rate.'

Nicholson and Fulrath [24] have described a 'differential thermal calorimeter' for measuring enthalpies of ceramic powders. Their design provides for calibration of the calorimeter with electric heat pulses. A novel embodiment of the principles of power-compensated d.s.c. was reported by Hill and Slessor [25]. In their unique design, the signal from a differential thermocouple between the sample and reference cells causes a heating lamp to pivot and turn toward the cell requiring the greater power.

Lagnier et al. [26] have described what appears to be an excellent differential scanning calorimeter which operates from 2–300 K. This is a very significant development, since no commercial instrument (to the author's knowledge) is capable of operating below about 100 K. The heat capacity of a sample of pure copper was measured with this d.s.c., and results were reported to be accurate to approximately ±1% from 2–125 K. Bonjour et al. [27] have recently reported an extension to even lower temperatures, describing an instrument for the range 0.3–10 K.

Some workers have modified the components of commercial instruments for specific applications. Wunderlich and Bopp [28] adapted a DuPont 900 DSC [29] for use at pressures up to 15 MPa and reported the pressure dependence of the melting points of various organic compounds. In a series of papers on high-pressure d.s.c., Kamphausen [30], Kamphausen and Schneider [31], Arntz [32], and Sandrock [33] have described modifications to Perkin-Elmer DSC-1B and DSC-2 instruments. They have also described [30] the use of thin-walled indium capsules to transmit the pressure of the surrounding gas to the sealed sample. No heat capacity data were reported, but results for enthalpies and temperatures of transition were given.

11.4.2.   Commercially Available Differential Scanning Calorimeters

Differential scanning calorimetry is almost unique among techniques for measurement of heat capacity in that the apparatus is commercially available. The two principal North American manufacturers who market d.s.c. instruments are Perkin-Elmer [5] and DuPont [29], although several European instruments, particularly those of Mettler [34], are easily available on this continent. This section will be a brief review of these commercial instruments. It is not intended to be comprehensive (particularly in the inclusion of every commercial instrument which may exist) and in no way should be construed as a recommendation by the author of one particular instrument over another.

As has already been discussed in Section 11.2.1, Perkin-Elmer [5] introduced the concept of 'power-compensated d.s.c.' and has incorporated it in their DSC-1, -1B, and -2 instruments. A schematic drawing of the sample holder assembly in the DSC-2 (the current instrument as of this writing) was given in Fig. 11.4. Pans containing the sample and reference materials are placed in nearly-identical platinum cups, which are then covered with platinum lids. The platinum cup as-semblies are symmetrically mounted in an aluminum block, which remains near the temperature of the cooling bath, no matter what the temperature of the sample cups. The operating range of the in-strument is nominally 100-1000 K. For operation above room temperature, the aluminum block is generally cooled with cooling water, while, for sub-ambient operation, the block can be cooled with dry ice, liquid nitrogen, or a small refrigeration unit. A purge gas is constantly passed through the aluminum block and over the sample cups. Various choices for purge gas are available, with dry nitrogen generally suitable for operation above 200 K. In previous years a programmable calculator option was available for digital data acquisition, and this has now been replaced by microprocessor and computer options in what is called the DSC-2C.

At the time of this writing, the current DuPont [29] instrument is the microcomputer-control-led DuPont 1090 Thermal Analysis System, incorporating the Model 910 DSC module. A cross-section of this module is shown in Fig. 11.6, which incorporates the basic measuring principle suggested by Boersma [3] (Fig. 11.3). In the DuPont module, pans containing sample and reference materials are placed on raised portions of a constantan disc, which forms one-half of the required thermo-couple junctions. The chromel wafers attached to the underside of the constantan platform form

DSC CELL CROSS-SECTION

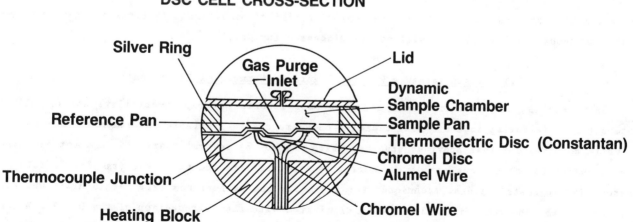

Figure 11.6. Cross-section of the d.s.c. cell in the DuPont 910 DSC System. Figure taken from DuPont Product Bulletin [29] and used by permission.

'area thermocouple' junctions, allowing the differential temperature [T_{rm}-T_{sm} of Eq. (15)] to be monitored.   Additional alumel wires permit the direct determination of the actual temperature (T_{sm}).   The circuitry of the instrument compensates for the R value [Eq. (15)] electronically to produce a 'constant calorimetric sensitivity,' and the primary signal output is obtained in units of power.   This differential power output can be plotted either vs. temperature (T_{sm}) or time. The operating range of the instrument is again 100–1000 K, but, unlike the cold block of the Perkin–Elmer DSC-2, the block of the DuPont 910 cell provides the heated environment for the samples.   A purge gas can be used, but operation down to 200 Pa is also possible.   In addition, DuPont markets a 'Pressure DSC' cell, which can be incorporated in the DuPont 1090 system and is capable of operation from 1 Pa to 7 MPa.   Lee and Levy [35] have recently presented a detailed analysis of the measuring principles of the DuPont system.

The Mettler [34] TA 3000 System, incorporating either the DSC 20 or the DSC 30 module, also uses the principle of 'heat-flux d.s.c.'   In a series of papers, van Dooren and Müller [36–39] have analyzed in great detail the operating characteristics of the Mettler instruments.   In addition, Mettler markets the TA 2000 C for simultaneous d.s.c. and thermogravimetric analysis, while an instrument for simultaneous d.t.a. and hot-stage microscopy has also been described by Mettler workers [40].

Although there are other less well-known American, European, and Japanese manufacturers, it is worth mentioning in particular the nature and capabilities of the SETARAM [41] instrument, which does not strictly use either of the 'power-compensated d.s.c.' or 'heat-flux d.s.c.' principles described earlier.   The SETARAM instrument has been described by Mercier [42] and incorporates a small-scale twin calorimeter of the Tian-Calvet type.   Unlike most other d.s.c.'s, access to the sample chambers, which are surrounded by multijunction thermopiles, is not obstructed during a measurement, permitting mechanical manipulation of the sample from the outside.   However, the output of this calorimeter is not adjusted electronically, and, thus, the calorimetric sensitivity varies with temperature.

Due to their excellent reputation among physical biochemists, the commercially available calorimeters of Privalov et al. [43] and of Microcal [44] deserve mention.   However, because they are optimized for high sensitivity in the region of biological interest (generally for liquid samples from about 273–373 K), they will not be discussed further.

11.4.3. Literature Reports Regarding Commercial Instruments

Section 11.9 will summarize work that has been reported by experimentalists specifically interested in measuring accurate heat capacity on commercially available instruments.   However, it is worth pointing out that previous studies, which may now be several years old, are not necessarily an indication of the quality of the currently available instruments.   For example, Barrall and Dawson [45] suggested several techniques for improving the performance of a Perkin-Elmer DSC-1B, but many of the observed difficulties were eliminated with the introduction of the DSC-2.   Mehta et al. [46] compared the performance of instruments from Perkin-Elmer, DuPont, and Mettler for the measurement of accurate heat capacity, but they acknowledged in a footnote that at least the DuPont instrument was not the model most recently introduced.   DuPont also claims that modifications in their current 910 DSC module have significantly reduced the difficulties discussed by Yuen and Yosel [47] regarding the importance of placement of sample pans in the earlier module.

Although manufacturers are still the best sources for detailed mechanical and electronic descriptions of their instruments, each quite naturally prefers the design adopted in their own system. Given this situation, independent reports by outside workers are often the best indicators of performance. However, with rapid electronic improvements and the constant introduction of new systems, older reports must be evaluated with caution.

11.5. ADVANTAGES AND DISADVANTAGES OF D.S.C. FOR HEAT CAPACITY MEASUREMENT

11.5.1. Advantages

The commercial availability of the above instruments for measurement of heat capacity ranks as one of the principal advantages of d.s.c. over other types of calorimetry. In a very few months, and for what should be considered a relatively modest cost, a prospective calorimetrist can go from an initial decision to measure heat capacity to the point where he is acquiring accurate thermodynamic data. The wide temperature range of applicability of the instruments, usually 100–1000 K, means that one instrument will suffice where two or three calorimeters of more classical design might be needed.

The second major advantage of d.s.c. over more classical techniques is that the sample size required is relatively small. Since accurate measurements can be made on approximately 0.01–0.03 cm^3 of sample, a wide range of materials which may be too expensive or too difficult to purify in larger quantities can be investigated.

Other advantages of d.s.c. generally center around the ease of sample loading, the speed with which data can be obtained, and the relative ease of obtaining or processing data by digital methods.

As will be shown in Section 11.9, heat capacity data which are accurate to approximately $\pm$2% from 120–220 K and approximately $\pm$1% from 220–1000 K can be obtained by careful application of d.s.c. techniques. A further advantage of d.s.c. is that, in certain instruments, measurements of almost this quality can be obtained on cooling as well as heating, providing a flexibility unmatched by most classical designs.

11.5.2. Disadvantages

Many of the disadvantages of d.s.c. stem, ironically, from the positive points that were discussed above. As Richardson [10] has observed, the very ease of obtaining data by d.s.c. can lead to work which is of questionable accuracy if the operator fails to observe many necessary and rigorous principles. A more classical scientist, who has painstakingly designed and built his own calorimeter, carefully loaded and sealed the sample, and assembled the many lead wires and vacuum cans that are involved in classical designs, will not usually proceed to make sloppy heat capacity measurements. Likewise, a prospective d.s.c. operator must force himself to truly think like a calorimetrist, not an 'analyst,' if he wishes to obtain accurate heat capacity results. A principal disadvantage of d.s.c. is that, because it is so easy to use, it is also very easy to abuse.

Other disadvantages of d.s.c. are more closely related to inherent limitations of current instruments. The temperature range below 100 K is inaccessible with present commercial equipment. Measurements of heat capacity in this region are necessary for the thermochemist to obtain the Third Law entropy of pure compounds at room temperature, and this is also the range most often explored by solid-state physicists. The work of Lagnier et al. [26] and Bonjour et al. [27], mentioned in Section 11.4.1, indicates that this range is definitely accessible by d.s.c. techniques, and it is hoped that instrument manufacturers will take up the challenge.

Finally, although there have been indications that heat capacity of accuracy even better than $\pm 1\%$ can be obtained in certain cases (see Section 11.9), it is extremely doubtful that the current generation of instruments can ever routinely approach the 0.05–0.20% accuracy traditionally obtainable by adiabatic calorimetry from 20 to at least 400 K.

11.6. EXPERIMENTAL PROCEDURE FOR MEASURING ACCURATE HEAT CAPACITY BY D.S.C.

11.6.1. Introduction

Section 11.6 will be a step-by-step description of the methods of heat capacity measurement by d.s.c., from the initial calibration of the instrument to the final acquisition of experimental heat capacity points. The methods described here should be equally applicable to both the 'power-compensated' and 'heat-flux' d.s.c. instruments, since both types of instruments, as discussed above, produce an ordinate signal in units of power and have a temperature-independent calorimetric sensitivity. In some instruments, the operator can select the abscissa of a data scan to be either the temperature [i.e., T_{sm} in Eq. (15)] or the time (t). We shall assume that the 'time' option has been selected in order to make the discussion general to all instruments, although the principles are easily translatable to the temperature case, if desired.

The basic method for heat capacity measurement by d.s.c. was originally described by O'Neill [48], and modifications and variations have been listed by Cassel [49]. This section will discuss the basic procedures involved, while Section 11.9 will deal in detail with the level of success that has been reported in the literature in measuring heat capacity accurately.

11.6.2. Instrument Calibration

A d.s.c. instrument must initially be calibrated for temperature and calorimetric sensitivity, and this calibration must be checked periodically. The operation manual for a given instrument will generally list step-by-step procedures for performing this calibration, so only a general summary need be given here.

Temperature calibration is accomplished using the transition points (usually melting points) of very pure materials, while the calorimetric sensitivity is determined using the enthalpy of transition (usually fusion) of these materials. Substances recommended for calibration have been described in a 'Catalogue of Physicochemical Standard Substances' [50], published in 1972, and, in North America, these materials are available from the U.S. National Bureau of Standards [51] (although not with the official certification of 'Standard Reference Material'). Some common calibration materials are often conveniently available from the instrument manufacturers themselves.

Values for the enthalpies of transition of most of the recommended materials have been measured by Gray [52], while Zeeb et al. [53] have recently selected 'best values' for indium and sodium nitrate. In addition, a program to provide a series of certified materials, including many organic compounds, is described by Andon and Connett [54] of the National Physical Laboratory in England. However, none of the above compilations list materials with recommended transition points below 320 K, so operation of a d.s.c. below this temperature, particularly in the sub-ambient region, often requires a careful selection of pure materials whose properties have been reported in the open literature. Water, mercury, and various organic compounds, such as cyclopentane, cyclohexane, and n-octane, have well-characterized transitions in the sub-ambient region and have been recommended for calibration [5]. A few materials with low-temperature transitions are also available from Ref. [51] above.

For temperature calibration, standards should be chosen to bracket the temperature range that will be used for measurement of heat capacity. Although this requires only two standards for calibration over a given range, it is wise to check the accuracy of intermediate temperatures with runs on additional standards. Likewise, although the enthalpy of fusion of a very pure sample of indium, 3283 $J \cdot mol^{-1}$ [53] is often recommended for calibration of the calorimetric sensitivity, it is again wise to check this calibration with a variety of materials having transitions which span the temperature range of interest. (If an even more detailed check is desired, the 'enthalpy method' (Section 11.6.4.3 below) could be used to determine the enthalpy of a standard material, such as α-Al_2O_3, over defined temperature intervals, e.g., 50 kelvins [55].)

Prior to performing the calibration, the operator should set up the instrument under conditions that will exactly match those that will be used for measurement of heat capacity. The type of cooling bath and the purge gas, once chosen, must remain the same, since these factors have a large effect on calibration. It is also desirable to use the same type of sample pans for calibration as will be used for subsequent measurements, although this is not always possible. For instance, reactive samples may require gold pans, while indium, which alloys easily with gold, must be encapsulated in an aluminum pan. However, as long as the type of pan remains the same (i.e., open vs. crimped vs. hermetically sealed), the switch from aluminum to gold should not present a major problem, as long as R'_s is reasonably constant from one pan to another and any further thermal resistances <u>within</u> the different pans are similar.

The importance of these considerations can be understood with reference to Eq. (14) for sample temperature lag, which was derived in Section 11.3. Even in the idealized analysis which leads to Eq. (14), three factors affect the lag of the sample temperature behind the indicated temperature: (1) the thermal resistance (R'_s); (2) the heat capacity of the pan plus sample (C_s); and (3) the scan rate (dT_s/dt). The procedure of calibrating the instrument for temperature actually involves identifying the program temperature with the actual sample temperature <u>for a given set of values</u> of R'_s, C_s, and dT_s/dt. Any subsequent change in these values produces a change in the sample temperature lag to a value other than that for which the calibration applies.

The precautions discussed above, of keeping coolant, purge gas, and type of sample pan constant, have the effect of minimizing variations in R'_s. In addition, the calibration should usually be performed at the same scan rate (dT_s/dt) as will be used in subsequent measurements, to eliminate the effect of this factor. (Alternatively, the actual dependence of temperature calibration on scan rate can be assessed experimentally, for a given set of conditions, by determining the apparent melting temperature of a pure material at different scan rates.) Finally, variations in the heat capacity of the sample and its pan (C_s) generally affect temperature lag only slightly. However, this is certainly not the case during a phase transition, when the effective heat capacity due to the latent heat of transition can be orders of magnitude greater than the contribution due to baseline heat capacity alone.

An excellent quantitative study of the effects of R_s', dT_s/dt, and C_s for a particular instrument has been reported by Richardson and Savill [12], who also studied the lag of the sample temperature itself behind that of the sample pan (due to a second resistance, e.g., R_s''), which was assumed to be nonexistent in the simplified analysis of Section 11.3.

With the above considerations in mind, the calibration procedure involves the analysis of a transition peak (often a fusion peak) such as that shown in Fig. 11.7, where the ordinate is power and the abscissa is either the monitored temperature (T_{sm}) or the time (t). A convenient, reproducible way to determine the transition temperature (T_{tr}) of the sample is to extrapolate the leading edge of the peak back to the baseline. The enthalpy of transition (ΔH_{tr}) is given by the area under the peak.

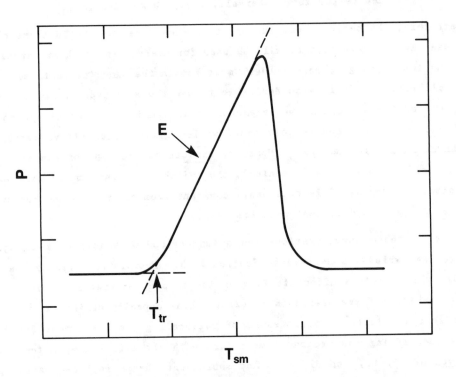

Figure 11.7.   Example of a d.s.c. chart recorder trace [differential power (P) vs. temperature (T_{sm})] for a first-order transition (e.g., fusion) in a very pure material. The observed transition temperature (T_{tr}) can be determined from the value of T_{sm} at the point where the dashed lines intersect. E is the slope of the 'leading edge' of the peak.

Although T_{tr} and ΔH_{tr} are the two values which are needed for calibration of the instrument in temperature and enthalpy, more detailed information can result from a further analysis of the peak in Fig. 11.7. When the sample is scanned in temperature, its scan rate (dT_s/dt) is initially equal to the programmed scan rate (dT/dt). However, when the sample reaches the transition point, its temperature (T_s) will remain constant while the latent heat of transition is absorbed, even though the temperature (T_{sm}) of the monitoring point will continue to rise. In power-compensated d.s.c., since $T_h = T_{sm}$, the ordinate is a measure of the heat flow between T_{sm} and T_s. Since the abscissa is a measure of the developing temperature lag ($T_{sm}-T_s$), the slope of the leading edge of the peak (vs. temperature, not time) is the inverse of the thermal resistance R'_s [this follows directly from Eq. (1)]. If time (t) is plotted as the abscissa instead of temperature (T_{sm}), the power-compensation circuitry ensures that, in principle, the two are directly related through the programmed scan rate (dT/dt). In heat-flux d.s.c., however, the ordinate is a measure of the heat flow between T_h and T_{sm}, not between T_{sm} and T_s, and these heat flows differ depending on the amount of heat flowing into the sample monitoring station [i.e., C_{sm} (dT_{sm}/dt)]. Although C_{sm} (dT_{sm}/dt) was assumed to cancel C_{rm} (dT_{rm}/dt) for measurement of heat capacity (see Section 11.3), this is no longer the case during a transition peak, since dT_{sm}/dt is not necessarily equal to dT_{rm}/dt. Since there is no direct power-compensation circuitry, dT_{sm}/dt will gradually assume values between the programmed scan rate (dT/dt) and zero, depending on the construction of a particular instrument, and no simple, general treatment can be given. However, an estimate of the value of R'_s for any type of d.s.c. instrument can still be obtained by one of a number of experimental methods (e.g., by determining the apparent transition temperature of a pure material as a function of scan rate and using Eq. (14) to evaluate R'_s). As mentioned earlier, such an investigation for a particular instrument is reported in Ref. 12.

In principle, the dependence of R'_s on temperature should be determined experimentally from a series of peaks which span the temperature range of interest. In practice, however, this type of detailed concern for sample temperature lag is usually not necessary for the measurement of heat capacity where no phase transition is present. As long as the calibration is done at the same scan rate and under the same operating conditions as will be used for the measurements, the sample temperature will generally lag the program temperature by no more than several tenths of a kelvin unless very high scan rates are used (>0.333 K·s^{-1}) [12].

11.6.3. Sample Encapsulation

Samples for heat capacity measurement by d.s.c. are generally encapsulated in sample pans which are available from the instrument manufacturer. Different versions of these pans can either be crimped shut or hermetically sealed, and pans made of aluminum, gold, platinum, steel, or graphite are often available. The physical and chemical characteristics of the sample will determine which type of pan should be chosen. Completely nonvolatile, nonreactive materials should be encapsulated in the 'standard' pans, since these permit a larger sample size than the sealed pans. However, materials which will develop a significant vapor pressure in the temperature range of interest must be hermetically sealed. In addition, it is often convenient to use the sealable pans for even nonvolatile samples if they are susceptible to reaction with air or moisture, in order to facilitate sample handling and storage. Aluminum pans will serve well for most materials, but particularly corrosive samples must be encapsulated in pans made of gold, platinum, or, occasionally, graphite.

Once the type of sample pan is chosen, it should be weighed carefully on a microbalance, filled with as much sample as possible, crimped or sealed, and then weighed again. The mass of the sample can be determined as the difference between the weighings, since, although proper correction should be made for buoyancy, it is rare that this effect will be significant compared to other measurement errors. The final weight of the full pan should be checked periodically, especially after runs above room temperature, to ensure that the sample is not being lost. Every effort should be made to enclose as much sample as possible in the pan in order to maximize the signal during measurement of heat capacity. With powdered samples, it is often helpful to use a small press to form pellets in order to increase the amount that can be enclosed in the pan.

It is often necessary to seal particular samples under something other than atmospheric pressure of room air. Because the sealing presses are generally small, it is usually possible to seal samples under an inert atmosphere in a glove bag or glove box. Furthermore, for some types of presses, a small enclosure can be constructed which will also allow pans to be sealed under less than atmospheric pressure of a given gas. Since many pans rupture or leak at internal pressures of 200–300 kPa, sealing at reduced pressures may be important, because the enclosed gas will be at two to three times its initial pressure if heated to 600–900 K. Depending on the particular design of the sample pan and of the d.s.c. instrument itself, this last consideration can also produce an adverse effect on the accuracy of the measurements themselves, even if the pan remains tightly sealed. In some cases, the internal pressure can cause the pan bottom to bulge outward, reducing the thermal coupling between the pan and its container (i.e., increasing R'_s) and, thus, increasing sample-temperature lag.

11.6.4. Measurement of Heat Capacity

11.6.4.1. General Considerations

Since the ordinate signal from a d.s.c. instrument is power (P in $J \cdot s^{-1}$), it can be divided by the programmed scan rate (dT/dt in $K \cdot s^{-1}$) to yield heat capacity directly (C_s in $J \cdot K^{-1}$). Two different methods of heat capacity measurement will be described in this section. In the 'scanning method' (Section 11.6.4.2), the ordinate values during a scan are converted directly to heat capacity values at the indicated temperatures. In the 'enthalpy method' (Section 11.6.4.3), all of the ordinate values for a scan between an initial and a final temperature are integrated to yield the total enthalpy change between the two temperatures.

For ease of description, the output will be depicted as it would appear on a strip-chart recorder, and all discussions will involve analysis of this recorded output. If the output is in fact available in digital form and can be analyzed by computer, the conversion to digital data manipulation will be obvious.

By either the scanning or the enthalpy method, the sequence of necessary runs is the same. Initially, an empty pan and lid is placed in the reference side of the d.s.c. and left there as ballast. (Actually, the lid can be removed, or extra lids added, in order to produce the desired signal level for subsequent runs. However, once a combination is decided upon, it must be left constant for all the remaining runs.)

The first run involves a scan with an empty pan in the <u>sample</u> side (in the remainder of this chapter, this run will often be referred to as the 'empty'). The pan used should be as similar as possible to the one that will contain the sample, although small differences in mass can be corrected for in subsequent data manipulation.

The second run is a scan of a sample pan containing a material whose heat capacity is accurately known. For work above room temperature, this material is usually α-Al_2O_3, sapphire. However, in the sub-ambient region, benzoic acid is often preferable because the heat capacity of sapphire varies rapidly with temperature in this region. Very accurate heat capacity data have been reported for both materials [56] and both are available as Standard Reference Materials [51]. For simplicity in the discussions, it will be assumed that sapphire is being used as the calibration material. The run on sapphire is used to determine its apparent heat capacity [$C_p(Al_2O_3$, obs)] as a function of temperature, and these values are then compared to the accepted literature values [$C_p(Al_2O_3$, lit)]. At each temperature a calibration factor (F) is calculated

$$F = C_p(Al_2O_3, \text{lit})/C_p(Al_2O_3, \text{obs}). \tag{16}$$

This factor (F) is actually the individual calibration constant of the calorimeter at each temperature and is used for all the runs to follow.

All subsequent runs are for the samples themselves, each of which should likewise be encapsulated in a pan similar to the empty pan. For each sample, the apparent heat capacity [$C_p(\text{sam}$, obs)] at each temperature is measured and then corrected by the calibration factor (F) for that temperature. The resulting value is taken as the experimental heat capacity of the sample [$C_p(\text{sam, exp})$]

$$C_p(\text{sam, exp}) = F [C_p(\text{sam, obs})]. \tag{17}$$

By this procedure, the d.s.c. instrument is calibrated with sapphire with a run which immediately precedes the sample runs and which is made under identical conditions over the identical temperature range. Since different instruments will have varying degrees of stability, it is impossible to say how often this calibrating run must be checked or repeated. However, it is often advisable to perform the sequence of empty, sapphire, and samples all in the same day in a given temperature range, to minimize the effects of instrument drift. If this procedure is adopted, the last 'sample' run of the day should always be made on a second material whose heat capacity is also accurately known from the literature as a final check on the results of that day. If possible, this material should have physical properties similar to the samples being investigated (i.e., a metal standard for metal samples, an organic for organics, a powder for powders, etc.). Since it is unlikely that the instrument could drift away from calibration after the sapphire run and coincidentally drift back in time for the final run, this last run on a material with known properties is an important safeguard for measuring accurate heat capacities by d.s.c.

11.6.4.2. Heat Capacity by the 'Scanning Method'

The 'scanning method' refers to heat capacity measurements wherein the ordinate values are directly converted to instantaneous values of heat capacity at the indicated temperatures. Any combination of scan rate and total temperature interval could conceivably be used, but rates of $0.083-0.333 \text{ K} \cdot \text{s}^{-1}$ and intervals of 30–100 kelvins are common. For the purpose of illustrating this discussion, a scan rate of $0.167 \text{ K} \cdot \text{s}^{-1}$ and an interval of 70 kelvins, between 400 and 470 K, will be assumed. Figure 11.8 depicts the sequence of runs schematically.

With the chosen amount of ballast in place in the reference side, the empty pan is loaded into the sample side, the sample holder is closed, and the sample and reference are heated to 400 K. When equilibrium is established, as indicated by a flat 'isothermal' baseline (i.e., the baseline at zero scan rate at a particular temperature), the operator begins the scan at the chosen rate, $0.167 \text{ K} \cdot \text{s}^{-1}$. There is initially a transient period, but, after a given amount of time which depends on the particular instrument, the system reaches a dynamic equilibrium. After seven minutes, the sample and reference reach 470 K, the scan is terminated, and, after another transient, the signal decays until it is again constant at the new 'isothermal' baseline value for 470 K.

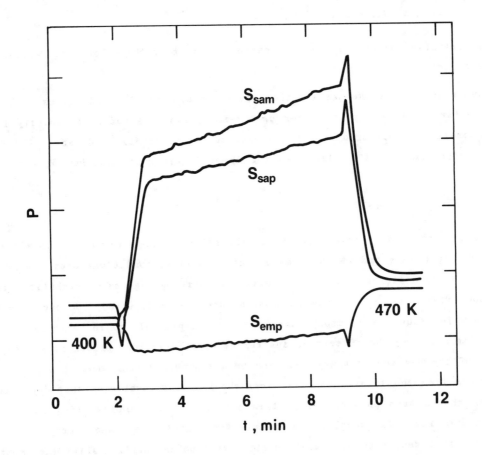

Figure 11.8. Typical chart recorder traces [differential power (P) vs. time (t)] for a heat capacity measurement by the 'scanning method.' For illustration, a temperature interval of 400–470 K and a scan rate of $0.167 \text{ K} \cdot \text{s}^{-1}$ have been chosen. S_{emp}, S_{sap}, and S_{sam} are the differential power signals for the empty, sapphire, and sample runs, respectively.

The calorimeter is cooled, the empty pan is removed from the sample side, and the pan containing the sapphire is loaded and heated to 400 K. Again, equilibrium is established at 400 K, the scan is made, and the calorimeter returns to equilibrium at 470 K. The calorimeter is again cooled, the sapphire removed, and the sample in its pan loaded and run. Although Fig. 11.8 shows a trace for only one sample, more than one can be run in sequence once the empty and sapphire runs have been completed.

Although it would be convenient if the three isothermal baseline values at 400 K were identical, and likewise those at 470 K, they usually are not, and Fig. 11.8 shows a possible set of signal traces. For a given temperature during the scan, the desired quantities are the differences in ordinate displacement between the sapphire and the empty, and between the sample and the empty, since these are proportional to the heat capacities of sapphire and sample, respectively. However, the true quantities are not simply $(S_{sap}-S_{emp})$ and $(S_{sam}-S_{emp})$, since correction must be made for the difference in isothermal baseline values between scans. There are several ways to do this, but one simple way is shown schematically in Fig. 11.9. The sapphire scan is used as an example, and the differences in baseline values have been exaggerated for clarity. For both the empty and the sapphire scans, the isothermal baseline is linearly interpolated between 400 and 470 K. These interpolated baseline values can be thought of as reflecting the power that would have been necessary to hold the calorimeter at the given temperature, even at zero scan rate. Then, at any temperature, the desired difference in ordinate values (D_{ord}) is given by

$$D_{ord} = (S_{sap}-I_{sap}) - (S_{emp}-I_{emp}) \qquad (18)$$

where S_{sap} and S_{emp} are the values of the ordinate signals, and I_{sap} and I_{emp} are the hypothetical isothermal baseline values, as indicated in Fig. 11.9. The quantity $(S_{emp}-I_{emp})$ can be either negative (as in Fig. 11.9) or positive, depending on the amount of ballast chosen for the reference side. The correct value of D_{ord} will still be calculated, however, as long as the subtraction is performed properly.

The apparent heat capacity of sapphire at a given temperature is obtained simply from the division of this ordinate difference (D_{ord} in $J \cdot s^{-1}$) by the selected scan rate (dT/dt in $K \cdot s^{-1}$), with a slight correction to be made for any difference in the masses of the empty pan and the pan containing the sapphire. From the known mass of the sapphire, the apparent molar heat capacity [$C_p(Al_2O_3$, obs) in $J \cdot K^{-1} \cdot mol^{-1}$] at each temperature is calculated and compared to the corresponding literature value [$C_p(Al_2O_3$, lit) in $J \cdot K^{-1} \cdot mol^{-1}$] in order to evaluate the calibration factor (F) in Eq. (16) at each temperature.

For each sample whose heat capacity is to be measured, the scan is analyzed exactly as described above (Fig. 11.9) for sapphire, in order to calculate the apparent heat capacity of the sample [C_p (sample, obs)] at each temperature. Multiplication by the factor (F), as shown in Eq. (17), yields the accepted value for the experimental heat capacity of the sample at the given temperature.

Figures 11.8 and 11.9 have shown the examples of a measurement covering the range 400–470 K. Assuming, for example, that the total temperature range of interest for a particular material is

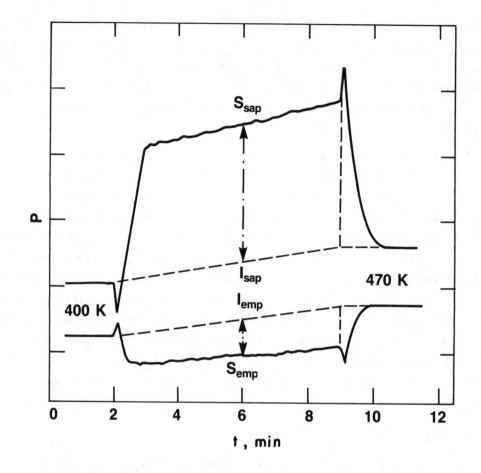

Figure 11.9.   Chart recorder traces [differential power (P) vs. time (t)] for the empty and sap-
                phire scans of Fig. 11.8. The differences for the isothermal baseline values at 400
                and 470 K have been exaggerated for clarity. S_{emp}, the power signal for the empty
                run; I_{emp}, the interpolated baseline for the empty run; S_{sap} and I_{sap} have analogous
                meanings for the sapphire run. Note that the 'isothermal' baseline values are
                interpolated between the initial and final <u>temperatures</u>, not <u>times</u>.

420-750 K, the sequence of runs described above is then repeated for a new temperature interval.
Since the first minute or so of a given scan is affected by the initial transient, additional runs
should be partially overlapped, both to cover the transient regions and to provide checks on the
reproducibility of the data. For the example given, the total range 420-750 K might be covered in
intervals of 400-470 K, 450-520 K, 500-570 K, and so on, with as many repeat or staggered runs as
might be felt necessary in a given situation.

    One way to cover any series of intervals such as those above would be to use a separate day
for each temperature range, with the sequence empty, sapphire, and as many samples as desired
being completed in a single day. Alternatively, the empty pan could be left in the calorimeter
and run over several temperature intervals before runs over the same intervals are done on the
sapphire and, finally, on the samples. As long as the instrument is monitored carefully (for instance, by including the 'sample' whose heat capacity is accurately known), the choice is largely
up to the operator. The latter procedure involves opening the calorimeter less often and, thus,
lends itself easily to automation.

11.6.4.3. Heat Capacity by the 'Enthalpy Method'

The 'enthalpy method' refers to heat capacity measurements wherein all the ordinate values during a scan are integrated to give the total enthalpy change (ΔH) for the given temperature interval (ΔT). The quantity $\Delta H/\Delta T$ is taken as the heat capacity at the midpoint of the temperature interval. Although this method naturally involves 'scanning' the sample in temperature, it more closely approximates the method of classical adiabatic calorimetry, since total enthalpy change is measured between two 'equilibrium' temperatures. Once again, a wide choice of scan rates and temperature intervals is available, but the temperature 'steps' are normally 5-20 kelvins, unlike the scans of 30-100 kelvins described in the previous method. Combinations such as 0.042 $K \cdot s^{-1}$ for 10 kelvins or 0.167 $K \cdot s^{-1}$ for 20 kelvins would presumably work well, but only the former, to the author's knowledge, has been tested thoroughly [57-60], and then only for one instrument.

No matter which scan rate is chosen, however, an important point must not be overlooked. Regardless of scan rate, temperatures in the enthalpy method are measured under static conditions (i.e., at zero scan rate). Thus, the original temperature calibration of the instrument described in Section 11.6.2 must also be done under conditions as nearly approaching zero scan rate as can be established. One way of doing this, as suggested by Flynn [14-16] and Richardson [10], is to step through the calibrating phase transition in very small temperature increments, waiting between each step. For a very pure material, almost all of the transition enthalpy will be recorded in only one interval, and the instrument can be calibrated to record this temperature properly. Alternatively, the calibration can be accomplished by recording the apparent transition temperature at a variety of scan rates and extrapolating to zero scan rate. In actual practice, it is often unnecessary to repeat this procedure every time a calibration is desired, if it can be established that calibration at one of the very slow scan rates available on a given instrument yields a temperature within a few-tenths kelvin of the temperature determined by the more elaborate procedure. In any event, the fact that the 'equilibrium' temperatures in the enthalpy method are measured under static conditions must not be overlooked.

The general description of the enthalpy method will follow that given by Mraw and Naas [57], although their method specifically referred to the use of a Perkin-Elmer DSC-2 with automated data acquisition. The present description should be quite general to any d.s.c. instrument, but, because of the amount of data manipulation required, some form of automation and digital acquisition will generally be necessary.

As an example, the temperature range 400-470 K will again be chosen, but this time it will be covered with seven scans of 10 kelvins each, at a scan rate of 0.042 $K \cdot s^{-1}$. A schematic of the measurement is shown in Fig. 11.10. With the ballast in place in the reference side, an empty sample pan is loaded into the sample side and brought to equilibrium at 400 K. The initial isothermal baseline value is recorded, and the first scan is begun. The value of the d.s.c. signal is recorded during the scan, which terminates four minutes later at 410 K. The signal is still monitored during the transient, until it reaches equilibrium at 410 K, whereupon the final baseline value is recorded. The isothermal baselines at 400 and 410 K are interpolated linearly to give the baseline values for temperatures within the interval. The ordinate values taken during

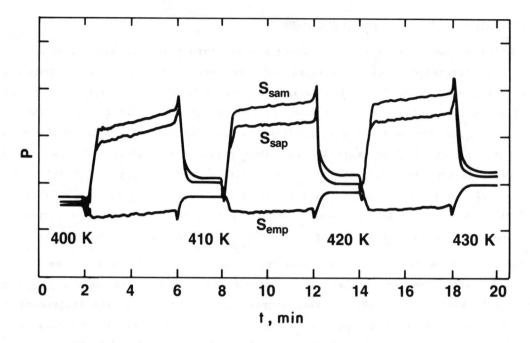

Figure 11.10.   Typical chart recorder traces [differential power (P) vs. time (t)] for heat capac-
                ity measurement by the 'enthalpy method.' For illustration, temperature steps of
                10 kelvins and a scan rate of 0.042 K·s^{-1} have been chosen. S_{emp}, S_{sap}, and S_{sam}
                are the differential power signals for the empty, sapphire, and sample runs,
                respectively.

and after the scan are then integrated, taking into account the values of the interpolated baseline, to yield the shaded area (H_{emp}) shown in Fig. 11.11. This value is recorded, and the next
scan, for the interval 410–420 K, is begun. The same calculation described for the first interval
is performed on all succeeding intervals, until an H_{emp} value has been calculated for each of the
seven intervals. The calorimeter is cooled, the empty is removed, and the pan containing the sapphire is inserted.

A series of seven scans on the sapphire between 400 and 470 K is performed in exactly the
same manner as for the empty, in order to measure the value H_{sap} (see Fig. 11.11) for each interval. When these values have been recorded, the enthalpy change for sapphire (ΔH_{sap}) in each
interval is evaluated by subtracting H_{emp} from H_{sap}

$$\Delta H_{sap} = H_{sap} - H_{emp} \tag{19}$$

including any necessary corrections for slight differences in pans. Again, it does not matter
whether H_{emp} is negative (as shown in Fig. 11.11) or positive. The observed molar heat capacity
[$C_p(Al_2O_3, obs)$] is then given simply by

$$C_p(Al_2O_3, obs) = \frac{\Delta H_{sap}}{n\Delta T} \tag{20}$$

where n is the number of moles of Al_2O_3 and ΔT is the temperature interval (in this case, 10
kelvins). This observed heat capacity is assigned to the midpoint of the heating interval and
compared to the literature value in order to calculate F for that interval, using Eq. (16).

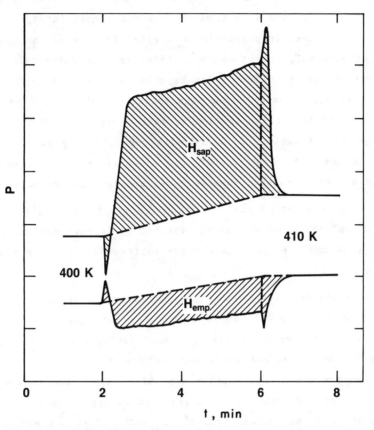

Figure 11.11.   Chart recorder traces [differential power (P) vs. time (t)] for the first of the empty and sapphire scans of Fig. 11.10.   The differences for the isothermal base-line values at 400 and 410 K have been exaggerated for clarity. H_{emp} and H_{sap} are the areas (proportional to enthalpy) under the curves for the empty and sapphire, respectively.

An exactly similar series of seven scans between 400 and 470 K is then performed on the sample in order to measure the observed heat capacity [C_p(sample, obs)] in each interval.   The observed value is corrected to the true experimental value by means of F, using Eq. (17) as described earlier.

The final result of a heat capacity determination by the enthalpy method is, thus, an experimental value for the heat capacity of the sample at the midpoint of each temperature interval. The total temperature range that can be covered in a single working day will depend on, among other factors, the size of the intervals, the scan rate, and the speed of data calculation between intervals.

11.6.4.4.   Comparison of Scanning Versus Enthalpy Methods

Careful work by a number of authors (see Section 11.9) has shown that accurate heat capacities can be measured by either of the two methods.   If neither method has yet been proven superior to the other, it may only be because every reasonable combination of scan rate and temperature interval has not yet been tested systematically for every instrument.   For instance, a criticism of the scanning method might be that the use of large temperature intervals allows the unseen 'isothermal' baseline to deviate too much from the assumed linear interpolation. However, since the scanning method could be used in principle on smaller intervals, the criticism should be directed at the choice of interval, not of method. Likewise, the relatively slow scan rates used for the enthalpy method in the past [57-60] can be criticized as having the effect of producing too small an ordinate signal. However, since the enthalpy method could be used at faster scan rates, this again is a criticism of the particular rate chosen, not of the method. For the time being, a particular operator's choice of method will generally depend on factors such as time available, number of samples to be investigated, and personal preference, rather than on the absolute accuracy attainable. However, some general considerations can be noted.

In either method, the ordinate signal is the primary experimental quantity. Thus, if the ordinate signal is correct, it should not matter whether it is recorded outright (scanning method) or integrated (enthalpy method), as long as the precautions discussed in preceding sections are observed in either case.

On the other hand, a similar statement cannot be made with regard to temperature measurement. In the scanning method, a well-behaved sample should lag the program temperature by an amount (usually small) that can be approximated using an empirically determined value of R'_s (Section 11.6.2). Even if the correction is not made, the assignment of the correct heat capacity to the (slightly) incorrect temperature is not usually a problem, since heat capacities generally vary only slowly with temperature. However, a particular sample may be encountered with a temperature lag much greater than would be calculated from the R'_s value discussed previously. This could be due to poor effective thermal conductivity within the sample pan itself, negating the previous assumption that sample material and sample pan are at the same temperature. Richardson and Savill [12], in studying the magnitude of temperature lag experimentally, have shown that a temperature error of about one kelvin exists even for solid discs of sapphire at a scan rate of 0.333 K·s^{-1}, and they further indicate that, in special cases, powdered materials may lag the calculated temperature by several kelvins. These authors have suggested that the 'tail' section at the end of a scan can be used as a measure of the actual sample temperature lag in individual cases. (It is important to note that, although the diagrams of Richardson and Savill seem to show rather alarming amounts of temperature lag, these are with respect to a calibration at zero scan rate for a sample of almost zero heat capacity. Calibration at the same scan rate to be used in subsequent measurements, as well as use of the 'time constant' ($R'_s C_s$) discussed in Section 11.3.4, would reduce the actual uncertainty considerably.) However, it is worth noting that the problem of sample temperature lag can be reduced considerably by use of the enthalpy method for heat capacity measurement described above. No matter how much the temperature of the sample differs from the program temperature _during_ the scan, the correct amount of enthalpy (ΔH) must still be supplied to heat the sample through the required temperature interval (ΔT). Since temperatures are measured under conditions of zero scan rate, the effect of sample thermal conductivity is almost completely eliminated, as long as the sample is truly at equilibrium at the initial and final temperatures. This similarity to classical adiabatic calorimetry may make the enthalpy method more suitable than the scanning method for a sample of very poor effective thermal conductivity. (In some cases, such a poor 'effective conductivity' might even be due to a very slow transition occurring in the sample.)

A possible exception to these general statements exists if a given sample pan suffers from so much temperature lag that it can be at a considerably lower temperature than the empty or sapphire-containing pan would be under the same conditions. For some instruments, it is possible that the total heat leak to the surroundings at a given temperature would be affected by this problem, so that even the ordinate value itself would be incorrect. Mraw and Keweshan [61] have noticed that inaccuracy in heat capacity measurements, even by the enthalpy method, appeared to be correlated with poor coupling of the sample pan to its sample holder. The reduced thermal coupling had been caused when the bottom of the pan containing the sample had become rounded after a high internal pressure was reached at high temperatures. Although the effect has not been proven unambiguously or quantitatively evaluated further, it was for this reason that it was recommended in Section 11.6.3 that high internal pressures be avoided.

11.6.4.5. Heat Capacity Measurements on Cooling

A very important aspect of differential scanning calorimetry is that heat capacity can be measured on cooling as well as on heating, although this may be more or less difficult to do, depending on individual instrument design. In principle, however, the procedures described above for either the scanning or the enthalpy method can be followed in the cooling mode exactly as in the heating mode.

However, there is difficulty with respect to temperature calibration of the instrument in the cooling mode. This is because most liquids, on cooling at moderate scan rates, will supercool, and thus the transition temperatures cannot normally be known to better than a few kelvins. This should be less of a problem for measurements to be done by the enthalpy method, since calibration can then be done while cooling either stepwise or at a very slow rate. Presumably, liquid samples (at least samples of pure metals) would crystallize fairly reproducibly if given enough time. In practice, the calibration can probably be established within one or two kelvins, and the behavior of sample temperature lag on cooling can be very crudely assumed to be the mirror image of that on heating. Although the more detailed method of determining sample temperature lag from the 'tail' area [12] should work on cooling as well as on heating, it would usually not be necessary to go to this length, since, as mentioned earlier, heat capacity generally varies slowly with temperature. As discussed in Section 11.8.3.1, errors involved in assigning the correct heat capacity to a slightly incorrect temperature are often only important below room temperature or in the vicinity of a phase transition.

11.7. MEASUREMENT OF ENTHALPY AND TEMPERATURE OF TRANSITION BY D.S.C.

11.7.1. Introduction

In this section, measurement of the characteristics of first-order phase transitions will be discussed only briefly, since most of the principles should already be evident from the discussions in previous sections. Determination of the enthalpy of transition involves measuring the area under the peak which results in the d.s.c. signal when the sample is scanned through the transition.   This can either be done with individual scans used only for the transition region (Section 11.7.2) or as part of a complete series of heat capacity measurements by the enthalpy method (Section 11.7.3).

The discussions in Sections 11.7.2 and 11.7.3 will deal only with first-order transitions. Gradual transitions or anomalies, including the glass transition, can usually be treated with the methods for heat capacity outlined in Section 11.6. The main concern will usually be to assign the temperature correctly in regions where the heat capacity is varying rapidly and where the sample temperature lag may be significant.

11.7.2. Individual Scans for the Transition Region

The first procedure for determining the enthalpy and temperature of transition for a sample is the same as is used for the standards in order to initially calibrate the instrument (Section 11.6.2). The sample is brought to equilibrium at some temperature below the transition and then heated through the transition at the same scan rate as was used for calibration.

The area under the peak yields directly the enthalpy of transition. In some cases, where the heat capacities of the low- and high-temperature phases are considerably different, there may be a shift in the baseline value from before to after the transition, causing some uncertainty as to how to interpolate the baseline for the transition region. Methods of dealing with this problem have been discussed by Dumas [22].

The temperature of the phase transition is determined from the following considerations. If the material has even a few-tenths percent impurity, it may display a transition peak, such as that shown in Fig. 11.12, which will not rise as sharply as did the peak for the pure standard shown in Fig. 11.7. If the observed slope (E_{obs}) of the leading edge of the transition peak is different from the slope of the transition peak of the standard material (E), the difference could be due to one or both of the following factors. Either (1) the temperature lag for this sample is different from that of the standard, or (2) the more gradual slope is only due to impurities, and the lag may be the same for the sample as for the standard. If the first explanation is correct, then the transition temperature should be given by T_{tr} in Fig. 11.12. If the second explanation is correct, the dashed line, representing the leading edge for the peak in the standard material, should be extrapolated from the peak maximum back to the baseline, giving T_{tr}' as the transition temperature. This latter construction reflects the fact that, at the peak maximum, the last bit of material is just completing the transformation, so that this temperature is least affected by any impurities. The leading edge of the standard peak (the dashed line) then describes how much the sample temperature is lagging the program temperature at that moment and allows the transition temperature to be calculated.

Since the decision between case 1 and case 2 in the preceding paragraph can rarely be made without a thorough understanding of all of the characteristics of the instrument and of the sample, it is fortunate that the difference between T_{tr} and T_{tr}' is seldom very large and, furthermore, that the portion of the difference due to temperature lag can be made smaller by

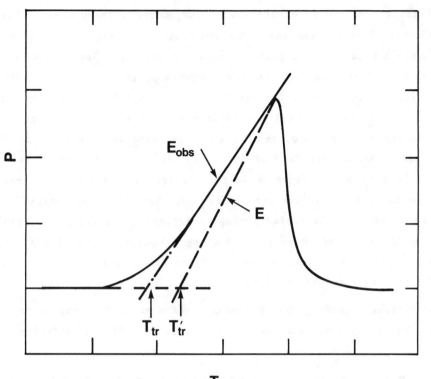

Figure 11.12. Example of a chart recorder trace [differential power (P) vs. temperature (T_{sm})] for a first-order phase transition in a sample under investigation. T_{tr}, transition temperature calculated using the observed leading edge of the transition peak (E_{obs}), dash-dot line); T'_{tr}, transition temperature calculated using the leading edge (E, dashed line) for the transition peak in the standard calibrating material (compare Fig. 11.7). See discussion in text.

choosing slower scan rates. However, the discussion points out once again that, as with measurements of heat capacity, the value of the enthalpy in d.s.c. is obtained directly and accurately, while the value of the temperature is only obtained secondarily and with some uncertainty.

Further mathematical analyses of the characteristics of transition peaks in d.s.c. can be found in many of the references [9–11, 13–22] described in Section 11.3.7.

11.7.3. Transition Scans During Continuous Heat Capacity Measurements

Rather than measure the heat capacity of a sample in one series of measurements and then the enthalpy and temperature of a phase transition in a separate scan, it is often more convenient to measure the properties of the phase transition as part of the continuous series of heat capacity measurements. The general difficulty with this is that the peak for a first-order phase transition would usually go completely off-scale at the high instrument sensitivities which are needed for determination of heat capacity.

Mraw and O'Rourke [60] have suggested a simple way around this problem, which can be used when heat capacities are measured by the enthalpy method of Section 11.6.4.3. In their method, heat capacities in intervals of 10 kelvins are run automatically by a progammable calculator system, using the highest sensitivity of the d.s.c. instrument. The procedure for enthalpy of transition involves simply interrupting the automatic programming sequence just prior to the particular 10-kelvin interval which will contain the phase transition. The instrument sensitivity is then manually reduced by a factor of ten, and the programming sequence is reinitiated. The enthalpy change, including the enthalpy of transition, for the next 10-kelvin interval is determined automatically in the usual fashion, but at the reduced sensitivity. When this interval is complete, the sequence is again interrupted, and the enthalpy value is manually multiplied by ten and re-entered into the proper memory location of the calculator. The instrument is returned to the highest sensitivity, the programming sequence is reinitiated, and heat capacity measurements proceed on the high temperature phase in the normal fashion. When the final output is obtained, the observed enthalpy change for the particular 10-kelvin interval containing the transition can be simply calculated from the $\Delta H/\Delta T$ which is printed along with the series of heat capacity values for the low and high temperature phases. Figure 11.13 shows the d.s.c. traces that were obtained for the enthalpy of fusion of acenaphthene [60].

However, one precaution regarding this procedure, which was not discussed in the earlier work [60], must be noted. Since the transition determination is so radically different from a normal

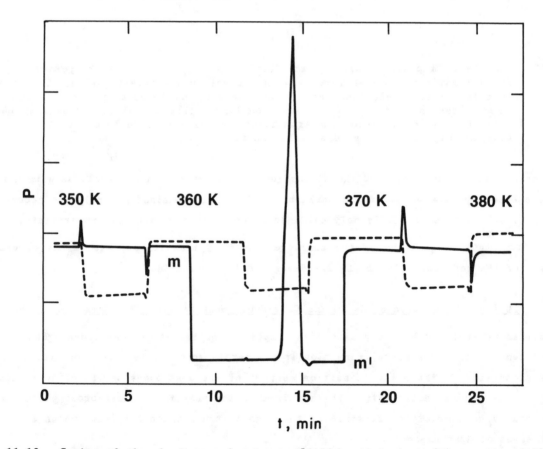

Figure 11.13. Copies of the chart recorder traces [differential power (P) vs. time (t)] obtained for the measurement of the enthalpy of fusion of acenaphthene [60]. The dashed curve is the d.s.c. signal for the empty runs, while the solid curve is the signal for the acenaphthene runs. At point m, the instrument sensitivity was manually reduced by a factor of ten, and then, at point m', it was returned to normal. See description in text.

heat capacity scan, the observed $\Delta H/\Delta T$ should _not_ be multiplied by the calibration factor (F) that is determined, using Eq. (16), from the heat capacity scan on the calibrating material. This is because the various factors which affect a heat capacity run to a given extent (e.g., nonlinearity of baseline or sample temperature lag) do not affect a transition scan to the same extent on a _percentage_ basis, and F, which is a percentage correction factor, is therefore not applicable. Thus, a different factor (e.g., F') must be evaluated from a similar type of transition scan on a material with a known enthalpy of transition. This factor can be evaluated as part of the instrument calibration procedure (Section 11.6.2), or it can be determined separately, when needed for a given application.

When the corrected enthalpy value is finally obtained for the transition interval, it is then treated exactly as it would be in classical adiabatic calorimetry. The heat capacities of the low and high temperature phases are extrapolated forward and backward, respectively, into the transition interval, and the 'baseline' heat capacity is subtracted from the total enthalpy to obtain the enthalpy of transition. This procedure does not suffer from the uncertainties in area determination that are caused in the method of Section 11.7.2 when the heat capacities of the low and high temperature phases are significantly different. In addition, because the scans just below the transition interval are done at the highest instrument sensitivity, pretransition effects can be readily assessed.

The temperature of the phase transition is determined from the transition peak exactly as was described in Section 11.7.2. However, if the instrument were calibrated for temperature at zero (or near-zero) scan rate, as described in Section 11.6.4.3, a small correction must be incorporated for the fact that the peak (Fig. 11.13) occurs while the sample is being scanned at a rate other than that for which the calibration was done. For the relatively slow rate that was used by Mraw and O'Rourke [60], this correction would often be no more than a few-tenths of a kelvin.

11.8. SOURCES OF ERROR IN DETERMINATION OF HEAT CAPACITY BY D.S.C.

11.8.1. Introduction

Many of the sources of error connected with the accurate determination of heat capacity by d.s.c., and the procedures for minimizing them, should already be apparent from the discussions in preceding sections. Random errors, such as those associated with instrument noise, will usually manifest themselves by producing scatter or irreproducibility in the results. Since these errors should be readily apparent to the operator, the remainder of this section will concentrate on the possibilities for systematic error, which could yield completely reproducible, but inaccurate, results. It will become apparent that, since d.s.c. is actually a relative method, which uses sapphire or some other standard as a calibrating material, many of the problems are reduced if conditions are rigorously matched for the runs on the empty, calibrant, and sample.

11.8.2. Errors Not Actually Due to the Instrument

As Richardson [10] has observed, many factors which might reduce the accuracy of heat capacity measurements by d.s.c. are not the fault of the calorimeter itself. For example, ancillary equipment used to record or process the output of the calorimeter must be at least as sensitive and accurate as the calorimeter signal itself. The sample must be weighed accurately on a sensitive balance, and correction in principle should be made for buoyancy, as well as for the heat capacity of any gas sealed within the pan. In addition, a run on a sample pan hermetically sealed in humid air may show a very minor heat capacity effect in sub-ambient measurements, due to the condensation (on cooling runs) or vaporization (on heating runs) of the water sealed within the pan. Because the heat capacity accuracy of present d.s.c. instruments is still near the ±1% level, these last three concerns are usually insignificant, but they could become more important if instrument accuracy should someday be improved.

A further problem concerning properties of the sample itself should be noted. If the sample is at all volatile in the temperature range of interest, it should, of course, be encapsulated in a hermetically sealed pan. Although volatility is usually a problem with liquids near their boiling points, many solids are also volatile, and others (such as hydrates, oxides, sulfides, etc.) generate a vapor as they convert to a different phase. When the vapor pressure of the sample, even in its sealed pan, has reached moderate values, usually less than 100 kPa, the enthalpy of vaporizing even a small amount of sample into the space within the sealed pan will contribute significantly to the apparent heat capacity of the condensed sample. The magnitude of the error will depend on the properties of the sample itself, as well as on the ratio of sample volume to space volume within the pan. For enthalpy of vaporization, simple calculations can show that, for many substances, the error for an 0.01 cm^3 sample in an 0.03 cm^3 pan will be approaching 1% as the vapor pressure approaches 50-100 kPa. Because of the gradual onset of this error, it can often go unnoticed until errors far greater than 1% have been reached. For these reasons, heat capacity measurements on a sample which generates a vapor should be terminated below the temperature of significant vaporization or else viewed with extreme caution. Although sample pans have been described [62,63] which are capable of withstanding high internal pressures, and in fact are available commercially [5], use of these pans for volatile materials in no way eliminates this problem.

11.8.3. Sources of Error in the Instrument Itself

11.8.3.1. Errors in Temperature Measurement

Preceding sections have dealt in detail with the fact that the sample temperature lags the monitored temperature in any d.s.c. instrument. Temperature calibration of the instrument under a particular set of conditions adjusts for this lag only for one particular scan rate and one particular value of the sample heat capacity.

The simplified case (no temperature lag within the pan), given by Eq. (14), can be described phenomenologically as follows. Even when operating conditions are maintained constant (keeping R'_s constant), samples of the same heat capacity (C_s) will lag differently depending on scan rate (dT_s/dt), and, likewise, samples of different heat capacity will lag differently even at the same scan rate. The importance of calibrating the instrument at the same scan rate as will later be used for the measurements has already been discussed, and this usually eliminates the larger of the two concerns. Concern over the second point can be reduced in a practical way by choosing the sample masses so as to closely match the heat capacities of all samples, including the calibrant, in a given temperature range. The problem of temperature lag can also be addressed, though this is rarely necessary, by the use of R'_s values, or by using the 'tail' area [12] as discussed in Section 11.6.

One exception to the normal lack of detailed concern as to the exact sample temperature occurs for measurements of heat capacity in the sub-ambient region, particularly below 200 K. When the heat capacity of a sample is changing rapidly with temperature, the assignment of the 'correct' heat capacity to the 'incorrect' temperature can be as serious as measuring the heat capacity incorrectly in the first place. Unfortunately, few detailed evaluations of commercial d.s.c. instruments in the sub-ambient region have been reported. Thus, previously reported experimental observations as to the values of R'_s, or the dependence of temperature on scan rate, which have generally been made for temperatures above 350 K, may not be valid below room temperature. Even if the temperature lag of a given sample could be made equal to the lag for the calibrating material, the problem is not eliminated by simple use of the correction factor (F), since dC_p/dT for the calibrant will not normally be the same as for the sample. This is why sapphire is not the best choice as a calibrant for low temperature measurements, since its rather large value of dC_p/dT is often very different from those of other samples.

In practice, it is usually sufficient to calibrate the instrument for temperature in the low temperature region as well as possible and to choose a calibrant for heat capacity which has as close a value of dC_p/dT to those of the samples as possible. Even reasonable care in following these procedures will usually bring heat capacity errors due only to temperature uncertainty down below the 1% level.

11.8.3.2. Errors in Baseline Interpolation

Both the scanning and enthalpy methods for measurement of heat capacity described in Section 11.6.4 use a linear interpolation of isothermal baselines to represent the unseen baseline during the scan. Even if this linear representation is not correct, no errors will be introduced as long as the true shape of the baseline remains constant. This is illustrated by the schematic representation in Fig. 11.14. Although the baselines are assumed to be given by the dashed lines in Fig. 11.14(a), the correct ordinate difference will still be obtained even if the true behavior is that of Fig. 11.14(b), since values for the empty pan are always subtracted from the values obtained in all subsequent runs.

In the preceding paragraph, it has been assumed that the isothermal baseline is constant at a given temperature, and this is usually the case. A further problem is encountered, however, when the isothermal baseline, even for a given temperature, drifts with time, which often happens at very low or very high temperatures. An example of this can be seen in the case where the slowly falling level of a bath of liquid nitrogen causes a slow change in the baseline value. However, as long as the temporal drift of the baseline remains constant throughout the entire temperature scan, only a small error will be introduced, since interpolation yields a value of the baseline which is very nearly correct.

11.8.3.3. 'Multiplicative' Errors

Many other possible measurement errors are reduced using the recommended procedure of calibrating with a known material prior to all the sample runs. The most obvious effect of this calibration is that, although many instruments are said to have 'temperature-independent calorimetric sensitivity,' this feature is not really necessary. The running of the heat-capacity standard is actually a direct calibration of the instrument in the <u>exact temperature range</u> where calibration is needed for the subsequent sample runs.

Other, less obvious multiplicative errors are also reduced with the calibration procedure. For example, even if the overall temperature calibration is fairly accurate, local fluctuations of a few-tenths of a kelvin might be possible, even for purely electronic reasons. Thus, in the enthalpy method for instance, what appears to be a 10-kelvin interval for a particular run may actually be a 9.8 or a 10.1-kelvin interval. This error will still be completely cancelled, since the $\Delta H/\Delta T$ of the empty, calibrant, and sample would all have the same error.

The importance of keeping the total sample heat capacity similar to that of the total calibrant heat capacity (by adjusting sample and calibrant masses) has already been mentioned in connection with sample temperature lag. Although detailed studies have not been reported, it is probable that this precaution also reduces many other possibilities for subtle, systematic errors.

11.8.3.4. Conclusion on Errors Due to the Instrument

The procedure of calibrating the instrument with a material of known heat capacity prior to the sample runs is valid only if all the relevant characteristics of the instrument are free

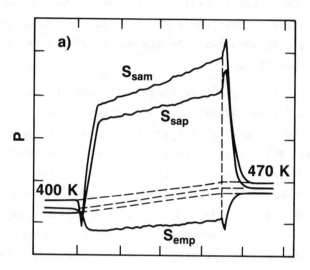

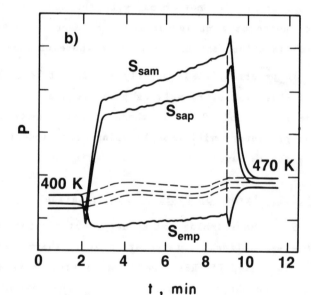

Figure 11.14. Chart recorder traces [differential power (P) vs. time (t)], such as those of Figs. 11.8 and 11.9, showing the possible behavior of the unseen isothermal baselines during a scan. S_{emp}, S_{sap}, and S_{sam} are the d.s.c. signals for the empty, sapphire, and sample runs, respectively. As shown by (a), linear interpolation of baselines is generally assumed; however, no error is introduced if the actual behavior is nonlinear [as in (b)], as long as the behavior is constant throughout all runs.

from significant drift throughout the runs.   As mentioned earlier, a final measurement should always be done on another material whose heat capacity is accurately known, as an excellent safeguard against unsuspected problems.   In general, there can be no substitute for careful, rigorous attention to detail when heat capacities of high accuracy are to be measured by d.s.c.

11.9. EXAMPLES OF ACCURATE HEAT CAPACITY INVESTIGATIONS BY D.S.C.

In recent years, various authors have reported results which indicate that heat capacity can be measured with an accuracy of approximately $\pm 1\%$ on commercial d.s.c. equipment.   These investigations have involved a comparison, for a given material, between the heat capacity values measured by d.s.c. and those determined in investigations by classical calorimetric techniques.   Since this section will merely summarize these reports, interested readers should consult the original references for further details.   In all cases, the scanning method of Section 11.6.4.2 was used, unless otherwise noted.

To the author's knowledge, O'Neill [48] was the first to demonstrate the accuracy of d.s.c. for measurements of heat capacity.   He reported heat capacities for graphite, diamond, silver nitrate, and gold in the temperature range 340–510 K and showed that the d.s.c. values agreed with literature values within ± 0.7–3.0%, depending on the sample.

Gaur et al. [64] measured the heat capacity of zinc between 410 and 520 K with a computer-interfaced d.s.c. and examined the effects of sample size, scan rate, and scan interval on the accuracy of the heat capacities measured.   They compared their d.s.c. results to literature values for zinc and found them accurate to ± 0.3–1.3%.   Krupka et al. [65] reported heat capacities for minerals from 350–1000 K, again with an accuracy of approximately $\pm 1\%$.

Clechet et al. [66] measured the heat capacity of aluminum (350–650 K) and of potassium nitrate (330–390 K) and tested the effects of rather high scan rates (0.53 K$\cdot$s^{-1}) and large scan intervals (100 kelvins).   They reported accuracies of ± 1–2%, depending on the particular method used.

Only a few authors have reported investigations intended to determine the accuracy of d.s.c. heat capacity measurements below 320 K.   Mraw and Naas [57] measured the heat capacity of FeS_2, pyrite, from 100–800 K by two separate d.s.c. methods: (1) a variation of the scanning method; and (2) the enthalpy method described in Section 11.6.4.3.   By either method, the d.s.c. heat capacity of pyrite was found to be accurate to approximately $\pm 1\%$ from about 220–700 K, with accuracy of ± 2–3% for temperatures below 200 K or above 700 K.   In other investigations on organic compounds, using the enthalpy method only, they showed that the heat capacities of cyclohexane [59] (120–325 K) and acenaphthene and n-heptadecane [60] (300–470 K) were again accurate to approximately $\pm 2\%$ from 120–220 K and $\pm 1\%$ above 220 K.   Wiedemeier et al. [67] measured the heat capacity of Sn_2S_3 from 110–610 K and found that d.s.c. values agreed ($\pm 2\%$) with the sum of the known heat capacities of SnS and SnS_2.   The comparison indirectly indicates the accuracy of the d.s.c. measurements on Sn_2S_3.

Despite the capability of d.s.c. for the measurement of heat capacity on cooling (Section 11.6.4.5), the absolute accuracy in this mode has rarely been investigated.   An exception is the work of Shu et al. [68], who determined the heat capacity of liquid and glassy selenium on cooling from 520-330 K. Comparison to literature could only be made for the liquid state at the upper temperatures and the solidified glassy state at the lower temperatures, but the comparison indicated excellent accuracy for the d.s.c. values ($\pm$1%). Mraw and O'Rourke [69] have also tested their enthalpy method in a cooling mode, but, although results on pyrite were encouraging ($\pm$1-2%) in certain temperature ranges, no systematic study has been done.

This section has concentrated on investigations intended to test the absolute accuracy of heat capacity measurements by d.s.c. It is worth noting, however, that it is possible to achieve an internal precision for a given scan which is much better than the overall absolute accuracy. This could be important when the location or characterization of a small thermal anomaly, barely above the baseline heat capacity, is the primary goal of the measurement. As only one example, Johnston et al. [70] defined a small thermal anomaly in $KFeS_2$ at 252 K with a precision of 0.1-0.2%, even though the overall accuracy of the values was only 2% by the particular method used.

11.10. ANALYSIS OF HEAT CAPACITY DATA BY D.S.C.

The experimental quantity determined in a d.s.c. heat capacity measurement can be either the heat capacity at constant pressure (C_p), i.e., for an inert, nonvolatile sample in an open pan, or the heat capacity under the saturation vapor pressure (C_{sat}), i.e., for a sample in a hermetically sealed pan. Since C_{sat} can almost always be identified with C_p for the type of measurements considered here, analysis of heat capacities from d.s.c. requires no special discussion. Once the experimental data have been obtained, the analysis of the results is exactly the same as it is for any C_{sat} or C_p data obtained by more classical methods. The only exception is that heat capacity points determined by the scanning method (Section 11.6.4.2) are measured 'instantaneously' and, thus, do not require curvature corrections. The enthalpy method (Section 11.6.4.3), since it simulates classical step-wise techniques, yields values of $\Delta H/\Delta T$ which would, in principle, require curvature corrections, though instances where this is important should be rare.

11.11. CONCLUSIONS

It is important to emphasize again that the very ease of using a d.s.c. should not be allowed to lead to careless use. The operator should practice all procedures with the utmost care and always be on the watch for unsuspected error, if data of the highest possible accuracy are desired.

The technique of differential scanning calorimetry has been proven suitable for the measurement of heat capacity values accurate to approximately $\pm$1% over a wide temperature range. These measurements can be made with samples of 10-100 mg on commercially available equipment, often with the aid of automation and computerization. The combination of these factors should allow d.s.c. to take its place legitimately among valuable calorimetric techniques for the acquisition of thermodynamic data. Careful, systematic studies with present equipment, or the introduction of improved instrumentation, may one day allow accuracies of a few tenths of a percent to be approached.

11.12. THERMAL ANALYSIS SOCIETIES AND JOURNALS

Many scientists involved in thermal analysis measurements, including measurements by d.s.c., belong to what is known as the International Confederation for Thermal Analysis (ICTA), an organization which, among other activities, tries to adopt rules for nomenclature and to establish recommended values for properties of standard materials. Recommendations for nomenclature were established in 1973 [71] and updated in 1979 [6]. As mentioned earlier, a 'Catalogue of Physicochemical Standard Substances' appeared in 1972 [50], which includes materials for temperature calibration in thermal analysis.

The North American Thermal Analysis society (NATAS) holds regular conferences, at which a large variety of topics in thermal analysis, including d.s.c., are discussed. In those cases where the methods of d.s.c. are specifically used for obtaining quantitative enthalpy or heat capacity data (as opposed to being used for 'thermal analysis' only), the results are often presented at the Calorimetry Conference, which holds yearly meetings in the United States or Canada. Papers on new instruments or methods, as well as theoretical treatments of d.s.c. principles, can often be found at either conference.

Two 'international' journals, the Journal of Thermal Analysis and Thermochimica Acta, are almost wholly devoted to thermal analysis techniques, applications, and results. However, the increased acceptance of d.s.c. in particular as a viable means of obtaining accurate calorimetric data has resulted in many papers appearing in other journals as well, notably the Journal of Chemical Thermodynamics.

11.13. ACKNOWLEDGMENT

The author thanks D. F. O'Rourke for a careful reading of the manuscript of this chapter and for many valuable comments and suggestions during its preparation. The valuable comments of J. E. Callanan, L. X. Finegold, and especially K. L. Churney are also gratefully acknowledged.

11.14. REFERENCES

1. Garn, P.D. and Menis, O., J. Macromolec. Sci. Phys., B13, 611-29, 1977.

2. Pope, M.I. and Judd, M.D., Differential Thermal Analysis, Heyden and Son, Ltd., London, England, 1977.

3. Boersma, S.L., J. Am. Ceram. Soc., 38, 281-4, 1955.

4. Watson, E.S., O'Neill, M.J., Justin, J., and Brenner, N., Anal. Chem., 36, 1233-8, 1964.

5. Perkin-Elmer Corp., Instrument Div., Norwalk, CT, USA, 06856.

6. International Union of Pure and Applied Chemistry (IUPAC), Pure Appl. Chem., 52, 2385-91, 1980.

7. McNaughton, J.L. and Mortimer, C.T., in IRS, Physical Chemistry Series 2, Vol. 10, Butterworths, London, England, 1975. Reprint available from Ref. 5 above.

8. Gill, P.S., Can. Res. Develop., 7, 23-6, 1974.

9. Gray, A.P., in Analytical Calorimetry (Porter, R.S. and Johnson, J.F., Editors), Plenum Press, New York, NY, 1968. Reprint available from Ref. 5 above.

10. Richardson, M.J., in <u>Compendium on Thermophysical Property Measurement Methods</u> (Maglic, K., Cezairliyan, A., and Peletsky, V.E., Editors), Vol. 1, Plenum Press, New York, NY, in press.

11. Baxter, R.A., in <u>Thermal Analysis</u> (Schwenker, R.F., Jr. and Garn, P.D., Editors), Vol. 1, Academic Press, New York, NY, 1969.

12. Richardson, M.J. and Savill, N.G., Thermochim. Acta, <u>12</u>, 213-20, 1975.

13. O'Neill, M.J., Anal. Chem., <u>36</u>, 1238-45, 1964.

14. Flynn, J.H., in <u>Analytical Calorimetry</u> (Porter, R.S. and Johnson, J.F., Editors), Vol. 3, Plenum Press, New York, NY, 1974.

15. Flynn, J.H., in <u>Status of Thermal Analysis</u>, NBS Spec. Publ. 338, U.S. Government Printing Office, Washington, DC, 119-36, 1970.

16. Flynn, J.H., Thermochim. Acta, <u>8</u>, 69-81, 1974.

17. Churney, K.L. and Garvin, D., 'A Third Appraisal of Methods for Estimating Self-Reaction Hazards,' National Bureau of Standards, NBSIR 80-2018 (DOT), 1980. (See appendices.)

18. Vallebona, G., J. Therm. Anal., <u>16</u>, 49-58, 1979.

19. Randzio, S.L. and Sunner, S., Anal. Chem., <u>50</u>, 704-7, 1978.

20. Heuvel, H.M. and Lind, K.C.J.B., Anal. Chem., <u>42</u>, 1044-8, 1970.

21. Smith, G.W., Thermochim. Acta, <u>21</u>, 431-5, 1977.

22. Dumas, J.P., J. Phys. D: Appl. Phys., <u>11</u>, 1-5, 1978.

23. Buckingham, M.J., Edwards, C., and Lipa, J.A., Rev. Sci. Instrum., <u>44</u>, 1167-72, 1973.

24. Nicholson, P.S. and Fulrath, R.M., J. Phys. E: Sci. Instrum., <u>3</u>, 351-5, 1970.

25. Hill, R.A.W. and Slessor, R.P., Trans. Faraday Soc., <u>65</u>, 340-8, 1969.

26. Lagnier, R., Pierre, J., and Mortimer, M.J., Cryogenics, 349-53, June 1977.

27. Bonjour, E., Lagnier, R., Locatelli, M., and Pierre, J., J. Calorim. Anal. Therm., <u>11</u>, 1-1-1/ 1-1-7, 1980.

28. Wunderlich, B. and Bopp, R.C., Therm. Anal., Proc. Int. Conf., 3rd 1971 (Wiedemann, H.G., Editor), <u>1</u>, 295-301, 1972.

29. DuPont Co., Scientific and Process Instruments Div., Wilmington, DE, USA 19898.

30. Kamphausen, M., Rev. Sci. Instrum., <u>46</u>, 668-9, 1975.

31. Kamphausen, M. and Schneider, G.M., Thermochim. Acta, <u>22</u>, 371-8, 1978.

32. Arntz, H., Rev. Sci. Instrum., <u>51</u>, 965-7, 1980.

33. Sandrock, R., Rev. Sci. Instrum., <u>53</u>, 1079-81, 1982.

34. Mettler Instrumente A.-G., CH-8606 Greifensee, Switzerland.

35. Lee, J.D. and Levy, P.F., North Amer. Thermal Anal. Soc., Proc. 11th Conf., 1981, in press.

36. van Dooren, A.A. and Müller, B.W., Thermochim. Acta, <u>49</u>, 151-62, 1981.

37. van Dooren, A.A. and Müller, B.W., Thermochim. Acta, <u>49</u>, 163-74, 1981.

38. van Dooren, A.A. and Müller, B.W., Thermochim. Acta, <u>49</u>, 175-84, 1981.

39. van Dooren, A.A. and Müller, B.W., Thermochim. Acta, <u>49</u>, 185-98, 1981.

40. Perron, W., Bayer, G., and Wiedemann, H.G., Therm. Anal., Proc. Int. Conf., 6th (Wiedemann, H.G., Editor, <u>1</u>, 279-84, 1980.

41. Societe d'Etudes d'Automatization de Regulation et d'Appareils de Mesures (SETARAM), Lyon, France.

42. Mercier, J., J. Therm. Anal., <u>14</u>, 161-71, 1978.

43. Privalov, P.L., Plotnikov, V.V., Filimonov, V.V., J. Chem. Thermodyn., <u>7</u>, 41-7, 1975.

44. Microcal, Inc., Amherst, MA, USA 01002.

45. Barrall, E.M., II, and Dawson, B., Thermochim. Acta, <u>8</u>, 83-92, 1974.

46. Mehta, A., Bopp, R.C., Gaur, U., and Wunderlich, B., J. Therm. Anal., 13, 197-203, 1978.

47. Yuen, H.K. and Yosel, C.J., Thermochim. Acta, 33, 281-91, 1979.

48. O'Neill, M.J., Anal. Chem., 38, 1331-6, 1966.

49. Cassel, B., 'New Techniques in DSC: Differential Heat Capacity Determinations for Maximum Accuracy,' presented at the Pittsburgh Conference on Analytical Chemistry, Cleveland, OH, 1974. Reprint available from Ref. 5 above.

50. International Union of Pure and Applied Chemistry (IUPAC), Pure Appl. Chem., 29, 597-616, 1972.

51. Office of Standard Reference Materials, Institute for Materials Research, National Bureau of Standards, Washington, DC, USA 20234.

52. Gray, A.P., Therm. Anal., Proc. Int. Conf., 4th, 1974. Reprint available from Ref. 5 above.

53. Zeeb, K.G., Lowings, M.G., McCurdy, K.G., and Hepler, L.G., Thermochim. Acta, 40, 245-9, 1980.

54. Andon, R.J.L. and Connett, J.E., Thermochem. Acta, 42, 241-7, 1980.

55. Churney, K.L., private communication.

56. Ginnings, D.C. and Furukawa, G.T., J. Am. Chem. Soc., 75, 522, 1953.

57. Mraw, S.C. and Naas, D.F., J. Chem. Thermodyn., 11, 567-84, 1979.

58. Mraw, S.C. and Naas, D.F., J. Chem. Thermodyn., 11, 585-92, 1979.

59. Mraw, S.C. and Naas-O'Rourke, D.F., J. Chem. Thermodyn., 12, 691-704, 1980.

60. Mraw, S.C. and O'Rourke, D.F., J. Chem. Thermodyn., 13, 199-200, 1981.

61. Mraw, S.C. and Keweshan, C.F., 38th Annual Calorimetry Conference, Williamsburg, VA, 1983.

62. Freeberg, F.E. and Alleman, T.G., Anal. Chem., 38, 1806-7, 1966.

63. Schouteten, C.J.H., Bakker, S., Klazema, B., and Pennings, A.J., Anal. Chem., 49, 522-3, 1977.

64. Gaur, U., Mehta, A., and Wunderlich, B., J. Therm. Anal., 13, 71, 1978.

65. Krupka, K.M., Robie, R.A., and Hemingway, B.S., Am. Mineral., 64, 86-101, 1979.

66. Clechet, P., Martelet, C., and Cau, N.D., Analusis, 6, 220-5, 1978.

67. Wiedemeier, H., Csillag, F., Gaur, U., and Wunderlich, B., Thermochim. Acta, 35, 187-96, 1980.

68. Shu, H.C., Gaur, U., and Wunderlich, B., J. Poly. Sci.: Poly. Phys. Ed., 18, 449-56, 1980.

69. Mraw, S.C. and O'Rourke, D.F., unpublished data.

70. Johnston, D.C., Jacobson, A.J., and Mraw, S.C., Solid State Commun., 44, 255-8, 1982.

71. International Union of Pure and Applied Chemistry (IUPAC), Pure Appl. Chem., 37, 439-44, 1974.

CHAPTER 12

Special Problems in Calorimetry of Radioactive Materials

M. G. CHASANOV, D. F. FISCHER, and L. LEIBOWITZ

12.1. INTRODUCTION

Of the 103 elements listed on the periodic table, approximately 20 are radioactive and also occur in significant quantities in nature (excluding the fission products and transuranics generated by a natural fission reactor such as at Oklo [1] in the Gabon). Since the discovery of radioactivity in 1896, by Henri Becquerel, scientific interest has spurred study of the properties of radioactive materials. The Manhattan Project and the subsequent availability of artificially radioactive materials led to development of a significant thermophysical data base for these elements; much of the early work from the Manhattan Project can be found in the books of Katz and Rabinowitch [2] and of Quill [3]. Nuclear reactor development programs have generated additional property data, especially for nuclear fuels, coolants, and structural materials. Calorimetric studies, mainly high-temperature work with fuel materials, have been of considerable interest in this later phase of the investigation of radioactive substances.

In performing calorimetry with radioactive materials, there are essentially two major problems: decay-heat effects and protection of the calorimetrist from exposure to the radioactivity. These aspects of calorimetry comprise the bulk of the following discussions.

12.2. CALORIMETRIC IMPLICATIONS OF RADIOACTIVE MATERIALS

12.2.1. Materials for Study

12.2.1.1. Naturally-Occurring Radioactive Materials

Those radioactive materials that occur in more than trace abundances in nature have, as one would expect, very long half-lives (greater than 10^9 years). The abundances of most of these isotopes, however, are low enough so that calorimetry with them constitutes no special problem. In the case of rhenium, the radioactive isotope (Re-187) has an abundance of nearly 63%; the major isotope (95.72%) of indium (In-115) is also radioactive. Th-232 and U-238 are virtually the only naturally-occurring species. However, the specific activities, curies per gram ($Ci \cdot g^{-1}$) of these last four isotopes are quite low: Re-187, 4×10^{-8}; In-115, 5×10^{-12}; Th-232, 1×10^{-7}; and U-238, 3×10^{-7}. For comparison, the specific activity of cobalt-60 is 1×10^3 $Ci \cdot g^{-1}$ (a curie corresponds to 3.7×10^{10} disintegration per second). In addition, the major radiations from these isotopes are not penetrating, consisting chiefly of β or α particles. Thus, heavy shielding is not necessary for handling these isotopes, even when daughters of their decay series are considered; however, protection against heavy-metal poisoning should be a consideration.

In effect here we have disregarded daughter products produced from the above isotopes by various decay chains, e.g., Po, Ra, Rn, Ac, Pa, and Tc. These daughters of the naturally-abundant radioactive isotopes are characterized by much shorter half-lives and concomitantly larger specific activities than those of the parent of the chain. A major consideration with such materials, most of which are bone-depositing elements, is the prevention of ingestion by the calorimetrist.

12.2.1.2. Fission Products and Other Artificially-Produced Radioactive Isotopes

Neutron-induced fission of heavy nuclear species, as is well known today, results in the production of fission fragments with atomic numbers ranging from 30 (zinc) to 66 (dysprosium). The fission products are radioactive species which, in turn, generally decay by beta emission to other species; associated with the decay of fission products is the emission of gamma radiation. These materials are characterized generally by short half-lives and large specific activities, e.g., La-140, half-life = 40.2 hours and an activity of about 5.6×10^6 curies per gram. Thus, if one were to work with these materials, adequate radiation protection for the experiments would be necessary; happily, the ordinary calorimetric objectives can usually be achieved by using the nonradioactive isotopes of the element to be studied. Self-heating studies for fission products, of course, necessitate use of the specific isotope.

The generation of radioactive isotopes by use of high-energy particle accelerators has populated the chart of the nuclides [4] with a multitude of radioactive species not found in nature. Calorimetric study of the properties of these materials is subject to the same considerations discussed above for naturally-occurring isotopes and for fission products.

12.2.1.3. Transuranic Materials

The elements in the periodic table with atomic numbers greater than 92 are the transuranics. They are artificially-generated, heavy metals resulting from neutron or charged-particle irradiation of uranium or other transuranium elements. They are generally long-lived when compared with many of the fission products; the radiations they emit are typically alpha and gamma, though some are beta emitters. However, in addition to being poisonous radioactive heavy metals requiring extreme caution in handling, many of the transuranics, especially Pu-239, are fissile and criticality considerations can necessitate special design of equipment utilizing these materials. Of course, the same must be said for studies involving highly-fissile isotopes of uranium as well.

12.2.2. Experimental Safety Considerations

12.2.2.1. Radiation Safety

The types of nuclear radiation encountered in calorimetry of radioactive materials are essentially alpha, beta, or gamma. These emanations are either 'hard' (gamma rays) or 'soft' (alpha and beta particles); protection of the experimenter from external exposure to these radiations depends on these characteristics. Alpha particles constitute no external hazard since they cannot

penetrate the outer layer of the experimenter's skin. Beta particles are more penetrating than
alpha particles, but shielding from them can readily be accomplished. It is the 'hard' gamma ray
that results in utilization of massive shielding to protect the experimenter.

    There are three general methods of protection against radiation: time, distance, and shield-
ing. The less time spent in the radiation field the better, and the further away from the radiation source the lower the dose to the experimenter ('inverse square law'). Shielding provides a
means of reducing the intensity of the radiation field by interposing absorbing materials between
the experimenter and the radiation source. The type of shielding employed depends on the radiation involved. Often, all three methods are called into play to provide adequate protection for
the experimenter.

12.2.2.2. Criticality Hazards

    Many of the actinides fission spontaneously and, in addition, can fission on capturing a neu-
tron; thus, if sufficient fissile material is present, a chain reaction can take place, releasing
lethal amounts of neutrons and gamma radiation. The mass of fissile material required to achieve
criticality, that is, the mass to support a self-sustaining fission chain reaction, depends on the
geometrical shape of the material and the presence of neutron reflecting materials. For example
[5], an unreflected sphere of uranium containing 93.5% U-325 has a critical mass of 48.6 kg of
U-235; a water-reflected sphere containing a water solution with 52 g U-235 per liter has a critical mass of only 0.84 kg. Clearly, if one needs to utilize fissile materials in a calorimetric
study, the possibility of a criticality hazard must be evaluated. In all likelihood, the amounts
of fissile material required for the calorimetry will not present criticality problems; e.g., no
criticality problem arises when less than 150 grams of plutonium are involved [6].

12.3. SPECIFIC-HEAT DETERMINATIONS

12.3.1. Decay-Heat Considerations

    Decay heat, that is, heating due to deposition of energy in the sample by the radiation
emanating from it, is chiefly of interest in three situations: (1) gross heating by large amounts
of radioactive isotopes – such as fission products after reactor shutdown, (2) small amounts of
heat release during low-temperature calorimetry of radioactive materials, and (3) the quantitative
assay of radioactive materials. As pointed out earlier, one would not choose to add the complications of handling radioactive substances to calorimetric operations unless it were unavoidable –
as, for example, when dealing with the transuranic elements.

    Decay-heat measurements for fission products are of considerable technical importance in
reactor design and in reactor-safety analyses. While theory can predict the heating expected due
to decay of fission products (e.g., Way and Wigner [7]), verification of theory by measurement has
always been necessary. In the case of fission-product decay from irradiated uranium, Day and
Cannon [8] of the Manhattan Project reported their measurements in 1944. They dropped an active
slug of uranium into a Dewer flask filled with water and followed the temperature change using a
Beckmann thermometer; the system was surrounded by a water bath that had its temperature manually

adjusted to match that of the calorimeter. The uranium slug was lowered into the water flask via a long rope and the whole assembly was located in the corner of a 'hot' laboratory behind an 8-in. lead shield.

Johnston [9], in 1965, published a calorimetric determination of fission-product heating from a plutonium fuel. His work was done using a microcalorimeter with samples containing about 40 mg of irradiated PuO_2 contained in two concentric aluminum capsules. There are no indications in the paper, however, of what precautions were taken in handling the system. He did conclude that gross fission-product decay heating in fast-fissioned Pu-239 is quite similar to that for thermally-fissioned U-235.

Low-temperature heat-pulse calorimetry for radioactive samples is described in a paper by Trainor et al. [10]. They point out, for example, that Pu-242 generates about 0.01 mW·g^{-1} which at 4 K is about an order of magnitude greater than the heater power for a heat-pulse experiment. The authors compensate for this by electronically regulating the sample environment so that the heat leak from the sample is essentially equal to the self-heating involved. Their experiment was carried out inside a glove box in order to confine the actinides they were working with.

These examples are for extreme situations; in most calorimetric studies the amounts of radioactive material are small enough so that the decay heat is of minor importance. Decay heat has also been used as a calorimetric tool for accurate determination of an isotope's half-life. The half-life of polonium has been determined from its decay heat by means of differential calorimetry [11] and a Bunsen ice calorimeter [12]. The two values obtained agreed within ±0.1%. More recently, the half-life of Pu-239 was determined by calorimetric means [13,14].

Decay-heat measurements have long been used to provide a quantitative assay of radioactive materials.   In fact, calorimeter measurements were first used in 1903 by Pierre Curie and A. Laborde [15] to demonstrate that nuclear decay is fundamentally different from any physical or chemical phenomenon known at that time. Rutherford and Barnes [16], also in 1903, determined the heat produced by 1 g of radium in equilibrium with its short half-life daughters. At the present time, calorimetry is a principal method for the quantitative assay of plutonium [17].

12.3.2. Remote Operation of Calorimeters

12.3.2.1. Glove Boxes

The characteristics of the transuranic elements are such that calorimetry involving them must be performed in a controlled environment both to protect the experimenter and to prevent spread of contamination to the surroundings. One approach is to use a completely closed system which permits direct access by the experimenter - the glove box. Glove-box design varies from facility to facility and is very dependent on the purpose of the box; however, there are some general features common to all of them. Some of these common characteristics will be discussed here. Detailed considerations of design philosophies can be found in many references; the material presented below is based on the works of Steindler [18] and of Appleton and Dunster [6].

The dimensions of the glove box should be such that the experimenter, working through the glove ports, can reach any position in the box; rather than limiting the dimensions of the box, this requirement dictates an adequate number of well-placed glove ports. Wood, plastics, aluminum, and steel have been used in the construction of frames for glove boxes; mild steel, painted to provide corrosion resistance, has proved to be satisfactory and relatively inexpensive. The floors and some sides of the glove boxes are normally fabricated of heavy-gauge sheet metal or metal plate. Transparent panels for the glove boxes are of two types: those containing glove ports (6-8 in. dia.) and those used only for lighting and visibility. The glove boxes that the authors have used had safety-glass windows installed in automobile windshield-type sealing strips. It should be kept in mind that the structural strength of the transparent panels is diminished by the glove-port holes. Other materials that have been used for transparent panels in glove boxes include tempered plate glass and plastics such as Lucite.

Glove ports made of plastic are inserted in the holes in the transparent panel; the holes are oval or round. Synthetic rubber gloves, held in place on the port by rubber O-rings or metal clamping rings, are usually used. Figure 12.1 is a photograph of the type of plutonium glove box in use in the Chemical Technology Division at Argonne National Laboratory.

Transfer of material in and out of the glove box constitutes a major source of possible dispersal of contamination. One of the most widely-used transfer methods involves dielectrically-sealed plastic pouches, typically 20 mil polyvinyl chloride. The exact transfer techniques vary from facility to facility, but, typically, three heat seals are made for out-of-box transfer and the plastic is cut along the central seal. Transfer into the box is simpler, involving only one seal (four layers of plastic). Details of such transfer procedures can be found in Ref. [19]. The boxes used by the authors had transfer ports (for dielectric-seal use) of 22 1/2-in. diameter. Small item transfer into the box can also be accomplished by passing special cans (sphincter cans) through gasket-sealed openings [20].

A common ventilation system for glove boxes is untreated laboratory air on a 'once-through' basis. Air is drawn into the box through filters and exhausted air from the box goes through a system of prefilters and filters to reduce spread of contamination. Often, normal laboratory air is not suitable for the glove-box operations; dried air, inert gas-dry air mixtures, or completely inert atmospheres are often employed. However, the operation of the ventilation system is significantly complicated by such requirements. Design of the ventilation system must allow for sufficient air flow for operational purposes while maintaining a negative pressure differential of the order of 0.2 to 1 in. of water. The system must be capable of handling a sudden surge of air in the event of ruptured gloves or panels.

It is obvious from the above descriptions of the glove box and its complications that calorimetry in such an environment becomes difficult and often tedious. Clearly, the calorimetric equipment will probably require modification for glove-box operation; often, the glove box itself will require special design to accommodate the calorimeter and ancillary equipment. In the authors' experience, it takes a minimum of three times as long to conduct a study in a glove box compared to working in the open. An example of a calorimetric apparatus operated in a dry-air glove box is given by Engel et al. [21]; heat of combustion studies in a once-through nitrogen glove box are discussed by Johnson et al. [22]. Additional recent examples of such operations can be found in the bibliography given in the Appendix at the end of this chapter.

Figure 12.1.   Cenham glove box used at Argonne National Laboratory.

12.3.2.2. Hoods

Hoods have long been used for chemical operations involving radioactive materials; however, since they are more likely to lead to spread of contamination than a closed box (the glove box described above), most operations involving the transuranics are performed in the closed system. Nevertheless, because of the difficulties in operating a massive high-temperature drop calorimeter in a glove box, the present authors successfully utilized a 'walk-in' hood for studies involving plutonium-containing reactor fuels. For a detailed discussion of the hood facility and calorimetric operation in it, the reader is referred to Ref. [23]. Some important features of this facility will be summarized below.

The need for accurate sample positioning in the drop calorimeter used by the authors required the kind of hands-on capability that would be very difficult to obtain in a glove box (see Fig. 12.2). The drop calorimeter system was enclosed in a 10 x 7 x 8-ft.-high aluminum walk-in hood. The front face of the hood was fitted with two 4-ft.-wide overlapping sliding doors, each of which had a 3-ft.-square Lucite window. The window in the door nearest the equipment was fitted with two neoprene arm-length gloves.

In the side wall of the hood nearest the equipment, an additional Lucite window (23 x 16 in.) was located. To the right of this window, two neoprene gloves (needed for equipment manipulations during an experiment) were located. Another window of the same size was situated above a pyrometer housing located on the side wall. To the right of this window, two more neoprene gloves provided access to components inside the hood. Two Lucite windows on top of the hood gave good lighting by means of fluorescent fixtures.

The flooring in the hood was painted with a liquid-tile paint. All seams along the flooring and the aluminum sides of the hood, plus all visible cracks and openings, were sealed with a silicon-rubber adhesive. The interior of the hood was maintained at a slightly negative pressure, even when one of the sliding doors was two-thirds open.

Radiation-monitoring probes were located outside the hood; one of the probes could be swung and extended inside the hood, if necessary. Protective clothing, such a one-piece paper suit, plastic boots and covers, gloves, and a plastic hood were worn by the experimenter before entering the hood. The plastic hood was attached by plastic tubing to a filtered breathing-air supply. The breathing-air supply flowed continuously and maintained a positive pressure under the head hood.

The walk-in hood, in this case, provided very reliable containment for plutonium-containing materials, good access to the experimental equipment, and easy breathing and good maneuverability plus alpha-radiation protection for the experimenter. The cost of putting the calorimeter system into a walk-in hood was considerably less than that of constructing a glove-box calorimeter.

12.3.2.3. Caves

Caves or 'hot cells' provide protection for the experimenter primarily against penetrating radiation from gamma emitters; if the isotopes being handled also are alpha emitters, a high

Figure 12.2.   Experimenter positioning sample capsule in high-temperature drop-calorimetric system inside walk-in hood.

degree of containment must also be provided in order to prevent spread of contamination. However, as indicated in an earlier section, calorimetry of fission products is usually not performed unless decay-heat determinations are to be conducted. The complications of operation of calorimeters inside cave areas using remote-handling devices such as manipulators are such that either highly-automated calorimetric systems must be utilized or else the degree of precision required for the calorimeter must be low enough to allow a relatively crude device to be operated in the cave. A brief description of a 'hot cell' should make the problems quite clear. The following discussion is based on the reports of Kelman et al. [19] and Moe et al. [24].

Figure 12.3 shows a hot cell in operation; this is a system designed to handle up to 1000 curies of 1 MeV gamma radiation. Quantities of isotopes up to 10 curies can be handled in a smaller facility, but the use of manipulators is still involved. Typically, the shielding walls of a cave would range from several inches of steel to three or four feet of concrete. Handling of samples or equipment in the facility is done via master-slave manipulators. Viewing within these

Figure 12.3. Cave with zinc bromide windows and master-slave manipulators.

facilities is provided by either windows, periscopes, mirrors, or television. Typically, the hot-cell windows consist of inner and outer walls (~1-in. thick) of lead glass filled with a $ZnBr_2$ solution; thus, the windows themselves are several feet thick in most caves. Ventilation in the caves is typically filtered air, with the cave maintained at negative pressure. Cell ventilation becomes more complex in the event that complete closure is required for alpha-emitting isotope containment. Transfer of materials in and out of the cave is accomplished with pouches, bags, transfer drawers, shielded casks, etc.

It is not surprising that only a few examples of calorimetry inside caves were found in our review of the literature; it is not a congenial environment for the type of careful manipulation of equipment necessary in precision calorimetry.

12.3.3. Safeguards

The utilization of fissile material in scientific experimentation carries along with it a responsibility to safeguard the material from diversion to nonauthorized uses – such as weapons. In the United States detailed Federal regulations are in force to account for and control these substances ('Special Nuclear Materials'). Title 10 of the Code of Federal Regulations, Section 70 (10CFR70) deals at considerable length with these requirements for accountability of special nuclear materials in excess of one kilogram. 10CRF73 details the requirements for physical protection of facilities that have five kilograms or more of special nuclear materials; they are considerable! More specific information can be found in the U.S. Nuclear Regulatory Commission's Regulatory Guides–Division 5, Materials and Plan Protection. Therefore, it should be kept in mind that extended calorimetric investigation in the U.S. involving special nuclear materials can be complicated by the necessities of compliance with Federal regulations.

12.3.4. Instrument Standardization and Radiation Damage

The experimenter who must perform calorimetry utilizing any of the remote operations described above in Section 12.3.2 would be wise to keep as much of his instrumentation as possible outside the enclosure system utilized. Standardization and repair of such instrumentation via glove ports or manipulators can become almost an impossible task. Some instrumentation can be especially susceptable to radiation, and this must be kept in mind when operating instrumentation in high-intensity radiation fields; an example of this is transient-pulse behavior and permanent radiation damage in semiconductor materials exposed to gamma radiation [25]. Cables and wiring can also be adversely affected by high-intensity radiation fields.

Of special importance to the experimenter who measures sample temperatures via optical pyrometry is the possible change in optical properties of glasses as a result of radiation damage [26]. Clearly, the transmission properties of windows and prisms must be carefully followed during experiments at high-radiation levels so that erroneous temperatures are not recorded.

12.3.5. Decontamination of Equipment and Facilities

A well-designed experiment involving radioactivity must include consideration of the decontamination of the equipment and its fate at the end of the study; advanced planning for decontamination can be of great aid in the event of an incident during the course of experimental studies as well. In the U.S., Federal regulations dealing with handling of radioactive materials detail requirements for packing, transportation, and permissible disposal of the wastes from decontamination of such experimental operations. The decontamination procedures themselves are essentially determined by the experimenter. A good discussion of appropriate decontamination techniques can be found in Ayres' Decontamination of Nuclear Reactors and Equipment [27].

The decontamination processes employed to handle the experimental equipment include treatment of surfaces with cleaning solutions (including ultrasonic cleaning), abrasive blasting, jet cleaning, or simply (if the activity level is low enough) placing the equipment in suitable containers and disposing of it in toto. These decontamination processes will result in two types of wastes for disposal: dry and wet. Since calorimetry will normally involve only small amounts of radioactive materials, the wastes generated would, for the most part, be considered as low-level; however, the presence of greater than 10 nanocuries of transuranics per gram of waste would remove the waste from the low-level category [28].

Most dry low-level wastes are eventually disposed of by shallow land burial, although intermediate treatment, such as incineration, may be involved as well. Wet wastes would usually first be reduced in volume (when possible) and then immobilized in cement or some other solidifying medium, with ultimate disposal via burial. However, in the U.S. those wastes containing greater than 10 nanocuries of transuranics (TRU) per gram of waste have to be separated from the non-TRU waste and stored in above-ground DOE facilities for eventual disposal in a Federal repository, although such a repository does not yet exist.

Therefore, in some instances, one may find that decontamination and clean-up of a facility used for transuranic calorimetry may constitute the major problem of the study.

12.4. REFERENCES

1. International Atomic Energy Agency, Symposium on the Oklo Phenomenon, Ribreville, Gabon, 1975, Vienna, 1975.

2. Katz, J.J. and Rabinowitch, E., The Chemistry of Uranium, National Nuclear Energy Series, Manhattan Project Technical Section, Division VIII - Vol. 5, McGraw-Hill Book Co., New York, NY, 1951.

3. Quill, L.L. (Editor), The Chemistry and Metallurgy of Miscellaneous Materials, Thermodynamics, National Nuclear Energy Series, Manhattan Project Technical Section, Division IV - Plutonium Project Record, Vol. 19B, McGraw-Hill Book Co., New York, NY, 1950.

4. Walker, F.W., Kirouac, G.J., and Rourke, F.J., Chart of the Nuclides, 12th Ed., Knolls Atomic Power Laboratory, 1977.

5. Clark, H.K., Handbook of Nuclear Safety, DP-532, E.I. DuPont and Co., Jan. 1961.

6. Appleton, G.J. and Dunster, H.J., Recommended Practice in the Safe Handling of Plutonium in Laboratories and Plants, AHSB(RP) R6, United Kingdom Atomic Energy Authority, 1961.

7. Way, K. and Wigner, E.P., Phys. Rev., 73, 1318, 1948.

8. Day, R.A. and Cannon, C.V., Paper 41, Direct Calorimetric Study of Fission Product Decay in Active Slugs (Coryell, C.D. and Sugarman, N., Editors), National Nuclear Energy Series, Manhattan Project Technical Section, Division IV - Plutonium Project Record, Vol. 9, McGraw-Hill Book Co., New York, NY, 1951.

9. Johnston, K., J. Nucl. Energy Parts A1B, 19, 527, 1965.

10. Trainor, R.J., Knapp, G.S., Brodsky, M.B., Pokorny, G.J., and Snyder, R.B., Rev. Sci. Instrum., 46, 1368, 1975.

11. Beamer, W.H. and Easton, W.E., J. Chem. Phys., 17, 1298, 1949.

12. Ginnings, D.C., Ball, A.F., and Vier, D.T., J. Res. Natl. Bur. Stand., 50, 75, 1953.

13. Seabaugh, P.W. and Jordan, K.C., Int. J. Appl. Radiat. Isotopes, 29, 489, 1978.

14. Gunn, S.R., Int. J. Appl. Radiat. Isotopes, 29, 497, 1978.

15. Curie, P. and Laborde, A., C.R. Hebd. Seances Acad. Sci., 136, 673, 1903.

16. Rutherford, E. and Barnes, H.T., Nature, 68, 622, 1903.

17. American National Standard Calibration Techniques for the Calorimetric Assay of Plutonium Bearing Solids Applied to Nuclear Materials Control, ANSI N15.22-1975.

18. Steindler, M.J., Comments on the Handling of Plutonium, ANL-6021, Argonne National Lab., June, 1959.

19. Kelman, L.R., Wilkinson, W.D., Shuck, A.B., and Goertz, R.C., The Safe Handling of Radioactive-Pyrophoric Materials, ANL-5509, Argonne National Lab., December, 1955.

20. Metz, C.F., Analytical Chemical Laboratories for Handling Plutonium, Paper A/Conf. 15/P/533, Second United Nations International Conference on the Peaceful Uses of Atomic Energy, June, 1958.

21. Engel, T.K., Jordan, K.C., Otto, G.W., and Scott, D.M., Rev. Sci. Instrum., 35, 875, 1964.

22. Johnson, G.K., Van Deventer, E.H., Kruger, O.L., and Hubbard, W.N., J. Chem. Thermodyn., 1, 89, 1969.

23. Fischer, D.F., Leibowitz, L., and Chasanov, M.G., A High-Temperature Calorimeter System for Use with Materials Containing Plutonium, ANL-7896, Argonne National Lab., December, 1971.

24. Moe, H.J., Lasuk, S.R., and Schumacher, M.C., Radiation Safety Technician Training Course, ANL-7291, Argonne National Lab., September, 1966.

25. Glasstone, S. and Dolan, P.J., The Effects of Nuclear Weapons, 3rd Ed., U.S. Department of Defense and U.S. Department of Energy, 1977.

26. Kircher, J.F. and Bowman, R.E., Effects of Radiation on Materials and Components, Reinhold Publishing Corp., New York, NY, 1964.

27. Ayres, J.A. (Editor), Decontamination of Nuclear Reactors and Equipment, The Ronald Press Co., New York, NY, 1970.

28. Kibbey, A.H. and Godbee, H.W., A State-of-the-Art Report on Low-Level Radioactive Waste Treatment, ORNL/TM-7427, Oak Ridge National Lab., September, 1980.

12.5. APPENDIX: A BIBLIOGRAPHY OF CALORIMETRY USING RADIOACTIVE MATERIALS

In this appendix are listed some calorimetric studies of radioactive materials; this is not an exhaustive bibliography. The references cited represent fairly recent studies culled from several computerized information data bases. It is expected that answers to many specific technical questions that have arisen as the reader examined the preceding sections will be found in these papers.

A. Fission Products

1. Osborne, D.W., Schreiner, F., Otto, K., Malm, J.G., and Selig, H., 'Heat Capacity, Entropy, and Gibbs Energy of Technetium Hexafluoride Between 2.23 and 350 K; Magnetic Anomaly at 3.12 K; Mean Beta Energy of Technetium-99,' J. Chem. Phys., 68, 1108-18, 1978.

2. Rau, F. and Wenzl, H., 'Measurement of Stored Energy Release and Specific Heat of Neutron-Irradiated Metals with a Differential Heat Flux Microcalorimeter,' Z. Angew. Phys., 27, 346-54, 1969.

3. Stewart, G.R. and Giorgi, A.L., 'Specific Heat of A-15 and bcc Molybdenum-Technetium
    (Mo0.4Tc0.6),' Phys. Rev. B, 17, 3534-50, 1978.

4. Stewart, G.R. and Giorgi, A.L., 'Specific Heat of bcc Molybdenum-Technetium (Moi-xTcx),'
    Phys. Rev. B, 19, 5704-10, 1979.

B. Thorium and Uranium and Their Alloys and Compounds

1. Alles, A., Falk, B.G., Westrum, E.F., Jr., Grønvold, F., and Zaki, M.R., 'Actinoid Pnic-
    tides. I. Heat Capacities from 5 to 950 K and Magnetic Transitions of Triuranium Tetra-
    arasenide (U3As4) and Triuranium Tetraantimonide (U3Sb4). Ferromagnetic Transitions,' J.
    Inorg. Nucl. Chem., 39, 1993-2000, 1977.

2. Armbruester, H., Franz, W., Schlabitz, W., and Steglich, F., 'Transport Properties, Sus-
    ceptibility and Specific Heat of Uranium-Aluminum (UA12),' J. Phys. Colloq., 4, 150-1,
    1979.

3. Balankin, S.A., Skorov, D.M., and Yartsev, V.A., 'Method of Determining the Thermophysi-
    cal Properties of Reactor Materials at Elevated Temperatures,' At. Energy, Ser. 41, Issue
    No. 4, 271-3, 1976.

4. Bastide, J.P., Loriers, C., Massat, H., and Vodar, B., 'Measurement of the Specific Heat
    of Metals Under High Pressure by an Impulsion Method. Application to Cerium and Uranium
    Between 0 and 100 kbar at 293 Degree K,' 4th Conf. Int. Thermodyn. Chim. (C.R), Ser. 9,
    90-7, 1975.

5. Blaise, A., Lagnier, R., Mulak, J., and Zolnierek, Z., 'Magnetic Susceptibility and Heat
    Capacity Anomalies of Uranium Hydroxide Sulfate (U(OH)2SO4) at 21 K,' J. Phys., Coloq.
    (Orsay, Fr.), No. 4, 176-8, 1979.

6. Blaise, A., Troc, R., Lagnier, R., and Mortimer, M.J., 'The Heat Capacity of Uranium
    Monoarsenide,' J. Low Temp. Phys., 38, 79-82, 1980.

7. Blaise, A., Lagnier, R., Troc, R., Henkie, Z., Markowski, P.J., and Mortimer, M.J., 'The
    Heat Capacities of Uranium (IV) Arsenide and Thorium Arsenide,' J. Low Temp. Phys., 39,
    315-28, 1980.

8. Clifton, P.G., 'Adiabatic Calorimetric Enthalpy Measurements for Alpha-, Beta-, and Gam-
    ma-Uranium and the Heats of the Transformations Alpha..far..Beta and Beta..far..Gamma,'
    LA-4521, 1970.

9. Elenbaas, R.A., Schinkel, C.J., and Swakman, E., 'Magnetic Susceptibilities and Specific
    Heats of Cerium-Thorium Alloys,' J. Phys. F, 9, 1261-9, 1979.

10. Fredrickson, D.R. and Chasanov, M.G., 'Enthalpy of Uranium Dioxide and Sapphire to 1500
    deg.K by Drop Calorimetry,' J. Chem. Thermodyn., 2, 623-9, 1970.

11. Flotow, H.E., Osborne, D.W., Lyon, W.G., Grandjean, F., Fredrickson, D.R., and Hastings,
    I.J., 'Heat Capacity and Thermodynamic Properties of Triuranium Silicide from 1 to
    1203 K,' J. Chem. Thermodyn., 9, 473-81, 1977.

12. Flotow, H.E. and Osborne, D.W., 'Heat Capacities and Thermodynamic Properties of Thorium
    Hydrides (ThH2 and ThH3.75) to 1000 K,' J. Chem. Thermodyn., 10, 537-51, 1978.

13. Hall, R.O.A. and Mortimer, M.J., 'Low-Temperature Specific Heat Anomalies in Polycrystal-
    line Alpha-Uranium,' J. Low Temp. Phys., 27, 313-16, 1977.

14. Hall, R.O.A., 'The Low-Temperature Specific Heat of Psuedo-Single Crystal and Polycrys-
    talline Alpha-Uranium,' Conf. Ser.-Inst. Phys., 37 (Rare Earths Actinide, 1977), 60-5,
    1978.

15. Harding, J.H., Masri, P., and Stoneham, A.M., 'Thermodynamic Properties of Uranium Diox-
    ide: Electronic Contributions to the Specific Heat,' J. Nucl. Mater., 92, 73-8, 1980.

16. Hein, R.A. and Flagella, P.N., 'Enthalpy Measurements of Uranium Dioxide and Tungsten to 3260 deg.K,' GEMP-578, 1968.

17. Inaba, H., Ono, S., and Naito, K., 'A Phase Transition in Doped Uranium Oxide (U4O9-y). I. Heat Capacity Measurement,' J. Nucl. Mater., 64, 189-94, 1977.

18. Lagnier, R., Wojakowski, A., Suski, W., Janus, B., and Mortimer, M.J., 'The Low Temperature Specific Heat of Uranium Susquisulfide,' Phys. Status Solidi A, 57, K127-K132, 1980.

19. Lyon, W.G., Osborne, D.W., Flotow, H.E., and Hoekstra, H.R., 'Sodium Uranium (V) Trioxide, NaUO3: Heat Capacity and Thermodynamic Properties from 5 to 350 K,' J. Chem. Thermodyn., 9, 201-10, 1977.

20. McLellan, R.B., 'High-Temperature Thermodynamic Properties of the Uranium Carbide UC1.fwdarw.2-Phase,' J. Phys. Chem. Solids, 40, 353-4, 1979.

21. Murabayashi, M., Takahashi, Y., and Mukaibo, T., 'High-Temperature Heat Capacity of Uranium Monosulfide by the Laser Flash Method,' J. Nucl. Mater., 40. 353-4, 1971.

22. Nakamura, J., Takahashi, Y., Izumi, S., and Kanno, M., 'Heat Capacity of Metallic Uranium and Thorium from 80 to 1000 K,' J. Nucl. Mater., 88, 64-72, 1980.

23. Satterthwaite, C.B. and Miller, J.F., 'Specific Heat and Superconductivity of Thorium Hydride (Th4H15) and Thorium Deuteride (Th4D15),' Proc. 14th Int. Conf. Low Temp. Phys., 2, 101-4, 1975.

24. Smith, T.F. and Phillips, N.E., 'Low-Temperature Heat Capacity Measurements at High Pressure,' Colloq. Int. Cent. Nat. Rech. Sci., No. 188, 191-5, 1970.

25. Stephans, H.P., 'Determination of the Enthalpy of Liquid Copper and Uranium with a Liquid Argon Calorimeter,' High Temp. Sci., 6, 156-66, 1974.

26. Stephens, H.P., 'Determination of the Enthalpy and Specific Heat of a (Uranium, Zirconium) Carbide-Graphite Reactor Fuel Material to 2000 K,' Proc. Symp. Thermophys. Prop., 7, 351-7, 1977.

27. Sterritt, D.E., Lalos, G.T., and Schneider, R.T., 'Specific Heat Ratio of Uranium Hexafluoride Measured with a Ballistic Piston Compressor,' Nucl. Technol., 25, 150-65, 1975.

28. Stewart, G.R., Giorgi, A.L., and Krupka, M.C., 'The Specific Heat of Yttrium Thorium Carbide (Y0.7TH0.3C1.58),' Solid State Commun., 27, 413-16, 1978.

29. Takahashi, Y., 'Heat Capacity Measurements of Nuclear Materials by Laser Flash Method,' J. Nucl. Mater., 51, 17-23, 1974.

30. Takeshita, T., Gschneidner, K.A., Jr., Thome, D.K., and McMaster, O.D., 'Low-Temperature Heat-Capacity Study of Haucke Compounds Calcium Nickel, Yttrium Nickel, Lanthanum Nickel, and Thorium Nickel (CaNi5, LaNi5, and ThNi5),' Phys. Rev. B, 21, 5636-41, 1980.

31. Yokokawa, H. and Takahashi, Y., 'Calorimetric Study on the Electronic Structure in Uranium Monophosphide-Uranium Monosulfide System,' J. Phys. Chem. Solids, 40, 603-12, 1979.

C. Transuranium Elements and Their Alloys and Compounds

    i. Neptunium

       1. Arkhipov, V.A., Gutina, E.A., Dobretsov, V.N., and Ustinov, V.A., 'Enthalpy and Specific Heat of Neptunium Dioxide in the 350-1100 deg. K Range,' Radiokyimiya, 16, 123-6, 1974.

    ii. Plutonium

       1. Affortit, C., 'Contribution to the Study of the Specific Heat of Actinide Compounds,' CEA-R-4266, 1972.

2. Affortit, C. and Boivineau, J.C., 'Application of Pulse Heating to the Measurement of Specific Heats of Refractory Compounds. Application to Uranium and Plutonium Oxides,' Bull. Inform. Sci. Techn. Commis. Energ. At., No. 180, 51-4, 1973.

3. Danan, J., 'Study of the Electronic and Magnetic Properties of Some Actinide Compounds by Specific Heat Measurements,' CEA-R-4453, 1973.

4. Danan, J., Conte, R.R., and DeNovion, C.H., 'Apparatus for Specific Heats Measurements Between 1.5 and 310 deg.K,' Bull. Inform. Sci. Tech. Commis. Energ. At., No. 180, 29-32, 1973.

5. Fischer, D.F. and Leibowitz, L., 'Enthalpy of Uranium-Plutonium Carbide from 298 K to the Melting Point,' J. Nucl. Mater., $\underline{67}$, 244-8, 1977.

6. Fukushima, S., Tokai, I., Abe, J., and Kurihara, M., 'Construction and Performance Test of the Apparatus for Measuring Thermal Properties of Uranium-Plutonium Mixed Carbide,' JAERI-M-8299, 1979.

7. Haines, H.R., Hall, R.O.A., and Lee, J.A., 'The Specific Heat of Plutonium Carbide (PuC1-x, Pu2C3 and Pu3C2) Between 10-300 K,' Proc. 2nd, Int. Conf. Electron. Struct. Actinides, 349-55, 1977.

8. Haines, H.R., Hall, R.O.A., Lee, J.A., Mortimer, M.J., and McElroy, D., 'The Low Temperature Specific Heats of Three Plutonium Carbides,' J. Nucl. Mater., $\underline{88}$, 261-4, 1980.

9. Hall, R.O.A., Lee, J.A., Martin, D.J., Mortimer, M.J., and Sutcliffe, P.W., 'Heat Capacity of Plutonium Nitride at Low Temperatures,' J. Chem. Thermodyn., $\underline{10}$, 935-40, 1978.

10. Leibowitz, L., Fischer, D.F., and Chasanov, M.G., 'Enthalpy of Uranium-Plutonium Oxides: $(U_{.8}Pu_{.2})O_{1.97}$ from 2350 to 3000 K,' J. Nucl. Mater., $\underline{42}$, 113-6, 1972.

11. Oetting, F.L., 'The Chemical Thermodynamics of Nuclear Materials. IV. The High-Temperature Enthalpies of Plutonium Monocarbide and Plutonium Sesquicarbide,' J. Nucl. Mater., $\underline{88}$, 265-72, 1980.

12. Osborne, D.W. and Flotow, H.E., 'Half-Life of Plutonium-242 from Precise Low-Temperature Heat Capacity Measurements,' Phys. Rev. C, $\underline{14}$, 1174-8, 1976.

13. Rose, R.L., Robbins, J.L., and Massalski, T.B., 'Enthalpy and Specific Heat of a Series of Plutonium-Gallium Alloys at Elevated Temperatures,' J. Nucl. Mater., $\underline{75}$, 98-104, 1978.

14. Sandenaw, T.A., 'Evidence for Order-Disorder in Plutonium Carbide (PuC0.80) at Low Temperatures Through Specific-Heat Measurements,' J. Nucl. Mater., $\underline{57}$, 145-50, 1975.

15. Thayer, W.L. and Robbins, J.L., 'Enthalpy and Heat Capacity of Liquid Plutonium and Uranium,' UCRL-76537, 1975.

iii. Americium

1. Meuller, W., Schenkel, R., Schmidt, H.E., Spirlet, J.C., McElroy, D.L., Hall, R.O.A., and Mortimer, M.J., 'The Electrical Resistivity and Specific Heat of Americium Metal,' J. Low Temp. Phys., $\underline{30}$, 561-78, 1978.

2. Smith, J.L., Stewart, G.R., Huang, C.Y., and Haire, R.G., 'Superconducting Critical Field and Low Temperature Heat Capacity of Americium,' J. Phys. Colloq. (Orsay, Fr.), No. 4, 138-9, 1979.

APPENDIX A

Materials of Construction
in High-Temperature Calorimetry

C. R. BROOKS and E. E. STANSBURY

1. INTRODUCTION

There are a number of factors which must be considered in choosing materials for the components of a high-temperature calorimeter system. In this section, we examine in some detail those choices which are associated with specific properties or construction techniques. Attention is given to the temperature dependence of relative properties and to comparing these properties for different types of materials. Representative values are presented, which are generally adequate for design purposes. If specific or more accurate data are required, detailed literature searches or actual measurements may be justified, as in many cases the values depend upon the purity and the processing methods employed in manufacturing the materials. Some specific sources of data are listed in Refs. [1-5].

We first list three general classifications of properties and then examine specific properties in detail in separate sections which follow.

(1) Thermal, chemical, and mechanical stability. The absolute upper temperature limit of use of a material is determined by the onset of its melting, but the acceptable limit is a considerably lower temperature. Limits are imposed by decomposition upon heating, excessive reaction with the environment, and loss of mechanical properties. Consideration must be given also to vaporization, which affects not only loss of the material but contamination of other components. At contacting surfaces of dissimilar materials interdiffusion may result in low melting phases, or a phase may form which is brittle, causing mechanical problems. Similar problems relate to components which must be disassembled later. These must be made of materials which will not bond at their contacting surfaces. The material must have sufficient high-temperature strength, including creep strength, to support any mass attached to it, as well as its own mass.

(2) Heat transfer properties. Thermal conductivity and surface emittance as a function of temperature are important physical properties governing heat transfer and, hence, are of prime importance in determining how well adiabatic conditions can be approached.

(3) Fabricability. Many calorimeter components are intricate in design and must be fabricated to exact dimensions. Most metallic materials can be machined, although some of the refractory metals (e.g., molybdenum) may require special consideration. Also, most metallic materials can be purchased in a variety of shapes (e.g., tubes, rods, etc.) which minimizes machining. Most ceramics can only be shaped by grinding, but they also can be purchased in a variety of shapes. Experience has shown that, in general, fabrication is not a significant limitation in calorimeter design and construction.

2. MECHANICAL PROPERTIES

In designing and constructing calorimeters, the mechanical stresses on the components must be assessed in order to choose materials of sufficient strength, and in order to size properly the components.

The mechanical stresses on the calorimeter components arise from two sources. One is due to gravitational forces on the components (i.e., the components have to support their own weight) and the other to gravitational forces on parts of the calorimeter that these components are supporting (e.g., the wires which support the calorimeter cell). The other source of stress is thermal expansion of components under conditions where the component is not free to expand or contract, but instead is restricted by contact with other components having a different expansion coefficient. An example is metal wire inside ceramic tubing, where the lower thermal expansion of the ceramic upon heating restricts the expansion of the wire, which has a higher expansion coefficient.

The response of materials to loads can be quite complicated and difficult to predict accurately. However, examination of some simple mechanical properties gives quite useful guidelines for the choice of materials and for sizing the components. For the support of a uniaxial tensile load, the stress at the temperature of interest which will initiate plastic deformation (the yield strength) is the best mechanical property to use. This property is determined from a uniaxial tensile test at this temperature which takes a relatively short time (e.g., five minutes) to perform. However, a property which is easier to obtain from the test is the tensile strength (sometimes called the ultimate strength, or the ultimate tensile strength), and is defined as the maximum load observed during the tensile test divided by the original cross-sectional area. Data of tensile strength are frequently easier to locate in the literature than the yield strength. The tensile strength, although higher than the yield strength (by about 20% typically) can be used for initial materials selection.

Figure A1 gives the approximate tensile strengths as a function of temperature for several commercially-pure metals. Since the actual values depend upon a number of metallurgical factors (e.g., grain size) relating to specific lots of material, these data should be taken only as representative. The strength decreases rapidly with increasing temperature; above 1300 K the high-strength metals are restricted essentially to the refractory metals, Mo, W, Ta, and Re. Generally, all pure metals can be strengthened by alloying and by controlling the heat treatment of the alloys. Figure A2 shows tensile strength-temperature curves for a few alloys. Alloys having good strength to about 1100 K are the austenitic stainless steels (approximately Fe-18% Cr-8% Ni), nickel-base alloys like Inconel (typically Ni-14% Cr-6% Fe) and the Ni-rich, Ni-Cr alloys (the Nimonic alloys). However, the strength of all of the alloys in Fig. A2 decreases drastically as 1300 K is approached, so that above this temperature the refractory metals (or their alloys) are the strongest.

Ceramics are generally stronger than metallic materials providing the loading is compressive. Figure A3 shows the compressive and tensile strength of several ceramics. Although the tensile strength is only about 1/10 of the compressive strength, it is in the same range as the metals and

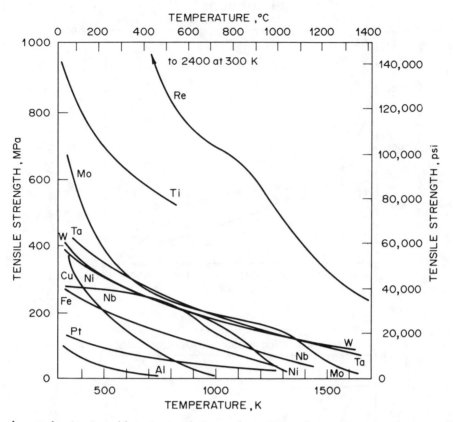

Figure A1. Approximate tensile strength as a function of temperature for several metals.

alloys. But, generally ceramics are not plastic even at high temperature, and, as a consequence, overloading results in fracture with little or no plastic deformation. Thus, their use in simple tensile loading must be considered cautiously, but in compressive loading they are quite superior to metallic materials.

To illustrate the utilization of these strength data, consider a case where it is desired to have a calorimeter component (e.g., a cell) suspended by a wire of as small a diameter as possible to minimize heat exchange between the components. Let the component weigh 150 g and the maximum temperature be 1300 K. Using the tensile strength data in Figs. A1 and A2, the minimum required diameter of the wire has the values in Table A1. The diameters shown are of sufficient size to be available commercially, and are of sufficient diameter to be manageable during installation.

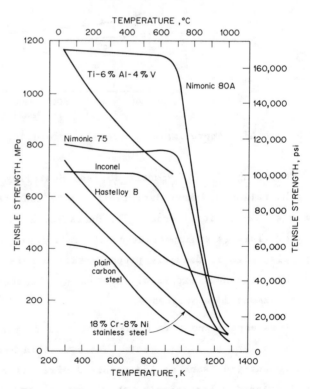

Figure A2. Approximate tensile strength as a function of temperature for several alloys.

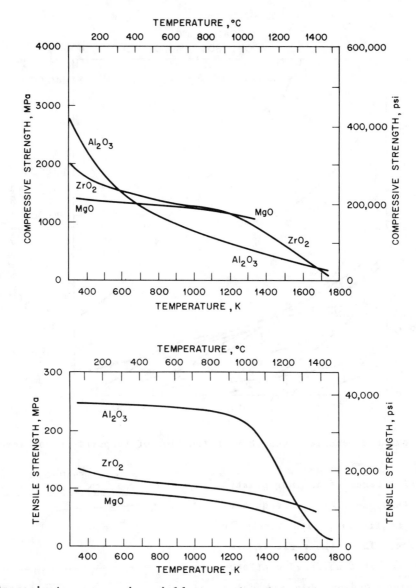

Figure A3. Approximate compressive yield strength and tensile strength of three ceramics.

Comparing Ni, Mo, Nimonic 80, and Hastelloy B, we see that the smallest wire is attained using Hastelloy B. More importantly, note that this material will allow a considerable reduction in the cross-sectional area. The Hastelloy B wire has an area 10 times smaller than that of the Ni wire. This is an important consideration as the rate of heat exchange by conduction along this supporting wire is directly proportional to this area, and, hence, using Hastelloy B allows considerably less heat exchange. Of course, the thermal conductivities must be considered also, and this is examined in Section 4.

It is emphasized that these mechanical properties are very sensitive to chemical composition, grain size, and heat treatment. The curves in Figs. A1-A3 are only approximate. Thus, in calculating quantities such as the required wire diameter to support a known load, a 'safety factor' should be used. For example, the diameter of the wire may have to be tripled from the calculated value.

Table A1. The Diameter of a Wire Required to Support a Load of 150 g at 1300 K, Based on the Tensile Strength Data in Figures A1 and A2

Material	Tensile Strength at 1300 K (MPa; psi)	Wire Diameter (cm)	Cross-Sectional Area (cm^2)
Al	not usable at 1300 K		
Cu	not usable at 1300 K		
Pt	20; 3,000	0.030	7.1 x 10^{-4}
Fe	28; 4,000	0.026	5.3 x 10^{-4}
Ni	28; 4,000	0.026	5.3 x 10^{-4}
Nb	60; 9,000	0.017	2.2 x 10^{-4}
W	130; 19,000	0.012	1.1 x 10^{-4}
Ta	140; 20,000	0.012	1.1 x 10^{-4}
Mo	160; 23,000	0.011	9.5 x 10^{-5}
Re	520; 75,000	0.006	2.8 x 10^{-5}
Inconel	40; 6,000	0.020	3.1 x 10^{-4}
Fe–18% Cr–8% Ni	55; 8,000	0.018	2.5 x 10^{-4}
Nimonic 75	70; 10,000	0.017	2.2 x 10^{-4}
Nimonic 80	100; 15,000	0.013	1.3 x 10^{-4}
Hastelloy B	280; 40,000	0.008	5.0 x 10^{-5}

Strength data obtained by a short-time (e.g., five minutes) test may not represent the response of the material to load for longer times. For example, if the yield strength of a material is 140 MPa, and the tensile load to be imposed on a component of this material is 70 MPa, then the component is supposedly loaded elastically, and the short-time yield strength indicates that the component should safely support the load. However, at temperatures of approximately 60% the absolute melting temperature, plastic deformation will be observed, the length will increase with time, and eventually fracture will occur. This plastic deformation under a stress less than the short-time yield strength is called creep.

Figure A4 shows typical creep curves [6]. The process is usually characterized by the time to fracture the test specimen, called the rupture life, and by the slope of the creep curve in the linear stage (second stage), called the creep rate. This linear region usually accounts for most of the change in dimensions and, therefore, is more important. Representative creep properties of several metals and alloys are given in Table A2.

Comparing Figs. A1 and A2 with the data in Table A2 shows that those metals and alloys with higher tensile strength also have better creep properties. The creep process is closely related to self diffusion, so the process is accelerated exponentially with increasing temperature. Also, the higher the stress, the higher the creep rate and the lower the rupture life. As a general rule, creep properties govern mechanical performance at temperatures above approximately two-thirds of the absolute melting temperature. Thus, in the higher temperature ranges of adiabatic calorimetry ceramics have considerably superior creep strength over the metallic materials.

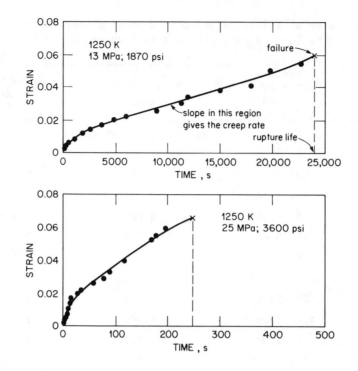

Figure A4. Creep curves for platinum at 1250 K under two different stresses (adapted from [6]).

Table A2. Approximate Creep Rate and Stress Rupture Data for Some Metals and Alloys

Material	Temperature (K)	Stress (MPa; psi) for 100 Hour Rupture Life	Stress (MPa; psi) for Given Creep Rate (%·hr^{-1})
Mo	1260	150; 22,000	----
Mo-25% W	1260	500; 73,000	----
W	1300	165; 24,000	----
Ta	1300	70; 10,000	----
Inconel 718	980	415; 60,000	----
Inconel 800	1260	170; 25,000	12; 1,800 for 0.1
Ni	670	----	70; 10,000 for 10^{-4}
Fe-18% Cr-8% Ni	1260	18; 2,650	8; 1,200 for 0.1

To illustrate a calorimeter design problem involving creep, consider a wire supporting 150 g on a diameter of 0.048 cm; the stress is 8.2 MPa (1200 psi). Let the temperature be 1260 K and the material 316 stainless steel. At 1260 K, 316 stainless steel has typically a rupture life of 1000 hours for a stress of 18 MPa (2650 psi), and for a stress of 8.2 MPa (1200 psi), the creep rate is 0.01%·hr^{-1}. Thus, under the load being considered (8.2 MPa), the wire will not break in 100 hours. The elongation of the wire in 100 hours is (0.01)(100) = 1%. If the wire is original-ly 2.5 cm long, and the design of the calorimeter is such that a shift of the component being sup-ported by the wire of more than 0.0025 cm is undesirable, then the wire should not be used for more than 100 hours while supporting this 150 g load.

Creep properties are quite sensitive to metallurgical variables, so the data in Table A2 are representative only. In using creep data, the uncertainty in the reliability of the data must be taken into account by use of a safety factor. Stress-rupture data are more common than creep rate data, as they are easier to obtain experimentally. These data allow an estimation of the life in creep of the component, but as it is based on rupture life, this is an optimistic estimate. If dimensional changes need to be estimated, then creep rate data are required, and they are considerably more scarce than stress-rupture data.

Closely related to the creep phenomenon is that of stress relaxation. Consider two dissimilar materials in intimate and rigid contact being heated. Due to their different coefficients of thermal expansion, they will generate stresses in each other, which may be sufficient to induce plastic deformation or even fracture in one or both of them. If the stresses generated are elastic, then, once high temperature is reached, stress relaxation occurs as the material undergoes creep. Thus, eventually the internal stress is relaxed to an insignificant value. However, then upon cooling, the uneven contraction generates the stresses again, and now because the temperature is decreasing, stress relaxation becomes less likely, and the stresses can lead to excessive plastic deformation or fracture.

An example where this may be a problem is in attempting to maintain a tight surface contact between two components by using, say, a bolt. At low temperature the bolt is tightened to elastically stress it and the components and provide a tight interface. To maintain this tightness upon heating, the bolt material is chosen with a coefficient of expansion less than that of the two components which it is holding together. Thus, upon heating, the stress in the bolt and components should increase with temperature and the joint become even tighter. However, at high temperature, if the components or the bolt undergo creep, the stress relaxes. Then upon cooling, as the components contract more than the bolt, the joint may lose its tightness.

Stress relaxation data, except for some high-temperature bolting alloys, are very scarce. Since the same structural processes control stress relaxation and creep, a guideline is to use materials only below about two-thirds of their absolute melting temperature.

3. THERMAL EXPANSION

The linear thermal expansion coefficients for several metals, alloys, and ceramics are shown in Fig. A5. Vitreous quartz has the lowest coefficient. The Fe-36% Ni alloys (e.g., Invar) have a low coefficient to about 500 K, but their value above this temperature is typical of that of nonrefractory metals. The refractory metals have coefficients below those of common ceramics. A new alloy, Inconel 907 [7], has a coefficient that is similar to that of the refractory metals up to about 700 K, but increases somewhat above this temperature. Lava, which has a low coefficient, is an important machineable ceramic used in calorimeter systems (see Section 8).

Differences in coefficients generate thermal stresses between materials which are bonded or otherwise have restricted movement. For example, upon heating from 300 to 1300 K a platinum wire 0.032 cm in diameter which is in a 0.032 cm diameter hole in Al_2O_3, a compressive stress is generated by the differential expansion which will greatly exceed the yield strength of the platinum. At 1300 K, the considerably higher strength of the Al_2O_3 prevents its plastic deformation, resulting in plastic deformation of the platinum. In this example, the materials have relatively

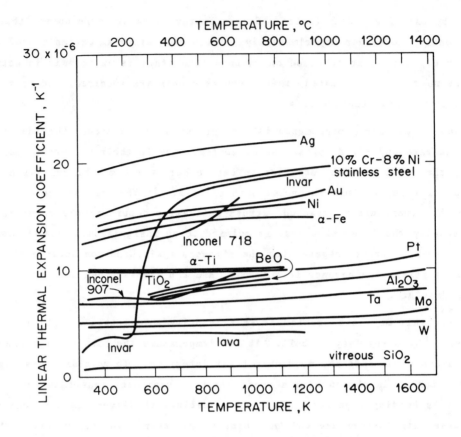

Figure A5. Approximate linear thermal expansion coefficients for several materials.

similar coefficients (Fig. A5). For heater wires such as Ni-Cr alloys, the thermal stresses will
be greater. Thus, plastic deformation, including creep of wires and other metallic components
confined by ceramics, is inevitable. This can lead to the fracture of metallic components that is
observed to occur after repeated heating (i.e., cycling) from 300 K to high temperature.

4. HEAT TRANSFER PROPERTIES

In the design of adiabatic shields and cells, heat transfer is an important consideration.
Since the calorimeter systems are usually evacuated, discussion is restricted to heat transfer by
solid conduction and by radiation, with emphasis on the importance of thermal conductivity and
emittance.

Thermal diffusivity (usually given as $cm^2 \cdot s^{-1}$) is the thermal conductivity λ divided by the
volumetric specific heat (e.g., $J \cdot cm^{-3} \cdot K^{-1}$). This specific heat has about the same value for most
construction materials of interest (i.e., it is about 0.2-0.3), and, thus, the thermal diffusivity
and thermal conductivity are approximately directly proportional to each other. Since at moderate
temperatures, thermal conductivity is generally more easily measured than diffusivity, and, since
more data are available for conductivity, material comparisons as they relate to heat transfer are
frequently made in terms of the conductivity. In general, high thermal conductivity is a major
factor minimizing temperature gradients of the surface of shields and on the surface of the sample

and cell.   In contrast, for the necessary connecting wires from the sample or cell to the surroundings, materials of low conductivity are desired to minimize heat leakage.

The thermal conductivity for several materials is shown in Fig. A6.   The superiority of Ag,
Cu, and Au for high conductivity requirements is evident.   Both Cu and Ag can be used to about
1100 K in vacuum, but above this temperature caution must be exercised because of vaporization
(see Section 7).   We have used Cu and Au to 1300 K.   Gold has the additional advantage of its
inertness, but its cost precludes its wide use in components of significant size.

The ceramic materials have conductivities an order of magnitude or more lower than the metallic materials.   However, for high-temperature calorimetry the exact choice of ceramics lies not in
their thermal conductivity, but in their electrical resistance, which is examined in Section 5.

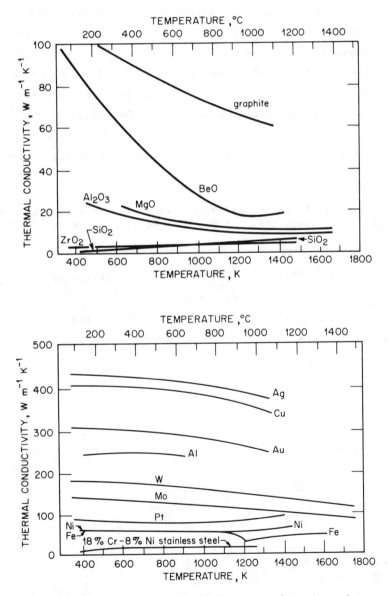

Figure A6.   Approximate thermal conductivity as a function of temperature for
            several metallic and ceramic materials.

Table A3 compares the conductivity and diffusivity of six materials and gives the ratio of the conductivity and diffusivity relative to Ag. These ratios emphasize the range of relative values and the close correspondence between thermal conductivity and thermal diffusivity. Note that Au and Cu have about the same diffusivity, although Cu has a higher conductivity.

Radiation heat exchange is related to the fourth power of the temperature of each surface, the surface radiation properties, and the configuration of the surfaces. If a small body at temperature T_1 is surrounded by a larger body of area A_2 at temperature T_2, then the rate of heat exchange $\dot{Q}$ is given by

$$\dot{Q} = \varepsilon_1 A_2 \sigma (T_1^4 - T_2^4) \tag{1}$$

where ε_1 is the emittance (see below) of the small body and σ is the Stefan–Boltzmann constant ($5.67032 \times 10^{-8} \ W \cdot m^{-2} \cdot K^{-4}$). The surfaces intrinsically radiate energy, the rate depending on the surface condition (e.g., roughness of the surface) which influences the value of the emittance. The governing material property is the emissivity if the surface is perfectly smooth; otherwise, it is referred to as the emittance, which is considered here. The radiation has a spectrum of wavelengths, such that the term total emittance is used to encompass all wavelengths. Some approximate values of emittance are tabulated in Table A4. The sample or cell should have a low emittance and, therefore, a polished surface is usually desirable. However, with any material having some initially prepared surface, it must be kept in mind that the surface microtopology may

Table A3. Approximate Values of the Thermal Diffusivity for Six Materials. Note that Cu and Au have Similar Diffusivities but Somewhat Different Conductivities.

Material	Temperature (K)	λ Thermal Conductivity ($W \cdot m^{-1} \cdot K^{-1}$)	c_p Specific Heat ($J \cdot mol^{-1} \cdot K^{-1}$)	α Thermal Diffusivity ($cm^2 \cdot s^{-1}$)
Al_2O_3	1300	10	130	0.02
Fe–18% Cr–8% Ni Stainless Steel	1300	30	37	0.06
W	1300	150	28	0.5
Au	1300	250	32	0.8
Cu	1300	350	32	0.8
Ag	1200	380	32	1.2

Material	α/α_{Ag}	λ/λ_{Ag}
Al_2O_3	0.02	0.03
18–8 Stainless Steel	0.05	0.08
W	0.4	0.4
Au	0.7	0.7
Cu	0.7	0.9
Ag	1	1

Table A4. Approximate Emittance of Several Materials. Although a Temperature is Listed
for the Emittance, Generally the Temperature Variation is Within a Factor of
Two from 300 to 1300 K.

Material	Temperature (K)	Emittance
Al (polished)	700	0.05
Al (oxidized)	700	0.1
Cu (polished)	350	0.02
Cu (oxidized)	700	0.6
Au (polished)	700	0.03
Fe (polished)	500	0.05
Fe or steel (oxidized)	1300	0.09
Ni (polished)	700	0.07
Ni (oxidized)	700	0.4
Fe-18% Cr-8% Ni stainless steel	700	0.4
Pt (polished)	700	0.1
Ag (polished)	700	0.02
W	1300	0.1
Mo	1300	0.1
Ta	1700	0.2
Graphite	1300	0.5
Quartz	1000	0.9
Ceramics	1300	0.8

change irreversibly upon heating to high temperature (e.g., facetting due to vaporization may
occur) or that variable amounts of oxidation may occur, both of which will significantly increase
the emittance. In this respect, gold and platinum are preferred materials because of their surface stability.

Radiation impinging on a surface is either absorbed, transmitted, or reflected. We will ne-
glect transmission since it is not significant in most calorimeter designs. The relative amounts
of energy absorbed and reflected are related to the absorptivity and reflectivity, which refer to
perfectly smooth surfaces [8]. Reflectance and absorptance refer to real surfaces. Since absorptance is proportional to the emittance (Kirchoff's law), low emittance means a high reflectance,
which is desired for a sample or cell surface; the shield surface should also have high reflectance, which means a shield with a low emittance also. Thus, for the shields, a polished and
unoxidized surface is required.

In some cases a high absorptance is required to enhance the heat transfer from a heater to
the calorimeter or a shield. These surfaces can be roughened, oxidized, or coated with a nonreac-
ting layer to increase the emittance (absorptance).

5. ELECTRICAL RESISTIVITY

The electrical resistivity of materials of construction must be considered in the choice of
heater materials and in evaluating electrical leakage through ceramics and other electrical
insulators.

Figure A7 shows the electrical resistivity as a function of temperature for metals, alloys,
and ceramics. The resistivity of metallic materials enters into the design of calorimeter heaters

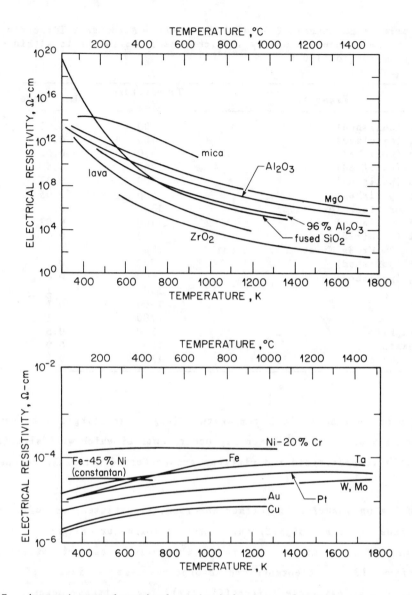

Figure A7. Approximate electrical resistivity as a function of temperature for several ceramic and metallic materials.

and shield heaters. The pure metals have large changes in resistivity (e.g., Cu changes by about 600% from 300 to 1300 K) compared to alloys such as Ni-20% Cr whose resistivity increases by only 10% over this temperature range. Since relatively constant energy transfer to the calorimeter from the heater is generally desired, use of Ni-20% Cr heaters accomplishes this within 10% using constant-voltage or constant-current power supplies. The Cu-Ni alloys in the range of about 45% Ni ('Constantan') have almost constant resistivity from 300 to about 800 K, but practical use is limited to 600 K. The property of low temperature coefficient of resistivity is usually desired in shield heaters because it simplifies design of power supplies.

The resistivity of ceramics is many orders of magnitude above that of the metallic materials. However, unlike metals, the resistivity decreases with increasing temperature. Since the ceramics are frequently chosen for their electrical insulating properties, problems due to increased electrical leakage at high temperatures must be considered.

Mica has a high resistivity (Fig. A7), although the exact values depend strongly upon the type of mica. It can be shaped somewhat to conform to desired geometries, but it is usable only up to about 1000 K. Al_2O_3 and MgO are the commonly-used ceramics where high electrical resistivity is required. They can be purchased in many shapes and with nearly theoretical density. Also, they can be obtained with high purity (99% or better), which is important as most common impurities lower the resistivity. For example, compare the curve in Fig. A7 for Al_2O_3 and for the 96% Al_2O_3. Lava is an important material that can be machined (see Section 8), then fired to give a ceramic material with the resistivity shown in Fig. A7.

The decrease in resistivity of ceramic insulating materials as temperature increases leads to errors in calculating the energy dissipated by heaters and in temperature measurement using either resistance thermometers or thermocouples. Wires, the calorimeter, and shields irregularly contact ceramics which, as the resistivity decreases upon heating, gives rise to complicated current leakage between components. Since these components (calorimeter and shields) are frequently grounded for signal stability, current leakage through ground loops is frequently a major source of error. Following are examples of leakage paths and estimates of the associated errors.

Consider the heater configuration shown in Fig. A8 where a heater lead passes through the sample (or the cell wall) and is insulated from the sample by a ceramic tube. Assume perfect interface contact between the tube, the wire, and the sample. For a tube 4 cm in length, an inside diameter of 0.02 cm, and an outside diameter of 0.05 cm, the inside and the outside surface areas A are each approximately 0.25 cm^2. With a wall thickness of 0.025 cm, the resistance R across the wall is given by

$$R = \rho \ (t/A) \simeq \rho \ 0.1 \ \Omega. \qquad (2)$$

If the tube is pure Al_2O_3 at 1300 K, with $\rho \simeq 10^7$, the resistance is about $10^6 \ \Omega$. The effective electrical circuit is that shown in Fig. A9, for which we need to determine the current leakage across the ceramic tube to ground. The power to the sample is required to calculate the specific heat and is based on the current obtained by measuring the voltage drop across a standard resistor. In this case, not all of this measured current passes through the heater; some is shunted across the ceramic wall to ground, and it is necessary to ensure that this current leakage is negligible.

The fraction of the current shunted is $i_2/(i_2+i_1)$, or $(1/R_2)/(1/R_1+1/R_2) = R_1/(R_1+R_2)$. Using $R_2 = 10^6 \ \Omega$ for Al_2O_3 from the earlier estimate and 10 Ω for the heater resistance, the

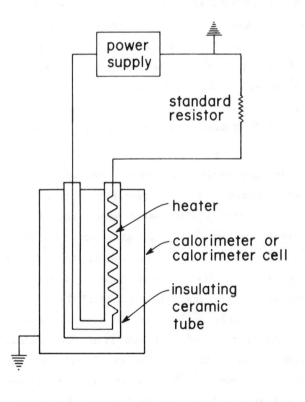

Figure A8. Schematic diagram of a typical adiabatic calorimeter heater circuit.

fraction of the current shunted is 10^{-5} or 0.001%, or 10^{-5} A for a 1 A heater current. Even with 96% Al_2O_3 ceramic with a resistivity of about 10^5 $\Omega \cdot$cm, the error is only 0.01%. Therefore, at 1300 K, this leakage is negligible; the calculation does point out that ceramics with resistivities much less than 10^5 $\Omega \cdot$cm can lead to measurable error. Also, the error obviously increases at temperatures higher than 1300 K.

Electrical leakage to thermocouples is a much more serious problem than that associated with sample heaters. If the leakage from a heater wire is along a thermocouple wire between components and, particularly to ground, an error signal is generated. This is depicted in Fig. A10. If the resistance of the thermocouple wire from the junction to the ground point is 3 Ω, and the shunted current is 1 A, then the voltage drop is (0.00001)(3) = 0.00003 V, or 30 μV, which is quite significant in introducing an error in the measurement of the temperature (~3 K for Pt–Pt10Rh thermocouples). These examples emphasize the importance, if not the necessity, of using high-purity Al_2O_3 or MgO for electrical insulating purposes at high temperatures.

Another serious electrical leakage problem is that caused by surface coatings on insulating ceramics, usually from metal vapor deposition. Consider the case in Fig. A11 where the free surface of the ceramic insulating tube for a specimen heater lead wire has been coated with a layer of metal. The deposited metal layer is a disk with a hole in it (for the wire), and current leakage is radially out from the contact region of the wire and the metal film to the grounded specimen. This is the same situation as that depicted in Fig. A9, and, with a heater voltage of 10 V, the thickness of the metal film such that the current leakage causes an error of 0.01% in the measured current can be calculated.

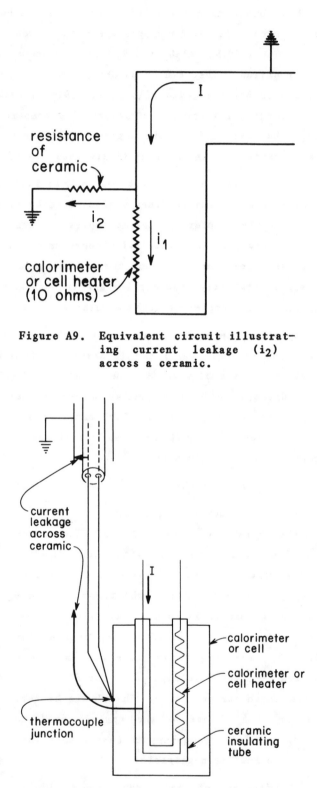

Figure A9.   Equivalent circuit illustrating current leakage (i_2) across a ceramic.

Figure A10.   Schematic diagram illustrating current leakage across insulating ceramic of a heater wire and flow along a thermocouple wire.

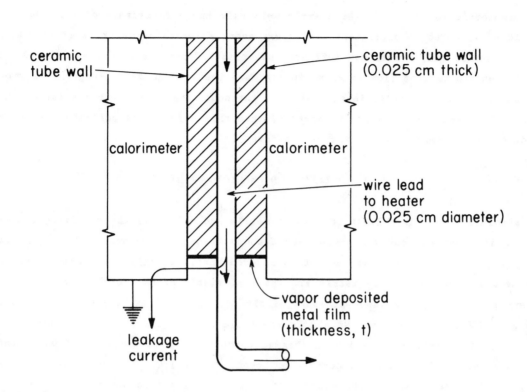

Figure A11.   Schematic diagram showing current leakage across a metal film vapor deposited on the exposed ceramic surface.

This layer is only a few atom layers.   Critical monitoring of this potential source of leakage is required for precision, high-temperature calorimetry.   We inspect such regions of the ceramics after every two or three operations to 1300 K.   If the ceramic appears discolored it is replaced, or the end broken off to expose fresh, clean ceramic.   We have found that if the ceramics are heated to about 1300 K in air they are reusable, although they may still be discolored.   To ensure that such films are not caused by organic matter present at 300 K prior to use in the calorimeter, we recommend use of nylon or rubber gloves to assemble these critical components, and that they be cleaned with a solvent, such as ether, after assembly.

6.   ELECTRON EMISSION

Another source of current leakage is electron (thermionic) emission from the heater wires. The current density i from the surface of a metal is given by

$$i \ (A \cdot cm^{-2}) = A \ T^2 \ e^{-53,400/T} \tag{3}$$

where A is a constant for each metal and ranges from about 10 to about 200.   For a metal such as nickel at 1300 K, the current density is about 10^{-10} A·cm^{-2}.   For a heater wire of diameter about 0.01 cm and a length of about 20 cm (hence, a surface area of about 0.8 cm^2), this gives a current of about 10^{-10} A.   If these electrons flow from the heater to the specimen, and then along one lead of a thermocouple to ground (just as in the example in Fig. A10), then an error is introduced in the thermocouple reading. If the thermocouple wire has a resistance of 3 Ω, the voltage drop is about 10^{-10} V, which is negligible. However, even if the calorimeter is at 1300 K, the heater wire will be hotter and, hence, may influence calorimeter operation. For example, if the wire is at 1570 K, then the current flow is about 10^{-7} A, and the voltage drop along the thermocouple wire is about 0.3 μV. This is sufficiently large to introduce a measurable error in the thermocouple signal. If the thermionic emission is erratic, then this effect will contribute to shield control noise and, hence, increase the scatter of the data.

7. THERMAL AND CHEMICAL STABILITY

The absolute temperature limit for use of metals and alloys is their melting temperature; for ceramics it is their melting temperature or decomposition temperature. These quantities are reasonably well-documented and will not be reproduced here. Clearly, the ceramics are more stable then the metals. The refractory metals are the most stable of the metallic materials, and these metals have sufficiently low vapor pressure (Table A5) to allow their use to quite high temperatures (e.g., 2300 K). In vacuum, Ni and Fe and their alloys generally show acceptable stability up to 1300 K. Cu, desired for its high thermal conductivity, is usable to 1300 K (the melting point temperature is 1356 K), but vaporization becomes a problem. Cu vapor condenses on the colder (but still hot) regions, causing contamination of ceramics (see Section 6). Ag (melting temperature 1233 K) is usable to 1100 K, possibly higher, until vaporization becomes a problem.

Scott [9] encountered problems with the resistance between the sample and the heater being lowered significantly after heating to above 900 K. He attributed this to vaporization of Zr from the Zr and Zr alloys on which he was measuring. He was able to minimize this effect by packing the internal heater hole with powdered, high-purity Al_2O_3. The main difficulty with this solution is that it reduces the heat transfer rate between the heater and the sample.

Table A5. Approximate Vapor Pressure at 1070 K for Several Materials

Material	Vapor Pressure (Pa)
Ag	7×10^{-3}
Cu	9×10^{-3}
Au	3×10^{-4}
Cr	3×10^{-4}
Fe	7×10^{-5}
Mo	3×10^{-14}
Ni	4×10^{-5}
W	$<10^{-9}$ (below 1970 K)
Pt	$<10^{-10}$
TiO_2	7×10^{-13}
SiO_2	1×10^{-3} (1770 K)
NiO	4×10^{-6}

We have also encountered indications that in vacuum above 900 K the vaporization of Cr from alloys containing this element may be a problem. We have had thermocouples show increasingly suspicious behavior as the duration of exposure to temperatures above 1000 K increases. There is some indication that this is associated with the reaction of Cr vapor with the Al_2O_3 ceramic, with the subsequent contamination of the thermocouple with Al. A solution to this problem has been use of pure Ni for the vacuum container and shields.

When it is required to place two dissimilar materials in contact (e.g., a thermocouple junction welded to a shield or a metal in contact with a ceramic), then the phase diagram should be examined to assist in predicting the formation of low-melting phases. For example, Fig. A12 shows that Cu and Ni are isomorphous, with the liquidus always above the melting point of copper [10]. Thus, Ni, Cu, and their alloys can be safely contacted with each other. On the other hand, care must be exercised to prevent Cu and Ag from contacting, as they form an eutectic system with an eutectic temperature of about 1070 K (Fig. A13) [10].

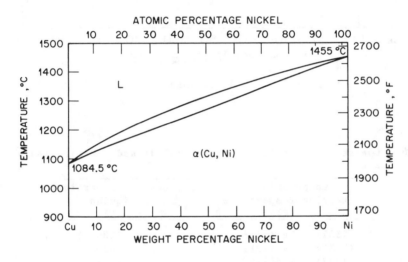

Figure A12. The Ni-Cu phase diagram [10].

Table A6 lists the lowest liquidus temperature for several metal and ceramic binary systems. Only those combinations likely to be encountered in calorimeter design are listed, and only if they have a liquidus significantly below the melting point of the lower melting component. To see how to use this information, consider welding a Pt-Pt:Rh thermocouple to titanium. In the table it is found that the liquidus for the Pt-Ti system is 1570 K, and thus, in use, the junction should not exceed this temperature.

In using alloys, multicomponent phase diagrams should be consulted. However, many diagrams have not been determined, so that to ensure stability between mated materials compatability tests should be conducted. The data for the binary systems can serve as a guide. For example, the Ni-Cr system has a low-temperature liquidus of 1618 K (Table A6), indicating that Cr lowers the melting point of Ni considerably. Thus, welding nichrome (e.g., Ni-20% Cr) wire to titanium may form a structure with a liquidus considerably lower than that predicted from the Ni-Ti phase diagram.

Table A6 also lists the liquidus of some ceramic binary systems. It is safest to use MgO and Al_2O_3. Care must be exercised to prevent SiO_2 from contacting FeO or Cu_2O.

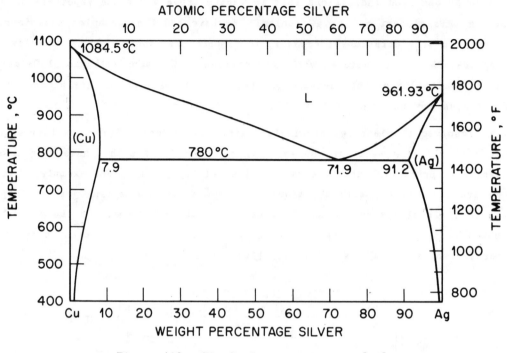

Figure A13. The Ag–Cu phase diagram [10].

Table A6. Lowest Liquidus for Several Metallic and Ceramic Binary Systems

Components and Melting Temperature (K)	Lowest Temperature Liquidus (K)
Au (1337) – Cu (1358)	1184
Au (1337) – Ge (1211)	629
Au (1337) – Ni (1728)	~1223
Au (1337) – Si (1687)	850
Ag (1235) – Cu (1358)	1053
Ag (1235) – Ge (1211)	924
Cu (1358) – Al (933)	821
Cu (1358) – Mn (1519)	1144
Cu (1358) – Si (1687)	1076
Cu (1358) – Ti (1945)	~1153
Cu (1358) – Ge (1211)	913
Al (933) – Si (1687)	850
Ni (1728) – Nb (2744)	1448
Ni (1728) – Mo (2896)	1593
Ni (1728) – Ta (3293)	1595
Ni (1728) – Ti (1945)	1216
Fe (1811) – Mo (2896)	~1723
Ni (1728) – Cr (2136)	1618
Co (1767) – Ti (1945)	1288
Cr (2136) – Ti (1445)	1673
Pt (2045) – Ti (1945)	~1573
Pt (2045) – Cr (2136)	1773
SiO_2 (1996) – FeO (~1623)	~1453
Cu_2O (1503) – SiO_2 (1996)	1333
SiO_2 (1996) – MgO (3093)	1816
SiO_2 (1996) – Al_2O_3 (2313)	~1943

8. FABRICABILITY

Most metals and alloys can be purchased in a wide variety of shapes; tubes, rods, and wires are readily available. They can be machined relatively easily into complex shapes and can be extensively plastically deformed.

One type of materials which is very promising for high-temperature use is sheathed wires, usually available as thermocouple wires. Wire inside a thin-walled tube is surrounded with powdered MgO for electrical insulation and the assembly swaged to final size. This construction is flexible and the inside wires are protected. Sizes are as small as 0.02 cm sheath diameter with 0.003 cm wire diameter, and wires and sheaths of a variety of materials (e.g., sheaths of stainless steel, Ta, Cu, and Al and the wires of Pt:Pt-13% Rh, Cu, etc.) are available.

The metallic materials have the advantage that they can be welded, which is very useful in construction of components. We routinely weld 0.02 cm diameter thermocouple wire to shields by discharge welding. Here a Cu electrode is placed on the wire and a force applied to make contact between the wire and the surface to which it is welded. Then a capacitor bank is discharged, sending a high current through the contacting interface. The contact resistance is sufficient to cause interface melting and welding. This procedure can be accomplished (with practice and patience) in rather remote locations. For example, we weld thermocouples onto the inner surface of a 4.4 cm diameter tube at a distance of about 80 cm from the end. A long Cu rod is used to press the thermocouple junction against the inside of the tube at the desired depth, and alignment prior to welding is achieved by use of a telescope.

This type of welding process is difficult to achieve if the surface has a low electrical resistivity, such as for Cu and Al. However, we do weld Pt-Rh thermocouples to Cu and Al. Also, we have successfully welded 0.008 cm diameter Pt wire to Al using a laser microwelder. Unfortunately, welded junctions may fracture after repeated heating and cooling of the calorimeter. This results from grain growth and diffusion during welding. There is no nondestructive method for testing the weld prior to use, other than careful inspection of the junction with a magnifying glass.

The influence of plastic deformation during fabrication on subsequent high-temperature mechanical behavior should be kept in mind when constructing parts for the calorimeter. For example, wire used to support components and sheets for shields must undergo various amounts of bending and twisting which is accompanied by extensive plastic deformation. Upon heating in the calorimeter, temperatures are frequently reached (e.g., about 1300 K for nichrome (Ni-20% Cr) and about 1100 K for platinum) at which the metal will recrystallize and eventually develop a relatively large grain size. This large grain size may be detrimental to creep and to creep-fatigue properties, lowering the strength.

The usable ceramics, such as high-purity Al_2O_3 and MgO, can be obtained in a variety of shapes, such as tubes and rods. They can only be shaped further by grinding. Some manufacturers will make items to specified shapes, but the cost may be prohibitively high. Also, there are available ceramic materials which are easily machined to desired shapes, which are then heated to obtain their functional properties. Noteworthy here is 'lava,' referred to in Section 5. It is quite machinable in the purchased form; it is hardened by heating slowly to 1300 K in air, then cooling slowly. We have used this material extensively to machine a variety of complicated shapes, including parts with threaded holes. Also, Corning Glass Company of Corning, New York sells a 'machinable' glass-ceramic material. It does not have to be fired, but is ready for use. The temperature limit of use is about 1300 K. However, we have had no experience with this material.

9. REFERENCES

1. Touloukian, Y.S. (Editor), Thermophysical Properties Research Literature Retrieval Guide, Plenum Publishing Corp., New York, NY, 2nd Ed., 1967, Supplement I, 1973, and Supplement II, 1979.

2. Ho, C.Y., Powell, R.W., and Liley, P.E., J. Phys. Chem. Ref. Data, Suppl. 1, 1974.

3. Clark, F.M., Insulation Materials for Design and Engineering Practice, Wiley, New York, NY, 1962.

4. Lyman, T. (Editor), Metals Handbook, 8th Ed., Vol. 1, Properties and Selection of Metals, American Society for Metals, Metals Park, OH, 1961.

5. Smithells, C.J., Metals Reference Book, Vol. 1 and 2, 4th Ed., Butterworths, Washington, DC, 1967.

6. Carreker, R.P., J. Appl. Phys., 21, 1289-96, 1950.

7. Smith, D.F. and Clatworthy, E.F., Metal Progress, 119, 32-5, 1981.

8. Love, T.J., Radiative Heat Transfer, C.E. Merrill Publishing Co., Columbus, OH, 1968.

9. Scott, J.L., 'A Calorimetric Investigation of Zirconium, Titanium and Zirconium Alloys from 60 to 960°C,' Dissertation, Univ. of Tennessee, 1957.

10. ASM, Metals Handbook, Vol. 8, American Society for Metals, Metals Park, OH, 1973.

APPENDIX B

Reference Materials for Calorimetry

R. K. KIRBY

1. INTRODUCTION

Information is provided in this appendix on the availability of reference materials (in the solid state) for use in calibrating or evaluating the performance of calorimeters. Certified Reference Materials are available from government bureaus in the United States [1], France [2], and USSR [3]. The Commission on Physicochemical Measurements and Standards of the International Union of Pure and Applied Chemistry (IUPAC) [4] recommends 10 different calibration and test materials for enthalpy measurements. Three of the materials recommended by IUPAC are also recommended by and available from the Calorimetry Conference.

A Certified Reference Material is a representative sample from a homogeneous lot of material that has been certified by an authorized agency to have a stated property such as enthalpy. The values certified are generally the best estimate of the 'true' values as determined by an accurate measurement technique. Certification generally includes the estimated uncertainties of the certified values.

Materials that are recommended by IUPAC can be obtained from a variety of sources as long as they meet the requirements specified by IUPAC. The recommended values for these materials are based on reliable experiments that are described in the open literature.

2. MATERIALS AVAILABLE FROM THE UNITED STATES

The following reference materials, Standard Reference Materials (SRM's) and Research Material (RM), are available from the U.S. National Bureau of Standards [1]: synthetic sapphire, molybdenum, polystyrene, polyethylene, and copper.

Synthetic sapphire, SRM 720, is sold as a 15-g sample and is in the form of small cylinders that were cut from rods grown by the Vernieul process. Values of enthalpy and heat capacity cover the range from 10 to 2250 K. These values are given in both tabular and equation format in a certificate provided with this Certified Reference Material. The heat capacity values are illustrated in Fig. B1 and some representative values are given in Table B1. The certified values of SRM 720 have an accuracy varying from 10% at 10 K, 0.1% at 70 K, 0.3% at 1200 K, to 0.5% at 2250 K. The measurements from 10 to 380 K were made with a vacuum adiabatic calorimeter, those from 273.15 to 1173 K with a Bunsen ice calorimeter, and those from 1173 to 2250 K with an adiabatic receiving calorimeter. Summary descriptions of these methods are also given in the certificate.

Molybdenum, SRM 781, is sold as both 3.2 and 6.4 mm diameter rods 100 mm in length. The rods were made by sintering high-purity (99.95 wt.%) powder, swaging, and grinding to final diameter and annealing at 1725 K under high vacuum. Values of enthalpy and heat capacity are certified from 273 to 2800 K and are provided in both tabular and equation format. The heat capacity values for this Certified Reference Material are also illustrated in Fig. B1 and representative values are given in Table B1. The certified values have an accuracy of less than 0.5% for temperatures up to 1200 K, 1.0% for temperatures between 1200 and 1850 K, less than 2% at 2000 K, and less than 3% at 2800 K. These measurements were made with a Bunsen ice calorimeter, an adiabatic receiving calorimeter, and with a high-speed pulse calorimeter at temperatures above 1500 K.

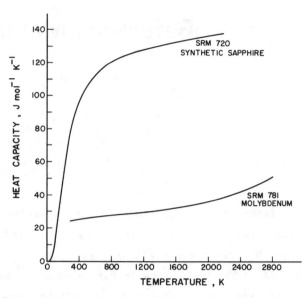

Figure B1. Heat capacity of two Certified Reference Materials that are available from the National Bureau of Standards.

Polystyrene, SRM 705, is a powder that is sold in 5-g amounts. The weight-average molecular weight of this polymer is about 184,000 g·mol^{-1}. Values of heat capacity, certified from 10 to 350 K, are provided in tabular form. Some representative values are given in Table B2. These values have an accuracy of less than 0.2%. Adiabatic calorimetry was used to measure the heat capacity.

Polyethylene, SRM 1475, is a powder that is sold in 50-g amounts. The weight-average molecular weight is about 52,500 g·mol^{-1}. Values of heat capacity are certified from 5 to 360 K as a

Table B1. Heat Capacity of SRM's 720 (Synthetic Sapphire) and 781 (Molybdenum)

Temperature (K)	SRM 720 (J·mol^{-1}·K^{-1})	SRM 781 (J·mol^{-1}·K^{-1})
10	0.009	–
20	0.073	–
40	0.697	–
70	4.59	–
110	16.35	–
160	35.95	–
220	57.95	–
290	77.2	23.8
370	92.0	24.8
470	103.6	25.6
600	112.6	26.5
800	120.1	27.4
1100	126.6	28.9
1600	133.4	32.5
2200	137.7	39.2
2800	–	51.6

function of density. Some representative values are given in Table B2. These values have an accuracy of 1% between 25 and 360 K. The heat capacity was determined with an adiabatic calorimeter.

Copper, RM 5, is not a Certified Reference Material but serves the same purpose as the Calorimetry Conference copper standard that is no longer available. RM 5 is a 99.999+ wt.% pure copper rod that has a diameter of 19 mm and a length of 120 mm. Recommended values of heat capacity cover the range from 0 to 300 K. Some of these values are given in Table B3.

Table B2. Heat Capacity of SRM's 705 (Polystyrene) and 1475 (Polyethylene)

Temperature (K)	SRM 705 ($J \cdot mol^{-1} \cdot K^{-1}$)	SRM 1475 ($J \cdot mol^{-1} \cdot K^{-1}$)
10	3.3	0.2
20	10.8	0.9
35	20.8	2.6
50	28.5	4.6
80	40.7	7.9
110	51.1	10.3
140	61.4	12.4
170	72.4	14.5
200	84.2	16.6
230	96.8	18.8
260	110.1	21.6
290	123.7	24.6
320	137.7	27.8
350	151.6	32.2

Table B3. Heat Capacity of RM 5 (Copper)

Temperature (K)	Heat Capacity ($J \cdot mol^{-1} \cdot K^{-1}$)	Temperature (K)	Heat Capacity ($J \cdot mol^{-1} \cdot K^{-1}$)
5	0.0094	70	10.9
10	0.0555	80	12.8
15	0.1835	100	16.0
20	0.462	125	18.7
25	0.963	150	20.5
30	1.69	200	22.6
40	3.74	250	23.8
50	6.15	300	24.5
60	8.60		

3. MATERIALS AVAILABLE FROM OTHER SOURCES

Certified Reference Materials that are available from other sources [5] include copper (350-1300 K) and platinum (298-2000 K) from France [2], and quartz (293-900 K), molybdenum (293-2600 K), and sapphire (293-2700 K) from USSR [3].

The calibration and test materials recommended by IUPAC [4] are as follows:

(1) A highly pure grade of synthetic sapphire—recommended values of enthalpy and heat capacity from 10 to 2000 K.

(2) High-purity platinum—values of enthalpy and heat capacity from 298 to 2000 K.

(3) 99.999+ wt.% pure copper—values of heat capacity from 1 to 25 K.

(4) Reagent grade succinic acid ($C_4H_2O_4$)—values of heat capacity from 5 to 320 K.

(5) 99.997 mol.% pure benzoic acid ($C_7H_6O_2$)—values of heat capacity from 10 to 340 K.

(6) 99.997 mol.% pure 2,2-dimethylpropane (C_5H_{12})—values of heat capacity from 4 to 139 K and 142 to 254 K (values near the solid-solid transition at 140.5 K are not recommended).

(7) 99.5 wt.% pure molybdenum—values of enthalpy and heat capacity from 273 to 2500 K.

(8) 99.985 mol.% pure naphthalene ($C_{10}H_8$)—values of heat capacity from 10 to 350 K.

(9) 99.999 mol.% pure diphenyl ether ($C_{12}H_{10}O$)—values of heat capacity from 10 to 300 K.

(10) 99.997 mol.% pure n-heptane (C_7H_{16})—values of heat capacity from 10 to 182.6 K (n-heptane is liquid at temperatures above 182.6 K).

Three materials are available from the Calorimetry Conference [6]. These are benzoic acid as a 75-g sample, synthetic sapphire as a 70-g sample, and n-heptane in a 100 mL ampoule. Both the benzoic acid and sapphire are in powder form and the recommended values range from 14 to 1200 K and 0 to 1200 K, respectively. The recommended values for the solid form of n-heptane range from 0 to 182 K.

4. REFERENCES

1. National Bureau of Standards, Office of Standard Reference Materials, Washington, DC 20234.

2. Services des Materiaux de Referance, 1 rue Gaston Boissier, 75015 Paris, France.

3. Gosstandart of the USSR, 9 Leninsky Prospekt, 117049 Moscow, USSR.

4. Recommended Reference Materials for Realization of Physicochemical Properties: Enthalpy, J.D. Cox, Pure Appl. Chem., 40(3), 399-450, 1974.

5. Directory of Certified Reference Materials (CRM): Sources of Supply and Suggested Uses, ISO/REMCO Directory, 1982. Prepared and disseminated by International Organization for Standardization, Case Postale 56, CH-1211, Geneve 20, Switzerland.

6. Contact the Office of Standard Reference Materials, B316 Chemistry Building, National Bureau of Standards, Gaithersburg, Maryland 20899, U.S.A.

Presentation of Thermophysical Data in the Scientific and Technological Literature

E. F. WESTRUM, JR.

1. INTRODUCTION

The performance of a thermophysical study does not end with the completion of the experimental work nor even upon the reduction of the data to smooth values at rounded temperatures. In order that the data themselves may be of value to the reader and be embodied in the scientific and technological literature by the process of critical evaluation, it is important that the author provide the reader with certain information on the definition of the systems studied, tests made on reference substances, methods used to test the reliability of the results, qualitative descriptions of the type of measurements made and apparatus used, of the experimental procedures, and of the performance of the measurement itself. In an earlier era, most thermophysical measurements were made by a group of experimentalists who identified themselves as either physical or chemical thermodynamicists. At the present time, however, much thermophysical data are taken by people whose primary allegiance is to some other field of science but who need the data for their own technological, biological, geological, or other applications. Because of the interdisciplinary nature of the production and utilization of such data, special care must be taken to ensure that appropriate information is provided for the benefit of the reader or the critical evaluator.

2. THE GUIDE FOR DATA PRESENTATION

Fortunately, the details of such information have been summarized in the 'Guide for the Presentation in the Primary Literature of Numerical Data Derived from Experiments' [1,2]. This document, devised by ICSU-CODATA (International Council of Scientific Unions - Committee on Data for Science and Technology), has been translated into a number of foreign languages for the convenience of authors, editors, and referees. These languages are: French [3], Spanish [4], Russian [5], Swedish [6], Japanese [7], and most recently, Chinese [8].

The application of the information contained in this document is intended to ensure that the reader (or the critical evaluator) of the scientific or technological presentation can understand the quantitative data, can assess their values and accuracy, and can recalculate the results when values for auxiliary data change. When it becomes necessary for the reader to compare the results of several studies or to reinterpret data, he may need to know whether the author paid attention to details that have since been realized to be important or whether his technique could have detected the existence of more recently revealed phenomena. If these questions cannot be answered, later workers and evaluators may assign very little weight to the results. Another frequently very awkward process is the interpretation of the precision (i.e., the reproducibility of the data) and the inaccuracy, which estimates the overall reliability of the measurements. Information provided by the author is very often important here.

Such a guide provides journal editors and referees with a set of consistent, considered criteria for judging the completeness and acceptability of papers insofar as the reporting of the numerical quantities is concerned. The recommendations reflect the experience of the data evaluator; hence, adherence to them will permit evaluators to consolidate the author's results with existing data and to facilitate their incorporation in critical compilations.

3. THE GUIDE FOR THERMODYNAMIC DATA

Another document more specifically designed for thermochemical and thermophysical data is 'A Guide to Procedures for the Publication of Thermodynamic Data.' This document, which predates the guide referred to in Section 2, has been widely approved by the International Union of Pure and Applied Chemistry (IUPAC), by national and international conferences, and by national bodies on experimental thermodynamics and calorimetry. The definitive text has been published in English [9]; verbatim publication has been made elsewhere also in English [10-15], in Japanese [16], in French [17,18], and in Russian [19]. Since differential scanning calorimetry and differential thermal analysis are closely related to the present discussion, it should be noted that a subdisciplinary guide concerning recommendations for the presentation of thermal analysis also exists [20].

4. REPOSITORIES FOR THERMOPHYSICAL DATA

In setting forth the important aspects of numerical data presentation so as to save the data and to promote the usefulness of the quantitative results of thermophysical measurements, there is, of course, an apparent conflict between the recommendations for provision of appropriate adjuvant data and the editors' usual exhortation to authors for brevity in their papers. The needs of the general reader can often be met by a brief article not containing the detailed information prescribed in the appropriate guides, but those of the specialists and of the evaluators cannot be thus provided. Although these recommendations call for more—but not much more—detail than is commonly provided, they do not exceed what appears in the better papers today. The needed statements may be terse and factual.

The ideal situation would be to have all of the relevant information in the published article. However, if this is not practical, then the supplementary material may be published in an auxiliary publication—submitted to the editor together with a shorter manuscript—and placed either in a suitable depository (repository) service or published as microform together with the article. In any event, the details must be made available to the scientific public from some source other than the author. The means of obtaining such auxiliary information may be clearly stated in the publication.

A single example of such a repository service is provided by the National Auxiliary Publications Service in the United States. This is accessible to journals published not only in the United States but in other countries and in other languages. The author simply includes a reference in the paper's bibliography indicating the nature of further details that are available in a supplementary publication and references this among the items in the bibliography. For a small fee, the reader may obtain either a microform or a full-scale xerographic copy of the material in question.

5. REFERENCES

1. CODATA Bulletin No. 9, 'Guide for the Presentation in the Primary Literature of Numerical Data Derived from Experiments,' Report of the CODATA Task Group on Publication of Data in the Primary Literature, September 1973.

2. Westrum, E.F., Jr., et al., 'Guide for the Presentation in the Primary Literature of Numerical Data Derived from Experiments,' Unesco-UNISIST Report, Paris, France, 1974.

3. Westrum, E.F., Jr., et al., 'Guide pour la Presentation dans la Litterature Primaire des Donnees Numeriques Deduites des Experiences,' Unesco-UNISIST Report No. SC. 74/WS/19, Paris, France, 1974.

4. Westrum, E.F., Jr., 'Guia para la presentatacion en la literatura primaria de los datos numericos derivados de esperimentos,' UNISIST, Unesco House, Paris, 1975.

5. Westrum, E.F., Jr., et al., 'Rekomendatsii po Predstavlenio i Pervichnoi Literature Chisennix Eksperimental 'nix Dannix,' Fundamental 'nie Konstanti Fiziki i Khimii, $\underline{1}$, 13–22, 1975.

6. Westrum, E.F., Jr., et al., 'Vagledning for presentation av numeriska data fran experiment i kalltidskrifter,' Kungl. Vetenskapsakademien, 8 pp., 1975.

7. Westrum, E.F., Jr., et al., Butsuri, 3D, $\underline{1}$, 17–21, 1975.

8. Westrum, E.F., Jr., et al., Chemistry (Hua Xue Tong Bao), $\underline{10}$, 55–7, 1981.

9. Westrum, E.F., Jr., et al., Pure Appl. Chem., $\underline{29}$, 397–407, 1972.

10. Westrum, E.F., Jr., et al., At. Energy Rev., $\underline{9}$, 869–78, 1972.

11. Westrum, E.F., Jr., et al., J. Chem. Thermodyn., $\underline{4}$, 511–20, 1972.

12. Westrum, E.F., Jr., et al., Indian J. Chem., $\underline{10}$, 51–5, 1972.

13. Westrum, E.F., Jr., et al., CODATA Newsletter, $\underline{8}$, 4–8, 1972.

14. Westrum, E.F., Jr., et al., Indian J. Phys., $\underline{12}$, 51–5, 1972.

15. Westrum, E.F., Jr., et al., J. Chem. Eng. Data, $\underline{18}$, 3–6, 1973.

16. Westrum, E.F., Jr., et al., 'A Guide to Procedures for the Publication of Thermodynamic Data' (in Japanese), The Society for Calorimetry and Thermal Analysis, 1–9, Nov. 1971.

17. Westrum, E.F., Jr., et al., 'Recommandations pour la Publication des Données Thermodynamique,' Bulletin de la Société Chimique de France, Special No., 1–9, 1973.

18. Westrum, E.F., Jr., et al., J. Chim. Phys., $\underline{69}$, 17–22, 1973.

19. Westrum, E.F., Jr., et al., Zh. Fiz. Khim., $\underline{47}$, 2459–64, 1973.

20. McAdie, H.G., Anal. Chem., $\underline{39}$, 543, 1967.

Subject Index